Second Harmonic Generation Imaging

Series in Cellular and Clinical Imaging

Series Editor
Ammasi Periasamy

PUBLISHED

Coherent Raman Scattering Microscopy
edited by Ji-Xin Cheng and Xiaoliang Sunney Xie

Imaging in Cellular and Tissue Engineering
edited by Hanry Yu and Nur Aida Abdul Rahim

Second Harmonic Generation Imaging
edited by Francesco S. Pavone and Paul J. Campagnola

FORTHCOMING

The Fluorescent Protein Revolution
edited by Richard N. Day and Michael W. Davidson

Natural Biomarkers for Cellular Metabolism:
Biology, Techniques, and Applications
edited by Vladimir V. Gukassyan and Ahmed A. Heikal

Optical Probes in Biology
edited by Jin Zhang and Carsten Schultz

SERIES IN CELLULAR AND CLINICAL IMAGING

AMMASI PERIASAMY, SERIES EDITOR

Second Harmonic Generation Imaging

Edited by

Francesco S. Pavone
Paul J. Campagnola

CRC Press
Taylor & Francis Group
Boca Raton London New York

CRC Press is an imprint of the
Taylor & Francis Group, an **informa** business

CRC Press
Taylor & Francis Group
6000 Broken Sound Parkway NW, Suite 300
Boca Raton, FL 33487-2742

First issued in paperback 2019

© 2014 by Taylor & Francis Group, LLC
CRC Press is an imprint of Taylor & Francis Group, an Informa business

No claim to original U.S. Government works

ISBN-13: 978-1-4398-4914-9 (hbk)
ISBN-13: 978-0-367-37990-2 (pbk)

Library of Congress Cataloging-in-Publication Data

Second harmonic generation imaging / [edited by] Francesco S. Pavone, Paul J. Campagnola.
   p. ; cm. -- (Series in cellular and clinical imaging ; 3)
   Includes bibliographical references and index.
   ISBN 978-1-4398-4914-9 (hardcover : alk. paper)
   I. Pavone, Francesco S. II. Campagnola, Paul J. III. Series: Series in cellular and clinical imaging ; 3.
   [DNLM: 1. Microscopy, Fluorescence, Multiphoton. 2. Image Processing, Computer-Assisted. 3. Microscopy, Confocal. 4. Photons--diagnostic use. QH 212.F55]

616.07'58--dc23   2013014914

Visit the Taylor & Francis Web site at
http://www.taylorandfrancis.com

and the CRC Press Web site at
http://www.crcpress.com

Contents

SECTION I   Principles of SHG Imaging

SECTION II   Biomedical Imaging Using SHG

SECTION III   Applications of SHG

Series Preface

A picture is worth a thousand words. This proverb says everything. Imaging began in 1021 with use of a pinhole lens in a camera in Iraq; later in 1550, the pinhole was replaced by a biconvex lens developed in Italy. This mechanical imaging technology migrated to chemical-based photography in 1826 with the first successful sunlight-picture made in France. Today, digital technology counts the number of light photons falling directly on a chip to produce an image at the focal plane; this image may then be manipulated in countless ways using additional algorithms and software. The process of taking pictures ("imaging") now includes a multitude of options—it may be either invasive or noninvasive, and the target and details may include monitoring signals in two, three, or four dimensions.

Microscopes are an essential tool in imaging used to observe and describe protozoa, bacteria, spermatozoa, and any kind of cell, tissue, or whole organism. Pioneered by Antoni van Leeuwenhoek in the 1670s and later commercialized by Carl Zeiss in 1846 in Jena, Germany, microscopes have enabled scientists to better grasp the often misunderstood relationship between microscopic and macroscopic behavior, by allowing for study of the development, organization, and function of unicellular and higher organisms, as well as structures and mechanisms at the microscopic level. Further, the imaging function preserves temporal and spatial relationships that are frequently lost in traditional biochemical techniques and gives two- or three-dimensional resolution that other laboratory methods cannot. For example, the inherent specificity and sensitivity of fluorescence, the high temporal, spatial, and three-dimensional resolution that is possible, and the enhancement of contrast resulting from detection of an absolute rather than relative signal (i.e., unlabeled features do not emit) are several advantages of fluorescence techniques. Additionally, the plethora of well-described spectroscopic techniques providing different types of information, and the commercial availability of fluorescent probes such as visible fluorescent proteins (many of which exhibit an environment- or analytic-sensitive response), increase the range of possible applications, such as development of biosensors for basic and clinical research. Recent advancements in optics, light sources, digital imaging systems, data-acquisition methods, and image enhancement, analysis and display methods have further broadened the applications in which fluorescence microscopy can be applied successfully.

Another development has been the establishment of multiphoton microscopy as a three-dimensional imaging method of choice for studying biomedical specimens from single cells to whole animals with sub-micron resolution. Multiphoton microscopy methods utilize naturally available endogenous fluorophores—including NADH, TRP, FAD, and so on—whose autofluorescent properties provide a label-free approach. Researchers may then image various functions and organelles at molecular levels using two-photon and fluorescence lifetime imaging (FLIM) microscopy to distinguish normal from cancerous conditions. Other widely used nonlabeled imaging methods are coherent anti-Stokes Raman scattering spectroscopy (CARS) and stimulated Raman scattering (SRS) microscopy, which allow imaging of molecular function using the molecular vibrations in cells, tissues, and whole organisms. These techniques have been widely used in gene therapy, single molecule imaging, tissue engineering, and stem cell research. Another nonlabeled method is harmonic generation (SHG and THG), which is also widely used in clinical imaging, tissue engineering, and stem cell research. There are many more advanced technologies developed for cellular and clinical imaging, including multiphoton tomography, thermal imaging in animals, ion imaging (calcium, pH) in cells, and so on.

The goal of this series is to highlight these seminal advances and the wide range of approaches currently used in cellular and clinical imaging. Its purpose is to promote education and new research across a broad spectrum of disciplines. The series emphasizes practical aspects, with each volume focusing on a particular theme that may cross various imaging modalities. Each title covers basic to advanced imaging methods, as well as detailed discussions dealing with interpretations of these studies. The series also provides cohesive, complete state-of-the-art, cross-modality overviews of the most important and timely areas within cellular and clinical imaging.

Since my graduate student days, I have been involved and interested in multimodal imaging techniques applied to cellular and clinical imaging. I have pioneered and developed many imaging modalities throughout my research career. The series manager, Ms. Luna Han, recognized my genuine enthusiasm and interest to develop a new book series on Cellular and Clinical Imaging. This project would not have been possible without the support of Luna. I am sure that all the volume editors, chapter authors, and myself have benefited greatly from her continuous input and guidance to make this series a success.

Equally important, I personally would like to thank the volume editors and the chapter authors. This has been an incredible experience working with colleagues who demonstrate such a high level of interest in educational projects, even though they are all fully occupied with their own academic activities. Their work and intellectual contributions based on their deep knowledge of the subject matter will be appreciated by everyone who reads this book series.

Ammasi Periasamy, PhD
Series Editor
Professor and Center Director
W.M. Keck Center for Cellular Imaging
University of Virginia
Charlottesville, Virginia

Foreword

The recent growth of harmonic generation imaging field has been phenomenal. Harmonic generation imaging that typically includes both second- and third-order processes is an important branch of nonlinear optics and has a long illustrious history. Similar to most other nonlinear optical contrast mechanisms, the theoretical prediction of harmonic generation can be traced back to the theoretical work of Dr. Maria Göppert-Mayer in 1930s [1]. The experimental demonstration of second-harmonic generation requires high excitation light peak power and that became possible with the invention of the laser in the 1960s. Franken and coworkers first demonstrated second-harmonic generation in ruby giving a practical start to the field of nonlinear optics [2]. After this time, the importance of harmonic generation in laser physics and in the characterization of solid-state materials is well recognized with many exciting works in their respective fields. However, the use of harmonic generation to study biological specimens had a later start (Figure 0.1).* Some of the pioneering works included the generation of second harmonic from amino acid crystals in 1965 [3]. In 1971, Fine and coworkers demonstrated second-harmonic generation from biological specimens including that of collagen, one of the most important endogenous molecular structures that provide strong second-harmonic signal [4]. In late 1970s, Sheppard and his coworkers first realized that the nonlinear nature of harmonic generation process results in 3D localization of optical interaction at the focal point of a high numerical aperture objective. This insight resulted

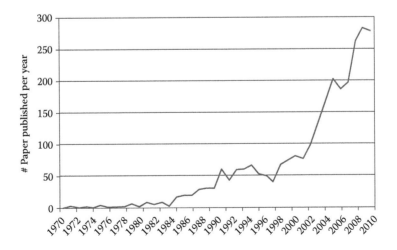

FIGURE 0.1   Journal publications in this field over the years.

---

* The chart is based on a PubMed search with keywords: "second harmonic generation" and "third harmonic generation."

in the first realization of a scanning second-harmonic generation microscope with 3D resolution [5,6]. Owing to laser limitations, this system was applied to the study of solid-state specimens but had not been used for imaging more delicate biological specimens. After 15 years, Freund and coworkers in 1986 demonstrated one of the first biological second-harmonic imaging experiments by mapping collagen structures in tendons although this work was performed at relatively low spatial resolution and without consideration of depth sectioning [7].

It is interesting to note that the number of publications of harmonic generation rises sharply starting in the late 1980s. This significant increase in the use of harmonic generation in biomedical research may be due to three factors. First, the introduction of reliable and robust femtosecond lasers, such as the titanium–sapphire laser, is probably an important factor. The availability of the laser sources that can be implemented in biological laboratories and clinics removes one of the major hurdles in the adoption of nonlinear optical imaging by biomedical researchers. Second, while harmonic generation microscopes have been developed for more than a decade, they were either applied for solid-state specimens or were producing relatively low-resolution images. In 1990, Denk, Webb, and coworkers first demonstrated two-photon fluorescence imaging in biological specimens and demonstrated exquisite 3D resolved imaging of developing embryos. This publication first captures the attention of biomedical researchers to the potential of nonlinear optical microscopy in general. The increasing interest of the biomedical field spurred the development of robust, convenient, and broadly available commercial nonlinear microscopes that can image based on both fluorescence and harmonic generation modalities. Third and possibly most importantly, a group of excellent biophotonic researchers entered the field of harmonic generation microscopy at that time and have successfully demonstrated that the harmonic generation signals can be correlated with tissue physiology and pathological states, have found a variety of other endogenous molecular structures that can also provide strong harmonic signals, and have developed contrast agents based on harmonic generation that are sensitive biochemical environment and are resistant to photobleaching. These successful pioneering works have firmly established harmonic generation imaging as one of the most important tools in the repertoire of microscopists. Many of this generation of pioneering researchers are the editors and authors of this excellent book.

In *Second Harmonic Generation Imaging*, the editors, Dr. Pavone and Dr. Campagnola, have assembled a team of authors who are all experts in the field of second-harmonic imaging. This volume started with a review of basic theories of harmonic generation and covers important topics such as instrument design and contrast agent development. With this foundation, this book covers many physiological systems where second-harmonic imaging provides important structural information, including that of the eyes, muscle, and skin. The applications of second-harmonic imaging in the study of different pathologies, such as cancer, cardiovascular diseases, and fibrotic diseases, are also described. Finally, this book covers more advanced topics such as holographic second-harmonic imaging and the development second-harmonic-based voltage-sensitive probes. In summary, this book provides an up-to-date coverage of our current understanding of harmonic imaging principles and also the most important biomedical applications of this technique.

References

1. M. Göppert-Mayer, Über Elementarakte mit zwei Quantensprüngen, *Ann Phys (Leipzig)* 5, 273–294, 1931.
2. P. A. Franken, A. E. Hill, C. W. Peters, and G. Weinreich, Generation of optical harmonics, *Phys Rev Lett* 7, 118, 1961.
3. K. E. Rieckhoff and W. L. Peticolas, Optical second-harmonic generation in crystalline amino acids, *Science* 147, 3658, 610–611, 1965.
4. S. Fine and W. P. Hansen, Optical second harmonic generation in biological systems, *Appl Opt* 10(10), 2350–2353, 1971.

5. J. N. Gannaway and C. J. R. Sheppard, Second harmonic imaging in the scanning optical micro-scope, *Opt Quant Electron* 10, 435–439, 1978.
6. C. J. R. Sheppard, J. N. Gannaway, R. Kompfner, and D. Walsh, Scanning harmonic optical micro-scope, *IEEE J Quant Electron* 13(9), D100–D100, 1977.
7. I. Freund and M. Deutsch, Second-harmonic microscopy of biological tissue, *Opt Lett* 11(2), 94, 1986.

Peter T. C. So
Department of Biological Engineering
MIT
Boston, Massachusetts

Preface

Second-harmonic generation (SHG) microscopy began to emerge as a high-resolution optical imaging modality in the late 1990s, about 10 years after two-photon excited fluorescence (TPEF) microscopy was pioneered by the laboratory of Watt Webb. Concurrent with two-photon microscopy, the initial suggestion for SHG imaging was made by Colin Sheppard. Although largely unnoticed by the biomedical optics community, Isaac Freund and coworkers published a fiber alignment study in rat-tail tendon using SHG polarization analysis, and while this was performed at low resolution (~50 μm), this landmark study captured the essence of using SHG to image collagen structure on a microscope. In an analog of surface SHG, in the mid-1990s, Leslie Loew and Aaron Lewis suggested the use of SHG to sense membrane potential, where the initial implementation used stage scanning at fairly low optical resolution.

The advent of two-photon microscopy on laser scanning confocal microscope platforms using commercially available mode-locked lasers set the stage for high-resolution SHG imaging. The first talks on SHG microscopy were given in the late 1990s at SPIE Photonics West and Focus on Microscopy. Concurrently, several labs published papers documenting the capabilities of the method. The development of SHG microscopy has taken two somewhat distinct paths: (i) exogenous staining with fluorescent dyes for imaging cell membranes, with the primary goal of achieving greater sensitivity for potential sensing and (ii) endogenous imaging of tissues. The latter has emerged as having greater utility, where the landscape of imageable harmonophores includes collagen in a wide array of tissues, acto-myosin in striated muscle and microtubules in live cells, particularly during cell division. As evidence of the growth of SHG microscopy, this modality has had dedicated sessions at national and international conferences for several years. Concurrently, papers utilizing SHG microscopy, especially for imaging collagen, have rapidly become more numerous. Additionally, third-harmonic generation (THG) together and in conjunction with SHG emerged as a new modality for imaging tissues with endogenous contrast. Several labs continue to specialize in the development of SHG techniques and analysis (and many have contributed chapters to this book), whereas many others now routinely use it as an additional contrast mechanism for imaging tissues.

The application of SHG in the field of medical diagnosis has opened up enormous possibilities. It has enabled the study of many pathologies related to collagen disorder in skin, such as keloids, keratoconus in cornea, liver fibrosis, reactive stroma evolution in methastasis, muscle dystrophy, and so on. There are ongoing efforts to improve instrumentation technology for practical clinical use. Furthermore, the coupling of SHG to other microscopy modalities, such as CARS (coherent anti-Stokes Raman scattering) or two-photon and FLIM (fluorescence-lifetime imaging microscopy), has shown to be very powerful for characterizing the morphochemistry of many tissues. Undoubtedly there are many other cellular features and tissue components yet to be detected and characterized. Motivated by our belief that the field still has a lot to discover, we decided to assemble a book dedicated to SHG and THG microscopy.

We have organized the book into three sections comprising of 18 chapters. The flow of topics is from fundamental SHG and THG theoretical concepts and instrumentation, followed by overviews of SHG capabilities and technology developments, and finally applications of the modalities, with an emphasis on imaging diseased states.

Section I discusses theoretical frameworks for SHG and THG microscopy. It also gives detailed instructions on how to build an SHG microscope, and provides in-depth comparisons with other nonlinear optical imaging modalities such as two-photon fluorescence and CARS microscopy.

Section II addresses the basic capabilities of SHG imaging of tissues and live cells, with chapters dedicated to imaging cell membranes, muscle, collagen in tissues, and microtubules in live cells, as well as emerging technology developments.

Section III focuses on imaging of diseased tissues, giving a timely overview of this exciting new frontier. This area is a natural target for SHG microscopy because numerous pathologies, including cancers, fibroses, autoimmune diseases, and connective tissue disorders, either result from or have changes in collagen content or organization. Historically, clinical imaging at the cellular and tissue level has been performed by pathologists on *ex vivo* biopsies removed by the surgeon. While histology remains the "Gold Standard" for pathologists, its interpretation remains highly subjective. Much of SHG research has focused on developing more quantitative, objective metrics. In this section, chapters are presented that describe efforts to use SHG to image a wide range of pathological conditions and diseases.

This book was developed for an audience of nonexperts in the field who need a tutorial, as well as experts looking for an up-to-date review of particular applications. We hope that this broad range of readers will find it useful.

MATLAB® is a registered trademark of The MathWorks, Inc. For product information, please contact:

The MathWorks, Inc.
3 Apple Hill Drive
Natick, MA 01760-2098 USA
Tel: 508 647 7000
Fax: 508-647-7001
E-mail: info@mathworks.com
Web: www.mathworks.com

Editors

Francesco S. Pavone obtained a PhD in optics in 1993 and spent two years as postdoctoral fellow at the Ecole Normale Superieure with the group of Claude Cohen Tannoudjy (Nobel Prize, 1997). He is currently a full professor at the University of Florence in the Department of Physics and at the European Laboratory for Non-Linear Spectroscopy (LENS), and group leader at the Biophotonics Laboratories. His research group is involved in developing new microscopy techniques for high-resolution and high-sensitivity imaging, and laser manipulation purposes. These techniques have been applied in single-molecule biophysics, single-cell imaging, and optical manipulation. He is also engaged in tissue imaging research, for which nonlinear optical techniques have been applied to skin and neural tissue imaging. He is the author of more than 100 peer-reviewed journal articles, has delivered more than 60 invited talks, and is on the editorial board of several journals. He coordinates various European projects and has organized several international congresses. He is also the director of the international PhD program at LENS.

Paul J. Campagnola obtained his PhD in chemistry from Yale University in 1992 after which he was a postdoctoral associate at the University of Colorado from 1992 to 1995. He was on the faculty in the Department of Cell Biology, Center for Cell Analysis and Modeling at the University of Connecticut Health Center from 1995 to 2010, having adjunct appointments in the physics department and biomedical engineering program. In 2010, he became an associate professor in the Departments of Biomedical Engineering and Physics at the University of Wisconsin—Madison.

His research focuses on the development of nonlinear optical spectroscopy and microscopy methods, with an emphasis on translational applications. One area of specialization has been the development of second harmonic generation microscopy and analysis/modeling tools for characterizing structural changes in the extracellular matrix (ECM) in diseased tissues (e.g., cancer and connective tissue disorders). A second area of research has developed multiphoton excited nano/microfabrication to create biomimetic models of the ECM. These efforts are directed at understanding cancer cell–ECM interactions in the tumor microenvironment as well as fabricating scaffolds for tissue regeneration.

He has published more than 50 peer-reviewed journal articles and several review articles and book chapters, and has delivered more than 70 invited talks. He serves on the editorial board for the *Journal of Biomedical Optics* and on numerous NIH and NSF review panels.

Contributors

Robert S. Balaban
Laboratory of Cardiac Energetics
National Institutes of Health
Bethesda, Maryland

Emmanuel Beaurepaire
École Polytechnique
CNRS–Inserm
Palaiseau, France

Christian Brackmann
Department of Physics
Lund University
Lund, Sweden

Edward B. Brown
Department of Biomedical Engineering
University of Rochester
Rochester, New York

Paul J. Campagnola
Department of Biomedical
    Engineering
University of Wisconsin—Madison
Madison, Wisconsin

Shean-Jen Chen
Department of Engineering Science
National Cheng-Kung University
Tainan, Taiwan

Szu-Yu Chen
Department of Electrical
    Engineering
National Taiwan University
Taipei, Taiwan

Xiyi Chen
LaserGen Corporation
Houston, Texas

Riccardo Cicchi
National Institute of Optics—National Research
    Council (INO—CNR)
University of Florence
Florence, Italy

Delphine Débarre
École Polytechnique
CNRS–Inserm
Palaiseau, France

Christian Depeursinge
École Polytechnique
Fédérale de Lausanne
Vaud, Switzerland

Christopher M. Dettmar
Department of Chemistry
Purdue University
West Lafayette, Indiana

Chen-Yuan Dong
Department of Physics
National Taiwan University
Taipei, Taiwan

Kevin W. Eliceiri
Department of Biomedical Engineering
University of Wisconsin—Madison
Madison, Wisconsin

Annika Enejder
Department of Chemical and Biological
    Engineering
Chalmers University of Technology
Gothenburg, Sweden

Steven C. George
Department of Biomedical Engineering
and
Department of Chemical Engineering and
    Materials Science
University of California, Irvine
Irvine, California

Xiaoxing Han
Institute of Optics
University of Rochester
Rochester, New York

Chiu-Mei Hsueh
Department of Physics
National Taiwan University
Taipei, Taiwan

Po-Sheng Hu
Department of Engineering Science
National Cheng-Kung University
Tainan, Taiwan

Patricia J. Keely
Department of Cell and Regenerative
    Biology
University of Wisconsin—Madison
Madison, Wisconsin

Alex C. Kwan
Helen Wills Neuroscience Institute
University of California, Berkeley
Berkeley, California
and
School of Applied and Engineering Physics
Cornell University
Ithaca, New York

Wen Lo
Department of Physics
National Taiwan University
Taipei, Taiwan

Bertrand M. Lucotte
Laboratory of Cardiac Energetics
National Institutes of Health
Bethesda, Maryland

Oleg Nadiarnykh
Department of Physics and Astronomy
VU University
Amsterdam, the Netherlands

Edward B. Neufeld
Laboratory of Cardiac Energetics
National Institutes of Health
Bethesda, Maryland

Mutsuo Nuriya
Department of Pharmacology
School of Medicine
Keio University
Tokyo, Japan

Nicolas Olivier
École Polytechnique
CNRS–Inserm
Palaiseau, France

Francesco S. Pavone
European Laboratory for Non-Linear
    Spectroscopy (LENS)
and
Department of Physics
University of Florence
Florence, Italy

Carolyn A. Pehlke
Spatiotemporal Modeling Center
University of New Mexico
Albuquerque, New Mexico

Seth W. Perry
Department of Biomedical Engineering
University of Rochester
Rochester, New York

Christopher B. Raub
Department of Bioengineering
University of California, Irvine
Irvine, California

Leonardo Sacconi
European Laboratory for Non-Linear
    Spectroscopy (LENS)
and
National Institute of Optics—National Research
    Council (INO—CNR)
University of Florence
Florence, Italy

Marie-Claire Schanne-Klein
École Polytechnique
CNRS–Inserm
Palaiseau, France

Caroline A. Schneider
The Alliance of Crop, Soil, and Environmental
    Science Societies (ACSESS)
Madison, Wisconsin

Etienne Shaffer
École Polytechnique
Fédérale de Lausanne
Vaud, Switzerland

Garth J. Simpson
Department of Chemistry
Purdue University
West Lafayette, Indiana

Chiara Stringari
Department of Biomedical Engineering
University of California, Irvine
Irvine, California

Ruth Sullivan
Research Animal Resources Center
University of Wisconsin—Madison
Madison, Wisconsin

Chi-Kuang Sun
Department of Electrical Engineering
National Taiwan University
Taipei, Taiwan

Hsin-Yuan Tan
Institute of Biomedical Engineering
National Taiwan University
Taipei, Taiwan
and
Department of Ophthalmology
Chang Gung University
Linko, Taiwan

Karissa Tilbury
Department of Biomedical Engineering
University of Wisconsin—Madison
Madison, Wisconsin

Bruce J. Tromberg
Beckman Laser Institute and Medical Clinic
University of California, Irvine
Irvine, California

Francesco Vanzi
European Laboratory for Non-Linear
    Spectroscopy (LENS)
and
Department of Evolutionary Biology
University of Florence
Florence, Italy

Rafael Yuste
Department of Biological Sciences
Howard Hughes Medical Institute
Columbia University
New York, New York

I

Principles of SHG Imaging

1

Theoretical Framework for SHG Microscopy: A Matrix-Based Approach

Christopher M.
Dettmar
Purdue University

Garth J. Simpson
Purdue University

The growing availability of turnkey ultrafast laser sources has fueled a revolution in optical microscopy, with second-harmonic generation (SHG) imaging emerging as a powerful probe for local order. The coherent nature of SHG results in high specificity for certain classes of ordered structures, often producing virtually no detectable background from disordered media. Consequently, image contrast is unique from that achievable using conventional optical methods. SHG has been used, and promises to provide continued insight, in the study of a wide variety of ordered systems, such as cells, organelles, tissues, complex surfaces, and other chiral materials as described in more detail in Parts II and III of this book.

The key objective of this chapter is to provide an initial foundation for quantitatively connecting SHG at the molecular level to the measured polarization-dependent image contrast in microscopy measurements. A general matrix-based mathematical approach is presented that removes some of the conceptual complexities of more conventional tensor-based methods. This approach has the additional advantage of providing simple extendibility for incorporating increasing levels of complexity. Furthermore, it is relatively simple to numerically invert to allow for either prediction or analysis. Several key limiting cases commonly arising in SHG microscopy measurements are considered within this framework to provide a reference point for more elaborate analyses.

1.1 Physical Overview

1.1.1 Linear Polarizability at the Molecular Level

Before tackling nonlinear optical effects, it arguably makes sense to describe linear optical effects within the same conceptual framework to better understand the close similarities. At the single molecule level, the linear polarizability describes the magnitude and phase of the dipole induced in the molecule by an AC or DC field. For an isolated molecule at optical frequencies, the linear polarizability describes optical scattering. When the density of scattering centers is high enough such that many are located distances much less than the optical wavelength, interference between the collective set of individual dipoles leads to directional reflection and refraction of light.

This linear polarizability can be qualitatively interpreted as the efficiency of "sloshing" the electron cloud of a molecule in a particular direction when driven by a polarization in a particular direction (not necessarily the same). Considering first a one-dimensional "molecule" as an illustrative example, the linear response is depicted in Figure 1.1. For a low applied field, $\vec{e}$, the polarization, $\vec{p}$, induced in the molecule will undergo harmonic oscillation, the magnitude of which is inversely related to the curvature of the surface. A high curvature corresponds to small charge displacement and induced dipole, and vice versa. (Although the potential energy surface holding the multi-electron cloud to the molecule is often not known, Taylor-series expansion about the zero-field energy minimum always produces a harmonic local potential close to the minimum sampled at low field strengths.) Formally, $\vec{P}$ is related to the driving field $\vec{e}$ by matrix multiplication, each element of which is inversely proportional to the curvature of the potential energy surface as in Equation 1.1. For a three-dimensional molecule, the polarizability is described by a 3×3 matrix, α, each ij element of which describes the magnitude and phase of the i-polarized dipole generated for a j-polarized driving field. Matrix multiplication of the polarizability matrix by the incident field describes the direction and magnitude of the induced polarization.[1]

$$\begin{bmatrix} P_x \\ P_y \\ P_z \end{bmatrix} = \begin{bmatrix} \alpha_{xx} & \alpha_{xy} & \alpha_{xz} \\ \alpha_{yx} & \alpha_{yy} & \alpha_{yz} \\ \alpha_{zx} & \alpha_{zy} & \alpha_{zz} \end{bmatrix} \cdot \begin{bmatrix} e_x \\ e_y \\ e_z \end{bmatrix} \qquad (1.1)$$

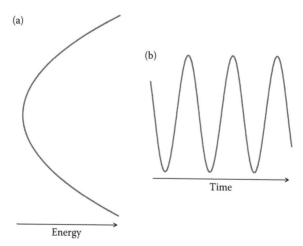

FIGURE 1.1 Diagram of the potential energy well of a one-dimensional molecule under the harmonic oscillator approximation (a) and the time-dependent trace of an oscillation within the well (b). This approximation is valid in the limit of low-energy driving fields.

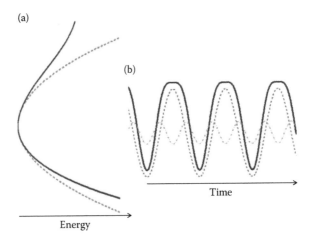

FIGURE 1.2 Diagram of an anharmonic oscillator potential energy well of a one-dimensional molecule (a, solid) with the harmonic oscillator overlaid for comparison (a, dashed) and the time-dependent trace of an oscillation within the well (b, solid). The oscillation can be largely recovered from a weighted sum of several harmonics (b, dashed, and dash-dotted).

1.1.2 Nonlinear Polarizability at the Molecular Level

Nonlinear optical interactions can be easily built into this same conceptual framework by consideration of an anharmonic oscillator.[2] From inspection of Figure 1.2 for a hypothetical one-dimensional molecule, introduction of anharmonicity in the potential surface will result in a temporal distortion in the induced polarization. For moderate driving fields, the distortion will be minimal, but will increase in significance with increases in the driving field and/or the degree of anharmonicity in the potential energy surface. These distortions in the time domain are recovered in the frequency-domain by including contributions from higher harmonics of the driving frequency. This effect is illustrated in Figure 1.2, where the addition of contributions at the doubled frequency recovers much of the distortion introduced by the depicted anharmonicity. From the Taylor-series expansion of the potential surface, the magnitude of the contribution at the doubled frequency scales with the square of the driving field.[2] Unlike the linear polarizability, the phase of the frequency-doubled contribution inverts if the orientation of the molecule flips. Consequently, the net polarization at the doubled frequency (and all higher even harmonics) sums to zero from an assembly of one-dimensional molecules with equal probability for both orientations.

1.2 Mathematical Formalism

1.2.1 The Molecular Tensor in the Local Frame

The molecular description of SHG becomes a bit more interesting for realistic molecules occupying three-dimensional space. The molecular nonlinear optical properties are described by a $3 \times 3 \times 3$ tensor, $\beta^{(2)}$, the magnitude of each β_{ijk} element describing the efficiency of generating i-polarized SHG for driving fields polarized along the j and k molecular axes. Fortunately, symmetry properties often reduce the number of parameters required to describe the molecular tensor well below the 27 possible. In this chapter, only systems of local uniaxial symmetry (e.g., membranes and fibers) will be considered for two key reasons: (i) they are among the most common and important classes of systems currently studied by SHG microscopy (confirmed by their ubiquitous appearance in other chapters), and (ii) they are significantly more concise to treat mathematically than the more general case of molecular assemblies of lower local symmetry.

Locally uniaxial systems include most naturally formed biopolymer assemblies studied in SHG microscopy measurements as well as lipid bilayers. In virtually all cases studying fibrous tissues, measurements are performed on assemblies of many fibrils (e.g., collagen, myosin, tubulin, cellulose, etc.), such that the ensemble fiber retains local uniaxial symmetry, even if the symmetry of the individual fibrils is lower as in the case of collagen and cellulose. In uniaxial systems, only four unique tensor elements are required to completely describe the polarization-dependent nonlinear optical properties within the local molecular-scale frame: β_{zzz}, $\beta_{zxx} = \beta_{zyy}$, $\beta_{xzx} = \beta_{yzy} = \beta_{xxz} = \beta_{yyz}$, and $\beta_{xyz} = \beta_{xzy} = -\beta_{yxz} = -\beta_{yzx}$, with the z-axis defined as the uniaxial axis.[3]

1.2.2 Experimental Observables and the Jones Tensor

To bridge the local tensor to the laboratory-frame measurements, it is useful to first consider the set of tensor elements accessible within the laboratory frame. Ultimately, the experimental observables will be dependent on both the intensity and polarization state of the exiting light. A concise means of representing the polarization state of coherent light is through a Jones vector, which is a unit vector describing the relative amplitude and phase of, in this case, the horizontal and vertical optical fields (indicated by the subscripts H and V, respectively).

$$\bar{e}_L^\omega = \begin{bmatrix} e_H^\omega \\ e_V^\omega \end{bmatrix}_L \tag{1.2}$$

The elements within the Jones vector can be real in the case of linearly polarized light or complex valued in the more general case of elliptical or circularly polarized light. The use of Jones vector representations is limited to characterizing light with a well-defined polarization state. In turbid matrices, or for partially depolarized light, Stokes vectors and Mueller matrices should be used instead,[4] which is beyond the scope of this work.

Every optical element that impacts the amplitude or polarization state of the light can be described mathematically by a corresponding Jones transfer matrix.[4,5] By matrix multiplication, all polarization-dependent observables can be uniquely predicted from precise knowledge of the initial Jones vector, provided the subsequent Jones transfer matrices are also known.

By extension to second-order nonlinear optics, an analogous Jones tensor can be generated.[6] This phenomenological 8-element $2 \times 2 \times 2$ tensor, 6 of which are unique in SHG, is effectively a nonlinear polarization transfer tensor, completely describing the expected polarization state generated experimentally for any combination of incident polarization states, both of which are typically coming from the same source in SHG. It serves an analogous role as the Jones matrix for reducing down the collective influence of a complex optical path to a single polarization transfer matrix. Since all possible experimental outcomes in an SHG measurement can be determined from detailed knowledge of the Jones tensor, it also contains all the information experimentally accessible in a single measurement configuration.

Mathematically, the link between the Jones vector of the incident light and the corresponding Jones vector of the doubled light can be written in terms of a matrix product.[7]

$$\bar{e}_L^{2\omega} = \begin{bmatrix} (e_H^\omega)^2 & 2e_H^\omega e_V^\omega & (e_V^\omega)^2 & 0 & 0 & 0 \\ 0 & 0 & 0 & (e_H^\omega)^2 & 2e_H^\omega e_V^\omega & (e_V^\omega)^2 \end{bmatrix}_L \cdot \begin{bmatrix} \chi_{J,HHH}^{(2)} \\ \chi_{J,HHV}^{(2)} \\ \chi_{J,HVV}^{(2)} \\ \chi_{J,VHH}^{(2)} \\ \chi_{J,VVH}^{(2)} \\ \chi_{J,VVV}^{(2)} \end{bmatrix}_L \tag{1.3}$$

In Equation 1.3, the six unique elements of the Jones tensor have been written as a vector, with the Jones vector of the doubled light given by the product of this vector with a matrix derived from the elements of the incident Jones vector for $\bar{e}_L^\omega$. This form is certainly not the only way to write this relationship, but the use of the vectorized form of the tensor has distinct practical advantages in subsequent steps.

Experimentally, image contrast in SHG microscopy depends on detection of either the intensities of the different polarization components or the fields themselves as functions of the incident polarization state(s). Precise knowledge of the Jones tensor allows prediction of the experimental observables within the image. Under appropriate conditions, the problem can also be inverted to recover the Jones tensor elements from polarization-dependent SHG measurements.[6b,7,8] The scope of the current work is focused primarily on providing a general framework for relating any experimental observables back to the Jones tensor and ultimately the local tensor introduced in the preceding section.

1.2.3 Case 1: Azimuthal Rotation Only

The strategy taken here to connect the local and laboratory responses will be to start with the simplest cases initially neglecting some of the more subtle interactions, then to systematically introduce increasingly general frameworks. Consistent with this approach, we will consider first the simplest case of plane-wave excitation of a point source located in vacuum.

Even within this simple context, the local frame is generally not perfectly oriented with the laboratory frame in SHG microscopy measurements. In most cases, the local uniaxial axis is tilted significantly from the horizontal and vertical image axes, with that orientation changing as a function of position within the field of view. Two obvious strategies emerge for interpreting polarization-dependent measurements in such instances: (1) mathematically rotate the molecular tensor to the laboratory frame to interpret the laboratory-frame measurements, or (2) rotate the laboratory frame optical fields to the local frame, then rotate the generated local polarization back to the laboratory frame. The two approaches are mathematically equivalent, but the first is arguably easier to implement with respect to book-keeping and will be the only one presented herein.

Considering a system with local uniaxial symmetry, only the two rotation angles ϕ and θ are required for projecting the local tensor onto the laboratory frame (Figure 1.3). In the case of fibers, the uniaxial axis runs parallel to the fiber axis. For surfaces and membranes, the uniaxial axis is normal to the surface. Often, the uniaxial axis lies close to the X–Y plane of the image, such that the rotation can be reasonably approximated as depending only on ϕ (i.e., assuming $\theta = \pi/2$). In this limiting case, the coordinate

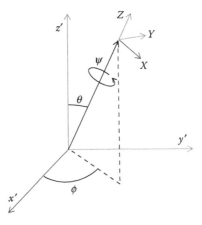

FIGURE 1.3 Depiction of the Euler angles involved in rotating a point from the XYZ coordinate system to the $x'y'z'$ coordinate system or *vice versa*. ϕ is the azimuthal angle, θ is the polar angle, and ψ is the twist angle.

transformation to express a local frame vector, v, or point in the laboratory frame, v', is performed using a simple rotation matrix.

$$\begin{bmatrix} v'_x \\ v'_y \end{bmatrix} = \begin{bmatrix} \cos\phi & -\sin\phi \\ \sin\phi & \cos\phi \end{bmatrix} \cdot \begin{bmatrix} v_x \\ v_y \end{bmatrix} \tag{1.4}$$

This expression can be rewritten more concisely.

$$\boldsymbol{v'} = R(\phi) \cdot \boldsymbol{v} \tag{1.5}$$

Transformation of a tensor is performed using the same basic approach, but with considerably more elements that must be rotated. If we consider first the most general set of eight nonzero possible elements within both the laboratory tensor (six of which are unique in SHG) and the molecular tensor when the z-axis lies within the X–Y plane, the rotation matrix is given by expanding the 2×2 rotation matrix in Equation 1.4 to an 8×8 matrix through two successive Kronecker products (i.e., multiplication of each element of one matrix by another matrix).

$$\bar{\chi} = [R(\phi) \otimes R(\phi) \otimes R(\phi)] \cdot \bar{\beta} \tag{1.6}$$

Explicit evaluation of this rotation matrix followed by simplification assuming local uniaxial symmetry for the case of SHG results in the following relatively concise 6×4 coordinate transformation matrix.

$$\begin{bmatrix} \chi^{(2)}_{J,HHH} \\ \chi^{(2)}_{J,HHV} \\ \chi^{(2)}_{J,HVV} \\ \chi^{(2)}_{J,VHH} \\ \chi^{(2)}_{J,VVH} \\ \chi^{(2)}_{J,VVV} \end{bmatrix}_L = \begin{bmatrix} 0 & 2g & g & i \\ 0 & f-h & -h & h \\ 0 & -2g & i & g \\ 0 & -2h & f & h \\ 0 & i-g & -g & g \\ 0 & 2h & h & f \end{bmatrix} \cdot \begin{bmatrix} \beta^{(2)}_{xyz} \\ \beta^{(2)}_{yyz} \\ \beta^{(2)}_{zyy} \\ \beta^{(2)}_{zzz} \end{bmatrix}_l \tag{1.7}$$

A concise shorthand notation has been introduced,[9] with each letter representing a different trigonometric function of an orientational angle (summarized below).

$$\begin{aligned} a &= \langle \sin\phi \rangle \\ b &= \langle \cos\phi \rangle \\ c &= \langle \sin^2\phi \rangle \\ d &= \langle \sin\phi\cos\phi \rangle \\ e &= \langle \cos^2\phi \rangle \\ f &= \langle \sin^3\phi \rangle \\ g &= \langle \sin^2\phi\cos\phi \rangle \\ h &= \langle \sin\phi\cos^2\phi \rangle \\ i &= \langle \cos^3\phi \rangle \end{aligned} \tag{1.8}$$

The expression in Equation 1.7 provides an initial handle to connect the local and laboratory frames in SHG microscopy measurements, offering a route for interpreting local structure and orientation information from polarization-dependent SHG measurements. Interestingly, all chiral-specific effects linked to the β_{xyz} and related tensor elements disappear in this orientation. Because the generated polarization is orthogonal to the two driving fields for β_{xyz} and its permutations, the only nonlinear polarization allowed is polarized along the direction of beam propagation (i.e., along z).

Using this set of expressions, the far-field polarization state of the doubled light expressed as a Jones vector can be related directly back to the azimuthal rotation angle of the sample within the field of view ϕ, the four unique elements within the local uniaxial tensor $\beta^{(2)}$, and the incident far-field fundamental polarization state through Equations 1.3 and 1.7.

However, the genesis of the laboratory frame tensor describing the experimental observables in such a relatively simple form required the use of several assumptions and approximations that may or may not hold in many microscopy measurements, including: (i) the assumption of the local z-axis lying within the image plane, (ii) the assumption of plane-wave excitation (as opposed to a tightly focused beam), (iii) the neglect of local field corrections from dielectric screening, and (iv) the assumption of a point SHG source, each of which is considered in subsequent sections.

1.2.4 Case 2: Combined Azimuthal and Polar Rotation

The situation becomes a bit more interesting when considering both polar (in and out of the image plane) and azimuthal (rotation within the image plane) orientation of the local z-axis, depicted in Figure 1.3. However, this case also represents the more common situation in practice, in which the local axis is not perfectly aligned within the image plane. Rotation of an object out-of-plane in a fixed laboratory reference frame can be treated by including a tilt angle rotation, θ, with the total rotation matrix given by the matrix product of the two individual operations.

$$R(\theta,\phi) = \begin{bmatrix} \cos\phi & -\sin\phi & 0 \\ \sin\phi & \cos\phi & 0 \\ 0 & 0 & 1 \end{bmatrix} \cdot \begin{bmatrix} \cos\theta & 0 & \sin\theta \\ 0 & 1 & 0 \\ -\sin\theta & 0 & \cos\theta \end{bmatrix} = \begin{bmatrix} \cos\theta\cos\phi & -\sin\phi & \sin\theta\cos\phi \\ \cos\theta\sin\phi & \cos\phi & \sin\theta\sin\phi \\ -\sin\theta & 0 & \cos\theta \end{bmatrix} \quad (1.9)$$

Working through the coordinate transformations using the equalities present for SHG of uniaxial assemblies $\beta_{yyz} = \beta_{yzy} = \beta_{xzx} = \beta_{xxz}$, $\beta_{zyy} = \beta_{zxx}$, and $\beta_{xyz} = \beta_{xzy} = -\beta_{yzx} = -\beta_{yxz}$ yields the following set of orientational angles connecting the local frame to the macroscopic observables.

$$\begin{bmatrix} \chi_{J,HHH} \\ \chi_{J,HHV} \\ \chi_{J,HVV} \\ \chi_{J,VHH} \\ \chi_{J,VVH} \\ \chi_{J,VVV} \end{bmatrix}_L = \begin{bmatrix} 0 & 2(Hi + Ag) & Hi + Ag & Fi \\ Db & Aa - 2Fh & -Fh & Fh \\ 2Da & -2Fg & Hh + Ai & Fg \\ -2Db & -2Fh & Hh + Af & Fh \\ -Da & Ab - 2Fg & -Fg & Fg \\ 0 & 2(Hf + Ah) & Hf + Ah & Ff \end{bmatrix} \cdot \begin{bmatrix} \beta_{xyz} \\ \beta_{yyz} \\ \beta_{zyy} \\ \beta_{zzz} \end{bmatrix}_{l,eff} \quad (1.10)$$

The same shorthand notation for the orientational averages is used, with capital letters indicating the trigonometric functions of the polar tilt angle, θ, and lowercase letters indicating the identical trigonometric functions of the azimuthal rotation angle, ϕ. Assuming a polar tilt angle of $\theta = \pi/2$, consistent with the local uniaxial z-axis oriented within the image plane, recovers the expression in Equation 1.7.

The expressions above provide a starting point for connecting local structure and orientation to the experimental observables measured in the laboratory frame. If one knows the molecular tensor *a priori*, it is in principle fairly straightforward to determine the most probable angles θ and ϕ for a given set of polarization-dependent measurements. Similarly, if one knows the orientation angles *a priori*, measurements of the laboratory-frame tensor elements can enable determination of the local tensor.

1.3 Special Considerations

To a first-order approximation, one might initially assume that the polarization state of the fundamental light within the focal volume will simply be equal to the polarization state of the incident light in the far field. However, this assumption neglects two key interactions: (i) the perturbations due to local field factors, which correct for the screening arising from the local dielectric medium in which the nonlinear optical source is embedded, and (ii) the influence of focusing on the local electric fields within the confocal volume. Both are considered here.

1.3.1 Corrections for the Local Dielectric Medium

For a point source embedded in a dielectric medium, the local fields experienced by the SHG-active source and detected from that source will be affected by the local dielectric properties. Introducing correction terms accounts for these perturbations. In the limit of a small source relative to the wavelength of light, the corrections are quite straightforward to apply, generally affecting only the relative magnitudes of the different polarization components. Lorentz corrections are arguably the simplest analytical expressions routinely used for handling such screening effects. The influence of the local field corrections can be introduced by rescaling each element.

$$\beta^{(2)}_{ijk \cdot eff} = L^{2\omega}_{ii} L^{\omega}_{jj} L^{\omega}_{kk} \beta^{(2)}_{ijk} \tag{1.11}$$

The Lorentz scaling factors can be written as elements of diagonal matrices of the following form:

$$L^{\omega} = \frac{1}{3} \begin{bmatrix} n_x^2 + 2 & 0 & 0 \\ 0 & n_y^2 + 2 & 0 \\ 0 & 0 & n_z^2 + 2 \end{bmatrix} \tag{1.12}$$

There are some lingering questions about the appropriate form for the local field correction matrix describing the doubled light $L^{2\omega}$,[10] but the most common approach in practice is arguably to use the same expression, but with the optical constants evaluated at the doubled frequency. The primary goal of this work is to define and describe a general framework, and we leave it up to the reader to decide the most appropriate specific model to use to treat the local field factors for the exiting beam should the use of the effective tensor prove insufficient for the analysis. In a uniaxial system (e.g., a thin fiber) with the *z*-axis defined as the unique axis, the local dielectric constant will generally be anisotropic, with $n_x = n_y = n_o$ and $n_z = n_e$ (i.e., the ordinary and extraordinary refractive indices of the fiber, respectively).

Most generally, the influence of the local field corrections can be incorporated mathematically by multiplication of the 27 element $\beta^{(2)}$ vector by a 27 × 27 matrix generated from the double Kronecker produce of the three *L* matrices.

$$L = L^{2\omega} \otimes L^{\omega} \otimes L^{\omega} \tag{1.13}$$

For SHG microscopy measurements of systems with local uniaxial symmetry with the z-axis defined as the extraordinary axis and using Lorentz scaling factors in Equation 1.12, only a few of these 27 triple products are required.

$$
\begin{bmatrix} \beta_{xyz} \\ \beta_{yyz} \\ \beta_{zyy} \\ \beta_{zzz} \end{bmatrix}_{l,eff} = \begin{bmatrix} L_{xx}^{2\omega}L_{yy}^{\omega}L_{zz}^{\omega} & 0 & 0 & 0 \\ 0 & L_{yy}^{2\omega}L_{yy}^{\omega}L_{zz}^{\omega} & 0 & 0 \\ 0 & 0 & L_{zz}^{2\omega}L_{yy}^{\omega}L_{yy}^{\omega} & 0 \\ 0 & 0 & 0 & L_{zz}^{2\omega}L_{zz}^{\omega}L_{zz}^{\omega} \end{bmatrix} \cdot \begin{bmatrix} \beta_{xyz} \\ \beta_{yyz} \\ \beta_{zyy} \\ \beta_{zzz} \end{bmatrix}_{l} \tag{1.14}
$$

For a point dipole embedded in an isotropic system, every tensor element is rescaled by the same constant term. In this limit, the Lorentz corrections influence only the overall intensity but not the polarization-dependence of the response (i.e., $\beta_{eff}^{(2)} = C \cdot \beta^{(2)}$). In the more general case of systems that are not point dipoles or are embedded in anisotropic media (e.g., fibers), the absence of off-diagonal elements within the local field factor matrices and L results in the rescaling of each element of the local $\beta^{(2)}$ tensor (i.e., $\beta_{ijk \cdot eff}^{(2)} \propto \beta_{ijk}^{(2)}$). As a result, all the equalities between the local tensor elements demanded by symmetry in uniaxial systems are still present after applying the local field corrections. Only the relative magnitudes (and possibly phases) of the effective tensor elements may be altered by the corrections. If the scientific objective is to quantitatively relate the laboratory measurements directly back to molecular-level properties, these corrections should be included. However, for many other objectives, such as imaging of cellular or tissue structure, they can simply be folded into the working definition of an effective local tensor and be otherwise neglected.

1.3.2 Incorporating the Influence of Focusing within the Paraxial Approximation

Before presenting methods for treating the local fields arising under tight focusing, it is useful to first develop a general framework that is valid in the limit of gentle focusing (i.e., with a numerical aperture (NA) < ~0.8). Within the paraxial approximation, the electric field for polarized light as a function of axial position z and radial distance from the optical axis ρ is given by the following expressions, calculated for a focused Gaussian beam.[5]

$$
\begin{bmatrix} E_x(\rho,z) \\ E_y(\rho,z) \end{bmatrix} = \begin{bmatrix} e_{x0} \\ e_{y0} \end{bmatrix} C^{\omega}(\rho,z) \tag{1.15a}
$$

$$
C^{\omega}(\rho,z) = \frac{W_0}{W(z)} \exp\left[-\frac{\rho^2}{W^2(z)}\right] \exp\left[-i\left(kz + k\frac{\rho^2}{2R(z)} - \tan^{-1}\left(\frac{z}{z_0}\right)\right)\right] \tag{1.15b}
$$

$$
R(z) = z\left(1 + \left(\frac{z_0}{z}\right)^2\right) \tag{1.15c}
$$

$$
W(z) = W_0\sqrt{1 + \left(\frac{z}{z_0}\right)^2} \tag{1.15d}
$$

$$
z_0 = \frac{\pi W_0^2}{\lambda} = \frac{\lambda}{\pi\theta_0^2} \tag{1.15e}
$$

In the above set of equations, W_0 is the beam waist, z_0 is the Rayleigh length describing the axial extent of the focal volume, θ_0 describes the angular spread of the incident light (related to the numerical aperture), and k is the wavevector, given by $2\pi n/\lambda$.[5] This expression differs from plane-wave propagation by two notable effects. First, there is a z-dependent Gaussian radial scaling of the field amplitude, the width of which is minimized for $z = 0$ (corresponding to the focal plane). Second, there is an overall additional phase change of π that varies smoothly upon traversing the focal volume from the $\tan^{-1}(z/z_0)$ term in the exponent, which is known as the Guoy phase shift.

Although the amplitude of the field changes significantly across the focal volume from Equation 1.15, both the x- and y-polarization components are equally affected by the change. As such, the normalized Jones vector describing the local polarization state remains identical for each individual position across the focal volume, scaled in magnitude and absolute phase by the axial and radial position. Since the polarization state of light described by Jones vectors only includes relative phase, the change in absolute phase between the fundamental and SHG across the focal volume is inconsequential in the limit of a thin sample. Consequently, the polarization transfer matrices connecting the local Jones vector to the far-field Jones vector and vice versa are simply given by identity matrices and can be largely ignored, with one caveat. This analysis *only* holds in the limit of thin or point samples relative to the focal volume. Because z_0 depends on wavelength, the absolute phase between the fundamental and the second-harmonic beams can shift across the focal volume. For samples in which the SHG source extends axially across distances approaching or exceeding the Rayleigh length, coherent interference between the SHG generated at each axial slice can influence the detected intensity and polarization, described in more detail in Section 1.3.3.

Formally, this outcome can be expressed as a Jones matrix by recasting Equation 1.14 using a position-dependent complex scaling factor, $C^\omega(\rho,z)$. Since the formalism introduced herein is based on expressing the local tensors within the laboratory frames, it is arguably most convenient to express the problem solving for the far-field polarization state, rather than in the local frame.

$$\frac{1}{C^\omega(\rho,z)}\begin{bmatrix} 1 & 0 & 0 \\ 0 & 1 & 0 \end{bmatrix}\begin{bmatrix} E_x(\rho,z) \\ E_y(\rho,z) \\ E_z(\rho,z) \end{bmatrix}^\omega = \begin{bmatrix} e_{x0} \\ e_{y0} \end{bmatrix}^\omega \tag{1.16}$$

If it is assumed that the detected SHG radiated by the sample can also be reliably modeled as a Gaussian beam, the paraxial approximation yields an analogous expression connecting the far-field and local field components for the SHG.

$$\frac{1}{C^{2\omega}(\rho,z)}\begin{bmatrix} 1 & 0 & 0 \\ 0 & 1 & 0 \end{bmatrix}\begin{bmatrix} E_x(\rho,z) \\ E_y(\rho,z) \\ E_z(\rho,z) \end{bmatrix}^{2\omega} = \begin{bmatrix} e_{x0} \\ e_{y0} \end{bmatrix}^{2\omega} \tag{1.17}$$

From inspection of Equations 1.16 and 1.17, all the field components are equally scaled in both amplitude and phase within the paraxial approximation in the limit of a thin SHG-active source. Assuming the SHG source mirrors the square of the driving fundamental field, $W_o^{2\omega}$ and $\theta_o^{2\omega}$ are both reduced accordingly. Interestingly, $z_o^{2\omega}$ is unchanged. Consequently, the situation is even simpler to treat mathematically than the local field correction factors described in the preceding section. In studies focused on interpreting polarization-dependent effects, the overall scaling factors for treating the perturbations to the polarization-dependence upon focusing simply disappear upon normalization!

1.3.3 Influence of Tight Focusing on the Local Polarization: Beyond the Paraxial Approximation

Under conditions of tight focusing, the local fields experienced within the confocal volume are generally distinct from those introduced in the far field. The reasons for this effect are depicted qualitatively in Figure 1.4. In brief, the high angles of incidence arising when using high NA objectives introduce field components in all three directions even if starting with linearly polarized light. The most common strategies developed for modeling the local fields generally consider the focusing of a Gaussian spatial profile for the far-field beam. In this limit, the local electric fields are given by integrals over Bessel functions.

$$
\begin{bmatrix} E_x \\ E_y \\ E_z \end{bmatrix}_{l,0} = \begin{bmatrix} -i(I_0 + I_2\cos(2\phi)) & -iI_2\sin(2\phi) \\ -iI_2\sin(2\phi) & i(I_0 + I_2\cos(2\phi)) \\ -2I_1\cos(\phi) & -2I_1\cos(\phi) \end{bmatrix} \begin{bmatrix} e_H \\ e_V \end{bmatrix}_L = G(r,\phi,z)_0 \begin{bmatrix} e_H \\ e_V \end{bmatrix}_L \tag{1.18a}
$$

where

$$
I_0(r,z) = \int_0^{\alpha} \cos^{\frac{1}{2}}\theta \, \sin\theta(1 + \cos\theta) J_0(kr\sin\theta)\exp(ikz\cos\theta)\mathrm{d}\theta \tag{1.18b}
$$

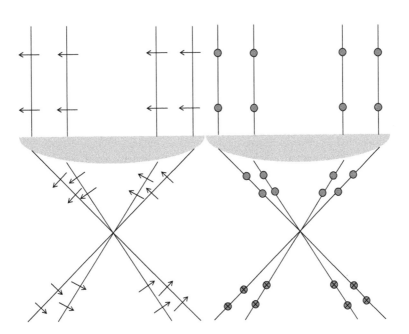

FIGURE 1.4 Graphical depiction of the origin of *z*-polarized incident field from beam polarized in the *x* direction. Upon focusing, the path of the beam is bent, converting some of polarization component in the plane of bending into the *z* direction. (left) Rays in the plane perpendicular to the polarization will not have this effect (right) and rays between these two extremes will produce some *y*- and *z*-polarization (not pictured).

$$I_1(r,z) = \int_0^\alpha \cos^{\frac{1}{2}}\theta \sin^2\theta \, J_1(kr\sin\theta)\exp(ikz\cos\theta)d\theta \qquad (1.18c)$$

$$I_2(r,z) = \int_0^\alpha \cos^{\frac{1}{2}}\theta \sin\theta(1-\cos\theta) J_2(kr\sin\theta)\exp(ikz\cos\theta)d\theta \qquad (1.18d)$$

$r = (x^2 + y^2)^{1/2}$, α is the aperture half-angle, k is the wave vector ($2\pi\, n/\lambda$), and the J_n terms are Bessel functions of the first type with order n.[11] The subscript of 0 on the cylindrical coordinates indicates the particular location of the SHG-active point source. Unlike in the paraxial case, the local fields change significantly as a function of position close to the focal point, such that the polarization components experienced by the local SHG-active source can also change while imaging.

The calculations shown in Equation 1.18 for the optical fields present within the volume of a focused Gaussian beam have interesting implications for polarization-dependent measurements in nonlinear optical imaging. For an x-polarized incident light, the y- and z-components of the optical field calculated using Equation 1.18 exhibit nodes along the optical axis. Even for high-NA excitation, the field strengths are at least two orders of magnitude lower than the x-polarized fields (for x-polarized incident light) in that region, and may safely be assumed to be negligible. Interestingly, the z-polarized fields exhibit two side lobes of opposite polarity positioned along the x-axis, such that objects adjacent to the beam center can couple to these additional field components. These field components drop off quickly from a maximum intensity of 0.25 relative to the maximum intensity of E_x for an NA of 1.4 to 0.05 for an NA of 0.8 and are therefore only generally significant when using oil or water immersion objectives. For measurements of moderate precision (~1–2 significant figures), the use of the much simpler paraxial approximation expressions can still be reasonably accurate for measurements obtained with an NA < 1. However, explicit numerical calculations may be necessary for interpreting polarization-dependent data obtained at higher NA.[12]

As expected, explicit calculations of the local fields also generate an accumulated Guoy phase-shift of 180° upon traversing the focal volume, qualitatively similar to that described within the paraxial approximation detailed in Section 1.3.2.

The implications of nonnegligible z-polarized field components are tractable, but nontrivial, since the emerging SHG light can now couple through a whole new set of local frame tensor elements containing z indices. Indeed, all 27 possible nonzero laboratory-frame tensor elements can now contribute to the detected response. To systematically build up to this most general cases, we will first consider SHG from a point source, or more precisely, a source that is thin axially relative to the wavelength of light (i.e., <~100 nm in thickness along the z-axis). Starting with this limit has the distinct advantage of largely eliminating the need to consider complicating interference effects, described in greater detail in Section 1.4. In this limit, the local electric field experienced at any specific location within the vicinity of the confocal volume can be related back to the incident polarization through the polarization transfer matrix described in Equation 1.18. Unlike a plane wave, the local fields change significantly as a function of position close to the focal point, such that the polarization components experienced by the local SHG-active source can also change while imaging.

Just as within the paraxial approximation in Section 1.3.2, the matrix in Equation 1.18 needs to be inverted to isolate the Jones vector and allow the local tensor to be expressed within the laboratory frame. Since G is a 3×2 matrix and inversion requires a square matrix, the simplest way to perform the inversion numerically is to left multiply both sides of the equation by G^T to recover an invertible matrix on the right, then left-multiply both sides by $(G^TG)^{-1}$ to recover the identity matrix on the right side of the equality. The local field vector is then multiplied by $\Gamma \equiv (G^TG)^{-1}G^T$, which is a 2×3 matrix. This same set of operations was performed implicitly in Section 1.3.2, but with a much simpler analytical outcome. The full 8 × 27 matrix to bridge the Jones tensor and the local effective tensor is then given by the Kronecker product of three such inverted matrices, two for the driving fields and a third calculated for the doubled light.

$$\Gamma(r,z,\varphi) = \Gamma^{2\omega} \otimes \Gamma^{\omega} \otimes \Gamma^{\omega} \tag{1.19}$$

The Kronecker products in Equation 1.19 will collectively produce an 8 × 27 matrix for relating all 27 elements of the effective laboratory frame tensor back to the experimentally measured polarization-dependent Jones tensor observables. The influences from local orientation (Sections 1.2.3 and 1.2.4) and local field factors (Section 1.3.1) can be systematically incorporated into this same mathematical framework.

$$\begin{bmatrix} \chi_{J,HHH} \\ \chi_{J,HHV} \\ \chi_{J,HVH} \\ \chi_{J,HVV} \\ \chi_{J,VHH} \\ \chi_{J,VVH} \\ \chi_{J,VHV} \\ \chi_{J,VVV} \end{bmatrix}_L = \Gamma(r,z,\varphi) \cdot R(\theta,\phi) \cdot \bar{\beta}_{l,eff} \tag{1.20}$$

The corresponding 27 × 27 matrix R can be easily generated by taking Kronecker products of the rotation matrix in Equation 1.9; $R(\theta,\phi) = R(\theta,\phi) \otimes R(\theta,\phi) \otimes R(\theta,\phi)$.

However, the matrix for $G^{2\omega}$ (and correspondingly, $\Gamma^{2\omega}$) has not yet been determined in the preceding analysis. This matrix effectively bridges the emission from the source polarization to the detected intensity and polarization state following the collection optics. Explicit values for calculating the matrix elements can be generated based on Green's function propagations of the nonlinear source polarization integrated over the focal volume and propagation of the source through standard matrices for describing the collection lens.[13] As in the case of the excitation fields, they are numerically calculable provided that the source polarization as a function of position is known *a priori*.

The good news is that this collective approach provides a very general toolkit for predicting the anticipated far-field experimental observables for a known source of arbitrary shape, arbitrary local tensor, and arbitrary incident far-field polarization state. However, the problem is not trivial to invert in order to recover the local source polarization from the detected intensity, since the matrices used are generated based on numerical integration of a very well-characterized source. In other words, if all the interesting properties of the sample are known *a priori*, the matrix for $G^{2\omega}$ can be explicitly calculated, but the more practical problem of relating a relatively small number of experimental observations back to the local structure is not as trivial beyond the paraxial approximation.

Since G^{ω} and $G^{2\omega}$ depend on position within the focal volume, the 8 × 27 matrix Γ also changes as a function of source location. The region immediately adjacent to the focal point will be considered explicitly, selected by the nature of its practical significance for SHG imaging. Within one beam waist of the focal point, both the z-polarized and y-polarized incident fields experience nodes and correspondingly low amplitudes. In this region of highest intensity, the local fields remain almost exclusively x-polarized for an x-polarized incident field. Consequently, provided the source is both thin and located close to the focal point, the validity of the plane-wave approximation for describing the polarization is recovered. There will be some uncertainty in the absolute phase of the exiting beam depending on the axial location of the sample relative to the focal point due to the Guoy phase shift, but the polarization state is a measure of relative phase and will not generally be impacted. Although the overall intensity drops quite rapidly with distance from the optical axis, the z-component of the fundamental field can be as much as 2/3 of the x-polarized component about two beam waists from the optical axis. Depending on the location of the SHG source within the focal volume, these z-polarized contributions may be relatively important, allowing access to tensor elements forbidden by the paraxial approximation.

1.4 Extension beyond the Thin Sample Limit

In all the preceding sections, the formalism has been limited intentionally to thin or point SHG sources. Extension to sources that are comparable to or greater than the wavelength of light in dimension requires revisiting some of the simplifying approximations related to both the local field corrections and the influence of the electric fields upon tight focusing. Using a Green's function approach, in which each location within the focal volume serves as a driven radiation source,[11,14] mathematical models have been generated for describing the predicted net radiation patterns expected for sources of different size, shape, and position relative to the focal volume, as described Section 1.3.3. In effect, the net emission arises from the collective interference between all the oscillating source polarizations within the excitation volume.

However, a basic understanding of the key interactions driving many of the interesting interference effects in SHG microscopy can arguably be most easily illustrated by first considering a Gaussian beam within the paraxial approximation. For simplicity, it will be assumed that the sample is uniform and semi-infinite (i.e., with a thickness significantly greater than z_0 and a width greater than W_0). In this limit, coherences that dictate the measured SHG responses are driven primarily by two key interactions: (i) the coherence length dictated by dispersion l_c and (ii) the coherence length dictated by the Guoy phase shift l_c^G.

1.4.1 Phase Considerations from Dispersion

The forward coherence length l_c^f depends on the presence of dispersion within the sample and is largely independent of focusing. Considering a plane wave propagating through a semi-infinite medium as illustrated in Figure 1.5, the difference in refractive index between the two wavelengths results in a phase-walk between the SHG generated previously and the next infinitesimal contribution generated by another slice. The phase-walk results in a transition from constructive to destructive interference, and oscillations in the net SHG intensity, referred to in bulk crystals as Maker fringes.[15] The expression for the forward coherence length is given by considering the distance required for a phase shift of π between the SHG and the square of the fundamental beam. This forward coherence typically ranges from 5 to 15 µm in tissues, and 30 µm in water.[16]

$$l_c^f = \frac{\lambda}{4} \cdot \frac{1}{\left|n^{2\omega} - n^{\omega}\right|} \tag{1.21}$$

For an SHG beam propagating in the backward (epi) direction, the phase walk between the fundamental and the SHG occurs much more quickly, over distances much less than the wavelength of light. In this case, the backwards coherence length is given by the following expression:

$$l_c^b = \frac{\lambda}{4} \cdot \frac{1}{\left|n^{2\omega} + n^{\omega}\right|} \tag{1.22}$$

In the absence of birefringence, all tensor elements are simply rescaled by the same factor in both the forward and backward directions, such that the coherence length from dispersion affects only the intensity and not the polarization state of the detected SHG.

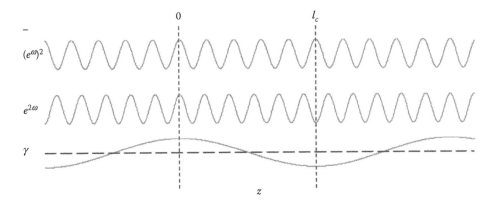

FIGURE 1.5 Difference in refractive index of ω and 2ω will give rise to a phase walk between the incident field and the SHG generated at a point z. This phase walk can result in interference between SHG generated in two different points along z. γ shows the oscillation between 100% constructive and 100% destructive interference as a function of distance between the points of generation of 2ω. l_c is the distance between these points. The frequency of γ will be proportional to the difference (or sum in the epi direction) of refractive indexes.

Although birefringence within the sample can impact both the forward and backward coherence lengths, the effect is most significant in the forward direction due to the difference appearing in the denominator. If the presence of birefringence can further reduce the denominator, the efficiency of SHG can improve dramatically. Quantitatively incorporating the role of birefringence into the most general forms for the analysis of semi-infinite media is beyond the scope of this work. However, full analysis is often not required. In many instances, the polarization dependence is dominated by the effective tensor contribution corresponding to the longest coherence length when measured in the limit of gentle focusing. For a uniaxial sample such as a collagen fiber, polarization along the uniaxial (fiber) axis is defined as the extraordinary axis and polarizations orthogonal to the unique axis are defined as ordinary. Under normal dispersion conditions, the polarization combination with the greatest coherence length will typically arise with the fundamental polarized along the unique (extraordinary) axis and the SHG polarized perpendicular to it (positive birefringence), or vice versa (negative birefringence). The latter leads to enhancement of the β_{zxx} tensor element, such that it can often be reasonably assumed to dominate the measured polarization-dependent response. Formally, this case corresponds to Type I phase matching. In the former case (positive birefringence), the β_{xzz} tensor element would be optimally enhanced, but is symmetry forbidden in uniaxial systems. The $\beta_{xxz} = \beta_{xzx}$ tensor elements provide the closest match and experience the greatest enhancement, corresponding to Type II phase matching. If the birefringence is sufficiently high that the coherence length for the Type I or Type II cases approaches or exceeds the Rayleigh length z_0, the one unique "phase-matched" tensor element can be reasonably assumed to dominate the polarization dependence of SHG. However, even under ideal phase-matching conditions, which never occur in tissues due to dispersion and randomness inherent in biological samples,[17] the dominance of a single element can be tempered significantly as the NA increases due to effective reduction of the net coherence length from the Guoy phase shift.

1.4.2 The Guoy Phase Shift

Unlike the coherence length contributions described by dispersion, the contributions from the Guoy phase shift are specific to focused beams such as those employed in microscopy measurements. Because the image inverts upon passing through the focal volume, the sign for each of the far-field polarization components also inverts. However, the SHG response scales with the square of the fundamental field, which does not invert. Consequently, the Guoy phase shift results in a net 180° phase walk between the doubled light and the square of the fundamental beam. This sign inversion arises smoothly across the focal volume manifest as a phase shift described by the $e^{-\tan^{-1}(z/z_0)}$ term derived from Equation 1.15. The Guoy phase term is summed in the exponent, such that it can be treated additively to other phase-shift contributions, including those arising from dispersion. The Guoy coherence length is given by the following expression derived assuming a Gaussian exiting SHG beam with a beam waist in the focal plane matched to the square of the fundamental beam waist.

$$l_c^G = 2z_0 \tan\left(\frac{\pi}{2}\right) = \infty \qquad (1.23)$$

Because the Guoy phase shift scales as a tangent function, it only approaches, but does not reach, a total phase shift of π. Therefore, by the conventional definition, the coherence length is ∞ for SHG (this is not the case for THG) but the span over which the Guoy phase shift arises is impacted by the degree of focusing. Therefore, we introduce the half-coherence length, corresponding to the distance producing a phase change of 90°.

$$l_c^{G/2} = 2z_0 \tan\left(\frac{\pi}{4}\right) = 2z_0 \qquad (1.24)$$

Because the Guoy phase length depends on z_0, higher NA objectives produce smaller values of $l_c^{G/2}$. In the backward direction (epi), the coherence length given by Equation 1.22 can be reasonably expected to be much shorter than the contributions from the Guoy phase shift and dominate. However, in the transmission direction if $l_c^{G/2}$ is significantly smaller than the forward coherence length, it can potentially dominate the net coherence length. Since the Guoy phase term affects each tensor element equally, it results only in a rescaling of each tensor element in the limit of comparatively long coherence lengths from dispersion.

The net coherence length of the sample in the forward direction is given by the sum of the contributions from normal dispersion and the Guoy phase shift. Over distances of $l_c^{G/2}$ spanning the focal volume, the Guoy phase shift varies approximately linearly with distance, just as l_c^f and l_c^b. In the limit of low NA, the Guoy half-coherence length approaches infinity and the dispersive/birefringent terms dominate. Conversely, under relatively tight focusing conditions and in the limit of weak dispersion/birefringence, the Guoy half-coherence length can be expected to dominate, impacting the magnitude and phase of each tensor element equally. In this latter limit, the rescaled $\beta_{eff}^{(2)}$ tensor is recovered for describing the net polarization-dependent SHG properties of the sample. Because the Guoy phase shift often brings the system close to destructive interference when focusing by inducing an initial shift of π, its presence can make the measurements much more susceptible to perturbations from even relatively weak dispersion, which can push the system to destructive interference more quickly than for plane waves. Some knowledge of the optical constants of the sample together with the beam path and the S/N of the measurements can be helpful for deciding whether one or both contributions need to be explicitly considered when interpreting image contrast and polarization-dependent measurements.

Finally, it should be noted once more that this analysis fundamentally relies on the validity of the paraxial approximation, which may not hold quantitatively when using high NA objectives. Under such conditions, the paraxial approximation provides only a qualitative means for aiding in the interpretation of image contrast, but should be used cautiously for quantitative analysis. The limits within which the paraxial approximation can still be used for describing polarization-dependent effects in

SHG microscopy depend largely on the precision of the experimental measurement. As described in Section 1.3.3, additional z-polarized field contributions can be as much as ~25% of the X and Y polarized contributions when using high NA oil- or water-immersion objectives, but drop off quickly as the NA decreases.

Although the influence on polarization is often minimal, the Guoy phase shift does significantly impact the scaling of the SHG intensity with the sample volume. For thin samples much smaller than the focal volume, the SHG intensity scales with the square of the number density. However, the detected intensity departs significantly from this quadratic scaling as the thickness increases due to the collective coherent interferences from dispersion, birefringence, and the Guoy phase shift. The length scale over which this asymptotic approach arises is dictated in the transmitted direction by the degree of focusing through the competition between changes in z_0 and $l_c^{G/2}$ versus those arising through dispersion from l_c^f.

1.5 Summary

The primary objective of this chapter is to provide an initial conceptual and mathematical framework for interpreting image contrast in SHG microscopy measurements, focusing on simplified methods valid in key limiting conditions. A matrix-based approach was presented for recasting the tensor mathematics into a more intuitive form and providing expandability to systematically include increasing levels of complexity in both the sample and the analysis. Limits of thin and thick samples were considered, along with effects in the limits of gentle and tight focusing. A key emphasis was centered on treating polarization-dependent measurements in SHG microscopy. This introductory chapter was designed to serve as a general starting point before advancing to some of the specific applications described in subsequent chapters.

Acknowledgments

This work was supported as part of the Center for Direct Catalytic Conversion of Biomass to Biofuels, an Energy Frontier Research Center funded by the U.S. Department of Energy, Office of Science, Basic Energy Sciences under Award # DE-SC0000997. GJS also acknowledges support from NIH R01.

References

1. Long, D. A., *The Raman Effect. A Unified Treatment of the Theory of Raman Scattering by Molecules*. John Wiley and Sons: New York, 2002.
2. (a) Shen, Y. R., *The Principles of Nonlinear Optics*. John Wiley & Sons: New York, 1984; (b) Boyd, R. W., *Nonlinear Optics*. 2nd ed.; Academic Press: Amsterdam, 2003.
3. (a) Campagnola, P. J., Millard, A. C., Terasaki, M., Hoppe, P. E., Malone, C. J., Mohler, W. A., Three-dimensional high-resolution second-harmonic generation imaging of endogenous structural proteins in biological tissues. *Biophys. J.* 2002, *82*(1), 493–508; (b) Chu, S.-W., Tai, S.-P., Sun, C.-K., Lin, C.-H., Selective imaging in second-harmonic-generation microscopy by polarization manipulation. *Appl. Phys. Lett.* 2007, *91*(10), 103903; (c) Cheng, X. G., Gurkan, U. A., Dehen, C. J., Tate, M. P., Hillhouse, H. W., Simpson, G. J., Akkus, O., An electrochemical fabrication process for the assembly of anisotropically oriented collagen bundles. *Biomaterials* 2008, *29*(22), 3278–3288.
4. Azzam, R. M. A., Bashara, N. M., *Ellipsometry and Polarized Light*. Elsevier: Amsterdam, 1987.
5. Saleh, B. E. A., Teich, M. C., *Fundamentals of Photonics*. John Wiley and Sons: New York, 1991.
6. (a) Plocinik, R. M., Simpson, G. J., Polarization characterization in surface second harmonic generation by nonlinear optical null ellipsometry. *Anal. Chim. Acta* 2003, *496*, 133–142; (b) Plocinik, R. M., Everly, R. M., Moad, A. J., Simpson, G. J., A Modular ellipsometric approach for

mining structural information from nonlinear optical polarization analysis. *Phys. Rev. B* 2005, *72*, 125409.

7. (a) Begue, N. J., Everly, R. M., Hall, V. J., Haupert, L., Simpson, G. J., Nonlinear optical Stokes ellipsometry. 2. Experimental demonstration. *J. Phys. Chem. C* 2009, *113*(23), 10166–10175; (b) Begue, N. J., Moad, A. J., Simpson, G. J., Nonlinear optical Stokes ellipsometry. 1. Theoretical framework. *J. Phys. Chem. C* 2009, *113*(23), 10158–10165.

8. (a) Polizzi, M. A., Plocinik, R. M., Simpson, G. J., Ellipsometric approach for the real-time detection of label-free protein adsorption by second harmonic generation. *J. Am. Chem. Soc.* 2004, *126*, 5001–5007; (b) Begue, N. J., Simpson, G. J., Chemically selective analysis of molecular monolayers by nonlinear optical Stokes ellipsometry. *Anal. Chem.* 2010, *82*(2), 559–566.

9. Davis, R. P., Moad, A. J., Goeken, G. S., Wampler, R. D., Simpson, G. J., Selection rules and symmetry relations for four-wave mixing measurements of uniaxial assemblies. *J. Phys. Chem. B* 2008, *112*, 5834–5848.

10. (a) Kaatz, P., Donley, E. A., Shelton, D. P., A comparison of molecular hyperpolarizabilities from gas and liquid phase measurements. *J. Chem. Phys.* 1998, *108*(3), 849–856; (b) Wortmann, R., Bishop, D. M., Effective polarizabilities and local field corrections for nonlinear optical experiments in condensed media. *J. Chem. Phys.* 1998, *108*(3), 1001–1007.

11. Yew, E. Y. S., Sheppard, C. J. R., Effects of axial field components on second harmonic generation microscopy. *Opt. Express* 2006, *14*(3), 1167–1174.

12. (a) Axelrod, D., Carbocyanine dye orientation in red cell membrane studied by microscopic fluorescence polarization. *Biophys. J.* 1979, *26*(3), 557–573; (b) Zyss, J., Chemla, D. S., Nicoud, J. F., Demonstration of efficient nonlinear optical crystals with vanishing molecular dipole moment: second-harmonic generation in 3-methyl-4-nitropyridine-1-oxide. *J. Chem. Phys.* 1981, *74*, 4800–11.

13. Wang, X. H., Chang, S. J., Lin, L., Wang, L. R., Huo, B. Z., Hao, S. J., Vector model for polarized second-harmonic generation microscopy under high numerical aperture. *J. Opt.* 2010, *12*(4), 8.

14. Cheng, J. X., Xie, X. S., Green's function formulation for third-harmonic generation microscopy. *J. Opt. Soc. Am. B-Opt. Phys.* 2002, *19*(7), 1604–1610.

15. Maker, P. D., Terhune, R. W., Nisenoff, M., Savage, C. M., Effects of dispersion and focusing on the production of optical harmonics. *Phys. Rev. Lett.* 1962, *8*(1), 21.

16. Kim, B. M., Eichler, J., Da, S. L. B., Frequency doubling of ultrashort laser pulses in biological tissues. *Appl. Opt.* 1999, *38*, 7145–7150.

17. (a) Mertz, J., Moreaux, L., Second-harmonic generation by focused excitation of inhomogeneously distributed scatterers. *Opt. Commun.* 2001, *196*(1–6), 325–330; (b) LaComb, R., Nadiarnykh, O., Townsend, S. S., Campagnola, P. J., Phase matching considerations in second harmonic generation from tissues: Effects on emission directionality, conversion efficiency and observed morphology. *Opt. Commun.* 2008, *281*(7), 1823–1832.

2

How to Build an SHG Apparatus

Riccardo Cicchi
National Institute of Optics—National Research Council (INO—CNR)

Leonardo Sacconi
European Laboratory for Non-Linear Spectroscopy (LENS)

National Institute of Optics—National Research Council (INO—CNR)

Francesco Vanzi
European Laboratory for Non-Linear Spectroscopy (LENS)

University of Florence

Francesco S. Pavone
European Laboratory for Non-Linear Spectroscopy (LENS)

University of Florence

Over recent years, second-harmonic generation (SHG) microscopy has been well established as a microscopic imaging modality used in biophysics, biomedical optics, and biophotonics applications. It has already been widely demonstrated that SHG is a powerful technique for both cellular and tissue imaging. However, a commercial microscope exclusively dedicated to SHG imaging is still lacking. Very often, SHG microscopes used in laboratory research are adapted from commercial (confocal or multiphoton) microscopes. However, for some applications, *ex novo* design is required to fit the technical specifications of the experiments to be performed. Several technical and experimental aspects have to be considered when designing a custom SHG microscope. This chapter aims to provide a technical guide to the researchers who want to build their own custom SHG microscope. This chapter is divided into three main parts. Starting from the possible solutions to be adopted for the mechanical system (Section 2.1), we then focus our attention on the most critical points of the experimental setup: the scanning (Section 2.2) and the detection (Section 2.3) systems. In each part of this chapter, we describe the most common configurations used in an experimental setup, highlighting their advantages and disadvantages, and focusing the attention of the reader on the most crucial aspects to be considered during both design and development. Finally (Section 2.4), we provide a short description of typical optical schemes used in SHG microscopy.

2.1 Mechanical System

In this section, we describe the mechanical elements required for building an SHG microscope. Starting from the optical table, we then focus our attention on the different solutions adaptable for the microscope stand, highlighting advantages and drawbacks of custom solutions in comparison to commercially available systems.

2.1.1 Optical Table

The mechanical structure of a custom SHG microscope must accomplish several functions. First of all, it has to be able to damp vibrations and avoid any external mechanical perturbation that can affect the measurements. Damping is in general obtained by installing the whole system on an antivibration optical table, which commonly consists of four independent legs that support the weight of a metallic horizontal breadboard. The legs have a hollow core in which a fluctuating actuator can move vertically driven by air under pressure provided by an external hydraulic circuit. Recently, optical table producers have developed a new damping technology based on piezo-electric actuators which are providing enhanced stability for applications requiring further damping. With this improvement, they offer three possible solutions for active vibration isolation: air-pressure damping, piezo-electric damping, and also hybrid systems. Air-pressure damping is generally sufficient for optical imaging applications, while piezo-electric damping is more suitable for applications requiring higher stability, such as electron or near-field microscopy.

The horizontal breadboard is made of a damping inner structure sandwiched between two stainless-steel plates with threaded holes (typically either ¼-20 or M7 thread) on one side in order to allow the positioning of opto-mechanical components. The two stainless-steel plates are a common feature for all the optical tables, while the inner structure varies among different models. There are several choices for the inner structure, including composite laminate, solid material, and lightweight honeycomb. The choice of construction depends on the type and size of the application, where sizes are typically 120×180 cm^2, 240 cm, or 300 cm (48×72, 48×96, or $48 \times 120''$). The optical surfaces with highest performance are honeycomb core tables, which offer rigidity with a light weight and the possibility to be extendable up to very large work surfaces. This is in general the most suitable solution for optical imaging applications, such as an SHG microscopy.

2.1.2 Microscope Stand

The microscope stand (intended as the microscope body) provides support to all the optical elements required for conventional optical microscopy. The most important elements are: the objective lens, the tube lens, the focusing system, bright field, and/or fluorescence illumination sources, the eyepiece, CCD mount, and the translational stage on which the sample is mounted. To give to the user the possibility of adding additional light sources and/or detectors for fluorescence or other imaging techniques, microscopes are equipped with additional access ports through which it is possible to couple and/or decouple light to/from the optical axis. Modern commercial microscopes offer solutions that are ergonomic in terms of comfort and versatile in terms of capability to be connected with add-on equipments (laser light, scanning head, detectors, filters) for various imaging techniques. Anyway, a commercial microscope is not always the best solution to be adopted for SHG imaging. In fact, while offering a large set of advantages in terms of simplicity and versatility, they are generally designed for imaging samples mounted on a standard microscope slide, hence they could have some limitations in imaging *in vivo* or massive *ex vivo* samples. The choice of the microscope body should thus be driven not only by the optical imaging technique to be employed, but also by the type of sample the researcher plans to analyze. Further, commercial microscope stands are generally designed for wide-field fluorescence imaging, so that different excitation modalities (such as laser scanning; see Section 2.2) and/or peculiar detection geometries (such as forward detection; see Section 2.3.2.1) cannot be easily implemented. Here, we provide two alternative solutions to the researchers who want to build their own custom microscope body without purchasing a commercial product.

2.1.2.1 Custom Microscope Stand

In this section, we show an example of a microscope stand built around a custom-made mechanical structure, which is basically equivalent to a commercial microscope skeleton. According to the description offered by Capitanio et al. (2005), a custom structure made of ErGaAl (erbium, gallium, aluminum) alloy can be used to build an optical microscope with high mechanical stability. In detail, the mechanical structure is composed by three $250 \times 250 \times 30$ mm³ platforms held by four stainless-steel columns with 25 mm diameter, 500 mm height (see Figure 2.1).

The lower platform serves as a basis for the whole microscope. The middle platform serves as a basis for the sample stage. Two manual translators, allowing gross movements of the sample in the XY plane, and a piezoelectric translator, allowing fine movements, can be mechanically coupled to the middle platform. The objective lens is positioned along the optical axis by means of two orthogonal ErGaAl brackets fixed to the same platform. The higher platform serves as a basis for the detection system and the support structure of the collection optics. The condenser lens is mounted on a support which can be translated vertically to allow mounting and removal of the sample. The microscope was mounted with an inverted geometry aimed at being used in molecular and cell biology imaging application. Both upright and inverted geometry can be realized starting from a custom mechanical body. In general, inverted or upright geometry is chosen according to the samples to be imaged: upright geometry is in general more comfortable for imaging thick massive samples, while inverted geometry is more indicated for imaging thin samples.

A schematic drawing of the experimental apparatus, mounted with inverted geometry, is depicted in a 3D rendering in Figure 2.1. This solution offers the great advantage of being relatively inexpensive (about few thousands of euros) and completely customizable according to the requirements of the experiment, but it also has some drawbacks. In particular, commercial microscopes offer a better comfort for the user and do not require any alignment procedure. Moreover, the technical complexity of such a project and the time consumed in designing, realizing, and mounting the instrument could

FIGURE 2.1 A 3D-rendering of a custom mechanical structure in ErGaAl alloy used to build an optical microscope. The platform at the bottom serves as a basis for the microscope, the middle platform sustains the objective and the microscope slide support. The upper platform hosts the condenser and the detection system. (From Capitanio, M., Cicchi, R. and Pavone, F. S. 2005. *Eur. Phys. J. B*, 46: 1–8. Reprinted with kind permission of *European Physical Journal B* (EPJ).)

represent an obstacle for the researcher. An alternative solution, with less complexity, much less time-consuming, upright geometry, and with the potential to be customizable is the use of a vertical optical breadboard, as described in the following section.

2.1.2.2 Optical Breadboard Solution

To have a customizable microscope without the complexity related with the solution presented above, an optical breadboard can be used as a microscope stand. For example, the solution presented by Cicchi et al. (2008) consists of a two-photon fluorescence/SHG custom-made microscope in which all the optics are fixed onto a stainless-steel honeycomb breadboard vertically mounted on an antivibrating optical table. A picture of this experimental setup is shown in Figure 2.2. The breadboard can host all the microscope components, including scanning head, scanning lens, tube lens, objective, sample stage, optical filters, and detectors.

The breadboard allows mounting all the optical components required for the experiment with easy interchangeability of the optical configuration. In comparison with the previously described solution, realizing such instrument is much simpler because the number of mechanical parts is drastically reduced, while the customizability remains unchanged. The main drawback of this solution is that it is harder to optically align all the system on a vertical plane.

An alternative solution is represented by mounting the microscope onto a horizontal optical breadboard, which offers an easy alignment procedure but it has the drawback of making an *in vivo* measurement on cells or tissues extremely hard because the sample has to be placed vertically. Both problems (ease of alignment and sample positioning) can be circumvented by aligning the system with the breadboard placed horizontally and then, once aligned, mount it in vertical onto the optical table. A couple of mirrors placed very close to the breadboard is strongly recommended in this configuration in order to have all the degrees of freedom on beam tilting required for an optimal optical coupling of the laser beam with the vertical optical system.

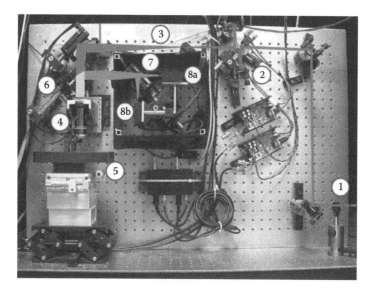

FIGURE 2.2 The figure shows the laser beam [in dark gray, coming from the bottom right side of the panel (1)] that is first scanned by two galvanometric mirrors (2), then expanded by a telescope (3), and finally focused by the objective (4) onto the specimen (5). The emitted light (light gray) is separated from the exciting beam by a first dichroic mirror (6) and then split by a second dichroic mirror (7) in two distinct chromatic components (TPEF and SHG). Two PMTs detect the split emissions (8a,b). (From Cicchi, R. et al. 2008. *Appl. Phys. B*, 92: 359–365; Allegra Mascaro, A. L., Sacconi, L. and Pavone, F. S. 2010. *Front. Neuroenerg.*, 2: 21.)

2.2 Laser Scanning System

In this section, we are going to describe one of the most important parts of an SHG microscope: the laser scanning system. Starting from basic principles, we then focus our attention on the different elements constituting the scanning system (Sheppard et al., 1977, Denk et al., 1990). Then, a brief description of two particular optical configurations for improving the acquisition speed in an SHG microscope is provided: a multifocal SHG microscope (Bewersdorf et al., 1998, Buist et al., 1998, Sacconi et al., 2003) and a digital holographic SHG microscope (Masihzadeh et al., 2010, Shaffer et al., 2010).

2.2.1 Basic Principles of Laser Scanning

In a conventional wide-field microscope, the entire specimen is bathed in light from a light source and the image can be viewed directly by eye or projected onto an image capture device (CCD). In contrast, the method of image formation in a laser scanning microscope is fundamentally different. Illumination is achieved by scanning one (or more) focused beam(s) of laser light across the specimen and by collecting the outcoming light (fluorescence or SHG) point-by-point with a high-sensitivity detector (Sheppard et al., 1977). The basic principles used for performing such scanning are illustrated in Figure 2.3a. As shown in Figure 2.3, a collimated laser beam propagating parallel to the optical axis is focused by a lens in a point geometrically defined by the intersection of the optical axis with the focal plane. When the collimated laser beam is tilted at an angle α with respect to the optical axis, the focal point is located within the focal plane at a distance d from the optical axis. Such distance d is related to the tilting angle α and to the focal length of the lens f by the following equation:

$$d = f \tan \alpha \tag{2.1}$$

For small tilting angles, Equation 2.1 can be approximated with the following form (with α expressed in radians):

$$d = f \alpha \tag{2.2}$$

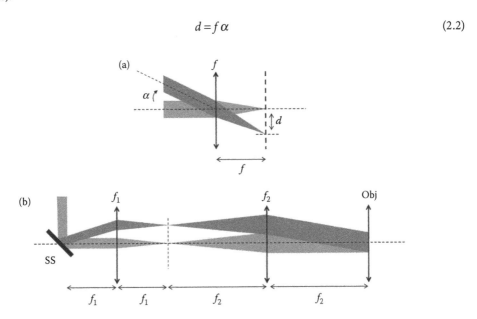

FIGURE 2.3 (a) Working principle of a lens with focal length f. (b) Schematic path of a beam propagating through a typical telescopic system composed of two lenses with focal lengths f_1 and f_2, respectively. In the figure SS indicates the scanning system, while Obj indicates the objective lens.

According to these equations, a beam focus can be scanned across the focal plane by tilting the laser beam with respect to the optical axis. This is in general achieved by using a scanning system, which allows tilting the laser beam from the optical axis (see Section 2.2.1 for a detailed description).

In a typical experimental setup (Denk et al., 1990), a telescope is inserted between the scanning system and the focusing lens (objective lens). This telescope has several functions. First of all, it allows re-conjugating the output of the scanning system with the input of the focusing objective lens. To maintain the laser beam centered on the microscope objective back aperture during scanning, it is important to follow the optical configuration shown in Figure 2.3b. Another crucial role of such a telescope is to magnify the laser beam diameter in order to fill the objective back aperture and take full advantage of the numerical aperture (NA) offered by the objective lens (see Section 2.2.5 for a detailed description). Finally, the telescope de-multiplies (by a factor equal to the magnification) the angle spanned by the laser beam during scanning, enhancing the scanning sensitivity in the objective focal plane. The magnification of the beam diameter and the de-multiplication factor for the tilting angle are related to the focal lengths of the two telescope lenses (f_1 and f_2), as follows:

$$\frac{w_o}{w_i} = \frac{\alpha_i}{\alpha_o} = \frac{f_2}{f_1} \tag{2.3}$$

where w_o and w_i are the output and input beam diameter, respectively; while α_i and α_o are the input and output laser beam tilting angles, respectively. The value of the tube lens focal length (f_2) should be chosen according to the objective lens used. The most important objective lens producers use tube lenses of different focal lengths: 200 mm for Nikon, 160 mm for Zeiss, 210 mm for Olympus. Although the best choice is to use a tube lens with a focal length optimal for the objective lens to be used in order to minimize aberrations, the values can be slightly changed in a custom-made setup without any dramatic effect on the image quality.

2.2.2 Laser Source

SHG is a second-order optical process so that the transition probability is extremely low with respect to other optical linear processes. To excite such a process and to detect a nonnegligible amount of SHG signal, an extremely high spatial and temporal density of photons is required in the focal volume. Such high photon density is realized spatially by using a high-NA objective lens (as it will be explained in the Section 2.2.5) and temporally by using femtosecond-pulsed laser sources. In this section, we describe the laser sources most commonly used for SHG imaging.

Among femtosecond-pulsed lasers, the most commonly used in SHG imaging is the Ti:sapphire oscillator which is emitting, depending on the brand and on the model, pulses in the range of 80–200 fs width at a repetition rate of several tens of MHz (in general between 50 and 100 MHz). Moreover, the Ti:sapphire laser has a typical emission of up to 5 W tunable in the 700–1000 nm spectral range, corresponding to the range in which biological molecules have a significant cross-section for SHG. All these features make Ti:sapphire the ideal laser source for an SHG microscope. The main drawback is represented by the cost, which is presently high (typically more than 100 kEuro).

Another femtosecond-pulsed source that can be used for SHG imaging is the Cr:forsterite laser, which is emitting pulses with width and repetition rate comparable to those of the Ti:sapphire, but in a different wavelength range (typically between 1150 and 1300 nm) and with a reduced power (typically between hundreds of mW and a 1–2 W). Even though the Cr:forsterite laser emission spectrum is not the most suited for SHG imaging (because of the reduced SHG cross-section of biological molecules in this wavelength range), this laser has the advantage of providing a source not only for second- but also for third-harmonic generation microscopy (Barad et al., 1997, Oron et al., 2004, Débarre et al., 2006). Moreover, the cost of such a source is approximately one-half the cost of a Ti:sapphire source.

An alternative solution with a lower cost is represented by femtosecond-pulsed fiber lasers. These compact and robust lasers provide a pulsed emission comparable with that of the Ti:sapphire in terms of pulse width and repetition rate, but they cannot be tuned. In general, they are preferably used in applications in which tunability is not required. Their cost is in the order of 20 kEuro and their maximum power is in the order of W.

Although high repetition rate sources (50–100 MHz) are most commonly used for SHG microscopy, amplified systems with a repetition rate in the range of few kHz and pulse width in the range of tens of fs can also be used (Theer et al., 2003, Qiao et al., 2008). These systems have the advantage of providing shorter pulses with corresponding wider spectral range and hence the capability to excite multiple fluorophores at the same time. On the other hand, since potential phototoxic effects depend on the 4th–5th or higher power of the excitation intensity peak, while the signal depends on the 2nd power, these systems can cause stronger unwanted side effect with respect to a corresponding high repetition rate source providing the same average light power.

2.2.3 Laser Power Adjustment

Femtosecond-pulsed laser sources used for SHG microscopy commonly provide a power output higher than that required for safely imaging biological samples and very often they do not allow a fine regulation of the emitted mean power. For this reason, a proper attenuation system is required for regulating the amount of optical power used for imaging. Laser power can be adjusted by using a neutral density filter ring with variable optical density: the filter can be rotated manually or with a motorized rotational mount allowing power remote control. The use of such a system could have problems with extremely high-power sources damaging the filter coating and it offers a limited dynamic range (most common filters have an optical density varying between 0.1 and 2, and hence a dynamic range of about 0.01–0.9). An alternative is offered by polarization optics. Laser output is very often polarized so that a half-wave-plate can be used in combination with a polarization beam splitter cube to adjust the laser power. The optical power transmitted through the polarization beam splitter cube depends on the rotation angle of the wave plate. As for the neutral density filter of the above, the wave plate can be rotated manually or with a motorized rotational mount to be controlled in remote. All these optical elements have to be chosen for sustaining the high intensity of femtosecond pulses with particular attention to the beam splitter cube. For this purpose, Glan-Laser or Glan-Taylor beam splitter cubes are suggested because they can be used with high peak power and have a higher extinction ratio for the unwanted polarization, allowing to obtain an extended dynamic range of about 0.001–1.

2.2.4 Scanning Head

The scanning head is one of the most important parts of an SHG microscope. It allows to rapidly scan the laser beam focus across the specimen under investigation. Laser beam tilting required for scanning is generally achieved by using galvanometric mirrors (galvo-mirrors), acousto-optic deflectors (AODs) or polygonal rotating mirrors. In this section, after having introduced the most popular scanners used in SHG microscopy, we describe four alternative solutions to be adopted for the scanning head: one commercial, and the other three custom-made based on galvo-mirrors, AODs, or polygonal mirrors.

2.2.4.1 Scanners

Laser scanning microscopy has proven to be a useful tool for examining cells and tissues. However, many interesting biological processes occurring on the millisecond timescale cannot be revealed by laser scanning microscopy because the imaging speed is limited by the bandwidth of the scanners used. Various approaches can be adopted in order to overcome this limitation and have a faster scanning system.

- *Nonresonant galvanometer-based scanners:* The most common scanners in laser scanning micros-copy consist in a couple of small mirrors mounted onto two galvanometer-based scanners (see Figure 2.4). The galvanometers are basically made by a magnet and a current-driven coil allowing a mirror tilting proportional to the injected current in the coil. Very often, they are equipped with a position sensor and a closed-loop servo system. They offer good precision and high-reso-lution scanning. For scanning the specimen with a raster waveform, the faster scanner is driven with a saw tooth (or triangular) waveform and the slower scanner with a linear decreasing ramp. However, the acquisition speed is limited by the bandwidth of the faster scanner. Typically, the scan rate of the saw tooth is several microseconds per pixel. For example, a 10 µs pixel dwell time translates to scanning time of more than 2.5 s for a 512 × 512 pixels image. A much faster acquisi-tion speed (typically up to 30 frames/s) can be obtained by using resonant galvanometer-based scanners.

- *Resonant galvanometer-based scanners:* When scanning a raster with nonresonant galvanometers, the scanning speed has to be reduced below an upper limit in order to avoid extreme accelera-tion and deceleration of the scanner when inverting scanning directions (at the image edge). This is the main phenomenon limiting the scanning speed. To overcome this limitation and achieve extremely fast scanning speed, it is very important to always maintain the scanner acceleration within an upper limit. This can be obtained by driving the faster scanner with a resonant sinusoi-dal waveform instead of a saw tooth. The advantage consists in the fact that a sinusoidal waveform can be used at a very high scan rate without having the problem of extremely high accelerations at image edges. Obviously, the pixel clock has to be modified according to the scanner speed, which is varying along the sinusoidal profile. In fact, due to the nonlinearity of the speed of a scanner driven with a sinusoidal waveform, the image pixels are scanned at the highest speed in the image center, with the speed decreasing when moving to the image edges. The simplest option for com-pensating scanning nonlinearity is to limit the scanning range to that portion of the oscillation where the velocity of the scanner is almost linear, which occurs over a region of approximately 70% of the total scanned field. Unfortunately, this solution reduces the amount of time that the emission signal can be collected and does not prevent overexposure of the sample regions falling out of the recorded area. An alternative option is represented by correction; in fact, image distor-tion introduced by nonlinear scanning is predictable and can be corrected by either hardware or software solutions. Both solutions rely on knowing the position of the faster galvanometer-based scanner. In this way, the resonant galvanometer position signal can be used as a master oscillator to which all the other timing components can be synchronized.

FIGURE 2.4 Picture of a typical commercial dual-axis mirror assembly. In particular, the picture refers to the GVS002 scanning galvo-mirror system produced by Thorlabs. (Courtesy from Thoralabs.)

- *AODs scanners:* Additional scanner technology capable of video rates acquisition speed includes instruments equipped with AODs. AODs utilize ultrasonic waves to generate pressure zones in a crystal that can diffract or deflect incident laser light at an angle that varies with the acoustical frequency (Bragg diffraction). These solid-state devices benefit from having few moving mechanical parts and negligible inertia, enabling the generation of highly accurate sawtooth raster scans having almost instantaneous flyback. Furthermore, AOD scanners can produce user-defined deflections to generate scanned regions of interest. The disadvantages of AOD microscopes are based on the dispersive properties of these devices, which are suited only for controlled passage of monochromatic light. Although these problems have been solved in several experimental instrument designs (Bullen et al., 1997, Sacconi et al., 2008, Reddy and Saggau, 2005), the concept of using AOD scanners has not been popular with the microscope manufacturers. A possible reason could be due to the fact that these systems need bright specimens to work at their best.
- *Polygonal mirror scanners:* An alternative scanning solution able to reach extremely high scanning speed consists in substituting the faster galvanometer scanner with a polygonal mirror scanner. The polygon scanning mirror is simply a prism with polygonal basis having mirrors placed onto the external faces. A laser beam hitting on the external face of a rotating polygonal mirror translates into a beam deflection with a sawtooth profile. Deflection of the light beam with a rotating polygon mirror system is a mature technology that uses optically simple, nondispersive surfaces to create an output beam having a basic sawtooth raster similar to conventional video scanning. However, as explained in Section 2.2.4.5, the use of polygonal mirror requires a complex optical and electronic design to be implemented in an instrument.

2.2.4.2 Commercial Scanning Head

A commercial scanning head is in general made by a pair of galvo-mirrors placed orthogonally to each other, so that one is scanning along the x-axis, and the other along the y-axis. Very often, commercial scanning heads (see Figure 2.4) are realized by placing the two galvo-mirrors very close to each other, so that the pivot point (the conjugate point of the intersection between the optical axis and the back focal plane of the objective) is located in between the two scanners.

This configuration can potentially create a nonuniform illumination when scanning a large field of view, because the positioning of the two galvo-mirrors does not allow a perfect pivoting. (A perfect pivoting is obtained by placing both scanners in a pivot point.) However, for implementation in a commercial microscope system, commercial scanning heads have the big advantage of being easily connected to a dedicated port of the microscope stand. Moreover, commercial scanning heads are in general digitally driven and also monitoring of the galvo-mirrors position is based on digital technology, so that it can be easily interfaced to other electronic devices.

2.2.4.3 Custom Galvo-Mirrors-Based Scanning Head

To overcome the problem of imperfect pivoting of the laser beam described above, two different solutions can be used, based on lenses or spherical concave mirrors.

The first configuration (see Figure 2.5a) is based on the optical relaying between galvo-mirrors by means of lenses. It consists in placing a telescope between the two galvo-mirrors in order to put both of them in a pivot point. Following an optical scheme like the one shown in Figure 2.5a, the first galvo-mirror is placed along the optical axis before the first lens, at a distance from the lens equal to its focal length; the second galvo-mirror is placed along the optical axis, after the second lens, at a distance from the lens equal to its focal length. This optical configuration requires an extended space, but it provides a "perfect" optical system. The second configuration (see Figure 2.5b) is quite similar to the first, but it uses concave spherical mirrors, instead of refractive lenses. In fact, it is known that a concave spherical mirror works as a lens with focal length equal to one-half the mirror curvature. The two spherical mirrors are placed in a symmetric geometry in between the two galvo-mirrors maintaining each of them at a focal distance from the first or the second spherical mirror.

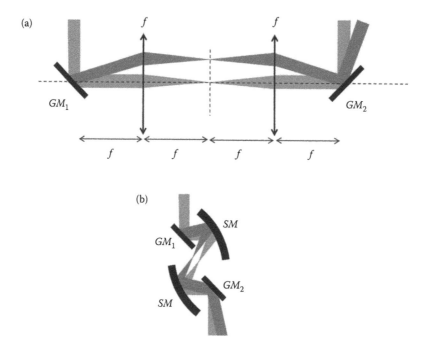

FIGURE 2.5 (a) Schematic sketch of a lens-based scanning head for relaying the two scanners. With this configuration, both scanners are placed in a conjugate plane with respect to the objective back-focal plane. (b) Schematic sketch of a mirror-based scanning head. In this configuration, the mirrors are slightly tilted in order to allow the positioning of scanners in conjugate planes with respect to the objective back focal plane.

The two spherical mirrors have to be rotated of a small angle along a vertical axis in order to make space for placing the two galvo-mirrors. In fact, in an ideal geometry, the two galvo-mirrors should be located exactly in the same point. (The ideal geometry is obtained with the beam orthogonal to the mirrors surface. This can be obtained only by placing both scanners in the common focus of the two mirrors.) The rotation of the two spherical mirrors allows to slightly shift in opposite directions the points in which the galvo-mirrors have to be placed (as in Figure 2.5b where the scanners are displaced from the common focus of the two mirrors). This optical configuration can be mounted in a set-up with limited available space but it could introduce a slight astigmatism in the laser beam, affecting the spatial resolution of the microscope.

2.2.4.4 Custom AODs-Based Scanning Head

The major limiting factor of galvanometric scanners is the scanning time. For this reason, the optical recording of fast physiological events (of the order of ms) is generally possible only in a single position of the field of view by using a line scanning procedure. On the other hand, in principle, the optical measurement of time-dependent processes at selected locations does not require the production of images at all. Instead, more time could be spent collecting as many photons as possible from selected positions, where the image plane intersects the biological objects of interest. Unfortunately, standard galvo-mirrors require about 1 ms to reach and stabilize a new position; therefore, even implementing a line scan approach on a selected sample region, the sampling frequency is always below 1 kHz, limiting the recording of fast physiological processes (such as, e.g., action potentials). On the other hand, high-speed scanning of a set of points within a plane can be achieved with two orthogonal AODs). In an AOD, a propagating ultrasonic wave establishes a grating that diffracts a laser beam at a precise angle, which can be changed within a few microseconds (see Figure 2.6).

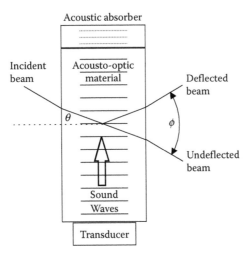

FIGURE 2.6 Diagram showing the principle of operation of an acousto-optic light-beam deflector. The diagram defines the deflection angle ϕ used in this chapter.

The AOD makes use of the acoustic frequency f dependent diffraction angle, where a change in the angle $\Delta\phi$ can be expressed as a function of the change in frequency Δf:

$$\Delta\phi = \frac{\lambda}{v}\Delta f \qquad (2.4)$$

where λ and v are the acoustic wavelength and velocity of the acoustic wave, respectively.

The first implementation of a high-speed, random-access, laser-scanning fluorescence microscope configured to record fast physiological signals from small neuronal structures with high spatiotemporal resolution has been presented by Bullen et al. (1997). Recently, Sacconi et al. (2008) combined the advantages of SHG with an AOD-based random-access (RA) laser excitation scheme to produce a new microscope (RASH) capable of optically recording fast (~1 ms) membrane potential (Vm) events. The RASH is based on a custom-made upright scanning microscope. The spatial distortion of the laser pulse in the AODs is known to affect the radial and axial resolutions of the microscope.

To compensate for the larger dispersion due to two crossed AODs, an acousto-optical modulator (AOM) placed at 45° with respect to the two axes of the AODs (see Figure 2.7) can be used. To compensate for the spatial distortion at the center of the field of view (F0), the AOM frequency can be fixed at a value given by F0 · √2 (Reddy et al., 2005). Clearly, the propagation direction of the ultrasonic wave

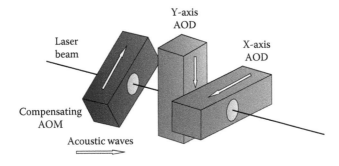

FIGURE 2.7 Scheme showing how the AOM is inserted at 45° before the two AODs. The propagation direction of the ultrasonic waves in each AOD is indicated by arrows.

in the AOM must propagate in the opposite direction with respect to the sum of the two waves in the AODs.

The RASH microscope is capable of collecting a field of view of $150 \times 150 \ \mu m^2$ with a radial spatial resolution of ~800 nm. The commutation time between two positions in the focal plane is of the order of 4 μs. RASH microscopy can be used to record fast physiological events from multiple positions with near simultaneous sampling. The RASH system AODs can be rapidly scanned between lines drawn in the membranes of neurons to perform multiplex measurements of the SHG signal monitoring Vm at the selected locations.

2.2.4.5 Scanning with a Polygonal Mirror

Polygonal mirrors have been utilized in several past commercial and experimental laser scanning instruments, but are currently not being implemented by the microscope manufacturers. In practice, the use of polygon mirrors requires considerably more complex optical component design, and the units are prone to miniscule variations in reflectivity and angle with respect to the axis of rotation. Known as pyramidal errors, these angular differences produce beam fluctuations that must be optically corrected. Due to the fact that the number of polygon facets must be proportional to the total number of raster scan lines, specialized mirrors must often be fabricated in order to build laser scanning instruments. Polygons having 15, 25, or 75 sides must rotate at 63,000, 37,800, or 12,600 revolutions/min, respectively, in order to generate video rate scanning at 15,750 lines/s. These high speeds require specialized bearings, further complicating instrument design. For these reasons, polygon mirror-based SHG microscopes are rare and generally relegated to special interest projects (Veilleux et al., 2008).

2.2.5 Microscope Objective Lens

In a laser scanning SHG microscope, the excitation objective determines the spatial resolution of the microscope. In wide-field microscopy, the spatial resolution achieved with an objective lens is limited by the diffraction limit of the wavelength used. In a nonlinear laser scanning microscope, the spatial resolution corresponds to the excitation volume, which is smaller than the diffraction limit (see Figure 2.8a) because of the nonlinear optical properties of the excitation process. Although coherent optical processes are not in general characterized by a PSF, it is useful to approximate the spatial resolution of an SHG microscope with the PSF associated with fluorescence excited at the same wavelength. By using this approximation, we can evaluate the best spatial resolution achievable with an SHG microscope. The spatial radial (ω_{xy}) and axial (ω_z) resolution for an SHG laser scanning microscope can be approximated by the following two relationships, respectively (Zipfel et al., 2003b):

$$\omega_{xy} = \begin{cases} \dfrac{0.320\lambda}{\sqrt{2}NA} & NA \leq 0.7 \\[2ex] \dfrac{0.325\lambda}{\sqrt{2}NA^{0.91}} & NA > 0.7 \end{cases} \tag{2.5}$$

$$\omega_z = \frac{0.532\lambda}{\sqrt{2}} \left[\frac{1}{n - \sqrt{n^2 - NA^2}} \right] \tag{2.6}$$

where λ is the excitation wavelength used, n is the refractive index of the immersion medium, and NA is the objective numerical aperture.

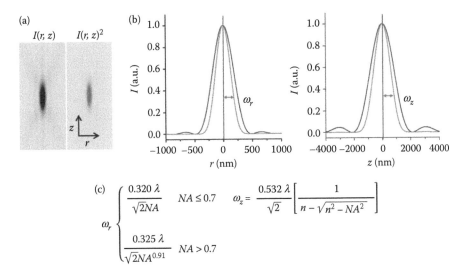

$$
\omega_r \begin{cases} \dfrac{0.320\,\lambda}{\sqrt{2}NA} & NA \le 0.7 \\[4mm] \dfrac{0.325\,\lambda}{\sqrt{2}NA^{0.91}} & NA > 0.7 \end{cases} \qquad \omega_z = \dfrac{0.532\,\lambda}{\sqrt{2}}\left[\dfrac{1}{n - \sqrt{n^2 - NA^2}} \right]
$$

FIGURE 2.8 (a) Axial and later views of intensity point spread function ($I(r,z)$ in black) in the excitation volume and squared point spread function ($I(r,z)^2$ in gray). Squaring the $I(r,z)$ results in minimal wings relative to center. (b) Radial (on the left) and axial (on the right) profiles of $I(r,z)$ (black line) and $I(r,z)^2$ (gray line). Both axial and lateral PSFs are intended for two-photon fluorescence and give the best approximation for the PSFs associated with SHG imaging. The simulated profiles have been obtained using the following parameters: $NA = 1$, $\lambda = 800$ nm, $n = 1.33$ (water).

Spatial resolution is not the only feature to be considered when choosing the excitation objective. Several specifications have to be taken into account when choosing an objective lens:

- *Numerical aperture:* As a general rule, the NA of the excitation objective should not be <0.5, in order to provide a photon density in the focus sufficient for two-photon excitation of the specimen. Higher NA objectives are preferable, since they increase both microscope spatial resolution and detected signal intensity (i.e., signal to noise). As a drawback, high-NA objectives correspond in general to higher magnification and, hence, to smaller fields of view. Furthermore, high-NA objectives require oil or water immersion. They could also introduce additional spurious effects in the polarization when using a polarization-scanning equipment, because of the phenomenon of scrambling occurring with very high NAs.
- *Objective transmission:* The spectral range commonly used in an SHG microscope is in the near infrared. The excitation objective should have a good transmittance in the excitation wavelength range used (typically 700–1000 nm).
- *Working distance:* Very often high NA objectives have working distances limited to 100–200 µm, imposing a limit on deep imaging inside a biological tissue. The working distance of the objective should be chosen according to the morphology of the sample to be investigated. For example, for *in vivo* deep tissue imaging, long working distances are required. A water immersion objective with NA ranging from 0.8 to 1 and a working distance ranging from 2 to 3 mm is a good solution for this purpose.
- *Back aperture:* The size of the objective back aperture has to be chosen carefully and considering the dimension of the exciting laser beam. In fact, it is very important that the excitation beam slightly overfills the objective back aperture. In this way, all the NA of the objective can be used, maximizing the resolution achievable with the objective (Figure 2.9a). The focal spot size critically depends on the ratio of the beam size and the objective back aperture, as shown in the insets of Figure 2.9. On the other hand, while overfilling guarantees the NA of the objective, it also reduces transmitted power.

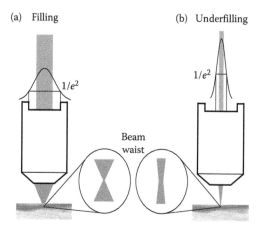

FIGURE 2.9 Beam size adjustment relative to the objective's back aperture. (a) Filling of the back aperture with an expanded beam. (b) Underfilling of the aperture, the transmitted power is maximum but the spatial resolution is reduced. The beam waist size in the focal volume is reproduced in the inset.

2.2.6 Polarization Scanning

An important feature in an SHG microscope is represented by the capability to vary the polarization of the exciting laser beam (Stoller et al., 2002a,b, Tiaho et al., 2007, Matteini et al., 2009, Nucciotti et al., 2010). In fact, the SHG intensity in far-field strongly depends on the mutual orientation between the molecular intrinsic hyperpolarizability and the direction of oscillation of the exciting field (Mertz and Moreaux, 2001). By using this property, it is possible to measure the angular distribution of hyper Rayleigh scattering (HRS) emitters in the focal volume, which is in turn related to molecular structural organization of the specimen under investigation. The measurement is usually performed by acquiring multiple images with different polarization orientations. The result is an image in which a polarization profile is embedded in each pixel. Such a profile contains the information on the angular distribution of HRS emitters within the focal volume. A fitting procedure by using a proper theoretically calculated function allows to extract the parameters describing the structural organization within each pixel (Sun, 2005, Plotnikov et al., 2006, Tiaho et al., 2007, Nucciotti et al., 2010).

The polarization scanning equipment is usually placed in a section of the microscope in which the beam is collimated, as close as possible to the objective lens. As a general rule, the equipment for polarization scanning comprises three main parts (Figure 2.10a).

- A first *polarizer* inserted in the beam path in order to make the polarization linear. In fact, even if the laser output is polarized, the beam travels through several optics which have detrimental effects on the polarization. A good linear polarization has to be obtained in order to provide the required input condition for the following component: a quarter wave plate.
- A *quarter wave plate* is inserted in the optical path and properly oriented for generating a circular polarization.
- The last element is a *rotating polarizer* for scanning the sample with a linear rotating polarization. The polarizer can be rotated by means of a rotating opto-mechanical component equipped with a stepping or a DC motor.

It has to be noted that, if higher polarization scanning speed is required, the last element (the rotating polarizer) can be substituted with an active element such as an electro-optical modulator (EOM) which is providing a much faster response with respect to a rotating polarizer, which offers a scanning speed limited by its inertia. It is also important to note that the light beam coming out from the quarter wave plate must have a perfect circular polarization in order to acquire accurate polarization SHG profiles.

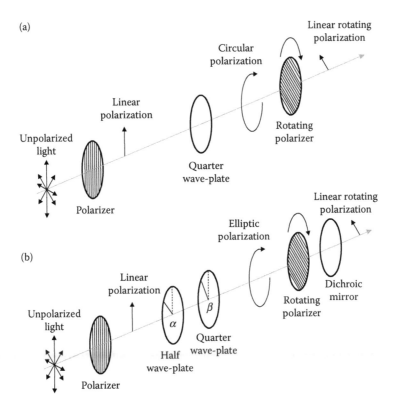

FIGURE 2.10 (a) Schematic sketch of a polarization scanning equipment to be used in an SHG microscope. (b) Schematic sketch of a polarization scanning equipment with compensation of the polarization ellipticity introduced by the dichroic mirror. The half-wave plate and the quarter-wave plate are rotated in order to compensate the phase retardation introduced by the dichroic mirror.

In fact, ellipticity produces a fluctuation in the intensity after the second polarizer, affecting the polarization scanning measurement. Further, an intensity fluctuation in the excitation beam will affect the detected SHG polarization profile introducing an intensity noise which is double with respect to the excitation intensity fluctuation, because of the nonlinearity of the optical process generating second-harmonic light.

All this equipment has to be placed immediately before the objective lens without any other optical component having the potential to affect the polarization of the beam. This geometry could represent a problem when a dichroic mirror is placed in front of the objective lens. In fact, in backward detection geometry (see Section 2.3.2.2), a dichroic mirror and the detector have to be placed as close as possible to the objective in order to maximize the detected signal. The coating of the dichroic mirror affects the polarization of the laser beam, generally producing elliptic polarization. A possible solution has been proposed by Chou et al. (2008). The procedure consists of placing an additional half-wave plate in the optical path. The half-wave plate and the quarter-wave plate should be rotated in order to introduce a compensation for the phase retardation introduced by the dichroic mirror, so that the output beam is polarized with linear rotating polarization (Figure 2.10b). This solution allows a compensation for a well-defined direction of the laser polarization, whereas it has to be modified while scanning the laser polarization. So, a synchronized rotation of both polarizer and quarter wave plate should be used. An alternative and much simpler approach has been used in the cases in which the sample itself can be rotated about the optical axis, while keeping a constant linear polarization of the incident laser beam (Hsieh et al., 2010).

2.2.7 A Multifocal SHG Microscope

To overcome the limitations in scanning time required for imaging applications (see also Section 2.2.4), an SHG microscope can be fitted with a multifocal excitation scheme, allowing fast scanning of a selected field of view by multiple laser beamlets.

A multifocal SHG microscope is built in the same manner of a multifocal multiphoton microscope (MMM) with the only difference of detecting SHG light instead of two-photon fluorescence. The differences reside in the detection system and in particular in the optical filtering, while the excitation system is exactly the same. The multifocal scheme can be implemented using microlens arrays: the scanning is achieved by a combining a microlens array and galvo-mirrors (Buist et al., 1998). An effective speed of 225 frames/s of MMM imaging was achieved by rotating a microlens disk (Bewersdorf et al., 1998). The system combining microlenses and galvo-mirrors (Buist et al., 1998), however, suffered from high loss of incident power (around 75%). The system utilizing rotation of a microlens disk also incurred in unavoidable loss of power in beam expansion and optical train (Bewersdorf et al., 1998). In addition, it suffered from nonuniformity of probe intensity such that the images were 50% less intense at the corners than in the center (Bewersdorf et al., 1998). Renormalization of intensities in the scan spots of multispot grid may be operated by nonlinear scaling of image intensity. Nevertheless, accomplishment of uniformity in foci intensity via optical conditioning of the beams is definitely a far more attractive approach as compared to image-processing software solutions. A promising solution with a better compactness and simplicity is represented by the use of a miniature diffractive optical element (DOE) in tandem with galvo-scanners to produce near multispot grids with high diffraction efficiency and provide a high degree of uniformity in foci intensity, resulting in high theoretical diffraction efficiency and an extremely small array uniformity error (Sacconi et al., 2003).

Figure 2.11 shows a scheme of the optical setup of an MMM microscope using a DOE. The expanded beam of the excitation laser source illuminates the DOE that generates an $n \times n$ scanning grid at the focal

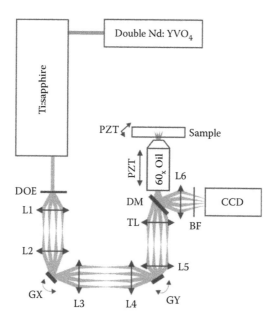

FIGURE 2.11 Schematic of the optical system of an MMM microscope. L1 and L2 are telescopic lens pair (2× magnification) that pivots the grid on the first scanner (GX). L3 and L4 are telescopic lens pair that pivots the grid on the second scanner (GY). L5 and TL (tube lens) form a telescopic lens pair (4× magnification) that pivots the grid on the back-focal plane of the objective. DM and BF are the dichroic mirror and the blocking filter, respectively. (From Sacconi, L. et al. 2003. Multiphoton multifocal microscopy exploiting a diffractive optical element. *Opt. Lett.,* 28: 1918–1920. With permission of Optical Society of America.)

plane of the microscope objective. As we can see from the scheme, the DOE is placed in a conjugate optical plane with respect to the back focal plane of the objective and to the two galvo-mirrors. Three telescopes are optically relaying the DOE with the X scanner (L1,L2), one scanner with the other (L3,L4), the scanning head with the objective (L5,TL). With this configuration, while a single beam of the grid has the same profile than in a single-beam scanning microscope, the grid profile is focused on both scanners and on the objective back focal plane, in order to provide a beam spacing in the focal plane and hence to scan adjacent points. The spacing of the beams in the grid is set by the magnification of the telescope formed by the tube lens and the microscope objective. To provide simultaneous acquisition of multiple scanned points, the emitted signal is detected by means of a CCD camera or a photomultiplier array. By means of this multispot arrangement for scanning operation, SHG images can be recorded with the same resolution and the same SNR in a time which is reduced by a factor n^2 with respect to the equivalent single-beam scanning.

2.2.8 An SHG Holographic Microscope

As shown in the previous sections, in the most general case, SHG microscopy is based on the scanning of femtosecond-pulsed beams, tightly focused by high numerical aperture objectives. However, this is not the only optical configuration that can be used when designing an SHG microscope. In fact, another relevant solution can be adopted taking advantage of the coherent nature of the optical process generating second harmonic. Such alternative is represented by holographic second-harmonic generation microscopy (Shaffer et al., 2010, Masihzadeh et al., 2010).

The combination of SHG microscopy and holographic imaging acquisition provides a powerful combination for improving imaging speed. The principle of operation is quite similar to digital holographic microscopy; the only difference is that second harmonic, instead of the scattered light, interferes and creates the hologram. Since holography encodes 3D optical field information in single 2D images, it enables the sample to be imaged in three dimensions at vastly improved speeds (typically in the order of ms/scan). The optical system used is analogous to a Mach–Zender interferometer (see Figure 2.12). A beam-splitter cube (BS) divides the excitation beam in two separate arms: the reference arm (R) and the

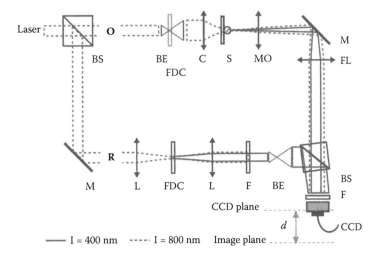

FIGURE 2.12 Experimental setup schematics for an SHG holographic microscope. BS, beamsplitter cube; BE, beam expander; C, condenser lens; S, sample; MO, microscope collection objective; M, mirror; FL, field lens; F, filter; L, lens and FDC; nonlinear crystal. **O** designates the object arm, while **R** designates the reference arm. d is the hologram reconstruction distance. (From Shaffer, E., Marquet, P. and Depeursinge, C. 2010. Real time, nanometric 3D-tracking of nanoparticles made possible by second harmonic generation digital holographic microscopy. *Opt. Express*, 18: 17392–17403. With permission of Optical Society of America.)

object arm (O). In the reference arm, the beam passes through a lens (L) that focuses into a nonlinear crystal for generating second harmonic. SHG light is then collected by a second lens (L), filtered with an optical filter (F) and expanded with a beam expander before passing through a second beam-splitter cube (BS). In the object arm, the laser beam is expanded with a beam expander (BE) and collected by a condenser lens (C) that illuminates the sample (S). Second-harmonic light generated in the specimen is then collected by the microscope objective lens (MO) and directed into the second beam-splitter cube. Light coming from the reference arm interferes with light coming from the object arm and forms an interference pattern, which is detected by a high-sensitivity CCD in the Fourier plane.

The detected hologram can be used for reconstructing the specimen in three dimensions. In fact, the position of a single emitter in the three-dimensional object space can be determined by measuring the phase shift between the reference and the object beam (see Figure 2.13a). The position of the emitter along the optical axis affects the phase of the SHG light, so that a measurement of the phase shift can be used to determine the axial position of the emitter. An alternative method proposed is based on the measurement of the hologram reconstruction distance. A schematic of the principle of operation of this method is provided in Figure 2.13b. The position of the emitter along the optical axis in the object space affects the hologram reconstruction distance in the image space, so that the measurement of the distance d can be used to determine the axial position of the emitter. Although, the holographic approach offers high 3D speed capabilities, the absence of a focused exciting beam requires samples with large SHG cross-section (typically in the order of 10^3–10^5 GM). These extremely large SHG cross-sections are

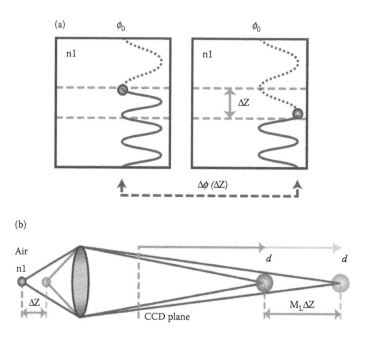

FIGURE 2.13   (a) The drawing explains how the observed SHG phase depends on the position of the emitter along the optical axis. The dotted line represents the laser light incident on the sample and the solid line represents SHG light. The horizontal dashed lines indicate the axial position of the scatterer (upper line in the left panel, referring to the reference arm; lower line in the right panel, referring to the object arm). The axial position of the emitter is determined directly from the phase of the generated second-harmonic field. (b) The drawing explains the relation between the axial position of the SHG emitter (in the object space) and that of its image plane (in the image space). The axial position of the emitter is determined from the hologram reconstruction distance d that brings the image into focus. (From Shaffer, E., Marquet, P. and Depeursinge, C. 2010. Real time, nanometric 3D-tracking of nanoparticles made possible by second harmonic generation digital holographic microscopy. *Opt. Express*, 18: 17392–17403. With permission of Optical Society of America.)

generally obtained by using the so-called SHRIMP (second-harmonic radiation imaging probes). These SHRIMPs are generally nanocrystals that can be functionalized for selective labeling of biological samples. The SHG properties of several kinds of nanocrystals have already been reported: ZnO (Prasanth et al., 2006), $KNbO_3$ (Nakayama et al., 2007), $BaTiO_3$ (Hsieh et al., 2009). For example, a measurement of the SHG cross-section for $BaTiO_3$ nanocrystals gave a value in the order of 10^4 GM (Hsieh et al., 2009). It has to be considered that typical cross-sections for two-photon fluorescence probes, for example, EGFP, are in the order of tens of GM (Heikal et al., 2001).

2.3 The Detection System

In this section, we analyze different optical detection modalities and geometries. In detail, we describe two detection geometries: forward detection and backward detection. The former is the most used, since SHG is mainly emitted in the forward direction. The latter is commonly used when imaging a sample with thickness large enough to prevent forward detection. In the last part, several optical filtering and detector solutions will be reviewed.

2.3.1 Optical Detection Modalities

There are two different detection modalities in laser scanning microscopy: de-scanned and non-de-scanned. The de-scanned mode is analogous to the confocal geometry: SHG light passes through the scanning head before being decoupled from the laser path and detected. In the non-de-scan mode, SHG light is decoupled from the laser path immediately behind the objective. The main difference consists of the fact that in de-scan mode the focused SHG light into the detector maintains the same position during scanning while in non-de-scan mode the focused SHG light scans across the detector. For deep tissue imaging, the non-de-scan solution (allowing the closest proximity of the detector to the objective lens) is essential to minimize the loss of multiple-scattered SHG photons from the tissue.

2.3.2 Detection Geometry

2.3.2.1 Forward Detection

Forward detection is the most common detection geometry used in SHG microscopes. Such detection geometry is preferably used with thin samples, such as cells (Campagnola et al., 1999, 2001), microtubules (Dombeck et al., 2003), cellular membranes (Bouevitch et al., 1993, Peleg et al., 1999, Moreaux et al., 2000a,b, Sacconi et al., 2005), and muscle fibers (Both et al., 2004, Boulesteix et al., 2004, Plotnikov et al., 2006, Nucciotti et al., 2010). In fact, as known from SHG theory (Moreaux et al., 2001, Mertz et al., 2001), SHG is a coherent process preserving both energy and momentum. Therefore, SHG is mainly directed in the same direction as the exciting wave vector. As shown in Figure 2.14a, the ratio between forward and backward emitted SHG is ranging from 1 to 10^{12}, depending on the angle between the single HRS dipolar axis and the optical propagation axis (Zipfel et al., 2003a). More in detail, as shown in Figure 2.14b, the forward emitted SHG light has a characteristic spatial pattern with two lobes forming an angle with the optical axis θ_{peak}. This angle is related to the phase anomaly and hence to the NA of the excitation objective (Moreaux et al., 2000b).

In agreement with this description, several considerations must be done regarding the collection objective lens. In principle, the ideal optical geometry is realized using a collection objective with the same NA of the excitation objective. If we are using a collection objective with lower NA, a substantial loss of SHG photons will occur. On the other hand, in a real experimental setup, it is preferable that the NA of the collection objective be slightly higher than the excitation one, so as to prevent photons leakage due to a inevitable misalignment of the objectives confocality during experiment (e.g., in axial scanning).

This optical configuration is maximizing the detected signal. As a general rule, objective lenses guarantee higher collection as the NA increases, so that the optimal solution should be, in principle, to

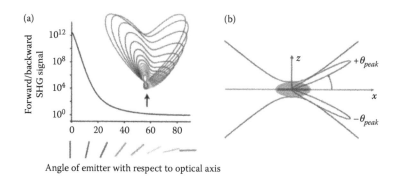

Angle of emitter with respect to optical axis

FIGURE 2.14 (a) SHG forward/backward intensity ratio versus the angle between the SHG emitter and the optical axis. Expected angular distribution of the emitted SHG for different angles between the emitter and the optical axis (indicated by the arrow). (From Zipfel, W. R. et al. 2003a. Live tissue intrinsic emission microscopy using multiphoton-excited native fluorescence and second harmonic generation. *Proc. Natl. Acad. Sci. USA*, 100: 7075–7080. Copyright (2003) National Academy of Sciences, USA.) (b) An excitation beam propagating in the *x* direction and polarized along the *z* axis is focused onto the specimen. Phase matching between the SHG and the excitation field causes the SHG radiation to be emitted with two lobes in the forward direction with axes oriented at an angle $\pm\theta_{peak}$ with respect to the optical axis. (From Moreaux, L., Sandre, O. and Mertz, J. 2000b. Membrane imaging by second-harmonic generation microscopy. *J. Opt. Soc. Am. B*, 17: 1685–1694. With permission of Optical Society of America.)

use an objective with the highest possible NA. Following the same considerations reported in Section 2.2.5, several specifications have to be taken into account when choosing a collection objective lens. For example, the objective should have optimal transmittance for SHG light without considering the transmittance in the NIR. However, a higher NA generally corresponds to a shorter working distance. The working distance of the objective, on the other hand, is a crucial parameter to be considered, since it limits the maximum thickness of the specimen that can be imaged.

2.3.2.2 Backward Detection

Backward detection is also possible in an SHG microscope, since, in specific conditions, there is a non-negligible amount of SHG signal emitted in the backward direction. This geometry does not allow detecting as strong signals as in the forward geometry, because the amount of SHG backward emitted is generally much lower than the forward emission. However, for some particular samples such as massive specimens (Guo et al., 1999, Provenzano et al., 2006, Pfeffer et al., 2007, Cicchi et al., 2010) or living subjects (Konig and Riemann, 2003, Llewellyn et al., 2008), forward detection is intrinsically prevented and only backward detection is possible. Furthermore, biological tissues are highly scattering media so that the forward-emitted SHG can be detected in backward geometry by taking advantage of multiple scattering events. The amount of backscattered signal generated by multiple scattering is extremely low: in general, it is ranging from 5% to 20% of the forward-emitted signal (Nadiarnykh et al., 2010). In this condition, the backward geometry is strongly limiting the strength of the detected signal, even if imaging of biological samples is still possible. In the backward modality, the excitation/collection objective should have optimal transmittance both for NIR and SHG light. If it is not possible to satisfy both conditions, as a general rule, it is preferable to optimize the SHG transmittance.

2.3.2.3 Combined Forward–Backward Detection

In the same application, a combined forward–backward detection could be required. Such a configuration is based on the detection of both forward and backward scattered light by means of two separated detectors. The net effect is:

- An increase of the detected signal because the collection cone is not limited to one objective lens but is extended to two objectives.

- The capability to estimate the mean orientation and/or the spatial distribution of the emitters contained in the focal volume with respect to the optical axis by measuring the ratio between forward and backward scattered light (LaComb et al., 2008, Nadiarnykh et al., 2010). In fact, as reported in Figure 2.13, the front-to-back emission ratio (F/B) depends on the angle between the emitter and the optical axis with a monotonic behavior.

2.3.3 Optical Filtering

Optical filtering is a crucial point for an SHG microscope, since the power density of the excitation is extremely high in comparison to the signals to be detected. Filtering is commonly achieved by using three optical elements:

- A *dichroic mirror* for diverting from the laser path the SHG light to be detected.
- In principle, the dichroic mirror needs to be chosen with a cutoff wavelength comprised between the excitation wavelength and the SHG wavelength. As a general rule, the dichroic mirror has to be placed as close as possible to the collection objective. This serves to minimize the distance between the detector and the objective, therefore maximizing the detected signal (Oheim et al., 2001, Svoboda and Yasuda, 2006). Furthermore, in backward detection, dichroic filters with a thin substrate are recommended to minimize the wave front distortions in the excitation beam.
- A *short-pass filter* for blocking spurious contamination of the excitation light in the detection path.
- Such filter has to be placed in the detection path independently if the microscope works with backward or forward detection. Its function is to remove spurious laser light from the detection path. The optical density at the laser emission wavelength should be at least 6–7, in order to avoid detection of undesired NIR photons. On the other hand, the filter transmission at the SHG spectral range should be as high as possible (generally more than 90%). The most common laser (e.g., Ti:sapphire laser) blocking filters for SHG microscopy have a cutoff wavelength between 650 and 700 nm, in order to block the laser light ($\lambda > 700$ nm) and transmit both SHG and fluorescence light. In fact, while in principle a shorter cutoff wavelength can be chosen for an SHG microscope, it could be of interest for the researcher to be able to detect also two-photon fluorescence signals with the same instrument. Additional filtering for removing fluorescence and selectively detect SHG is performed by using another filter, as described below.
- A *narrow band-pass filter* selectively detects only photons involved in the SHG process.
- This filter is placed in the detection path to selectively detect only photons involved in the SHG process. Considering that the emitted SHG light has a wavelength of exactly one-half the excitation wavelength, the central wavelength of the bandpass is chosen at simply half the excitation wavelength. On the other hand, the optimal filter width depends on the spectral broadening of excitations laser pulses. For a Ti:sapphire laser, a dielectric bandpass filter with 20 nm FWHM is a good choice. Narrower filters, based on interference, can be used as well.

2.3.4 Detectors

This section describes the most common detectors that can be used in an SHG microscope and the features to be taken into account when choosing the detector. In particular, the detection modality, the spectral sensitivity, and the dimension of the sensitive area will be considered.

2.3.4.1 Detector Technology

The main technologies used for building detectors suitable for SHG microscopy are summarized below.

- *Photomultiplier tubes (PMTs):* The most used detector in SHG microscopy is the PMT. A conventional PMT is a vacuum device which contains a photocathode, a number of dynodes (amplifying stages) and an anode which delivers the output signal (Figure 2.15a). An electrical field accelerates

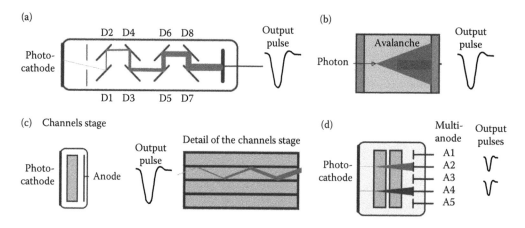

FIGURE 2.15   Sketch of various detectors: (a) Photomultiplier tube; (b) avalanche photodiode; (c) microchannel plate; (d) photomultiplier array. (From Becker, W. and Bergmann, A. *Tutorial—Becker & Hickl GmbH*. Available at: http://www.becker-hickl.com/literature.htm.)

the electrons from the cathode to the next dynodes and to the anode. In each dynode, the number of photoelectrons is amplified and the total gain reaches values of 10^6–10^8. A wide variety of photocathode and dynode geometries has been developed. The sensitivity of the PMT and its spectral response depend on the photocathode type, while the dynode geometry affects the overall gain and the temporal response of the detector.

- *Avalanche photodiodes (APDs):* The APD consists in a photodiode driven with a voltage close to or slightly above the breakdown voltage. In this way, when a photon is detected, the generated electron–hole pair creates an avalanche that can be detected. The quantum efficiency is extremely high in the green range of the spectrum (up to 0.7), but it drops in the blue region (0.2–0.4) (Cova et al., 1982). The sensitivity and the quantum efficiency are related by the following expression:

$$QE = S \frac{hc}{e\lambda} \tag{2.7}$$

where QE is the quantum efficiency, S the sensitivity, h the Planck constant, c the speed of light in vacuum, e the electron charge, and λ the wavelength.

- *Microchannel plates (MCPs):* An MCP is a detector with a working principle similar to the conventional PMT, but the photoelectron amplification is achieved in a channel with a conductive coating instead of in a multiple dynode stage, as depicted in Figure 2.15c. MCPs are the fastest detectors currently available. Moreover, the MCPs technique allows building position-sensitive detectors and image intensifiers. The sensitivity is comparable to PMTs.

- *CCDs and PMT arrays:* These array detectors are mostly used in combination with digital holographic equipment (CCD) and with multifocal setups (both CCD and PMT arrays). In general, the sensitivity and the noise level of a common CCD sensor are not suitable for the detection of weak SHG signals, so that an intensifying system is required to amplify the signal originated by detected photons. Very often, cooled intensified CCDs or electron-multiplying CCDs (EMCCDs) are used in both digital holographic and multifocal systems in non-de-scan mode. On the opposite, a PMT array (Figure 2.15d), although it does not provide enough resolution to be used with digital holographic equipment, can be used in MMM in de-scan mode (Kim et al., 2007, Martini et al., 2007). It offers the advantage of a very high sensitivity and fast response but can be used only in backward detection. A forward detection is possible as well, but it requires a second scanning head for de-scanning SHG light synchronously with the scanning of the beam focus.

2.3.4.2 Sensitive Area

As previously described, the optical detection of SHG signal can be performed in de-scan and non-de-scan mode. In de-scan mode, the stability of the focused SHG light allows the use of a detector with a very small sensitive area, such as an APD. On the other hand, in non-de-scan mode, it is necessary to use a detector with larger sensitive area, such as a PMT. The dimension of the scanned area depends on the magnification obtained with the objective and the focusing optics. Therefore, the choice of the sensitive area of the detector must be operated taking into account the maximum scanning area with the properties of the optical detection system. A small sensitive area can cause a limited field of view, due to dependence of the detected intensity on the position of the scanned beam. This effect is translated in a bright central region and darker sides in the images.

2.3.4.3 Electrical Detection Modality

The most important issue to be considered when choosing a detector for an SHG microscope is the modality in which the detector is going to be used: proportional mode or photon-counting regime. Each of these two modalities requires a detector with particular features in terms of temporal response and dynamic range. In an ideal detector (with quantum efficiency equal to 1), an output pulse is provided every time a photon interacts with the photocathode. In the real scenario, the quantum efficiency is <1, but the output pulse rate is still proportional to the rate of incoming photons (Figure 2.16).

- *Proportional regime.* If the pulse rate is equal or greater than the inverse of the pulse width, the detector pulses are superimposed resulting in a current signal, which is proportional to the intensity of the detected light (Figure 2.16a). The detection is obtained by integrating the output signal with a current integrator. The time constant of the circuit should be chosen according to the sampling rate. The proportional regime has to be used if a high light level is expected and a fast detector is not required.

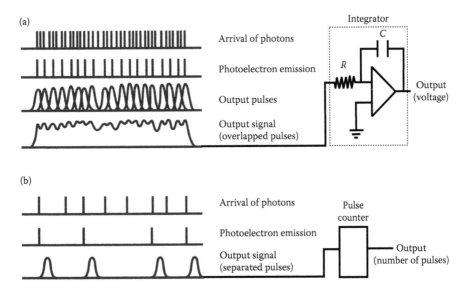

FIGURE 2.16 Detection modality with the corresponding acquisition principle. (a) Proportional regime: Output pulses are overlapped in time. The output voltage is proportional to the detected light intensity over the integration time. (b) Photon-counting regime: Pulses are well separated in time. The output is the number of pulses detected in the pixel dwell time.

- *Photon-counting regime.* On the opposite, if the pulse rate is much lower than the inverse of the pulse width, the pulses come out well separated in time and in this case the number of pulses is proportional to the intensity of the detected light (Figure 2.16b). The detection is obtained by counting the number of pulses with a digital counter. The photon-counting regime is most likely used in low light level and requires a fast detector with high gain.

As a general rule, proportional regime is most likely used in applications where the signals to be detected are high and a fast scanning speed is required; photon-counting regime is instead preferable in applications, where the signals to be detected are low and the required scanning speed is not high.

2.4 Optical Scheme of a Typical SHG Microscope

In this section, we conclude this chapter presenting the whole optical scheme of a typical SHG microscope. In detail, three representative optical configurations will be considered: the forward detection geometry (Figure 2.17) and backward detection geometry in both non-de-scan (Figure 2.18) and de-scan mode (Figure 2.19).

2.4.1 Forward Detection

This optical configuration is the most commonly used in SHG microscopy. This optical configuration is generally used in SHG imaging of nonmassive samples, such as cultured cells, thin tissue slices, and so on. Referring to Figure 2.17, the experimental setup is composed by:

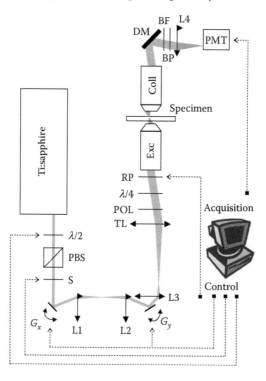

FIGURE 2.17 Optical scheme of a typical SHG microscope with forward detection geometry. Ti:sapphire: femtosecond laser source; $\lambda/2$: rotating half-wave plate; PBS, polarizing beam splitter cube; S, shutter; G_x, galvo-scanner x; L1, relay lens; L2, relay lens; G_y, galvo-scanner y; L3, scanning lens; TL, tube lens; POL, polarizer; $\lambda/4$, quarter-wave-plate; RP, rotating polarizer; Exc, excitation objective lens; Coll, collection objective lens; DM, dichroic mirror; BF, laser-blocking filter; L4, collection lens; BP, narrow band-pass filter; PMT, photomultiplier detector.

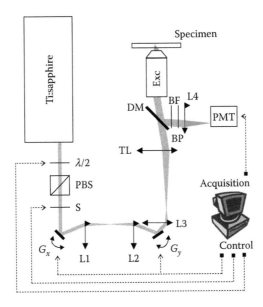

FIGURE 2.18 Optical scheme of a typical SHG microscope with backward detection geometry (non-de-scan mode). Ti:sapphire:femtosecond laser source; $\lambda/2$: rotating half-wave plate; PBS, polarizing beam splitter cube; S, shutter; G_x, galvo-scanner x; L1, relay lens; L2, relay lens; G_y, galvo-scanner y; L3, scanning lens; TL, tube lens; DM, dichroic mirror; BF, laser-blocking filter; L4, collection lens; BP, narrow band-pass filter; PMT, photo-multiplier detector; POL, polarizer; $\lambda/4$, quarter-wave plate; RP, rotating polarizer; Exc, excitation/collection objective lens.

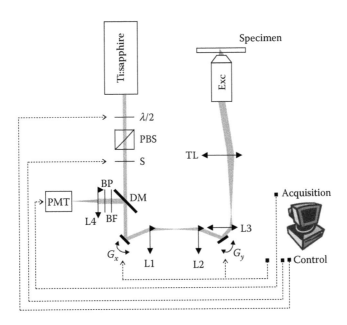

FIGURE 2.19 Optical scheme of a typical SHG microscope with backward detection geometry (de-scan mode). Ti:sapphire:femtosecond laser source; $\lambda/2$: rotating half-wave plate; PBS, polarizing beam splitter cube; S, shutter; DM, dichroic mirror; BF, laser-blocking filter; L4, collection lens; BP, narrow band-pass filter; PMT, photo-multiplier detector; G_x, galvo-scanner x; L1, relay lens; L2, relay lens; G_y, galvo-scanner y; L3, scanning lens; TL, tube lens; POL, polarizer; $\lambda/4$, quarter-wave plate; RP, rotating polarizer; Exc, excitation/collection objective lens.

- The laser source (Ti:sapphire) providing femtosecond pulses. See Section 2.2.2 for a detailed description of suitable laser sources for SHG microscopy.
- The power adjustment system (Section 2.2.3), made by a rotating half-wave plate ($\lambda/2$), (remotely controlled from PC), and a polarizing beam-splitter cube (PBS).
- The scanning system (Section 2.2.4), made with two orthogonal galvo-mirrors (G_x and G_y) optically relayed by a telescope (L1,L2) with magnification equal to 1. A detailed description of the basic principle of a scanning system is provided in Section 2.2.1.
- The telescope (L3,TL) for magnifying the beam size and de-multiply the scanned angle. See Section 2.2.1 for a detailed description of the function accomplished with this optical configuration. Generally, the magnification obtained ranges from 3 to 5.
- A polarization scanning system made with: a polarizer in order to make the polarization linear; a quarter wave plate for generating a circular polarization from a linear one; a rotating polarizer for illuminating the sample with a linear rotating polarization. See Section 2.2.6 for a detailed description of the polarization scanning system.
- The excitation objective lens (Exc). See Section 2.2.5 for a more detailed description.
- The collection objective lens (Coll). See Section 2.3.2.1 for a more detailed description of the forward detection geometry.
- The optical filtering system, composed of a dichroic mirror (DM), a laser-blocking filter (BF), and a narrow band-pass filter (BP). See Section 2.3.3 for a detailed description of the optical filtering system.
- The detection system (Section 2.3), constituted by a collecting lens (L4) and a PMT. It has to be noted that the focus of the SHG light detected by the PMT is scanned across the detector during scanning so that a detector with large sensitive area is necessary in this configuration, as previously described in Section 2.3.4.2.

2.4.2 Backward Detection

This optical configuration is generally used in SHG imaging of massive samples, such as bulk *ex vivo* excised tissues (biopsies) or *in vivo* experiments. The backward detection geometry can be performed in both non-de-scan and de-scan mode.

2.4.2.1 Non-De-Scan Mode

The main differences with respect to the previous detection scheme are:

- The absence of a collection objective. The excitation objective serves for both excitation and collection.
- The position of the detection system (L4, PMT) and the optical filtering system (DM, BF, BP). In this optical configuration, these two systems are placed immediately behind the excitation objective, as close as possible to the objective back aperture.

As in forward detection, in non-de-scan mode, the focus of the SHG light detected by the PMT is scanned across the detector during scanning, so that a detector with large sensitive area is necessary in this configuration, as previously described in Section 2.3.4.2.

It has to be noted that the backward scheme strongly limits the use of a polarization scanning system. In fact, if the polarization scanning system is placed before the dichroic mirror (DM), the birefringent coating of DM can affect the intensity uniformity during polarization scanning (see Section 2.2.6). On the other hand, if the polarization scanning system is placed between the dichroic mirror (DM) and the objective lens, the polarization optics will cause a drastic reduction of the detected SHG signal.

2.4.2.2 De-Scan Mode

The main difference with respect to non-de-scan mode consists in placing the optical filtering system (DM, BF, BP) and the detection system (L4, PMT) before the scanning system (G_x, L1, L2, G_y).

With this geometry, SHG light generated in the specimen travels back along the laser light path, passing through the scanning system before being decoupled from the laser path. In this way, the focus of SHG light detected by the PMT always occupies the same position (as in confocal microscopy), allowing the use of a detector with a very small sensitive area.

In this configuration, the distance between the detector and the collection objective drastically reduces the collection of multiple-scattered SHG photons, limiting *in vivo* applicability.

References

Allegra Mascaro, A. L., Sacconi, L. and Pavone, F. S. 2010. Multi-photon nanosurgery in live brain. *Front. Neuroenerg.*, 2: 21.

Barad, Y., Eisenberg, H., Horowitz, M. and Silberberg, Y. 1997. Nonlinear scanning laser microscopy by third harmonic generation. *Appl. Phys. Lett.*, 70: 922–924.

Becker, W. and Bergmann, A. Detectors for high-speed photon counting. *Tutorial—Becker & Hickl GmbH.* Available at: http://www.becker-hickl.com/literature.htm.

Bewersdorf, J., Pick, R. and Hell, S. W. 1998. Multifocal multiphoton microscopy. *Opt. Lett.*, 123: 655–657.

Both, M., Vogel, M., Friedrich, O., Von Wegner, F., Kunsting, T., Fink, R. H. A. and Uttenweiler, D. 2004. Second harmonic imaging of intrinsic signals in muscle fiber in situ. *J. Biomed. Opt.*, 87: 882–892.

Bouevitch, O., Lewis, A., Pinevsky, I., Wuskell, J. P. and Loew, L. M. 1993. Probing membrane potential with non-linear optics. *Biophys. J.*, 65: 672–679.

Boulesteix, T., Beaurepaire, E., Sauviat, M. P. and Schanne-Klein, M. C. 2004. Second-harmonic microscopy of unstained living cardiac myocytes: Measurements of sarcomere length with 20 nm accuracy. *Opt. Lett.*, 29: 2031–2033.

Buist, A. H., Muller, M., Squier, J. and Brakenhoff, G. J. 1998. Real time two photon absorption microscopy using multi-point excitation. *J. Microsc.*, 192: 217–226.

Bullen, A., Patel, S. S. and Saggau, P. 1997. High-speed, random-access fluorescence microscopy: I. High-resolution optical recording with voltage-sensitive dyes and ion indicators. *Biophys. J.*, 73: 477–491.

Campagnola, P. J., Clark, H. A., Mohler, W. A., Lewis, A. and Loew, L. M. 2001. Second harmonic imaging microscopy of living cells. *J. Biomed. Opt.*, 6: 277–286.

Campagnola, P. J., Wei, M. D., Lewis, A. and Loew, L. M. 1999. High-resolution nonlinear optical imaging of live cells by second harmonic generation. *Biophys. J.*, 77: 3341–3349.

Capitanio, M., Cicchi, R. and Pavone, F. S. 2005. Position control and optical manipulation for nanotechnology applications. *Eur. Phys. J. B*, 46: 1–8.

Chou, C. K., Chen, W. L., Fwu, P. T., Lin, S. J., Lee, H. S. and Dong, C. Y. 2008. Polarization ellipticity compensation in polarization second-harmonic generation microscopy without specimen rotation. *J. Biomed. Opt.*, 13: 014005.

Cicchi, R., Kapsokalyvas, D., De Giorgi, V., Maio, V., Van Wiechen, A., Massi, D., Lotti, T. and Pavone, F. S. 2010. Scoring of collagen organization in healthy and diseased human dermis by multiphoton microscopy. *J. Biophoton.*, 3: 34–43.

Cicchi, R., Sacconi, L., Jasaitis, A., O'Connor, R. P., Massi, D., Sestini, S., De Giorgi, V., Lotti, T. and Pavone, F. S. 2008. Multidimensional custom-made non-linear microscope: From ex-vivo to in-vivo imaging. *Appl. Phys. B*, 92: 359–365.

Cova, S., Longoni, A. and Ripamonti, G. 1982. Active-quenching and gating circuits for single-photon avalanche photodiodes (SPADs). *IEEE Trans. Nucl. Sci.*, NS29: 599–561.

Débarre, D., Supatto, W., Pena, A. M., Fabre, A., Tordjmann, T., Combettes, L., Schanne-Klein, M. C. and Beaurepaire, E. 2006. Imaging lipid bodies in cells and tissues using third-harmonic generation microscopy. *Nat. Methods*, 3: 47–53.

Denk, W., Strickler, J. H. and Webb, W. W. 1990. Two-photon laser scanning fluorescence microscopy. *Science*, 248: 73–76.

Dombeck, D. A., Kasischke, K. A., Vishwasrao, H. D., Ingelsson, M., Hyman, B. T. and Webb, W. W. 2003. Uniform polarity microtubule assemblies imaged in native brain tissue by second-harmonic generation microscopy. *Proc. Natl. Acad. Sci. USA*, 100: 7081–7086.

Guo, Y., Savage, H. E., Liu, F., Schantz, P., Ho, P. P. and Alfano, R. R. 1999. Subsurface tumor progression investigated by noninvasive optical second harmonic tomography. *Proc. Natl. Acad. Sci. USA*, 96: 10854–10856.

Heikal, A. A., Hess, S. T. and Webb, W. W. 2001. Multiphoton molecular spectroscopy and excited-state dynamics of enhanced green fluorescent protein (EGFP): Acid-base specificity. *Chem. Phys.*, 274: 37–55.

Hsieh, C. L., Grange, R., Pu, Y. and Psaltis, D. 2009. Three-dimensional harmonic holographic microscopy using nanoparticles as probes for cell imaging. *Opt. Express*, 17: 2880–2891.

Hsieh, C. L., Pu, Y., Grange, R. and Psaltis, D. 2010. Second harmonic generation from nanocrystals under linearly and circularly polarized excitations. *Opt. Express*, 18: 11917–11932.

Kim, K. H., Buehler, C., Bahlmann, K., Ragan, T., Lee, W. C. A., Nedivi, E., Heffer, E. L., Fantini, S. and So, P. T. C. 2007. Multifocal multiphoton microscopy based on multianode photomultiplier tubes. *Opt. Express*, 15: 11658–11678.

Konig, K. and Riemann, I. 2003. High-resolution multiphoton tomography of human skin with subcellular spatial resolution and picosecond time resolution. *J. Biomed. Opt.*, 8: 432–439.

Lacomb, R., Nadiarnykh, O. and Campagnola, P. J. 2008. Quantitative second harmonic generation imaging of the diseased state osteogenesis imperfecta: Experiment and simulation. *Biophys. J.*, 94: 4504–4514.

Llewellyn, M. E., Barretto, R. P. J., Delp, S. L. and Schnitzer, M. J. 2008. Minimally invasive high-speed imaging of sarcomere contractile dynamics in mice and humans. *Nat. Lett.*, 454: 784–788.

Martini, J., Andresen, V. and Anselmetti, D. 2007. Scattering suppression and confocal detection in multifocal multiphoton microscopy. *J. Biomed. Opt.*, 12: 034010.

Masihzadeh, O., Schlup, P. and Bartels, R. A. 2010. Label-free second harmonic generation holographic microscopy of biological specimens. *Opt. Express*, 18: 9840–9851.

Matteini, P., Ratto, F., Rossi, F., Cicchi, R., Stringari, C., Kapsokalyvas, D., Pavone, F. S. and Pini, R. 2009. Photothermally-induced disordered patterns of corneal collagen revealed by SHG imaging. *Opt. Express*, 17: 4868–4878.

Mertz, J. and Moreaux, L. 2001. Second-harmonic generation by focused excitation of inhomogeneously distributed scatterers. *Opt. Commun.*, 196: 325–330.

Moreaux, L., Sandre, O., Blanchard-Desce, M. and Mertz, J. 2000a. Membrane imaging by simultaneous second-harmonic generation and two-photon microscopy. *Opt. Lett.*, 25: 320–322.

Moreaux, L., Sandre, O., Charpak, S., Blanchard-Desce, M. and Mertz, J. 2001. Coherent scattering in multi-harmonic light microscopy. *Biophys. J.*, 80: 1568–1574.

Moreaux, L., Sandre, O. and Mertz, J. 2000b. Membrane imaging by second-harmonic generation microscopy. *J. Opt. Soc. Am. B*, 17: 1685–1694.

Nadiarnykh, O., Lacomb, R. B., Brewer, M. A. and Campagnola, P. J. 2010. Alteration of the extracellular matrix in ovarian cancer studied by second harmonic generation imaging microscopy. *BMC Cancer*, 10: 94.

Nakayama, Y., Pauzauskie, P. J., Radenovic, A., Onorato, R. M., Saykally, R. J., Liphardt, J. and Yang, P. D. 2007. Tunable nanowire nonlinear optical probe. *Nature*, 447: 1098–U8.

Nucciotti, V., Stringari, C., Sacconi, L., Vanzi, F., Fusi, L., Linari, M., Piazzesi, G., Lombardi, V. and Pavone, F. S. 2010. Probing myosin structural conformation *in vivo* by second-harmonic generation microscopy. *Proc. Natl. Acad. Sci. USA*, 107: 7763–7768.

Oheim, M., Beaurepaire, E., Chaigneau, E., Mertz, J. and Charpak, S. 2001. Two-photon microscopy in brain tissue: Parameters influencing the imaging depth. *J. Neurosci. Methods*, 111: 29–37.

Oron, D., Yelin, D., Tal, E., Raz, S., Fachima, R. and Silberberg, Y. 2004. Depth-resolved structural imaging by third-harmonic generation microscopy. *J. Struct. Biol.*, 147: 3–11.

Peleg, G., Lewis, A., Linial, M. and Loew, L. M. 1999. Nonlinear optical measurement of membrane potential around single molecules at selected cellular sites. *Proc. Natl. Acad. Sci. USA,* 96: 6700–6704.

Pfeffer, C. P., Olsen, B. R. and Légaré, F. 2007. Second harmonic generation imaging of fascia within thick tissue block. *Opt. Express,* 15: 7296–7302.

Plotnikov, S. V., Millard, A. C., Campagnola, P. J. and Mohler, W. A. 2006. Characterization of the myosin-based source for second-harmonic generation from muscle sarcomeres. *Biophys. J.,* 90: 693–703.

Prasanth, R., van Vugt, L. K., Vanmaekelbergh, D. A. M. and Gerritsen, H. C. 2006. Resonance enhancement of optical second harmonic generation in a ZnO nanowire. *Appl. Phys. Lett.,* 88:181501/1–181501/3.

Provenzano, P. P., Eliceiri, K. W., Campbell, J. M., Inman, D. R., White, J. G. and Keely, P. J. 2006. Collagen reorganization at the tumor–stromal interface facilitates local invasion. *BMC Med.,* 4: 38.

Qiao, L., Ni, J., Mao, Z., Wang, C. and Cheng, Y. 2008. Two-color two-photon excitation of indole using a femtosecond laser regenerative amplifier. *Opt. Commun.,* 282: 1056–1061.

Reddy, G. D. and Saggau, P. 2005. Fast three-dimensional laser scanning scheme using acousto-optic deflectors. *J. Biomed. Opt.,* 10: 064038.

Sacconi, L., D'amico, M., Vanzi, F., Biagiotti, T., Antolini, R., Olivotto, M. and Pavone, F. S. 2005. Second harmonic generation sensitivity to transmembrane potential in normal and tumor cells. *J. Biomed. Opt.,* 10: 024014.

Sacconi, L., Froner, E., Antolini, R., Taghizadeh, M. R., Choudhury, A. and Pavone, F. S. 2003. Multiphoton multifocal microscopy exploiting a diffractive optical element. *Opt. Lett.,* 28: 1918–1920.

Sacconi, L., Mapelli, J., Gandolfi, D., Lotti, J., O'connor, R. P., D'angelo, E. and Pavone, F. S. 2008. Optical recording of electrical activity in intact neuronal networks with random-access second-harmonic generation microscopy. *Opt. Express,* 16: 14910–14921.

Shaffer, E., Marquet, P. and Depeursinge, C. 2010. Real time, nanometric 3D-tracking of nanoparticles made possible by second harmonic generation digital holographic microscopy. *Opt. Express,* 18: 17392–17403.

Sheppard, C. J. R., Kompfner, R., Gannaway, J. and Walsh, D. 1977. Scanning harmonic optical microscope. *IEEE J. Quantum Electron.,* 13E: 100D.

Stoller, P., Kim, B. M., Rubenchik, A. M., Reiser, K. M. and Da Silva, L. B. 2002a. Polarization-dependent optical second-harmonic imaging of a rat-tail tendon. *J. Biomed. Opt.,* 7: 205–214.

Stoller, P., Reiser, K. M., Celliers, P. M. and Rubenchik, A. M. 2002b. Polarization-modulated second harmonic generation in collagen. *Biophys. J.,* 82: 3330–3342.

Sun, C. K. 2005. Higher harmonic generation microscopy. *Adv. Biochem. Eng. Biotechnol.,* 95: 17–56.

Svoboda, K. and Yasuda, R. 2006. Principles of two-photon excitation microscopy and its applications to neuroscience. *Neuron,* 50: 823–839.

Theer, P., Hasan, M. T. and Denk, W. 2003. Two-photon imaging to a depth of 1000 mm in living brains by use of a Ti:Al$_2$O$_3$ regenerative amplifier. *Opt. Lett.,* 28: 1022–1024.

Tiaho, F., Recher, G. and Rouede, D. 2007. Estimation of helical angles of myosin and collagen by second harmonic generation imaging microscopy. *Opt. Express,* 15: 12286–12295.

Veilleux, I., Spencer, J. A., Biss, D. A., Coté, D. and Lin, C. P. 2008. *In vivo* cell tracking with video rate multimodality laser scanning microscopy. *IEEE J. Quantum Electron.,* 14: 10–18.

Zipfel, W. R., Williams, R. M., Christie, R., Nikitin, A. Y., Hyman, B. T. and Webb, W. W. 2003a. Live tissue intrinsic emission microscopy using multiphoton-excited native fluorescence and second harmonic generation. *Proc. Natl. Acad. Sci. USA,* 100: 7075–7080.

Zipfel, W. R., Williams, R. M. and Webb, W. W. 2003b. Nonlinear magic: Multiphoton microscopy in the biosciences. *Nat. Biotechnol.,* 21: 1369–1377.

3

THG Microscopy of Cells and Tissues: Contrast Mechanisms and Applications

Nicolas Olivier
École Polytechnique
CNRS–Inserm

Delphine Débarre
École Polytechnique
CNRS–Inserm

Emmanuel Beaurepaire
École Polytechnique
CNRS–Inserm

3.1 Introduction

Third-harmonic generation (THG) and second-harmonic generation (SHG) are both coherent nonlinear processes that can be used for microscopy, and that can be produced and detected simultaneously using a single laser. However, they are sensitive to different electronic symmetries and obey different phase-matching conditions, so that THG and SHG images are usually both distinct and complementary. Since the principles and applications of SHG microscopy are discussed at large in other chapters, the aim of this chapter is to discuss the parameters that govern THG efficiency, and to present a few examples where combined THG–SHG–2PEF imaging provides complementary information on biological samples.

3.1.1 First Demonstrations of THG Microscopy

THG microscopy relies on the coherent nonlinear process of THG, in which three infrared photons (typically between 900 nm and 1.5 μm) are coherently scattered by a molecule to produce one harmonic photon, as illustrated in Figure 3.1.

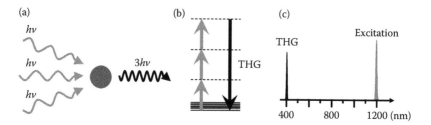

FIGURE 3.1 Schematic view of the THG process. (a) Three photons of energy $h\nu$ are coherently scattered to produce a harmonic photon of energy $3h\nu$. (b) Virtual energy levels of the molecule involved in the THG process. (c) Wavelength representation.

The first experimental demonstrations of THG in calcite [1,2], gases [3,4], and liquids [5,6] were performed shortly after the demonstration of SHG, and an accurate theory of THG with focused Gaussian beams was proposed shortly after [3,4]. Tsang et al. [7,8] later reported efficient THG at interfaces between two dielectric volumes.

THG microscopy was initially demonstrated by two groups at the end of the 1990s: Barad, Eisenberg, Horowitz, and Silberberg at the Weizmann Institute [9,10]; and Squier, Müller, Brakenhoff, and Wilson at UCSD (University of California, San Diego) [11,12]. The authors have shown that THG can be used as a contrast mechanism and allows structural imaging of several samples (root tips, algae, neurons, yeast cells, etc.) with micrometer resolution. Although the contrast mechanisms were not entirely characterized at this point, these pioneering studies showed that because of the phase shift experienced by focused beams, THG signals mainly originate from interfaces and sub-micron-sized structures, which were in good agreement with both the experimental demonstration of THG at interfaces [7] and the theoretical analysis performed by Ward and New [4] who had shown that there is no THG from a homogeneous isotropic medium and a maximum signal from structures approximately half the Rayleigh range of the focused excitation beam.

From these early studies, one can already point out important differences between THG and SHG microscopies:

1. THG is observed only at interfaces or inclusions, that is to say where the sample is heterogeneous at the scale of the wavelength.
2. However, THG is possible at the interface between two homogeneous isotropic media, and is not limited to organized structures like SHG is.

Therefore, the two modalities provide complementary information about the sample, which makes THG/SHG microscopy an interesting combination.

3.1.2 Chapter Outline

In this chapter, we will concentrate on the principles of THG microscopy. We will start by discussing the contrast mechanisms in the case of isotropic media, and present applications of combined THG/SHG imaging in developmental biology illustrating the complementarity between the two signals. In a second part, we will briefly discuss the mechanisms of THG from organized media, and illustrate this analysis with the case of THG/SHG imaging of the human cornea.

3.2 THG Microscopy of Isotropic Media

3.2.1 Introduction

The coherent nature of signal generation in THG/SHG microscopy is one fundamental difference with fluorescence microscopy, as it makes the geometrical structure of the sample become an important parameter: if the sample has an appropriate molecular and microscopic structure, constructive interference will increase the harmonic signal, whereas an inefficient geometry will result in destructive interference and weak or null signal.

We will in this first part consider the case of isotropic media, and a linearly polarized excitation (and therefore harmonic) beam. This approximation allows us to neglect the tensorial properties of THG and to first derive a simpler scalar description.

3.2.2 THG with Plane Waves (1D Case)

Before discussing microscopy, we first remind a few basic concepts of nonlinear optics. Coupling between the excitation electric field and the produced harmonic field is described by the nonlinear wave equation, which in the case of THG can be approximated as

$$\Delta E(3\omega) + \frac{n^2(3\omega)(3\omega)^2}{c^2} E(3\omega) = -\frac{(3\omega)^2}{\epsilon_0 c^2} \chi^{(3)}(3\omega;\omega;\omega;\omega) E^3(\omega) \tag{3.1}$$

To discuss the parameters governing THG efficiency, we first consider the case of a plane wave forward propagating along direction $z > 0$ on the optical axis. The fundamental and harmonic waves can be described by

$$E_\omega(z,t) = A_1 e^{i(k_\omega z - \omega t)} \tag{3.2}$$

$$E_{3\omega}(z,t) = A_3(z) e^{i(k_{3\omega} z - 3\omega t)} \tag{3.3}$$

with $k_\omega = (n_\omega \omega / c)$.

Under the assumption of the slowly varying amplitude approximation, the nonlinear wave equation can be written as

$$2ik \frac{\partial}{\partial z} E_{3\omega}(z) = -\frac{(3\omega)^2}{\epsilon_0 c^2} \chi^{(3)}(3\omega;\omega;\omega;\omega) A_1^3 e^{3i(k_\omega z)} \tag{3.4}$$

$$\Downarrow$$

$$\frac{\partial}{\partial z} A_3(z) \propto A_1^3 e^{i(\Delta k z)} \tag{3.5}$$

where $\Delta k = k(3\omega) - 3k(\omega) = 3(n_\omega - n_{3\omega})\omega/c$ is the phase difference between the propagating fundamental and harmonic beams, and is called the *wave vector mismatch*, here governed by the dispersion $(n_\omega - n_{3\omega})$ of the refractive index.

We can already extract a qualitative view of the THG process from Equation 3.5: (i) the TH amplitude is proportional to the cube amplitude of the fundamental, and (ii) it is modulated by the phase mismatch Δk between the co-propagating beams. This equation can be easily integrated, and the harmonic field after an interaction distance l can be written as

$$E(3\omega, z = l) \propto -A_1^3 \int_0^l e^{i\Delta k z} dz$$

$$\propto A_1^3 l \, \mathrm{sinc}(\Delta k l/2) \tag{3.6}$$

In the case of perfect phase matching ($\Delta k = 0$), the THG intensity increases quadratically with l. If there is dispersion ($n_\omega \neq n_{3\omega}$) and therefore phase mismatch, THG exhibits an oscillating behavior as a function of l (sinc function) with a characteristic period of $\pi/\Delta k$: the signal starts to decrease when the interaction length gets larger than this period.

This length is referred to as the *coherence length* (noted l_c) of this particular nonlinear interaction.

$$l_c^{FTHG} = \frac{\pi}{|\,k(3\omega) - 3k(\omega)\,|} \tag{3.7}$$

The case of backward (i.e., counter propagating) radiation is described by the same equation, except that we now have:

$$E_\omega(z,t) = A_1 e^{i(k_\omega z - \omega t)} \tag{3.8}$$

$$E_{3\omega}(z,t) = A_3(z) e^{i(-k_{3\omega} z - 3\omega t)} \tag{3.9}$$

We can perform the same analysis as in the case of forward propagation, except with $\Delta k^{BTHG} = k(3\omega) + 3k(\omega)$. The coherence length for backward THG is therefore defined as

$$l_c^{BTHG} = \frac{\pi}{|\,k(3\omega) + 3k(\omega)\,|} \approx \frac{\lambda}{12 n_\omega} \tag{3.10}$$

where the approximation holds when neglecting dispersion. With typical biological materials and wavelengths, $l_c^{FTHG} \approx 10\,\mu m$ and $l_c^{BTHG} \approx 80\,nm$, so that the backward coherence length is much smaller than the forward coherence length. This difference in direction between the wave vectors is responsible for the small amount of backward-emitted signals in harmonic generation microscopy, as we will confirm later.

This simple 1D geometry illustrates the importance of the relative phase between the fundamental and the harmonic field, which results in either constructive or destructive interference. However, in the case of microscopy, we need to consider a focused excitation beam. We will see that going from plane waves to focused beams not only changes the intensity distribution (and therefore the effective interaction length), but also changes the phase-matching conditions.

3.2.3 THG with Gaussian Beams

We refine the previous analysis by now considering a Gaussian excitation beam instead of a plane wave. Although the Gaussian beam model does not provide an accurate description of the phase and intensity distributions in the case of high numerical aperture (NA) focusing [13], it usually provides solutions that are in good qualitative agreement with nonlinear microscopy experiments, at least at moderate NAs.

The most obvious difference with the plane-wave case is the nonconstant intensity distribution. The main consequence is that the nonlinear interaction is confined within a finite effective interaction volume. However, the intensity distribution is not the only change: the field distribution near the focal point illustrated in Figure 3.2 also exhibits a progressive phase slippage along the optical axis z known as the Gouy phase shift [14] that results in an overall π radian phase difference between a plane wave and a focused beam. The Gouy phase shift is particularly significant for THG phase matching, and makes the phase-matching conditions derived for a plane wave invalid in the case of a focused beam.

There is no simple way to fully account for the phase and intensity dependence of THG, which is why researchers often rely on numerical simulations to study the influence of different parameters. However, simple models can be derived from the simulations to understand the key phenomena: in the case of the phase influence, one elegant model was proposed by Cheng et al. [15] in which they consider the interaction between a focused Gaussian excitation beam and a harmonic plane wave propagating along

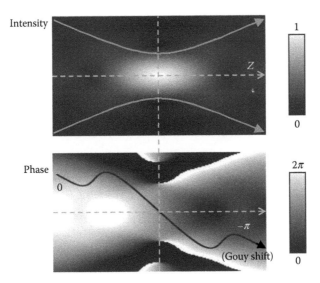

FIGURE 3.2 Intensity and phase distribution of a focused Gaussian beam. 2D distribution of intensity and phase near the focus. The phase propagation term along direction z has been subtracted. $-2\,\mu\text{m} < \rho < 2\,\mu\text{m}$, $-5\,\mu\text{m} < z < 5\,\mu\text{m}$, $\lambda = 1.2\,\mu\text{m}$.

the same axis. The projection on the propagation axis of the phase difference between the two beams can then be written as

$$\Delta k = k(3\omega) - 3(k(\omega) + k_G) \tag{3.11}$$

where k_G represents the Gouy phase shift and is negative. This yields an *effective* coherence length of

$$l_{c,eff}^{FTHG} = \frac{\pi}{|\,k(3\omega) - 3(k(\omega) + k_G)\,|} \approx \frac{\pi}{|\,3(k_G)\,|} \tag{3.12}$$

The right-side approximation in the equation above is valid in the limit of strong focusing. In that case, the Gouy phase rather than the material dispersion is the dominant factor defining $l_{c,eff}^{FTHG}$. This simple model provides a good estimate of the size of an heterogeneity yielding the maximum signal in THG microscopy. In the same spirit, we will now use numerical simulations of simple geometries to dissect the THG process.

3.2.4 Green's Function Formulation of Coherent Scattering

Throughout the remaining of this chapter, we will use the Green function formalism presented in Ref. [15]. We here briefly introduce this model (see also Appendix). For details about the numerical implementation, we refer the reader to Refs. [13,15,16]. The numerical simulation is performed as follows:

1. The focal field at the fundamental wavelength is calculated using Debye–Wolf integrals [13,17,18].
2. The distribution of the induced nonlinear polarization is calculated for a particular sample geometry.
3. The nonlinear polarization is propagated to the far field using Green's functions.

This model is quite general, and can be used to analyze complex geometries. We will however first consider simple geometries and discuss the relevant parameters.

3.2.5 Size Effects

We consider the case of a sphere of diameter d with a nonlinear susceptibility $\chi^{(3)} = 1$ centered at the focus, and embedded in a medium with the same index of refraction, but a null nonlinear susceptibility. We note that, because there is no far-field THG signal from an isotropic normally dispersive homogeneous medium [4], and because the propagation of the harmonic field is a linear process, it is equivalent to consider an object with a nonlinear susceptibility $\chi_1^{(3)}$ embedded in a medium with a nonlinear susceptibility $\chi_2^{(3)}$, or to consider an object with a nonlinear susceptibility $\chi_1^{(3)} - \chi_2^{(3)}$ embedded in a medium with $\chi_0^{(3)} = 0$.

We denote F-THG the harmonic signal detected in the forward direction ($z > 0$) and B-THG the harmonic signal detected in the backward direction ($z < 0$). Figure 3.3 illustrates the dependence of both F-THG and B-THG signals on the sphere size. As the sphere size increases from zero, both F-THG and B-THG signals first exhibit a coherent increase characterized by a quadratic intensity dependence on the sphere volume. Yet, as the size of the sphere further increases, destructive interferences start to occur in the backward direction, and the signal decreases. This is due to the large phase mismatch in that direction. The effective coherence length for B-THG can be approximated as

$$l_{c,eff}^{BTHG} = \frac{\pi}{3k_\omega + k_{3\omega}} \approx \frac{\pi}{6k_\omega} \approx \frac{\lambda}{12n_\omega} \tag{3.13}$$

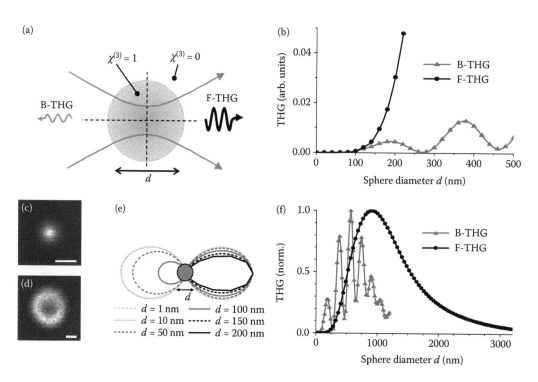

FIGURE 3.3 THG signal from the center of a sphere as a function of size. (a) Geometry considered. (b,f) Calculated B-THG and F-THG signal when the excitation beam is focused at the center of a sphere of variable diameter d. Conditions: $NA = 1.2$, $n_\omega = n_{3\omega} = 1.33$, $\lambda = 1.2$ μm. (c,d) Experimental THG images of 600 nm and 3 μm diameter polystyrene beads. The larger bead appears hollow. Scale bar = 2 μm, $NA = 0.8$, $\lambda = 1.2$ μm. (e) Calculated normalized emission patterns for several sphere sizes. (Reprinted from Débarre D, Olivier N, Beaurepaire E 2007. Signal epidetection in third-harmonic generation microscopy of turbid media. *Opt. Express* 15:8913–8924. With permission of Optical Society of America.)

which is the value we obtained by considering plane waves. Taking the same parameters as in the simulation, we find $l_{c,eff}^{BTHG} \approx 70\,nm$, which is consistent with the maximum signal value observed. The B-THG then exhibits damped oscillations with a period corresponding to twice the backward coherence length.

The F-THG signal has a simpler dependence: F-THG increases with bead size until destructive interference due to the Gouy phase shift (and, if present, to index mismatch between the fundamental and the harmonic field) becomes significant. F-THG from the sphere center then decreases to zero for large spheres, which can be considered as a homogeneous medium at the scale of the excitation volume. The effective forward coherence length can be expressed as

$$l_{c,eff}^{FTHG} = \frac{\pi}{\Delta k} = \frac{\pi}{3(k_\omega + k_g) - k_{3\omega}} \qquad (3.14)$$

where k_g is the linear phase variation induced by the Gouy phase shift. The forward coherence length was estimated in Ref. [15] as $l_{c,eff}^{FTHG} \approx 0.7\lambda$ in the conditions considered here, this value being close to the predicted maximum signal.

There are no visible oscillations on the F-THG signal because the excitation intensity at $z = 2l_{c,eff}^{FTHG}$ is negligible compared to its value at the focus. We note however that in highly dispersive media where the coherence length ($l_c = \pi c/3\omega(n_\omega - n_{3\omega})$) is smaller than the Rayleigh range of the excitation beam, oscillations would be present in the forward direction.

3.2.6 Influence of the Excitation NA

Owing to the dependence of the effective forward coherence length on the axial extension of the Gouy shift, the relative THG signal from objects of different sizes depends on the excitation NA. For details on the influence of the focusing condition in THG microscopy, we refer the reader to Ref. [20]. We only outline here the NA dependence in the case of spherical objects centered on the focus. The general rule is quite simple: reducing the NA results in an increased visibility of large structures relative to small structures. Changing the excitation NA therefore results in changes of the relative contrast between structures, as illustrated in Figure 3.4.

This phenomenon can be analyzed simply by considering a paraxial Gaussian excitation beam. The axial distribution of the Gouy phase shift can then be written as $arctan(z/b)$, where b is the confocal

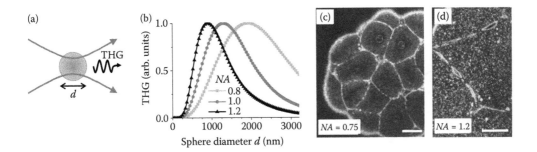

FIGURE 3.4 THG signal as a function of the NA. (a) Centered sphere geometry. (b) Calculated THG from a centered sphere as a function of size, for three different excitation NAs. (c,d) THG images of cells in a zebrafish embryo at cleavage stage recorded using $NA = 0.75$ and $NA = 1.2$. Scale bars: 50 µm (c), 30 µm (d). (From Olivier N et al. 2010. Cell lineage reconstruction of early zebrafish embryos using label-free nonlinear microscopy. *Science* 329:967–971. Reprinted with permission of AAAS.)

parameter. If we make the approximation that it varies in a linear manner between $z = -b$ and $z = +b$, we have an accumulated phase shift across the focus of

$$k_G \approx \frac{\pi}{4b} \tag{3.15}$$

and the coherence length can now be approximated as

$$l_{c,eff}^{FTHG} \approx \frac{4b}{3} \tag{3.16}$$

If we use the geometrical relation between the NA and the waist at the focus, we have

$$b \propto \frac{1}{NA^2} \tag{3.17}$$

so finally

$$l_{c,eff}^{FTHG} \propto \frac{1}{NA^2} \tag{3.18}$$

This result is consistent with the coherence lengths calculated in Figure 3.4: for example, the ratio between the size of the structures that yield a maximum signal for the NAs of 0.8 and 1.2 is equal to $\approx 2/0.9 \approx 2.2$, while the ratio between the squared NAs is equal to 2.25.

Figure 3.4 also illustrates the difference in the images obtained with two different excitation NAs in a zebrafish embryo. In this particular example, it is quite visible that, although higher NA imaging reveals smaller details, it results in decreased contrast of the large structures. In the lower resolution image, the cell contours are more clearly revealed than the smaller intracellular organelles, which make possible the algorithmic detection of cell shapes, as illustrated in Section 3.2.13.

3.2.7 Influence of Dispersion

To discuss the influence of the linear dispersion, we consider the slightly different geometry of a slab perpendicular to the z-axis. Figure 3.5 illustrates the influence of linear dispersion on the THG signal

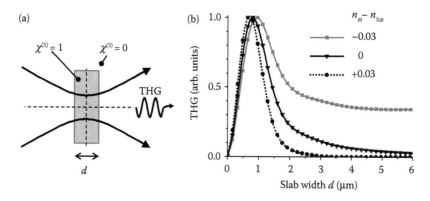

FIGURE 3.5 Influence of dispersion on the THG signal from a slab. (a) Geometry considered. (b) F-THG signal as a function of slab width in the case of negative, null, and positive dispersion. Conditions: $n_{3\omega} = 1.5$, $1.47 < n_\omega < 1.53$, $NA = 1.2$, $\lambda = 1.2$ μm.

as a function of the slab width. We notice that for all the dispersions considered we have a bell-shaped THG response, with a coherent increase followed by signal decrease due to the destructive interference caused by the Gouy phase shift. Yet, in the case of negative dispersion, the dispersion partly compensates the effect of the Gouy shift, and the destructive interference is not complete.

Interestingly, THG can be obtained from a homogeneous medium exhibiting strong negative dispersion, as demonstrated in the case of THG in gases [4]. However, this situation is usually not found in cells and tissues, so that THG microscopy typically requires the presence of an interface or an inclusion.

3.2.8 Interference Effects

More generally, when several objects, or objects of different sizes are present within the excitation volume, interference effects can result in increased or decreased THG intensity. This is illustrated in Figure 3.6a, where thickness variations at the tip of a pulled glass pipette produce constructive and destructive interference. In Figure 3.6b, distance variations between two lipid droplets result in a local THG over-intensity. We note that a similar effect has been described in SHG imaging of labeled vesicles [22].

3.2.9 Orientation Effects

In THG microscopy, phase matching is different in the transverse (x,y) and axial (z) directions. This results in different THG efficiencies for structures oriented along different directions. The geometry we consider here is an interface between a medium with a nonlinear susceptibility $\chi^{(3)}$ and another medium with the same linear index of refraction, but with a nonlinear susceptibility equal to zero. Figure 3.7d illustrates the different geometries used in this section: we consider interfaces along the x-, y-, and z-axis, and calculate the THG signal as a function of the angle it makes around its axis.

In the case of an interface along the optical axis z (Figure 3.7a), there is a weak though nonnegligible orientation dependence ($\approx 20\%$). This is related to the transverse asymmetry of a linearly polarized beam focused by a high NA lens, as will be illustrated later. More intriguing is the fact that far-field scattering occurs along two off-axis lobes. This phenomenon can be described analytically by considering 3D phase-matching conditions:

Coherent THG scattering is efficient only along the directions that fulfill the following vectorial phase-matching condition:

$$\left|3(\mathbf{k}_\omega + \mathbf{k}_G) - \mathbf{k}_{3\omega}\right| = 0 \tag{3.19}$$

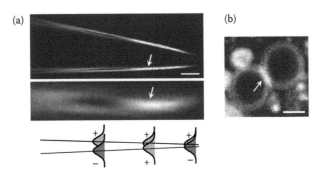

FIGURE 3.6 Examples of interference in THG images. (a) Original and zoomed THG images of the tip of a pulled glass pipette. (b) THG images of lipid droplets in a rat hepatocyte. Arrows indicate local over-intensities resulting from interference. Scale bars: 5 μm. (Adapted from Débarre D 2006. *Thèse de doctorat*. Ecole Polytechnique, Palaiseau.)

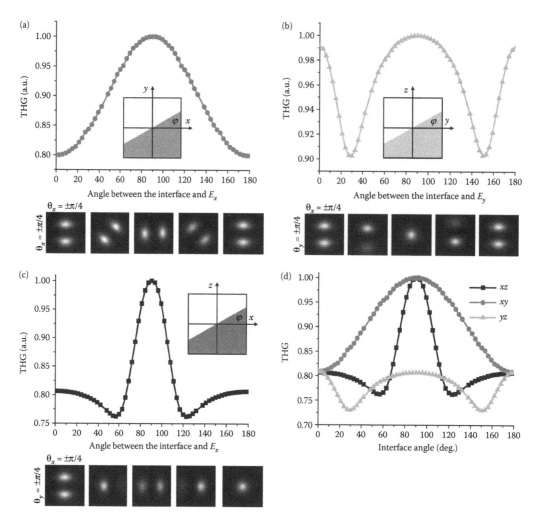

FIGURE 3.7 THG as a function of interface orientation, for an *x*-polarized beam focused by a *NA* =1.2 lens. (a) THG from an interface parallel to the *z*-axis, as a function of the angle between the interface and the *x*-axis. Underneath the curve are the far-field harmonic field distributions for various interface angles. (b,c) Similar calculations for interfaces rotating around the *x*- and *y*-axes. (d) Comparison between the different orientations.

where $\mathbf{k}_\omega$ is the wave vector of the fundamental wave, $\mathbf{k}_G$ represents the Gouy phase shift, and $\mathbf{k}_{3\omega}$ is the wave vector of the harmonic wave.

If we consider scattering with an angle θ relative to the optical axis, and neglect the index mismatch, we obtain the following phase-matching condition along the *z* axis:

$$\left|3(k_\omega + k_G) - \cos(\theta)k_{3\omega}\right| = 0 \tag{3.20}$$

implying that coherent scattering will occur at an angle θ such as

$$\cos(\theta) = 1 + \frac{3k_G}{k_{3\omega}} \tag{3.21}$$

In the conditions of the simulation ($n_\omega = n_{3\omega} = 1.33$, $NA = 1.2$), this equation yields $\theta \approx 34°$, which is close to the maximum value of the emission pattern.

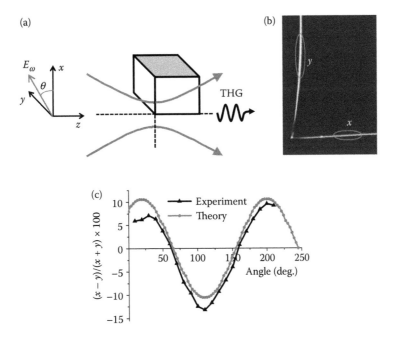

FIGURE 3.8 THG as a function of the polarization direction. (a) Geometry. (b) THG image of a coverslip showing xz and yz glass/water interfaces ($NA = 1.2$). (c) Ratio between the signal obtained from the xz-oriented interface and the signal from the yz-oriented interface as a function of the direction of the excitation polarization.

If we now consider the case of an interface along the x or y axis (Figures 3.7b and 3.7c), the emission diagrams evolve from a single-peaked emission in the case of an interface perpendicular to the optical axis, to the aforementioned symmetric double-peaked emission in the case of an interface parallel to the optical axis, with intermediate positions showing an asymmetric double-peaked emission.

This orientation dependence can be confirmed experimentally by imaging the top corners of a glass coverslip (Figure 3.8a). In this geometry, signals from glass/water interfaces with three different orientations can be detected. It would be complex to compare the signal obtained from the vertical interfaces and the horizontal one because of the presence of aberrations caused by the focusing on an interface. However, signals from the two vertical interfaces can be compared because the aberrations are similar in both cases. Figure 3.8b shows the ratio of THG intensities from the vertical interfaces as a function of the polarization of the excitation beam. This experiment is fully consistent with the theoretical prediction plotted on the same graph.

3.2.9.1 Influence of the NA on Orientation Effects

The relative visibility of structures with different orientations is also dependent on the excitation NA. Let us now consider a centered slab with a nonlinear susceptibility $\chi^{(3)}$ oriented along the axial or transverse directions. Figure 3.9 presents the corresponding calculated THG signal as a function of both the orientation and the size of the slabs.

The THG signal dependence as a function of the width of an xy-oriented slab is similar to the case of a centered sphere (see Section 3.2.5), because phase matching is also dominated by the Gouy phase shift. For the same reason, it is also strongly dependent on the excitation NA. In the case of an xz-oriented slab, THG efficiency is not dominated by axial phase matching and THG signal exhibits a lesser dependence on the NA. This implies that the relative signals from structures having different orientations change with the focusing conditions: THG is more orientation-dependent for lower NAs, as illustrated by Figure 3.9 which compares the orientation visibility ratio for NAs of 0.8 and 1.2.

.

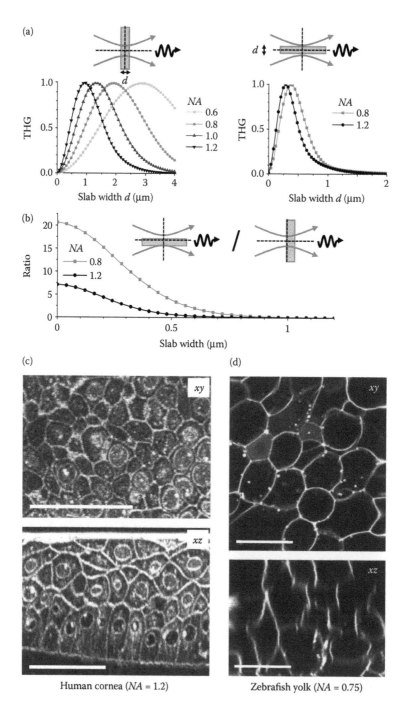

FIGURE 3.9 Influence of the NA in THG microscopy. (a) F-THG signal as a function of the width of a centered *xy*-oriented and *xz*-oriented slabs. (b) Ratio between the maximum signals obtained when scanning the focus across a *xz*-oriented and a *xy*-oriented slab, for two different excitations NAs. (c,d) Illustration: THG images recorded in a human cornea and in a zebrafish embryo. Scale bars: 50 μm. (From Aptel F et al. 2010. Multimodal nonlinear imaging of the human cornea. *Investigative Ophthalmology & Visual Science* 51:2459–2465; Olivier N et al. 2010. Cell lineage reconstruction of early zebrafish embryos using label-free nonlinear microscopy. *Science* 329:967–971. Reprinted with permission from ARVO and AAAS.)

More generally, the rule is that the orientation visibility ratio is large for structures that are small compared to the beam focus (see, e.g., Figure 3.9 in the case $NA = 0.8$). Calculations show that the orientation response becomes isotropic for objects larger than ≈ 500 nm using $NA = 1.2$, and for objects larger than ≈ 1.2 μm using $NA = 0.8$.

Above this "threshold," structures are equally visible in both orientations (although resolution is anisotropic); this is illustrated in Figure 3.9 in the case of the epithelium of the cornea. In contrast, below this size threshold, visibility strongly depends on object orientation, as illustrated in Figure 3.9 in the case of the yolk of the zebrafish embryo, where only axially oriented structures are visible on the *xz* reprojection.

3.2.10 Epidetection

As outlined in the previous sections, THG from dielectric volumes is predominantly forward directed. This is because of the large phase mismatch for backward radiation, preventing efficient phase matching for counter-propagating THG. This implies that THG microscopy is exclusively a transmitted-detection technique in transparent media.

However, when imaging a thick biological tissue, although coherent THG scattering initially occurs in the forward direction, incoherent scattering by the tissue can redirect some of the harmonic light toward the excitation objective. Since scattering of visible light by tissues is mostly forward directed, harmonic photons need to travel several scattering mean free paths in the tissue without being reabsorbed before they can travel back to the surface (see Figure 3.10). Epidetected THG imaging is therefore usually possible in scattering, weakly absorbing tissues [19], as illustrated in skin [25] and brain tissue [26].

Another possibility for epidetection is the detection of THG from efficient nano-sized volumes such as nanoparticles or metal surfaces [19,28], from which the THG radiation pattern is nearly dipolar, and therefore nearly as efficient in the forward and backward directions. Significant backward emission can also arise from periodically structured objects as will be discussed in Section 3.3.4.

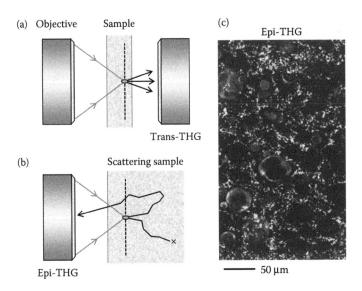

FIGURE 3.10 Epidetection of backscattered THG light. (a) In transparent samples, THG is co-propagating with the excitation beam and is most efficiently detected in transmission. (b) In thick scattering samples such as biological tissues, scattering can redirect a fraction of the THG light toward the excitation objective, so that epidetection is possible. (c) Epidetected THG image of rat lung tissue. (Reprinted by permission from Macmillan Publishers Ltd. *Nat. Methods.* Debarre D et al. Imaging lipid bodies in cells and tissues using third-harmonic generation microscopy. 3:47–53. Copyright 2006.)

3.2.11 Conclusion on Phase Matching in Isotropic Media

In conclusion, we summarize here the main characteristics of THG microscopy in isotropic media:

1. There is no THG from a homogeneous medium due to poor phase matching caused by the Gouy phase shift of the focused excitation beam. However, signal is obtained for inclusions that have a size close to the effective coherence length. Since this effective coherence length depends on the axial extent of the Gouy phase shift, changing the excitation NA changes the relative THG signals obtained from objects of different sizes.
2. Because of the asymmetry in both phase and intensity between the axial and lateral axes, the THG signal also depends on the orientation of structures. Axial structures are generally more visible than structures orthogonal to the optical axis.
3. THG is a forward-directed process, because THG coherent signal buildup needs phase matching over a significant distance. B-THG usually gives a negligible contribution to the THG signal in isotropic media. Epidetection is however possible for some geometries using backscattered light [19,26,29].

3.2.12 Sources of Contrast

Whereas the second-order susceptibility ($\chi^{(2)}$) is nonzero only in the case of anisotropic media, there is no such requirement for the third-order susceptibility ($\chi^{(3)}$), and most isotropic materials (such as glass and water) have a nonzero $\chi^{(3)}$. However, as previously mentioned, the parameter governing signal level in THG microscopy is not the absolute value of the $\chi^{(3)}$, but rather the spatial variations of $\chi^{(3)}$ around interfaces. The peak THG signal observed at an interface between two media scales as $\kappa \, | \, \alpha_1 - \alpha_2 \, |^2 < I_\omega^3 >$, where κ depends on geometry as discussed in the previous sections, I_ω is the excitation intensity, and $\alpha_1 - \alpha_2$ depends on the optical properties of the two media. Under high excitation NA phase matching is dominated by the Gouy shift and $\alpha \approx \chi^{(3)}$, whereas in the case of moderate focusing, $\alpha \approx \chi^{(3)}/(n_{3\omega}(n_{3\omega} - n_\omega))$ [30].

A few articles have reported measurements of $\chi^{(3)}$ of solvents [31,32] and biological liquids [30,33]. The general principle of these measurements is to measure the THG intensity at the interface between a known material (glass coverslip) and a liquid to be characterized. In particular, these studies have shown that lipids and water have very different $\chi^{(3)}(-3\omega;\omega,\omega,\omega)$ at 1.2 μm (Figure 3.11) [30].

$\lambda = 1.18$ μm		$\chi^{(3)}$ ($\times 10^{-22}$m^2 V^{-2})	$|\chi^{(3)} - \chi^{(3)}_{water}|^2$
Water	—	1.68 ± 0.08	0
NaCl 1M	Ions	1.79 ± 0.09	1.2×10^{-2}
Glucose 1M	Sugar	1.83 ± 0.08	2.2×10^{-2}
Glycine 1M	Amino acid	1.69 ± 0.13	1.0×10^{-4}
Triglycine 1M	Polypeptide	1.69 ± 0.12	1.0×10^{-4}
BSA 1mM	Protein	1.75 ± 0.13	4.9×10^{-3}
Triglycerides	Lipids	2.58 ± 0.5	0.81
Oil	Lipids	2.71 ± 0.5	1.06
BK7	Glass	2.79	1.2

FIGURE 3.11 Third-order nonlinear susceptibility $\chi^{(3)}$ of biological liquids. Lipids have a large non-resonant nonlinear susceptibility. Lipid/water interfaces provide strong $\chi^{(3)}$ contrast, making lipid structures visible in THG images of cells and tissues. The excitation wavelength is $\lambda = 1180$ nm. (Adapted from Débarre D and Beaurepaire E 2007. *Biophys. J.* 92:603–612.)

As a consequence, a lipid/water volume heterogeneity a few hundred nanometers in size is an efficient structure for THG by a focused beam. Consistently, lipid droplets have been identified as strong sources of contrast in THG images of cells and tissues [27]. For the same reason, THG has been reported as an effective means for imaging myelin in the central nervous system [26,29].

Other dense nonaqueous structures such as mineral deposits, or cell nuclei can usually be detected in THG images. In contrast, most aqueous solutions exhibit similar values of $\chi^{(3)}$ [30], implying that THG from cellular organelles is usually not concentration-sensitive in physiological conditions. We also point out that, since solvents are optically different from water, the relative visibility of cellular structures is affected by mounting, index matching, or optical-clearing agents [30,25].

Besides this basic contrast mechanism, the third-order nonlinear susceptibility of a medium is generally wavelength dependent. In particular, $\chi^{(3)}(-3\omega;\omega,\omega,\omega)$ is altered by 1-photon, 2-photon, or 3-photon absorption. THG can therefore be resonantly enhanced in absorbing structures. For example, Clay et al. [33] identified resonant contributions in the THG signal from hemoglobin, which contribute to the visibility of red blood cells in THG images. Chloroplasts in plant cells also exhibit strong THG [34], probably related to chlorophyll absorption properties. Along the same line, Bélisle et al. [35] took advantage of the resonance of hemozoin pigments at the harmonic wavelength to obtain sensitive detection of malaria-infected cells. Similarly, hematoxylin (an absorbing stain commonly used in histology) has been shown to enhance the contrast from cell nuclei [36].

3.2.13 One Application of THG–SHG Microscopy: Imaging Embryo Development

The previous sections discussed how THG microscopy highlights optical heterogeneities such as intracellular lipidic organelles. At the supra-cellular scale, THG microscopy provides a convenient way to image the structure of unstained tissues with 3D resolution, while being compatible with SHG or 2PEF imaging. One field of application that has been explored in recent years is the imaging of embryonic development in small animal models.

Embryo development involves the spatio-temporal coordination of large ensembles of morphogenetic processes including collective cell movements and cell divisions. Global imaging of morphogenesis in complex organisms with subcellular resolution is technically challenging because the shape and opacity of embryos hamper deep imaging. Nonlinear microscopy is attractive for embryo studies because it provides deep 3D imaging with reduced phototoxicity. Although 2PEF microscopy is the most commonly used technique for biology studies, it usually relies on fluorescent protein expression, which can be challenging to obtain in mutant embryos or at very early stages.

3.2.13.1 Multimodal Imaging of Zebrafish Embryonic Development

Chu et al. [37,38] demonstrated the possibility of imaging developing zebrafish embryos from the cleavage stage to the larva stage using THG microscopy without damaging the embryo. They showed that THG can be combined with SHG imaging of the muscle myofilaments during the larva stage, and even of the mitotic spindles forming during cell division.

This approach was used by Olivier et al. [21] in conjunction with an optimized scanning scheme to image all the cell divisions occurring during the first 3 h of development of the zebrafish embryo, and to reconstruct the cell lineage during the corresponding 10 division cycles.

Representative THG images of zebrafish embryo during divisions are presented in Figure 3.12. Strong signals are observed near the interface of dividing cells at post-cellularization stages. This signal reflects the presence of a sizable intercellular space, as corroborated by high-NA THG images of dividing cells showing locally double interfaces (Figure 3.4d). THG therefore highlights dividing and motile cells, and provides a direct visualization of cell morphology in the early zebrafish embryo. This remarkable contrast is compatible with automated cell contour detection [21]. These images also reveal the traffic of intracellular lipidic vesicles and the dynamics of the vitelline stores (yolk).

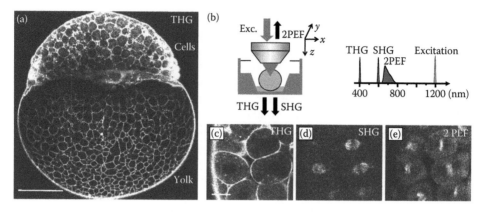

FIGURE 3.12 Multimodal nonlinear imaging of a zebrafish embryo. (a) THG image of a zebrafish embryo during initial cell divisions. Scale bar 200 μm. (b) Geometry and wavelengths involved for multimodal THG–SHG–2PEF imaging. (c–e) Simultaneous THG–SHG–2PEF image of dividing cells in a H2B-mCherry embryo. Scale bar 20 μm. (From Olivier N et al. 2010. Cell lineage reconstruction of early zebrafish embryos using label-free nonlinear microscopy. *Science* 329:967–971. Reprinted with permission of AAAS.)

Combined THG/SHG imaging can be used to detect a number of key events during cell divisions. SHG reveals the formation of mitotic spindles, providing a specific marker of cell divisions. The SHG intensity exhibits a nearly Gaussian temporal behavior with a maximum intensity corresponding to the anaphase, permitting the measurement of division timings and cell cycle duration with high precision.

Moreover, THG, SHG, and 2PEF signals from red fluorescent proteins can be produced and detected simultaneously. Figures 3.12c through 3.12e show an example of multimodal THG–SHG–2PEF imaging of H2B-mCherry mutant embryos, where the 2PEF signal is used to follow the chromatin distribution along with THG–SHG signals [21].

3.2.13.2 THG–SHG Imaging of Other Embryos

Similarly, THG–SHG microscopy has also been shown to be an effective approach for providing morphological images of developing embryos in other species: drosophila [39], mouse [40,41], *Caenorhabditis elegans* [42], and so on. We note that, unlike in zebrafish embryos, in these other species the THG images usually reveal the intracellular organelles rather than the cell contours.

3.3 THG Microscopy of Anisotropic Media

In the previous section, we discussed THG by simple isotropic media. In this section, we will consider more complex media. We will first discuss the sensitivity of THG to tensor symmetries and to the incident polarization, and discuss THG by anisotropic and birefringent media. We will then outline the possibility of quasi-phase matching from axially periodic samples. Finally, we will illustrate the application of THG–SHG imaging to a highly organized biological medium, namely the human cornea.

3.3.1 Field and Tensor Symmetries

The nonlinear susceptibility $\chi^{(3)}$ relating the induced third-harmonic polarization to the incident fields is generally a fourth-rank tensor. However, before discussing the $\chi^{(3)}$ tensor properties, we point out that the excitation field distribution is also intrinsically vectorial, even in the case of an incoming linear polarization. For example, the E_z component of the electric field of a tightly focused Gaussian beam may not be negligible, as illustrated in Figure 3.13.

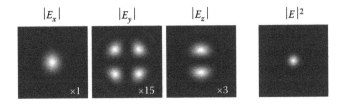

FIGURE 3.13 Vectorial aspect of the focused Gaussian beam. Calculated electric field at the focus of a Gaussian beam with initial *x*-oriented linear polarization. $-1\ \mu m < x,y < 1\ \mu m$. Conditions: $NA = 1.4$, $n = 1.5$.

For simplifying the discussion, we will however neglect this z polarization component in the remaining of this chapter. This approximation is valid for THG microscopy applications using moderate excitation NAs. We will therefore consider only plane linear and circular polarization distributions.

In a medium characterized by its third-order nonlinear tensor $\chi^{(3)}_{ijkl}(\mathbf{r})$, the excitation field induces a nonlinear polarization described by

$$P_i^{(3\omega)} = \sum_{j,k,l} \chi^{(3)}_{ijkl} E_j E_k E_l \tag{3.22}$$

In the case of a homogeneous isotropic medium, the $\chi^{(3)}$ tensor verifies [43]:

$$\chi^{(3)}_{ijkl} = \chi_0(\delta_{ij}\delta_{kl} + \delta_{ik}\delta_{jl} + \delta_{il}\delta_{jk}) \tag{3.23}$$

and we can express the nonlinear polarization induced by the exciting field E in Cartesian coordinates as

$$\mathbf{P}^{(3\omega)} = 3\chi_0 \begin{bmatrix} E_x(E_x^2 + E_y^2 + E_z^2) \\ E_y(E_x^2 + E_y^2 + E_z^2) \\ E_z(E_x^2 + E_y^2 + E_z^2) \end{bmatrix} \tag{3.24}$$

Several points can be noticed from this equation:

1. Although THG is a third-order process, P_y and P_z depend both linearly and nonlinearly on E_y and E_z. This means that one has to be careful before deciding to neglect the vectorial components of the exciting field. Of particular importance is the spatial overlap between the different polarizations. For example (see Figure 3.13), if we consider a focused Gaussian beam with initial linear polarization along x, the E_y and E_z components of the electric field have their maxima away from the optical axis, and so their overlap with E_x is small. This is one of the reasons why the paraxial scalar approximation works well for THG microscopy.
2. For an isotropic medium, if we consider a circular polarization in the paraxial approximation, we have $E_x = i.E_y$ so $P_x = P_y = 0$, that is, no third-order polarization. For that reason, no THG is obtained from isotropic media excited with circularly polarized light, even at interfaces [44]. One consequence is that THG with circular incident polarization can be used to discriminate between isotropic and anisotropic media. It also has implications in the case of combined THG/SHG imaging: linearly polarized excitation is generally preferred for general-purpose THG imaging, however, linear polarization results in orientation effects particularly pronounced in SHG images.

3.3.2 $\chi^{(3)}$ Tensorial Effects

We have seen that the absence of THG signal using circularly polarized light in isotropic media comes from the tensor symmetries. We will now consider anisotropic media, and first neglect refractive index dispersion and linear birefringence (i.e., $n_\omega = n_{3\omega}$ everywhere). However, the nonlinear tensor considered is not that of a homogeneous isotropic sample anymore.

3.3.2.1 Tensors and Nonlinear Polarization

The form of the isotropic tensor has been given in Equation 3.23. The case of the third-order nonlinear susceptibility for various symmetries has been studied by Butcher [45], expanded by Hellwarth [46], and can be found in a number of references, including Ref. [43].

For simplicity, here, we restrict our analysis to the case of a crystal with one main axis along the x-axis with a $C_{\infty v}$ symmetry. This tensor has three independent tensor elements:

$$
\begin{aligned}
\chi_\| &= \chi_{xxxx} \\
\chi_{xxyy} &= \chi_{xyyx} = \chi_{xyxy} = \chi_{xxzz} = \chi_{xzzx} = \chi_{xzxz} \\
\chi_{yyzz} &= \chi_{yzzy} = \chi_{yzyz} = \chi_{zzyy} = \chi_{zyzy} = \chi_{zyyz} \\
&= \chi_{yyxx} = \chi_{yxxy} = \chi_{yxyx} = \chi_{zzxx} = \chi_{zxzx} = \chi_{zxxz} \\
\chi_{zzzz} &= \chi_{yyyy} = 3\chi_{yyzz}
\end{aligned}
\tag{3.25}
$$

If we assume $\chi_\| = \chi_{xxxx}$, $\chi_{cr} = 3\chi_{xxyy}$, $\chi_\perp = 3.\chi_{yyzz}$, we can then express the nonlinear polarization as (still neglecting E_z):

$$
\mathbf{P}^{(3\omega)} =
\begin{bmatrix}
E_x(\chi_\| E_x^2 + \chi_{cr} E_y^2) \\
E_y(\chi_{cr}.E_x^2 + \chi_\perp . E_y^2) \\
0
\end{bmatrix}
\tag{3.26}
$$

Let us now consider an interface between such a medium and an isotropic medium ("air") sharing the same linear properties but with $\chi^{(3)} = 0$, and look at the influence of two parameters on the THG signal: the incoming polarization and the ratio between the tensor elements.

3.3.2.2 Interface between an Anisotropic Medium and Air

Isotropic medium, linear polarization

Figure 3.14 illustrates THG signal obtained in the classical situation [8,9,15] where the focus is z-scanned across the interface between an isotropic medium ($\chi^{(3)} = \chi_0$) and air ($\chi^{(3)} = 0$), where z denotes the distance between the interface and the focal plane.

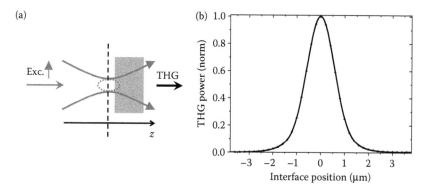

FIGURE 3.14 THG from an interface with linearly polarized light. (a) Geometry considered. (b) $C_0(z)$: numerical calculation of the THG signal obtained when the excitation beam is z-scanned across an xy interface between an isotropic medium an air. $NA = 1.2$, $\lambda = 1.18$ μm.

Anisotropic medium, linear polarization

We first consider the case of an interface between air and the uniaxial crystal, excited by a linearly polarized Gaussian beam. θ denotes the angle between the incident polarization and the axis of the crystal.

Calculations show that the shape of the curve given in Figure 3.14 is conserved for every combination of polarization angle and ratio between tensor elements (not shown).

Not unexpectedly, the signal is maximum when the polarization angle is along the axis of the crystal that has the largest tensor element, much like SHG from a collagen fibril is maximized when the excitation polarization and the fibril are co-aligned. This indicates that tensor anisotropy does not change the phase-matching conditions: there is no THG signal from a homogeneous *anisotropic* sample, and the signal from an interface is proportional to the signal expected from the interface between two isotropic media.

Anisotropic medium, circular polarization

In the case of a circularly polarized excitation, the THG signal scales as $((\chi_\| - \chi_{cr})^2 + (\chi_{cr} - \chi_\perp)^2)$. Figure 3.15 illustrates the THG signal as a function of the value of $\chi_\|$ and χ_{cr} (assuming $\chi_\perp = 1$). The degenerate situation $\chi_\| = \chi_{cr} = \chi_\perp$ corresponding to an isotropic medium yields, as expected, no THG signal.

3.3.3 THG from Birefringent Media

Birefringent media have polarization-dependent indices. When the material has a single axis of symmetry, the birefringence can be described by assigning two different refractive indices to the material for different polarizations: one along the ordinary axis (perpendicular to the axis of symmetry) and another along the extraordinary axis (parallel to the axis of symmetry).

If some tensor elements create a nonlinear field polarized perpendicularly to the excitation, we obtain a situation where the excitation propagates along the ordinary axis, while the harmonic field propagates along the extraordinary axis, thus producing an effective negative dispersion for the harmonic generation process.

THG in highly birefringent structures such as calcite crystals (calcium carbonate ($CaCO_3$)) or liquid crystals [48] has been reported by Oron et al. [44,49]. Calcite is one of the most birefringent structures found in biological samples, and has been extensively used in nonlinear optics. At a wavelength of 590 nm, calcite has ordinary and extraordinary refractive indices of 1.658 and 1.486, respectively.

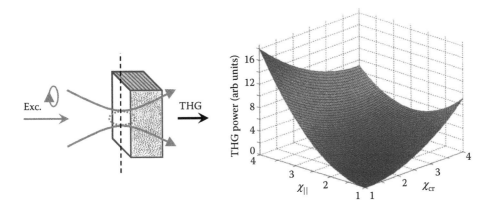

FIGURE 3.15 Sensitivity to $\chi^{(3)}$ anisotropy with circular polarization. (Left) Geometry considered. (Right) THG signal obtained with circular incident polarization at the interface between an anisotropic medium and air, as a function of the parameters $\chi_\|$ and χ_{cr} with $\chi_\perp = 1$ (see text). No signal is obtained in the case of an isotropic medium ($\chi_\| = \chi_{cr} = \chi_\perp = 1$). (Reprinted from Olivier N et al. 2010. Harmonic microscopy of isotropic and anisotropic microstructure of the human cornea. *Opt. Express* 18:5028–5040. With permission of Optical Society of America.)

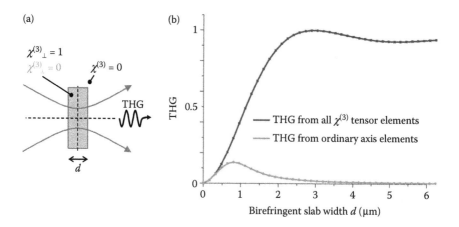

FIGURE 3.16 Calculation of THG from a birefringent medium. (a) Geometry. (b) Comparison between THG from a slab of calcite-like medium considering either only tensor elements in the ordinary–ordinary axis (gray) or also taking into account the ordinary–extraordinary elements (black) Conditions: $NA = 1.2$, $n_o = 1.66$, $n_e = 1.49$.

Figure 3.16 shows numerical simulations of THG from a slab of calcite-like medium considering an incoming beam polarized along the ordinary axis. The dispersion is neglected, and linear indices are taken to be $n_o = 1.66$ and $n_e = 1.49$. The gray curve corresponds to a calculation considering only the tensor elements that produce a harmonic field along the ordinary axis. The size response in this case is the typical single-peaked curve that reaches its maximum for a size approximately equal to the coherence length, and then decreases until it reaches zero. However, if we now take into account the tensor elements that induce a nonlinear polarization along the extraordinary axis (black curve), we see the negative dispersion compensates the Gouy phase shift and the THG signal increases until it reaches a plateau.

THG can therefore be observed from homogeneous birefringent media [44].

3.3.4 Quasi-Phase Matching

We consider here the possibility of quasi-phase matched THG from an axially periodic medium. Quasi-phase matching (QPM) is an idea developed in nonlinear optics [50,51], where the conversion efficiency is an important parameter, and stems from the consideration that it is sometimes easier to change the structure of the sample (typically a nonlinear crystal) than to change the structure of the excitation (usually a focused Gaussian beam). QPM is usually implemented using a stacked noncentrosymmetric media exhibiting a periodic arrangement of alternative perpendicular orientations, with a period corresponding to the coherence length of the nonlinear process. That way, interference is always constructive along the optical axis.

Here, we will not consider the usual arrangement, as it usually implies birefringence, but we will consider instead a medium that exhibits a sinusoidal variation of $\chi^{(3)}$ from 0 to 1 with a period δe, while keeping a constant linear index (see Figure 3.17a).

$$\chi^{(3)}(z) = \frac{1}{2}\left(1 + \sin\left(\frac{2\pi z}{\delta e}\right)\right) \tag{3.27}$$

The F-THG and B-THG obtained with a focused Gaussian beam from this geometry are shown in Figure 3.17b as a function of the period δe. The signal is here normalized to that obtained in the forward direction from a single interface perpendicular to the z-axis, a geometry that corresponds to the same

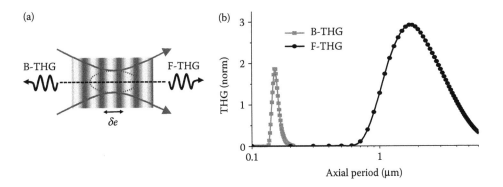

FIGURE 3.17 Quasi-phase matching. (a) Geometry considered: the medium exhibits a sinusoidal variation of $\chi^{(3)}$ along the optical axis. (b) Calculated F-THG and B-THG signal as a function of the axial period of the sample. (Reprinted from Débarre D, Olivier N, Beaurepaire E 2007. Signal epidetection in third-harmonic generation microscopy of turbid media. *Opt. Express* 15:8913–8924; Olivier N, Beaurepaire E 2008. Third-harmonic generation microscopy with focus-engineered beams: A numerical study. *Opt. Express* 16:14703–14715. With permission of Optical Society of America.)

number of emitters within the focal volume. This allows a direct comparison of the THG signal that is only due to the quasi-phase matching. In both cases, we notice a large enhancement of the harmonic signal for a particular sample periodicity. The theoretical value for the QPM resonance period can be expressed as

$$\delta e = 2.l_c \tag{3.28}$$

It can be noted that the B-THG resonant period corresponds closely to the theoretical value ($l_c^{BTHG} \approx 70\,nm$ and the maximum B-THG is obtained for $\delta e \approx 150$ nm), as the phase and intensity distribution can be approximated as constant at this scale.

While this geometry demonstrates that more complex phase-matching conditions can be obtained in organized media, it describes a hypothetical situation, since most organized media also have an organized microstructure which means that they exhibit some degree of anisotropy and tensorial aspects have to be taken into account.

3.3.5 Cornea Imaging

In this section, we discuss combined THG–SHG of the human cornea. We show how multimodal nonlinear imaging can provide multiscale complementary information about the structure of this highly organized biological medium.

3.3.5.1 Structure of the Cornea

The outermost part of the cornea is an epithelium consisting of 6–7 cell layers. Beneath this epithelium is a 500 μm-thick collagenous stroma. The stroma consists of layered 2 μm-thick collagen lamellae with alternate orientations. These lamellae themselves are made of aligned collagen fibrils.

3.3.5.2 THG and SHG Signals in the Cornea

An example of combined THG–SHG imaging is shown in Figure 3.18. Multiharmonic images provide a rich description of the lamellar organization of the intact stroma over its entire thickness. One interesting observation is that THG and SHG signals are very different and generally exhibit anticorrelated maxima, as illustrated in Figure 3.18c.

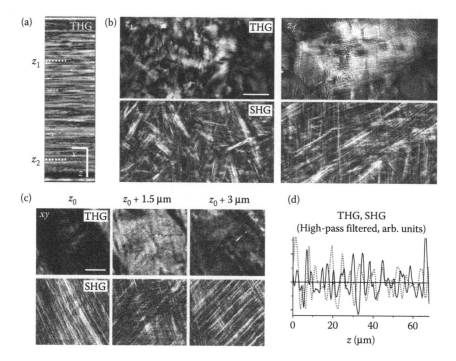

FIGURE 3.18 THG–SHG imaging of the human cornea. Forward-detected THG–SHG imaging of the stromal organization at different depths. (a) *xz* reprojection of a series of THG images recorded with the epithelium (top) facing the objective. $NA = 0.75$, scale bars 20 μm (*x*) × 100 μm (*z*). (b) *xy* images recorded at depths indicated in (a). Scale bar 100 μm. (c) THG and SHG images at three different depths illustrating the general anticorrelation of THG and SHG maxima. ($\Delta z = 1.5$ μm) Scale bar = 10 μm. (d) *z*-profiles through THG and SHG image stacks. (Aptel F et al. 2010. *Investigative Ophthalmology & Visual Science* 51:2459–2465; Olivier N et al. 2010. *Opt. Express* 18:5028–5040. Reprinted with permission from ARVO and OSA.)

1. In the outermost epithelium, THG images reveal the cellular structures (boundaries and organelles). In the organized collagen stroma, THG reveals the interfaces between the lamellae, that is, the anisotropy changes that create an effective $\chi^{(3)}$ heterogeneity. One striking feature of these images is that they reveal the different large-scale organizations of collagen lamellae at successive depths, as exemplified in Figure 3.18b. Large-scale (100s of μm) heterogeneity is more pronounced in the anterior stromal region, whereas the posterior stroma exhibits a more regular, long-range-stacked organization.

2. The corneal SHG signal originates from the collagen fibrils, and depends on the macro-molecular organization of the collagen. It is maximum in the middle of a layer, where the organization is the most crystal like. Since individual fibrils are not resolved because of their small diameter (35 nm) and dense packing, SHG images result from interference processes. Forward-SHG images exhibit striated features that reflect the orientation and distribution of the fibrils (see Figure 3.18b). Backward-SHG (B-SHG) images result from a shorter coherence length and appear as relatively uniform or speckle like [24].

3.3.5.3 Influence of the Polarization on THG Images

In the epithelium, THG images highlight the cell structures, whereas in the stroma they also reveal the anisotropy changes between collagen lamellae. These two types of THG signals have different origins and can be distinguished using polarization.

Figure 3.19 illustrates the influence of the polarization of the excitation in THG microscopy of the cornea. The THG image recorded with circularly incident polarization specifically detects the stroma

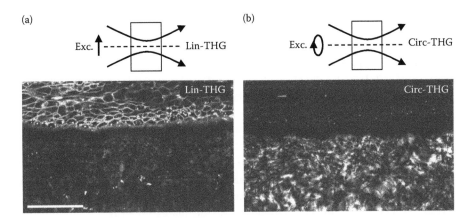

FIGURE 3.19 Influence of the excitation polarization on THG contrast. (a) THG image of a human cornea recorded with linear incident polarization. (b) THG image of the same area recorded with circular incident polarization. Scale bar = 50 μm. Conditions: $NA = 0.8$, $\lambda = 1.2$ μm. (Reprinted from Olivier N et al. 2010. Harmonic microscopy of isotropic and anisotropic microstructure of the human cornea. *Opt. Express* 18:5028–5040. With permission of Optical Society of America.)

signals, since the epithelium is made of isotropic media. In contrast, the image from the same area recorded with linear incident polarization also reveals the cellular structures with strong contrast.

In conclusion, the case of the cornea is an interesting illustration of the benefit of multimodal, nonlinear imaging: (i) THG and SHG images reveal different levels of organization of the microarchitecture of the corneal stroma; and (ii) Polarization can be used to detect anisotropy.

3.4 Conclusion

THG is a nonlinear imaging modality that uses a single excitation laser, so that it can be combined with SHG and 2PEF microscopy. The contrast mechanisms of THG and SHG are different, so that the two modalities provide different and complementary information. While SHG reveals a few number of organized structures with high specificity (fibrillar collagen, myofilaments, starch, astroglial fibers, polarized microtubule assemblies, etc.), THG microscopy reveals optical heterogeneities in a general manner. In cells, strong THG signals are observed from dense lipidic, mineralized, or absorbing structures. At the tissue scale, THG imaging provides rich 3D morphological information. One remarkable property of THG microscopy is that no signal is obtained from homogeneous media, implying that heterogeneities and interfaces are revealed with a good contrast. THG is particularly sensitive to structure size in the 100 nm–2 μm range.

THG imaging has been reported in a number of contexts including imaging developing embryos [37,39] brain tissue [26], neurons [10], hamster oral cavity [25], human skin [52], elastic fibers [53], lipid droplets [27], hemozoin pigments [35], and so on. THG from organized biological media have been reported from calcite [44], tooth dentin [54], and the human cornea ([24,47]). While few "real" applications have been reported to date (one example being the reconstruction of the early zebrafish development [21]), it is anticipated that THG microscopy will find many uses as an "add-on" imaging technique, owing to its versatility and its relatively straightforward combination with 2PEF–SHG microscopy.

Acknowledgments

We thank Marie-Claire Schanne-Klein for many discussions and comments, Maxwell Zimmerley for critical reading, and Jean-Louis Martin for constant encouragement. We are indebted to all the colleagues involved in the articles from which the illustrations were adapted. In addition, we thank Nadine Peyriéras and Louise Duloquin for preparing the zebrafish embryos, Laurent Combettes for the hepatocytes, and the French Eye Bank and the ophthalmology department at Hôtel-Dieu Hospital for the supply of human corneas. This work was supported by Centre National de la Recherche Scientifique (CNRS), Institut National de la Santé et de la Recherche Médicale (INSERM), Délégation Générale de l'Armement (DGA), Agence Nationale de la Recherche (ANR), and Fondation Louis D. de l'Institut de France.

Appendix

In this appendix, we summarize the two commonly used formalisms for calculating THG from various geometries. The first model assumes scalar Gaussian beam excitation, moderate focusing, axially symmetric samples, and provides some analytical results. The second model uses a Green's function formalism and is more general, but its implementation is more computation-intensive.

THG from an Axially Symmetric Sample with a Focused Gaussian Beam

Here, we follow the method used in Boyd's *Nonlinear Optics* [43]:

1. The fundamental ($n = 1$) and n-th harmonic beams are described as 2D (ρ,z) Gaussian beams:

$$A_n(\rho,z) = \frac{B_n}{1 + 2iz/b_n} e^{-r^2/w_n^2(1+2iz/b_n)} \qquad (3.29)$$

   where w_0 is the waist of the fundamental beam, $w_n^2 = w_0^2/n$, $b_n = k_{n\omega}w_n^2$ denotes the confocal parameter, and $k_{n\omega} = 2\pi n\omega/n_{n\omega}$ is the wavenumber.

2. We only consider interactions along the z-axis (i.e., slabs or interfaces), and the intensity of the fundamental beam is considered constant.

   The equation is modified to account for the z dependence of the harmonic signal:

$$A_n(r,z) = \frac{B_n(z)}{1 + 2iz/b_n} e^{-r^2/w_n^2(1+2iz/b_n)} \qquad (3.30)$$

$$= B_n(z) \cdot G_n(r,z) \qquad (3.31)$$

with $G_n(r,z) = 1 + 2iz/b_n e^{-r^2/w_n^2(1+2iz/b_n)}$.

The paraxial wave equation for n-th harmonic generation can then be written as

$$
\begin{aligned}
C_n G_1(r,z)^n &= \left(\nabla_r^2 + 2ik_{n\omega}\frac{\partial}{\partial z} \right)(B_n(z) \cdot G_n(r,z)) \\
&= B_n(z)\nabla_r^2(G_n(r,z)) + 2ik_{n\omega}B_n(z)\frac{\partial}{\partial z}(G_n(r,z)) + G_n(r,z)\frac{\partial B_n(z)}{\partial z} \\
&= G_n(r,z)\frac{\partial}{\partial z}(B_n(z))
\end{aligned} \qquad (3.32)
$$

since $G_n(r,z)$ is a solution to the paraxial wave equation.

Moreover, since

$$\frac{G_1(r,z)^n}{G_n r,z} = \frac{e^{i\Delta k z}}{(1 + 2iz/b)^{(n-1)}} \qquad (3.33)$$

This equation can easily be integrated, and we find

$$B_n(z) = B_0 \int_{z_0}^{z} \frac{e^{i\Delta k \zeta} d\zeta}{(1 + 2i\zeta/b_1)^{(n-1)}} \qquad (3.34)$$

$$= B_0 J_n(\Delta k, z_0, z) \qquad (3.35)$$

where J_n is called the *J*-Integral of the *n*-th order process.

In the case of THG, we have

$$B_3(z) = B_0 \int_{z_0}^{z} \frac{e^{i\Delta k \zeta} d\zeta}{(1 + 2i\zeta/b_1)^2} \qquad (3.36)$$

$$= B_0 J_3(\Delta k, z_0, z) \qquad (3.37)$$

This integral unfortunately cannot be integrated analytically in most cases, except in the case of an infinite homogeneous medium ($z_0 = -\infty$, $z = \infty$), in which case it yields [43]:

$$J_H^{(n)}(\Delta k > 0) = \frac{b}{2} \frac{2\pi}{(n-2)} \left(\frac{b\Delta k}{2} \right)^{(n-2)} e^{-b\Delta k/2} \qquad (3.38)$$

$$J_H^{(n)}(\Delta k \le 0) = 0 \qquad (3.39)$$

Despite the simplifying hypotheses, several important properties of THG with focused beams are described by this integral. In particular, it is found analytically that no THG is obtained from a homogeneous, normally dispersive or even nondispersive medium. In addition, the numerical integration of this integral in the case of slab and interface geometries yields results which are consistent with nonparaxial simulations.

However, the two main hypotheses do not hold in the case of nonaxially symmetric samples, and a more complex model must be used.

Green's Function Formalism for THG

The following model, which is used to represent a coherent nonlinear microscope, has been proposed by Cheng et al. [15,55]. Figure 3.20 illustrates the geometry and the notations used. It consists of three elements:

1. The excitation field near focus is calculated from the field distribution at the back aperture and from the objective properties.
2. The interaction between the fundamental field and the sample is described by the spatial distribution of a nonlinear tensor (the 4th rank $\chi^{(3)}$ tensor in the case of THG).
3. The propagation of the nonlinear polarization to the far field is calculated using Green's functions.

This model neglects the spectral width of the excitation pulses: only the central frequency is considered. Therefore, all effects related to the spectral phase of the excitation are neglected. This approximation is justified by the fact that chromatic effects are negligible over the extent of the excitation volume for the pulses typically involved in THG microscopy (100 fs duration, 5–10 nm bandwidth) in the

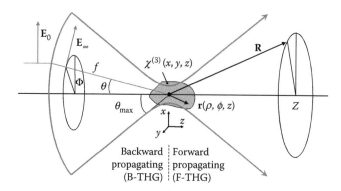

FIGURE 3.20   Geometry for modeling coherent microscopy. Illustration of the notations used in this section. See text. (Adapted from Novotny L, Hecht B 2006. *Principles of Nano-Optics.* Cambridge University Press; Cheng JX, Xie X 2002. *J. Opt. Soc. Am. B* 19:1604–1610; Olivier N, Beaurepaire E 2008. *Opt. Express* 16:14703–14715.)

absence of resonance effects. The linear index mismatches are also neglected, although they may have a significant influence on beam propagation.

Description of Focused Fields

The angular spectrum representation [13] is used to calculate the field distribution obtained by propagating an initial field distribution (amplitude and phase) at the back aperture of a microscope objective under a given set of focusing conditions (NA, index of refraction). In this representation, the field is described as a sum of plane waves with variable amplitudes and propagation directions, so that the excitation field near focus is calculated by propagating all the plane waves and then summing them up coherently. Assuming a homogeneous isotropic linear medium, we have

$$\mathbf{E}(\rho,\phi,z) = \frac{ik_\omega f e^{-ik_\omega f}}{2\pi}\int\limits_0^{\theta_m}\int\limits_0^{2\pi} e^{-ik_\omega z\cos(\theta)}e^{-ik\rho\sin(\theta)\cos(\Phi-\phi)}\sin(\theta)\mathbf{E}_\infty(\theta,\Phi)\,d\Phi\,d\theta \tag{3.40}$$

with

$$\mathbf{E}_\infty(\theta,\Phi) = (\cos\theta)^{1/2}\left[\mathbf{E}_0(\theta,\Phi)\cdot\begin{pmatrix}-\sin\Phi\\\cos\Phi\\0\end{pmatrix}\right]\begin{pmatrix}-\sin\Phi\\\cos\Phi\\0\end{pmatrix}$$
$$+\,(\cos\theta)^{1/2}\left[\mathbf{E}_0(\theta,\Phi)\cdot\begin{pmatrix}\cos\Phi\\\sin\Phi\\0\end{pmatrix}\right]\begin{pmatrix}\cos\Phi\cos\theta\\\sin\Phi\cos\theta\\-\sin\theta\end{pmatrix} \tag{3.41}$$

where $\mathbf{E}_0(\theta,\Phi)$ describes the field distribution at the back aperture of the objective, $k = k_\omega = 2\pi\omega/n_\omega$ is the wavenumber, f is the focal length of the objective, n_ω is the refractive index at frequency ω, (ρ,ϕ,z) are cylindrical coordinates near focus, and $\theta_{max} = \sin^{-1}(NA/n)$ is the maximum focusing angle of the objective, as illustrated in Figure 3.20.

This integral can then be integrated numerically on each point of a 3D grid representing the focal volume.

Calculation of the Nonlinear Polarization

The nonlinear polarization induced by the focused fundamental beam can then be expressed in each point of the grid as

$$P_i^{(THG)}(3\omega) = \sum_{j,k,l} \chi_{ijkl}^{(3)} \cdot E_j(\omega)E_k(\omega)E_l(\alpha) \tag{3.42}$$

where $\chi_{ijkl}^{(3)}$ is the third-order nonlinear tensor of the medium. In the case of isotropic media, it can be expressed as [43]

$$\chi_{ijkl}^{(3)} = \chi_0(\delta_{ij}\delta_{kl} + \delta_{ik}\delta_{jl} + \delta_{il}\delta_{jk}) \tag{3.43}$$

where $\delta_{ab} = 1$ if $a = b$, and $\delta_{ab} = 0$ if $a \neq b$.

The last step is to propagate all these sources of THG to the far field, taking into account their phase to determine the resulting interference.

Propagation to the Far Field Using Green's Functions

To calculate the far-field interference pattern created by the nonlinear polarization, each point of the grid is considered as a dipole which electric field can be computed using Green's function, and all the contributions of the sources located in the focal volume are then summed up in the far field.

The harmonic field originating from all positions $\mathbf{r}$ in the focal region and propagated to a position $\mathbf{R}$ in the collection optics aperture can be expressed as [13,15]

$$E_{FF}(\mathbf{R}) = \int_V \mathbf{P}^{(nl)}(\mathbf{r})\mathbf{G}_{FF}(\mathbf{R} - \mathbf{r})dV \tag{3.44}$$

where V spans the excitation volume and $\mathbf{G}_{FF}$ is the far-field Green's function:

$$\mathbf{G}_{FF} = \frac{exp(ikR)}{4\pi R}[\mathbf{I} - \mathbf{RR}/R^2] \tag{3.45}$$

where $\mathbf{R}$ is the coordinate of a point in the far field, $R = |\mathbf{R}|$ and $\mathbf{I}$ is the third-order identity tensor. To calculate the emitted THG intensity, $E_{FF}(\mathbf{R})$ is finally evaluated on a grid describing a 2D surface (usually corresponding to a solid angle defined by the detection NA) and all the intensities are added up.

References

1. Terhune RW, Maker PD, Savage CM 1962. Optical harmonic generation in calcite. *Phys. Rev. Lett.* 8:404–406.
2. Maker PD, Terhune RW 1965. Study of optical effects due to an induced polarization third order in the electric field strength. *Phys. Rev.* 137:A801–A818.
3. New GHC, Ward JF 1967. Optical third-harmonic generation in gases. *Phys. Rev. Lett.* 19:556–559.
4. Ward JF, New GHC 1969. Optical third harmonic generation in gases by a focused laser beam. *Phys. Rev.* 185:57–72.
5. Bey PP, Giuliani JF, Rabin H 1967. Generation of a phase-matched optical third harmonic by introduction of anomalous dispersion into a liquid medium. *Phys. Rev. Lett.* 19:819–821.
6. Kajzar F, Messier J 1985. Third-harmonic generation in liquids. *Phys. Rev. A* 32:2352–2363.
7. Tsang T 1995. Optical third-harmonic generation at interfaces. *Phys. Rev. A* 52:4116–4125.

8. Tsang T 1996. Third- and fifth-harmonic generation at the interfaces of glass and liquids. *Phys. Rev. A* 54:5454–5457.

9. Barad Y, Eisenberg H, Horowitz M, Silberberg Y 1997. Nonlinear scanning laser microscopy by third harmonic generation. *Appl. Phys. Lett.* 70:922–924.

10. Yelin D, Silberberg Y 1999. Laser scanning third-harmonic generation microscopy in biology. *Opt. Express* 5:169–175.

11. Squier J, Müller M, Brakenhoff G, Wilson K 1998. Third harmonic generation microscopy. *Opt. Express* 3:315–324.

12. Millard A et al. 1999. Third-harmonic generation microscopy by use of a compact, femtosecond fiber laser source. *Appl. Opt.* 38:7393–7397.

13. Novotny L, Hecht B 2006. *Principles of Nano-Optics.* Cambridge University Press, Cambridge, UK.

14. Gouy CR 1870. Sur une propriete nouvelle des ondes lumineuses. *Comptes rendus de l'Académie des sciences (Paris).* 110:1251–1253.

15. Cheng JX, Xie X 2002. Green's function formulation for third harmonic generation microscopy. *J. Opt. Soc. Am. B* 19:1604–1610.

16. Olivier N, Beaurepaire E 2008. Third-harmonic generation microscopy with focus-engineered beams: A numerical study. *Opt. Express* 16:14703–14715.

17. Wolf E 1959. Electromagnetic diffraction in optical systems. i. An integral representation of the image field. *Proc. Royal Soc. A* 253:349–357.

18. Richards B, Wolf E 1959. Electromagnetic diffraction in optical systems ii. Structure of the image field in an aplanetic system. *Proc. Royal Soc. A* 253:358–379.

19. Débarre D, Olivier N, Beaurepaire E 2007. Signal epidetection in third-harmonic generation microscopy of turbid media. *Opt. Express* 15:8913–8924.

20. Débarre D, Supatto W, Beaurepaire E 2005. Structure sensitivity in third-harmonic generation microscopy. *Opt. Lett.* 30:2134–2136.

21. Olivier N et al. 2010. Cell lineage reconstruction of early zebrafish embryos using label-free nonlinear microscopy. *Science* 329:967–971.

22. Moreaux L 2001. Coherent scattering in multi-harmonic light microscopy. *Biophys. J.* 80:1568–1574.

23. Débarre D 2006. *Thèse de doctorat.* Ecole Polytechnique, Palaiseau.

24. Aptel F et al. 2010. Multimodal nonlinear imaging of the human cornea. *Investig. Ophthalmol. Visual Sci.* 51:2459–2465.

25. Tai SP et al. 2006. *In vivo* optical biopsy of hamster oral cavity with epi-third-harmonic generation microscopy. *Opt. Express* 14:6178–6187.

26. Witte S et al. 2011. Label-free live brain imaging and targeted patching with third-harmonic generation microscopy. *Proc. Natl. Acad. Sci.* 108:5970–5975.

27. Debarre D et al. 2006. Imaging lipid bodies in cells and tissues using third-harmonic generation microscopy. *Nat. Methods* 3:47–53.

28. Chang CF et al. 2008. Cell tracking and detection of molecular expression in live cells using lipid-enclosed CdSe quantum dots as contrast agents for epi-third harmonic generation microscopy. *Opt. Express* 16:9534–9548.

29. Farrar M, Wise F, Fetcho J, Schaffer C 2011. *In vivo* imaging of myelin in the vertebrate central nervous system using third harmonic generation microscopy. *Biophys. J.* 100:1362–1371.

30. Débarre D, Beaurepaire E 2007. Quantitative characterization of biological liquids for third-harmonic generation microscopy. *Biophys. J.* 92:603–612.

31. Schins JM, Schrama T, Squier J, Brakenhoff GJ, Müller M 2002. Determination of material properties by use of third-harmonic generation microscopy. *J. Opt. Soc. Am. B* 19:1627–1634.

32. Barille R, Canioni L, Sarger L, Rivoire G 2002. Nonlinearity measurements of thin films by third-harmonic-generation microscopy. *Phys. Rev. E* 66:1–4.

33. Clay G et al. 2006. Spectroscopy of third harmonic generation: Evidence for resonances in model compounds and ligated hemoglobin. *J. Opt. Soc. Am. B* 23:932–950.

34. Müller M, Squier J, Wilson KR, Brakenhoff GJ 1998. 3D microscopy of transparent objects using third-harmonic generation. *J. Microsc.* 191:266–274.

35. Bélisle JM et al. 2008. Sensitive detection of malaria infection by third harmonic generation imaging. *Biophys. J.* 94:L26–L28.

36. Yu CH et al. 2008. Molecular third-harmonic-generation microscopy through resonance enhancement with absorbing dyes. *Opt. Lett.* 33:387–389.

37. Sun CK et al. 2004. Higher harmonic generation microscopy for developmental biology. *J. Struct. Biol.* 147:19–30.

38. Chu SW et al. 2003. *In vivo* developmental biology study using noninvasive multi-harmonic generation microscopy. *Opt. Express* 11:3093–3099.

39. Débarre D et al. 2004. Velocimetric third-harmonic generation microscopy: Micrometer-scale quantification of morphogenetic movements in unstained embryos. *Opt. Lett.* 29:2881–2883.

40. Hsieh CS, Chen SU, Lee YW, Yang YS, Sun CK 2008. Higher harmonic generation microscopy of *in vitro* cultured mammal oocytes and embryos. *Opt. Express* 16:11574–11588.

41. Watanabe T et al. 2010. Characterisation of the dynamic behaviour of lipid droplets in the early mouse embryo using adaptive harmonic generation microscopy. *BMC Cell Biol.* 11:38.

42. Tserevelakis G et al. 2010. Imaging *Caenorhabditis elegans* embryogenesis by third-harmonic generation microscopy. *Micron* 41:444–447.

43. Boyd R 2003. *Nonlinear Optics*, 2nd edition. Academic Press, San Diego, CA.

44. Oron D, Tal E, Silberberg Y 2003. Depth-resolved multiphoton polarization microscopy by third-harmonic generation microscopy. *Opt. Lett.* 28:2315–2317.

45. Butcher PN 1965. *Nonlinear Optical Phenomena*. Ohio State University, Columbus, OH.

46. Hellwarth R 1979. Third-order optical susceptibilities of liquids and solids. *Progr. Quantum Electron.* 5:1–68.

47. Olivier N, Aptel F, Plamann K, Schanne-Klein MC, Beaurepaire E 2010. Harmonic microscopy of isotropic and anisotropic microstructure of the human cornea. *Opt. Express* 18:5028–5040.

48. Yelin D, Silberberg Y, Barad Y, Patel JS 1999. Phase-matched third-harmonic generation in a nematic liquid crystal cell. *Phys. Rev. Lett.* 82:3046–3049.

49. Oron D et al. 2004. Depth-resolved structural imaging by third-harmonic generation microscopy. *J. Struct. Biol.* 147:3–11.

50. Armstrong JA, Bloembergen N, Ducuing J, Pershan PS 1962. Interactions between light waves in a nonlinear dielectric. *Phys. Rev.* 127:1918–1939.

51. Franken PA, Ward JF 1963. Optical harmonics and nonlinear phenomena. *Rev. Mod. Phys.* 35:23–39.

52. Tsai TH et al. 2006. Optical signal degradation study in fixed human skin using confocal microscopy and higher-harmonic optical microscopy. *Opt. Express* 14:749–758.

53. Sun CK et al. 2007. *In vivo* and ex vivo imaging of intra-tissue elastic fibers using third-harmonic-generation microscopy. *Opt. Express* 15:11167–11177.

54. Elbaum R et al. 2007. Dentin micro-architecture using harmonic generation microscopy. *J. Dentistry* 35:150–155.

55. Cheng JX, Volkmer A, Xie X 2002. Theoretical and experimental characterization of coherent anti-Stokes Raman scattering microscopy. *J. Opt. Soc. Am. B* 19:1363–1375.

4

SHG Microscopy and Its Comparison with THG, CARS, and Multiphoton Excited Fluorescence Imaging

Xiyi Chen
LaserGen Corporation

Paul J. Campagnola
*University of
Wisconsin—Madison*

4.1 Introduction

Second-harmonic generation (SHG) is a nonlinear effect that was discovered in 1961 by Franken, Hill, and Weinreich [1] where frequency-doubled 347 nm light was generated when intense 694 nm light from a ruby laser was focused on a quartz sample. Subsequently, with the development of short-pulse lasers that produce much higher instantaneous light intensity and improved methods to grow uniaxial birefringent crystals with large nonlinear susceptibility, SHG has been used as an effective approach for new wavelength generation. This greatly extends scientific researchers' capability as most primary laser sources are in the visible/near-infrared spectrum, whereas the optical absorption of most materials is at bluer wavelengths.

Soon after SHG was discovered in crystals, it began to find useful applications in biological research. In 1971, Fine and Hansen reported that SHG can be produced from collagenous tissues [2]. Freund then extended this by implementing SHG into a stage scanning microscope to image rat tail tendon at ~50 micron resolution. Subsequently, Lewis et al. showed that image cell membranes that had been labeled voltage-sensitive dyes produced SHG contrast [3]. Later, Campagnola et al. implemented SHG into a laser scanning microscope, first for cellular imaging and then for imaging tissues [4–6].

Owing to the nonlinear nature of harmonic generation, the SHG signal level is proportional to the square of the applied laser intensity, and thus it occurs predominately at the focus of a microscope objective, where the peak power is sufficiently high. This aspect gives SHG microscopy the powerful capability of intrinsic 3D sectioning and is shared by other nonlinear optical (NLO) microscopy methods that have been developed almost concurrently with SHG microscopy. The most well-known modalities include third-harmonic generation (THG), two-photon excited fluorescence (TPEF), sometimes referred to as multiphoton excited fluorescence (MPEF), and coherent anti-Stokes Raman scattering (CARS) microscopy [7]. They all fall into the overarching category of laser scanning microscopy and experimentally they are similar in many ways such as point scanning/detection and image formation. However, they are all based on distinct physical processes that lead to different contrast mechanisms. Their applications as imaging approaches are most often complementary, even though they can be readily incorporated into one multimodal imaging system due to their common methodology. In this chapter, we give a detailed comparison of these imaging methods. At the end of this chapter, some new developments of such non-linear microscopy will be briefly reviewed.

4.2 Physical Origin and Contrast Mechanism of SHG and the Other NLO Microscopies

When a dielectric medium is exposed to an oscillating electric field E, a charge oscillation within that medium will then be induced. Moreover, this charge oscillation has the capability of launching a secondary optical wave under certain conditions. All the NLO microscopic methods mentioned earlier involve the response of some biological material to an incident laser beam of high intensity, and emission of the induced new optical wave that carries signatures of the imaged material. In this section, we will discuss the nature of nonlinear interaction of light and matter that gives rise to the contrast mechanism for each of the NLO imaging techniques.

The material response to the applied electric field E is often described using polarization P according to the following relationship:

$$P = \chi^{(1)}E^1 + \chi^{(2)}E^2 + \chi^{(3)}E^3 + \cdots \tag{4.1}$$

where $\chi^{(n)}$ is the nth-order nonlinear susceptibility. The nonlinear effects are represented by the higher-order susceptibility ($n > 1$). The tensor $\chi^{(n)}$ decreases dramatically with increasing n, signifying that higher-order nonlinear processes are very weak responses to the driving optical waves. This also explains why NLO microscopy has only emerged as a standard tool in recent years, as the technology requires user-friendly high-repetition-rate ultrafast lasers, which only became commercially available in the mid-1990s. The first-order susceptibility, $\chi^{(1)}$, describing the linear response of the material to the optical field, is often invoked to explain linear (one-photon) absorption and the index of refraction. SHG is governed by $\chi^{(2)}$ along with two closely related nonlinear processes, sum frequency generation (SFG) and difference frequency generation (DFG). The third-order susceptibility, $\chi^{(3)}$, gives rise to THG, two-photon and three-photon absorption, CARS, and stimulated Raman scattering (SRS). The readers are referred to Refs. [8,9] for detailed discussion of all these different high-order susceptibilities.

4.2.1 SHG Photophysics

4.2.1.1 Molecular and Bulk Relationships

For SHG imaging, photon emission at the harmonic frequency 2ω is collected as a result of second-order nonlinear interaction of the fundamental optical wave (ω) and the nonlinear material. SHG is closely related to hyper-Rayleigh scattering (HRS), an incoherent second-order light scattering process generating a new wavelength at 2ω in isotropic bulk solutions with random molecular orientation. In a sense,

SHG is the coherent version of HRS. In both cases, molecules exposed to the electric field need to possess a dipole moment, which ensures that a harmonic optical wave component can be produced as a result of the nonsymmetrical oscillation of the electrons in response to the symmetrically oscillating driving wave. An additional noncentrosymmetric component can arise from molecular chirality [10–12]. The efficiency of HRS is characterized by the molecular first hyperpolarizability, β, which relates to the second-order induced dipole moment of the molecule as follows:

$$d^{(2)} = \beta E^2 \tag{4.2}$$

The second-order nonlinear polarization $P^{(2)}$ response of a material is simply the bulk representation of $d^{(2)}$, and the emitted SHG is a coherent addition of the HRS emission from the local molecular ensemble. The bulk property $\chi^{(2)}$ can be related to the molecular-level property β as below:

$$\chi^{(2)} = N_s \langle \beta \rangle \tag{4.3}$$

where N_s is the number of molecules involved for the coherent SHG generation and $\langle \beta \rangle$ is the orientational average of β. We want to point out that E, P, $d^{(2)}$ are all vectors, whereas β and $\chi^{(2)}$ are tensors, and thus $\langle \beta \rangle$ is zero for randomly orientated molecules such as in liquid solutions, and this type of material cannot produce SHG. HRS emits in all directions, and only when the molecules are packed in an organized way is $\langle \beta \rangle$ nonzero, and the frequency-doubled emission becomes predominant in a certain direction as a result of the collective response of all the molecules to the driving optical field. In summary, effective SHG requires that the molecules of the medium have a permanent dipole moment and nonzero hyperpolarizability, and, at the bulk level, the dipole moments be aligned as an organized array. In a microscope, the bulk level translates to the focal volume.

Even if a medium meets all of the aforementioned three requirements, SHG is typically still weak. One strategy to increase the SHG intensity is to utilize resonant enhancement. This can be done as illustrated in Figure 4.1. SHG can be viewed as a wave mixing process where two input photons are coupled to form one output photon via a virtual energy level. However, if the SHG excitation energy is resonant

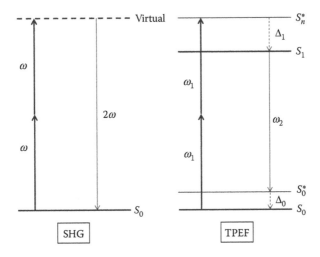

FIGURE 4.1 Energy diagram for SHG and TPEF. S_0 is the ground state; S_0^* is the vibrationally excited S_0 state; S_1 is the lowest singlet excited state; S_n^* ($n \geq 1$) is a vibrationally excited lowest or higher singlet excited state. The thick solid arrows refer to excitation and the thin solid arrows refer to emission. Δ_0 and Δ_1 represent relaxation that leads to heat dissipation into the solvent.

with a real excited state, the SHG intensity will be greatly enhanced. A simple two-state model indicates [4,13,14] that the hyperpolarizability can be written as

$$\beta \sim \frac{\omega_{ge} f_{ge} \Delta\mu_{ge}}{(\omega_{ge}^2 - \omega^2)(\omega_{ge}^2 - 4\omega^2)} \tag{4.4}$$

where ω_{ge} is the frequency that corresponds to the energy gap between the ground state (g) and the upper state (e), f_{ge} the oscillator strength that describes the optical transition probability from state g to state e, and $\Delta\mu_{ge}$ denotes the dipole moment change upon the optical transition. Equation 4.4 clearly indicates that when the excitation energy ($2\hbar\omega$) matches the energy difference the two states ($\hbar\omega_{ge}$), maximal hyperpolarizability is achieved, as the denominator goes to zero, which results in enhanced SHG signal. However, utilizing resonant enhancement will result in photobleaching in-plane. This condition is necessary for SHG imaging of dye-labeled membranes but is not typically employed for imaging proteins.

4.2.1.2 SHG Emission Directionality

An additional consequence of the coherent nature of SHG is the emission directionality and underscores a large difference relative to the incoherent process of TPEF. Neither single nor multiphoton excited fluorescence has a phase relationship to the excitation laser and is emitted over 4 pi steradians. In strong contrast, SHG has a phase relationship with the laser and a well-defined emission directionality. The directionality is explained in terms of phase-matching conditions, a requirement of momentum conservation of optical waves, in addition to the energy conservation requirement that the second harmonic has photon energy twice that of the fundamental. The momentum of an optical wave is $\hbar k$, where k is the wavevector. The direction of the k vector defines the optical wave propagation direction and its value is given by $k = (2\pi/\lambda) = (\omega/c)$, with λ, ω, and c being wavelength, angular frequency, and speed of light in vacuum, respectively. For efficient SHG in nonlinear crystals induced by a laser approximated as a plane wave ω, momentum conservation requires that $k_{2\omega} = 2k_{\omega}$, or $\Delta k = k_{2\omega} - 2k_{\omega} = 0$. This mandates that the second harmonic (2ω) follow the *forward* direction of the fundamental wave (ω) in the limit of perfect phase matching, as achieved in certain uniaxial birefringent crystals such as potassium dihydrogen phosphate (KDP) and β barium boron oxide (BBO).

However, for SHG microscopy with biological tissues, perfect phase matching is never achieved as no type I phase-matching conditions exist, that is, matching the refractive index of the ordinary wave of the SHG with the extraordinary wave of the fundamental. Thus, the minimum phase mismatch is the dispersion in refractive index between the fundamental and SHG wavelengths. Additionally, the molecules (e.g., collagen) are not perfectly aligned in a tissue. Thus, the strict phase-matching requirement of tissues is relaxed and nonzero Δk is allowed, in such a way that the SHG signal varies with it as follows:

$$I_{\text{SHG}} \propto \sin\left(\frac{m\Delta kL}{2}\right) \tag{4.5}$$

where m is an integer and L the coherence length [15,16]. While the SHG conversion efficiency decreases for nonzero Δk, *backward* SHG signal is also produced with this relaxed phase-matching condition due to the need to conserve momentum. Through the development of a general model of phase matching, we showed that smaller and larger values of Δk are associated with primarily forward and backward SHG, respectively [15].

At the same time, a laser beam focused by a microscope objective can no longer be treated as a plane wave, and it experiences a swift phase anomaly (Gouy shift) at the focus [17,18]. This reduces the effective optical momentum along the prorogation direction and further complicates the phase-matching conditions [19]. The direct effect of this is that the SHG beam profile does not match the cone-shaped fundamental beam that propagates from the focal point. Instead, the emission is a two-lobe profile so that the momentum conservation is satisfied with its intensity on the propagation axis being minimum [20]. The angle between the lobes increases at high numerical aperture.

4.2.2 TPEF Photophysics

When resonantly enhanced SHG occurs, another nonlinear process, two-photon induced fluorescence (TPEF), occurs concurrently, both leading to photon emission at wavelengths shorter than the applied optical wave. Owing to the Stokes shift in fluorescence, the TPEF emission must lie to the red of the SHG wavelength. Typically, fluorescence will be characterized by a broader spectrum ~30–100 nm FWHM, and can be readily separated from the SHG, allowing concurrent detection. Figure 4.2 shows a typical situation where the emission spectrum was measured when 880 nm light was used to excite the dye Di-6-ASPBS [21]. The sharp peak at 440 nm (twice the excitation frequency) is the photon emission due to SHG. The SHG bandwidth will be $1/\sqrt{2}$ of the fundamental laser, or for typical 100 femtosecond laser pulses, about 7 nm FWHM. The broad spectrum to the red of the SHG emission is the TPEF spectrum.

Figure 4.1 compares the energetics of TPEF to that of SHG. Whereas for SHG the upper state is generally a virtual state (it is a real state in the resonant enhancement scenario) and the emission photon carries away the total input photon energy ($2\hbar\omega$), for TPEF, the instantaneously excited state is a real excited state S_n^* ($n \geq 1$), from which the molecule then relaxes to the lowest excited state S_1 before it transits to a vibrationally excited ground state S_0^*. Figure 4.1 shows that the fluorescence emission photon energy ($\hbar\omega_{21}$) is less than the total excitation energy ($2\hbar\omega_1$), and the difference is due to the relaxation both from S_n^* to S_1 and from S_0^* to S_0, which ultimately results in heat dissipation into the surrounding environment. This is also revealed by Figure 4.2 where the total excitation energy ($2\hbar\omega_1$), which corresponds to the SHG wavelength at 440 nm, is higher than the TPEF emission photon energy ($\hbar\omega_2$) corresponding to the broad band centered around 580 nm [21]. Regardless of the excitation energy or photon absorption order, the molecules always relax to the lowest excited singlet state, S_1, from which fluorescence emission occurs. As a consequence, the TPEF spectrum must be identical to that of one-photon fluorescence. Moreover, for the case of resonance SHG, the SHG and TPE excitation spectrum would be essentially identical.

While photon emission of TPEF is not a coherent process, the initial two-photon absorption (TPA), however, is coherent. Nonintuitively, the absorption (excitation) process is actually a third-order nonlinear process ($\chi^{(3)}$), even though only two photons are absorbed. This is because it involves coherent interaction of two optical waves and a material wave (polarization) [8], in contrast to the second-order

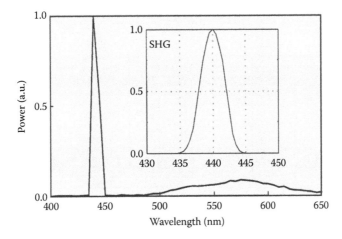

FIGURE 4.2 SHG and TPEF spectrum of Di-6-ASPBS excited by 880 nm. (Reproduced from Moreaux. et al. 2000., Membrane imaging by simultaneous second-harmonic generation and two-photo microscopy. *Opt. Lett.* 25:320–322. With permission of Optical Society of America.)

SHG nonlinear process. All third-order nonlinear processes can be viewed as four-wave mixing processes, and the material wave in the case of TPA is induced by two optical waves of the same frequency. TPA in semiconductors is among the most thoroughly studied in nonlinear optics [22]. The resonant condition (a stable upper state) is characterized by an imaginary $\chi^{(3)}$, while a real $\chi^{(3)}$ is attributed to the third-order nonlinear refraction, much in the same way that the real and imaginary components of the linear susceptibility $\chi^{(1)}$ are associated for linear refraction and absorption.

4.2.3 THG Photophysics

THG is another third-order nonlinear process, which involves a real $\chi^{(3)}$ susceptibility. Here, three fundamental photons at the frequency of ω interact with the material to generate one photon at the third-harmonic frequency, 3ω (Figure 4.3) [7,9]. For systems with inversion symmetry, $\chi^{(2)}$ is zero and third-order processes are the dominant nonlinearities. It should be pointed out that third harmonic (3ω) can also be generated involving a $\chi^{(2)}$ SHG process and a sequential SFG (also a $\chi^{(2)}$ process) of the SHG photon (2ω) and another fundamental photon (ω), for noncentrosymmetric systems where $\chi^{(2)}$ processes are allowed. However, for biological materials, the *direct* $\chi^{(3)}$ process prevails [7] and we use THG in this chapter to denote the direct $\chi^{(3)}$ nonlinear process shown in Figure 4.3, which can be easily understood in analogy to the $\chi^{(2)}$ SHG process.

THG is related to the nonlinear refractive coefficient n_2 according to the following equation

$$n_2 = \frac{3}{4\varepsilon_0 n_0^2 c}\Re e\left\{\chi^{(3)}\right\} \tag{4.6}$$

where ε_0 is the vacuum dielectric constant, n_0 is the linear refraction coefficient ($n_0^2 \propto \Re e\{\chi^{(1)}\}$), and $\Re e$ denotes the real component of a complex value. As a result, THG is sensitive to inhomogeneities such as aqueous medium interfaces and microstructures, where n_2 is mismatched [23]. Again, this reveals the similarity between n_2 and n_0 (or $\chi^{(3)}$ and $\chi^{(1)}$) for linear optics governed by the first-order susceptibility, where signals (reflections or scattering) are generated only at interfaces and inhomogeneities where there is a mismatch of the refractive index n_0. For laser scanning microscopy where a Gaussian beam is focused using an objective tightly onto an inhomogeneous sample, the third-harmonic power is given by

$$I_{3\omega} \propto \left(\delta\chi^{(3)}\right)^2\left(1 + \frac{4z_\omega^2}{b^2}\right)^{-1} \tag{4.7}$$

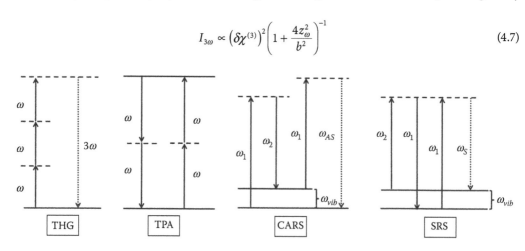

FIGURE 4.3 $\chi^{(3)}$ Nonlinear processes commonly used in NLO microscopy. Solid lines are real molecular states; dashed lines are virtual states. Solid arrows (ω, ω_1, ω_2) refer to interaction between the applied optical waves and the material; dotted arrows (3ω, ω_{AS}, ω_{SRS}) refer to emitted optical waves. ω_{vib} is the vibrational frequency of a molecular ground state.

where z_ω is the distance between the interface of a structure and the beam waist, $b = k_\omega \omega_0^2$ is the confocal parameter of the fundamental beam with a wave vector of k_ω and a waist radius of ω_0, and $\delta\chi^{(3)}$ is the difference of the third-order susceptibility values. Equation 4.7 clearly indicates that THG is strongest when the laser is focused at the interface ($z_\omega = 0$), where mismatch of $\chi^{(3)}$ is located. Barad et al. [23] demonstrated that a large third-harmonic signal can be generated when a femtosecond laser is focused at the interface of glass and index-matching oil, indicating that mismatched nonlinear index (n_2) is the physical origin of THG. We further stress that phase matching conditions are difficult to satisfy for direct THG with a strongly focused laser beam, as THG generated within homogeneous material vanishes due to Gouy phase shift near the laser focus [24,25].

Still, since all materials have some nonvanishing third-order susceptibility and biological systems are often composed of transparent microstructures interfaced with fluid (cytosol, body fluid, etc.), it is expected that THG can be an effective tool to map material distributions in cells and tissues.

4.2.4 CARS and SRS Photophysics

CARS and stimulated Raman scattering (SRS) are two other related third-order nonlinear optical processes, which have been recently developed for biomedical imaging [26,27]. In contrast to SHG, THG, and TPEF, their strength lies in their chemical sensitivity/selectivity.

Raman spectroscopy, together with infrared absorption spectroscopy, is sometimes termed as vibrational spectroscopy (Figure 4.4), as they measure vibrational "fingerprint" signatures of molecules. As an example, Figure 4.5 shows the Raman spectrum of the P22 virus, which reveals the power of vibrational spectroscopy for detecting specific biochemical molecules [28]. Thus, coupling vibrational spectroscopy with high-resolution microscopy is a powerful means to monitor the spatial distribution of proteins/nuclei acids in cells and tissues. Microscopy based on infrared absorption spectroscopy is not practically feasible because the laser excitation has to match the intrinsic vibrational frequency, demanding high tunability over a broad range of infrared wavelengths. Such lasers do not exist in typical lab settings and equivalent light sources are mostly confined to synchrotrons. Additionally, the long excitation wavelengths reduce the spatial resolution. Additionally, microscopy using absorption as a contrast mechanism has low sensitivity because generally the absorption introduces a small decrease of a relatively intense light signal (e.g., 3–6 orders of magnitude). By comparison, spontaneous Raman microscopy is background free as it detects a new wavelength. The other advantage of Raman microscopy is that the upper state of the excitation process is a virtual state, meaning that the excitation wavelength is not limited by the energy level system of the molecule under study. Thus, excitation

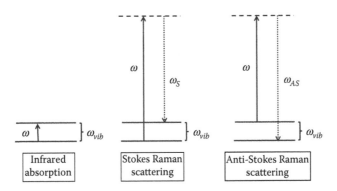

FIGURE 4.4 Energy diagrams for vibrational spectroscopic methods. ω is the excitation optical wave; ω_S and ω_{AS} are the Stokes and anti-Stokes Raman emission, respectively; ω_{vib} is a representative vibrational frequency of the molecule. The horizontal solid lines are stable vibrational energy levels and the horizontal dashed lines are virtual energy levels. The solid arrows represent excitation and the dashed arrows are spontaneous Raman emission.

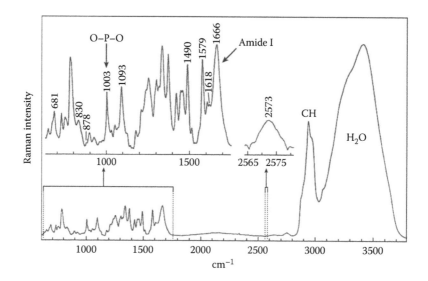

FIGURE 4.5 The Raman spectrum of the P22 virus showing characteristic vibrational frequencies observed in biological samples. Several vibrational modes of particular interest in vibrational microscopy are labeled. The O–P–O stretching vibration arises from the vibration of the DNA backbone. The amide-I band is characteristic of proteins and can be used to map out protein density. The CH-stretching band is typically used to image lipids. The H_2O-stretching vibrations of water are important for following water flow and density. (Adapted from Evans, C. L. and X. S. Xie. 2008. *Annu. Rev. Anal. Chem.* 1:883–909. With permission.)

wavelength can be in the visible range, leading to higher spatial resolution. Indeed, confocal spontaneous Raman microscopy has been previously developed to provide a "chemical" imaging tool with 3-D capability [29].

However, owing to the small scattering cross section ($\sim 10^{-30}$ cm² compared to $\sim 10^{-16}$ cm² for absorption of a typical dye), the spontaneous Raman signal is weak. This is especially true for anti-Stokes Raman scatter as it is usually much weaker than Stokes Raman scattering, because anti-Stokes Raman scattering occurs from vibrationally excited molecules (Figure 4.4), whose population is many orders of magnitude smaller than the molecules in the ground state. The relative population for molecules in the vibrationally excited and ground states can be estimated from the Boltzmann distribution law by $\sim e^{-\frac{\hbar \omega_{vib}}{RT}}$, where $\hbar \omega_{vib}$ is the energy gap between the two vibrational energy levels shown in Figure 4.4, and R and T are the gas constant and the absolute temperature, respectively. For the CH-stretching vibrational band that is typically chosen to image lipids, $\hbar \omega_{vib}$ is ~ 3000 cm⁻¹. At room temperature, RT corresponds to ~ 200 cm⁻¹. From this, it can be calculated that only 0.3 ppm of lipid molecules is in the CH-stretching vibrationally excited state. Apparently, to use the Stokes or anti-Stokes Raman scattering signal for imaging of living specimens, where fast acquisition is required, strong enhancement mechanisms such as CARS and SRS need to be implemented to increase the sensitivity by several orders of magnitude.

CARS introduces another optical wave, termed as pump wave ω_2 (Figure 4.3), to amplify the anti-Stokes Raman signal (ω_{AS}) induced by the Stokes wave ω_1. The two laser wavelengths are set so that their beat frequency ($\omega_1 - \omega_2$) matches the frequency of the Raman active vibration ω_{vib}. This activates an oscillating polarization with the vibrational frequency and a strong emission optical field at the anti-Stokes frequency $\omega_{AS} = (2\omega_1 - \omega_2)$. CARS is a vibrationally resonant third-order nonlinear process that occurs instantaneously, where the molecules do not gain energy from the optical waves, as they both start and end at the ground state. Assuming plane pump and Stokes waves, the anti-Stokes Raman signal intensity is given as [28]:

$$I_{AS} \propto (\chi^{(3)})^2 I_p^2 I_S \left(\frac{\sin(\Delta kz/2)}{\Delta kz/2} \right)^2 \qquad (4.8)$$

where I_P and I_S are intensities of the pump and Stokes waves, respectively, z is the sample thickness, and $\Delta k = k_{AS} - (2k_1 - k_2)$ is the wavevector mismatch. Equation 4.8 indicates that CARS signal is strongest when $\Delta k = 0$, a phase-matching condition similar to the one introduced for SHG, where the signal wave copropagates with the excitation laser.

When two laser beams of strong intensity are concurrently present as the sample with the frequency difference matching a Raman-active vibration for CARS generation, another $\chi^{(3)}$ process, SRS [9] also occurs simultaneously, which enhances the spontaneous Stokes Raman signal (ω_S) induced by the Stokes wave ω_1 (Figure 4.3). This has been successfully used in Xie's group as an alternative to CARS for vibrational imaging [27]. SRS can be viewed as a four-wave mixing process, where the Stokes Raman scattering, as described by the solid arrow and the dashed arrow on the right in Figure 4.3, is enhanced by the coherent interaction between the optical waves (the two solid arrows on the left in Figure 4.3) and the molecule (denoted by the energy levels). Figure 4.3 indicates that the photon energies of the two applied optical waves, ω_1 and ω_2, satisfy the energetics for both CARS and SRS ($\Delta\omega = \omega_1 - \omega_2 = \omega_{vib}$). A major difference is that CARS leads to a new wavelength generation at $\omega_{AS} = \omega_1 - \omega_{vib}$, while for SRS the emission frequency (ω_S) is the same as the stimulating frequency ω_2 resulting an intensity increase for the stimulating laser beam (ω_2; Raman gain) and a decrease of the Stokes beam (ω_1; Raman loss). This difference has a significant consequence for signal retrieval in the experimental design (see Section 4.3), as the CARS emission is a signal over zero background while the SRS signal has a strong background. However, CARS can also occur by means of a nonresonant $\chi^{(3)}$ process in any material, where a real upper vibrational state is absent, and thus this nonresonant contribution to ω_{AS} is manifested as an ever-present nonspecific background and unfavorably interferes with the resonant CARS signal. This has the result that the CARS spectrum is not equivalent to the Raman spectrum and phase retrieval efforts must be employed to extract the latter. On the other hand, the nonresonant background is not a problem for SRS because of the detection scheme, that only is sensitive to the upper vibrational state when $\Delta\omega$ matches ω_{vib} [27,30]. As an additional benefit over CARS, SRS detection results in the true Raman spectrum.

4.3 Instrumentation of NLO Microscopes

For all of the NLO effects introduced in the previous section, the intensity of the generated optical signal scales nonlinear with the input optical waves, that is, $I_{NLO} \propto I_\omega^2$, $I_{\omega 1}^2$, I_ω^3, $I_{\omega 1}^2 I_{\omega 2}$ and $I_{\omega 1}^2 I_{\omega 2}$ for SHG, TPF, THG, CARS, and SRS, respectively. Pulsed lasers with high instantaneous peak power, typically high-repetition-rate, mode-locked Ti:sapphire lasers, are used to drive the nonlinear effects. For tightly focused laser beams in the material, such relationships dictate that the nonlinear effects can be confined to the focal point only, where the peak power is most intense. Owing to the similar nature of nonlinear signal generation, the NLO microscopes share a common instrumental design with some subtle differences among themselves. Experimentally, all NLO microscopes are based on the idea of point illumination and detection of point scanning microscopy, which is also the approach used for confocal microscopy [31]. At the early phase of development, point scanning was implemented via sample scanning with an XYZ transitional stage, and this has switched to laser beam scanning in later years, which greatly increases the image acquisition speed, and microscopy based on this method is generally known as laser scanning microscopy (LSM).

Conceptually, laser scanning microscopy contrasts with the traditional wide field microscopy, where a region of interest is uniformly illuminated and the whole image is projected on a light-sensitive surface detector such as a CCD camera. Thus, laser scanning does not employ Kohler illumination, although the laser must be aligned with the Kohler image planes. While the imaging acquisition speed of LSM (typically one frame per second for nonresonant galvos) is inferior to the frame-based wide field micro-scopy, it offers other powerful imaging capabilities that would not be possible for wide field microscopy due to the different nature of nonlinear optical imaging contrast mechanisms described in this chapter.

Figure 4.6 displays an ideal schematic of a laser scanning NLO microscope showing the simplest scanning pattern between point 1 and point 2 with a forward and a backward detection channels which are optically symmetric with respect to the sample plane. The way the detection is set up differs from that of a confocal microscope, which has the detector located before the scanning mirror, so that the emitting ray reverses the propagation of the pump laser beam all the way back through the scan mirror and the "descanned" ray is then stationary and focused on the confocal pinhole before collection by the detector. The combination of descanning and a confocal pinhole is essential to achieve 3D sectioning in confocal fluorescence imaging, but is not necessary for NLO microscopes as there is no out-of-plane signal. Thus, essentially all NLO microscopes use nondescanned detection by placing the detectors in a return path where the signal does not traverse the scanning mirrors. Moreover, confocal detection is highly inefficient and nondescanned detection increases sensitivity by several factors due to the optical path alone. This sensitivity is increased further for scattering specimens.

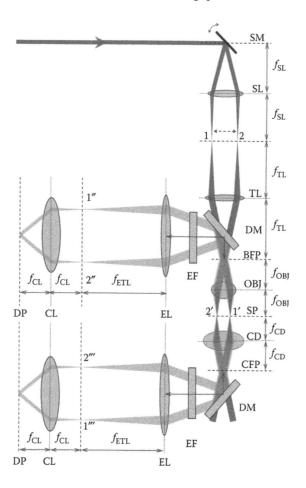

FIGURE 4.6 Laser scanning microscopy. Light paths are shown for two scan positions (1–1′–1″ (or 1‴)) and (2–2′–2″ (or 2‴)). The excitation laser beams are in dark gray and the emission light paths are in light gray. SM, scan mirror; SL, scan lens; TL, tube lens; DM, dichroic mirror; BFP, back focal plane of the objective (OBJ); SP, sample plane; CD, condenser; CFP, condenser focal plane; EF, emission filter; EL, emission lens; CL, collimation lens; DP, detection plane.

For laser scanning microscopy, the relative distance between the optical components is strictly regulated. In the ideal case shown in Figure 4.6, they are spaced by the focal length of the lenses except for the dichroic mirrors and emission filters, which are located in the infinity space. This arrangement is important to maintain both proper beam path (propagation directionality) and correct beam profile (size and divergence). The laser beam is expanded by the scan lens and the tube lens to fill the microscope objective back aperture for maximum resolution as the beam can then be focused into the tightest spot (1' or 2') at the sample plane. One crucial aspect is that when the scan mirror moves the laser beam between 1 and 2, the laser beam actually remains stationary at the position of the scanning mirror, the back focal plane of the objective and the condenser focal plane, and meanwhile, the emitting rays remain stationary at the objective (or condenser) focal plane and the detector. Such stationary status has to be maintained as the objective back aperture, the condenser aperture, and the detector optical window are beam-restrictive components and violation of this condition will result in the loss of field of view.

The NLO microscopes all have their own special characteristics, despite the common aspects outlined earlier, and this is reflected in the apparatus designs. SHG, TPEF, and THG microscopes all use single laser beam illumination and can be easily implemented as multiple modalities of one setup. However, they are different in the detected emission wavelength and switching modalities often involve only careful emission filter (EF in Figure 4.6) selection. With a pump laser beam at λ, SHG and THG require a narrow-band interference filter at $\lambda/2$ and $\lambda/3$, respectively, while TPEF uses a broad-band filter between $\lambda/2$ and λ or multiple narrower-band filters for the case of multiple fluorophores.

Because THG is a nonlinear process of higher order with a smaller interaction cross section, it often requires higher pump laser power, which may lead to undesirable sample damage. Additionally, the third harmonic of the Ti:sapphire output is usually deep in the UV (<333 nm) and can be absorbed and scattered by the sample, and cannot be efficiently transmitted by conventional glass optical components. All these make THG signal detection much harder. As a result, an optical oscillator (OPO) is commonly used to shift the pump laser wavelength toward infrared, which is less invasive [24,32], and this also moves the third harmonic toward the visible spectrum. The other laser source used for THG has been the Cr:fosterite laser at ~1240 nm [33].

CARS and SRS imaging are more complex than the previous modalities because in addition to proper optical filter selection they involve two different laser wavelengths, a pump beam at ω_1 and a Stokes beam at ω_2, which need to be combined spatially and before they are introduced to the microscope. Owing to the coherent nature of CARS and SRS, the pulses from the two lasers need to be synchronized so that both wavelengths can be present simultaneously at the same location in the sample to produce the nonlinear signal. For this purpose, either outputs from two synchronized Ti:sapphire lasers are used [34] or a single OPO (along with the pump laser) can be used to provide both wavelengths with the required timing [27]. Signal detection for CARS can still be accomplished in the same way as SHG except that now a narrow filter at a shorter wavelength than the input wavelengths is used ($\omega_{AS} = \omega_1 + \omega_{vib} > \omega_1 > \omega_2$).

SRS is very unique among the NLO methods under discussion in terms of signal detection. All these other NLO methods record the generated nonlinear signal at a new wavelength, and thus the signal is background free. By comparison, SRS deals with small intensity variation ($\sim 10^{-6}$) of the input laser beams (Figure 4.7), which means that the resulting SRS signal (Raman gain or loss) is a very weak signal on top of a large background. In such situations, it is necessary to use a lock-in amplification system for signal recovery [35]. Here, the intensity of one of the input laser beams is modulated with an optical chopper, at a reference frequency (~10 kHz) while the intensity of the other beam is monitored using a photodiode. Because the two laser beams are coupled at the sample via the SRS process, the intensity variation of the second laser beam also carries the reference frequency and can be retrieved by the lock-in amplifier, which is only sensitive to the signal at the reference frequency. One should note that the detector used for SRS imaging is different from those of the other NLO imaging techniques, which are almost always PMTs detecting low single levels on top of zero background. For SRS, a photodiode

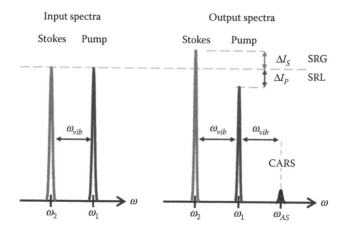

FIGURE 4.7   Input and output spectra of SRS and CARS: SRS leads to an intensity increase in the Stokes beam (SRG) and an intensity decrease in the pump beam (SRL); CARS generates new signal at the anti-Stokes frequency. (From Freudiger, C. W. et al. 2008. Label-free biomedical imaging with high sensitivity by stimulated Raman scattering microscopy. *Science* 322:1857–1861. Reproduced with permission of AAAS.)

is used to monitor the intensity of one of the input laser beams after it has gone through the sample. Lastly, CARS and SRS microscopes mostly employ picosecond lasers instead of the femtosecond sources mostly commonly used for SHG, THG, and TPEF. This is because the shorter pulses have bandwidths (~10 nm FWHM) broader than that of vibrational bands (<1 nm). For CARS, this greatly increases the ubiquitous nonresonant background, decreasing the attainable signal-to-noise ratio. This is not necessarily problematic for SRS, but still decreases selectivity in the crowded fingerprint region. Thus, while the use of longer pulses decreases the peak power and the resulting CARS and SRS signals, the trade-off is the increased sensitivity to specific bonds and better background rejection.

For all the NLO imaging techniques described earlier, the polarization of the excitation laser beam and/or the emitted light can be manipulated for special considerations. For example, the information content in SHG imaging can be enhanced by measuring the signal anisotropy measurement, where a linearly polarized laser is used to excite the sample, and the resulting SHG emission signals parallel and perpendicular to the excitation polarization are collected separately and can be further analyzed [36–38]. Other measurements show the dependence of the SHG signal on the laser polarization. Collectively, these measurements relate to the underlying matrix elements of $\chi^{(2)}$ [39]. For CARS, polarizers were used for both excitation laser beams (pump and Stokes) and emission light (CARS signal) to suppress the nonresonant background as a means to enhance SNR [26].

4.4 Representative Biological Applications of SHG Imaging and Its Comparison with Other NLO Techniques

4.4.1 Representative SHG Examples and Unique Aspects

Based on Equation 4.3, SHG requires that the target sample contain noncentrosymmetric molecules packed in a well-ordered manner. This attribute implies that SHG microscopy is sensitive to the bulk property (susceptibility) of the material $\chi^{(2)}$. In this context, let us consider the structure of collagen. Collagen is the most abundant protein in mammals and makes up 25–35% of the whole-body protein content. Collagen has a molecular weight about 300 kD and has dimensions of 300 nm in length and 1 nm in width. The triple helical collagen molecules are aligned with one another and crosslinked to form fibrils of 10–300 nm, which then aggregate into collagen fibers with a diameter on the order of a micron. Each of the α-chains (~300 nm long) within the triple helix is a polypeptide chain of high regularity, composed of repetitive –Gly–X–Y– sequences, where X and Y can be any amino acid, though X is commonly proline and Y hydroxyproline. This degree of order resembles that of a nonlinear crystal with periodically arrayed unicells. SHG has been shown to be sensitive to both the fibril and fiber levels of organization [38]. A broad range of collagen tissues have so far successfully imaged with SHG techniques, which includes self-assembled collagen gels [40], the epimyseum and perimyseum wrapping of muscle [6], bone [41], fish scale [6], skin [42], rat or mice tendon [43–45], cornea [46–49], and ovary [38]. We will discuss the unique application of SHG imaging for collagen tissues in more detail below with two examples, explicitly showing the structural sensitivity.

Cornea, the important organ of the visual system for organisms, is composed essentially of all type I collagen [50], and it has been studied intensely using SHG imaging [47–49]. Han et al. found that cornea collagen fibers generate strong SHG signal, and the emission almost completely follows the direction of the incident laser, consistent with the transparency of cornea. At the same time, SHG signal generated in the closely related sclera tissue (also largely composed of collagen fibers) is more backward directed than forward (Figure 4.8) [47]. Scanning electron microscopy indicates that corneal collagen fibrils have a very uniform diameter of ~25 nm, and are regularly packed into layers of extended structures (lamellae) resembling crystalline lattices much larger than λ_{SHG}. In contrast, scleral fibrils possess diameters that are widely distributed in the range of 25–300 nm, and present a random morphology of smaller structures [50]. Clearly, microstructures of the collagen tissues indicate that cornea resembles a crystal more than sclera does, and the general relaxed phase-matching conditions described earlier would predict predominantly harmonic generation in the forward direction, as is experimentally observed. In contrast, the smaller structures of sclera lead to a mixture of forward and backward components.

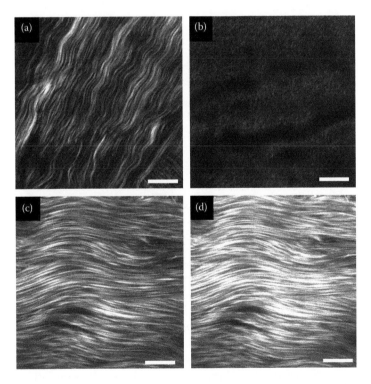

FIGURE 4.8 SHG imaging of collagen fibrils in forward and backward directions of cornea (a, b) and sclera (c, d). Bars: 10 μm. (Reproduced from Han, M., G. Giese, and J. F. Bille. 2005. *Opt. Express* 13:5791–5797. With permission.)

The underlying coherence of the SHG is the enabling factor in elucidating these structural details. Incoherent modalities such as fluorescence and reflectance confocal cannot reveal this level of detail. We note that while collagen does possess weakly imageable autofluorescence, the emission typically appears somewhat diffuse and is not necessarily reflective of the fibrillar structure.

An example demonstrating how the coherence of SHG can uniquely obtain information imaging collagen in diseased states comes from our group, where we compared the structure of normal and malignant ovarian tissues (almost all type I collagen). Sharp differences exist for collagen morphology in these tissues [38], where the malignant collagen fibers display higher regularity and are more densely packed along a predominant direction, while the normal collagen fibers are loosely distributed in all directions (Figure 4.9). This change in morphology was also evident in the directional data, which, along with Monte Carlo simulations, allowed us to deduce changes in the fibril size and distribution. Concurrently, SHG polarization anisotropy revealed changes in the fiber morphology. More details on this aspect are presented in Chapter 6. Thus, as is the case with cornea, the coherence enables subresolution information on the fibril assembly to be obtained.

Besides fibrillar collagen tissues, several other endogenous protein structures in biological systems have been investigated using SHG microscopy, where this modality produces unique contrast relative to the other NLO techniques. Detailed features of actomyosin structures in muscle tissue

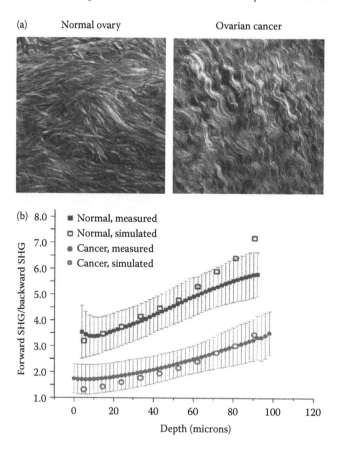

FIGURE 4.9 (a) Representative SHG images of collagen fibers in normal human ovary (left) and malignant ovary (right) with 890 nm excitation wavelength (field size = 170 μm). (b) Forward/backward SHG ratio versus imaging depth for normal and malignant ovarian tissues and Monte Carlo simulations using measured bulk optical parameters. (Adapted from Campagnola, P. 2011. *Anal. Chem.* 83:3224–3231.)

were revealed with SHG imaging in our lab [51], with the myosin thick filaments generating strong harmonic signal and actin filaments not producing contrast. This arises due to the strong and weak birefringence of myosin and actin, respectively. We also successfully acquired SHG images of microtubules in mitotic spindles and centrosomes [6]. Microtubules are polymers of α- and β-tubulin dimers of high polarity, with the α-subunit exposed at one end and β-subunit at the other. The SHG emission is much weaker compared to the coiled-coil myosin and collagen (~10–50-fold). It is also very interesting to note that SHG signal disappears at the midzone of mitotic spindles and the center of centrosomes, where microtubules are aligned antiparallel, so there is no net dipole and the harmonic emission from individual filaments interfere destructively as described by Equation 4.3. In contrast, fluorescence from GFP:tubulin was observed throughout these regions as no symmetry constraints exist.

The assembly of intermediate astroglial filaments from the central nervous system were also studied by SHG imaging by Cheng et al. [52]. However, the mechanism of the harmonic generation remains unknown, as the basic units of an intermediate filament are antiparallel tetramers, which possess an inversion symmetry [53], and no SHG should be expected. On the other hand, SHG imaging of a mouse ear skin containing hair follicles indicates complete absence of SHG signal from the keratin intermediate filaments of the hair, consistent with the molecular orientation of intermediate filaments [6]. We note that fluorescence and CARS/SRS are not constrained by these symmetry considerations. Indeed, we observed strong keratin autofluorescence.

4.4.2 TPEF and SHG Comparison

The symmetry constraints of SHG discussed earlier allow the extraction of structural details that cannot be obtained by other optical methods. But on the other side, it limits the scope of applicability for this technique. For example, in contrast to SHG, two-photon absorption is a noncoherent process, and it does not require the target sample to contain molecules with specific structural alignment, which enables TPEF microscopy to have broader applicability. For example, autofluorescence from endogenous biomolecules such as NADH, flavins, or tetinol is commonly used for TPEF imaging [54]. In addition to the ability to image endogenous species, TPEF has become widespread more due to the wide range of fluorophores with desired spectral and chemical properties as well as labeling essentially any protein of interest [55]. This can be achieved via the well-developed technology of immunostaining, or via the vast palette of fluorescent proteins [56].

Compared to SHG microscopy, TPEF has several other disadvantages. First is the lack of sensitivity to structural information, where fluorescent tags are reporters of protein assembly, whereas SHG visualized the proteins directly. Second, photoexcitation for TPEF is a real photon absorption process, where molecules are promoted to the electronic excited state. Heat release and photobleaching (in-plane) are two undesired consequences that lead to sample damage and image quality deterioration. For SHG, there is no energy exchange between light and the materials under study. Lastly, SHG imaging targets are endogenous protein structures (with the exception of dye-labeled membranes), which require no special sample preparation, while TPEF often involves adding external or genetically encoded fluorophores. Adding foreign species to the system often changes the property of the system. For example, immunostaining requires that the tissues or cells be fixed, which makes *in vivo* study impossible. Additionally, genetic reporters can alter the function of some proteins, especially those that actively polymerize/depolymerize.

4.4.3 THG and SHG Comparison

Unlike TPEF microscopy, THG shares similar structure sensitivity of SHG for biomedical imaging of tissues. However, THG and SHG operate on significantly different mechanisms. While SHG requires the target to be in a bulk noncentrosymmetric microenvironment, THG does not have this constraint. However, THG is not highly versatile as a change in refractive index is required when focused beams are used. This is because the signal arises from optical heterogeneities in the specimen where there are changes in the third-order nonlinear susceptibility, $\chi^{(3)}$. One point of comparison comes from imaging biological membranes with both SHG and THG. SHG images the two-dimensional membrane itself where dye molecules have assembled parallel to each other [16,21,57]. In contrast, THG does so by probing the volume around the membrane, where there is an interfacial change of susceptibility/refractive index [32,49]. The sensitivity of THG to membrane boundaries has been successfully utilized to image unstained whole zebrafish embryos with micrometer 3D resolution. This was used for cell lineage reconstruction over the first 10 cell division cycles, with SHG complementary imaging of mitotic spindles [49]. THG microscopy was also used to image lipid bodies in cells and tissues in real time [32]. The examples indicate that THG, like SHG, is a relatively noninvasive microscopic method that can be applied for *in vivo* imaging with 3D capability. However, the applicability is limited to regions where these is a change in refractive index, and imaging bulk homogeneous tissues is not possible.

4.4.4 CARS, SRS, and SHG Comparison

Like SHG and THG, CARS and SRS are also coherent processes, where the generated signal maintains a specific phase relationship with the incident waves. CARS and SRS are not limited by the noncentrosymmetry requirement of SHG microscopy. In fact, CARS and SRS differentiate target tissues according to their molecular vibrational signatures, and in principle, they have tremendous potential for biomedical imaging, as biological molecules (protein, DNA, lipid, etc.) all have distinct vibration spectra. This has been increasingly enabled by the availability of widely tunable pico/femtosecond lasers with difference in photon energy that can be tuned to match the vibrational frequencies of target molecules. For example, the versatility of label-free CARS and SRS microscopy has been demonstrated through imaging of DNA backbone by targeting the PO_2^- symmetric stretching vibrational mode [58], of plant cell walls using the 1600 cm^{-1} aryl ring stretching of lignin polymers [28], of omega-3 fatty acids, DMSO, and retinoic acid as demonstrations of drug delivery monitoring [27,59], and of elastin and collagen fibrils in Yorkshire pig carotid artery walls [60].

However, at the current stage of development, CARS and SRS are mostly used to image samples rich in lipids using the 2845 cm^{-1} CH$_2$ stretch mode. Examples have included lipid multilamellar vesicles [61], endothelium cells of carotid artery [60], lipid droplets in mouse adrenal cortical cells [62], mouse brain and ear skin [27], and the axonal myelin sheath from Guinea pigs [52,63]. This limitation is in part due to weak Raman cross section and the resulting high number of local oscillators that are needed to produce useful contrast. Moreover, the overall signal must be high enough to overcome the

TABLE 4.1 Comparison of Nonlinear Optical Microscopy Modalities

Method	Primary Information	Lasers	Concentration	Emission Direction	Applicability	Detection Wavelength
SHG	Structural/assembly	Fs Ti:sapphire	Quadratic	F/B Tissue dependent	Bulk tissues	Near UV-visible
THG	Structural/assembly	OPO, Cr:fosterite	Quadratic	F/B Tissue dependent	Interfaces in cells, tissue	Near UV
CARS	Chemical, mostly C–H lipids	PS + OPO	Quadratic	Primarily forward	Cells, tissues	Visible
SRS	Chemical, mostly C–H lipids	PS/fs + OPO	Linear	Forward (laser detection)	Cells, tissues	NIR pump or probe
Two-photon	Protein localization	Fs Ti:sapphire	Linear	4π	Cells, tissues	Visible

nonresonant background [28]. This requirement is most readily met by abundant C–H stretches. One such example showing this point comes from the imaging of collagen fibrils using CARS and SFG, where CARS had much inferior S/N than SFG (SFG is closely related to SHG and their S/N should be comparable) [60]. When used for imaging proteins, CARS and SRS methods directly report CH_2 content of the proteins. Wang et al. indicated that when CARS imaged both elastin and collagen fibers, simultaneously, the former showed higher contrast due to greater abundance of CH_2 groups. On the other hand, SFG imaging only recorded collagen fibers, where elastin was completely absent [60]. The more complex and expensive CARS/SRS microscopes have the advantage of a wider range of applications, but lack the capability of structural sensitivity of SHG/SFG. SHG/SFG imaging methods will be more valuable for disease diagnosis, where only change of protein conformations occurs without the variation of chemical composition. However, the modalities have great promise when used in conjunction as they provide complementary information. An overall summary of the attributes of all the NLO modalities is given in Table 4.1.

4.5 Summary

SHG and other nonlinear optical microscopies are all imaging methods whose development resulted from introducing nonlinear optics into laser scanning microscopy, of which multiphoton excited fluorescence imaging is the most straightforward conceptually. While all the NLO imaging methods have many similarities experimentally and their nonlinear nature enable their 3D capability to image biological samples, the unique aspects of SHG microscopy arise because of its sensitivity to protein microscale structures, especially to those of collagen fibrils/fibers. Given the numerous and diverse pathologies that have alterations in the collagen assembly, SHG has a high potential to develop into a clinical diagnostic imaging tool. However, the physics behind its structural sensitivity also limits its application generality.

The NLO imaging techniques are often incorporated into one microscope as complementary modalities to combine the merits of the different contrast mechanisms. This field is still undergoing rapid development, with the current technologies finding broader applications, with new techniques to improve the current technologies (such as NLO endoscopies, and manipulating laser polarizations for the benefit of information extraction or noise suppression), and sometimes with completely new technologies starting to emerge (e.g., stimulated emission microscopy). NLO microscopies have already significantly pushed forward scientists' ability to understand academic questions about cells and organisms, and will likely revolutionize clinical imaging approaches for disease diagnosis.

References

1. Franken, P. A., A. E. Hill, and G. Weinreich. 1961. Generation of optical harmonics. *Phys. Rev. Lett.* 7:118.
2. Fine, S. and W. P. Hansen. 1971. Optical second harmonic generation in biological systems. *Appl. Opt.* 10:2350–2353.
3. Bouevitch, O., A. Lewis, I. Pinevsky, J. P. Wuskel, and L. M. Loew. 1993. Probing membrane potential with non-linear optics. *Biophys. J.* 65:672–679.
4. Campagnola, P. J. and L. M. Loew. 2003. Second-harmonic imaging microscopy for visualizing biomolecular arrays in cells, tissues and organisms. *Nat. Biotechnol.* 21:1356–1360.
5. Campagnola, P. J., M. D. Wei, A. Lewis, and L. M. Loew. 1999. High resolution non-linear optical microscopy of living cells by second harmonic generation. *Biophys. J.* 77:3341–3349.
6. Campagnola, P. J., A. C. Millard, M. Terasaki, P. E. Hoppe, C. J. Malone, and W. A. Mohler. 2002. 3-Dimensional high-resolution second harmonic generation imaging of endogenous structural proteins in biological tissues. *Biophys. J.* 82:493–508.

7. Masters, B. R. and P. So. 2008. *Handbook of Biomedical Nonlinear Optical Microscopy*. Oxford University Press, New York, NY.

8. Shen, Y. R. 1984. *The Principles of Nonlinear Optics*. John Wiley and Sons, New York, NY.

9. Bass, M., G. Li, and E. V. Stryland. 2010. *Handbook of Optics, Volume IV—Optical Properties of Materials, Nonlinear Optics, Quantum Optics*. McGraw-Hill Professional, New York, NY.

10. Pena, A. M., T. Boulesteix, T. Dartigalongue, and M. C. Schanne-Klein. 2005. Chiroptical effects in the second harmonic signal of collagens I and IV. *J. Am. Chem. Soc.* 127:10314–10322.

11. Haupert, L. M. and G. J. Simpson. 2009. Chirality in nonlinear optics. *Annu. Rev. Phys. Chem.* 60:345–365.

12. Simpson, G. J. 2004. Molecular origins of the remarkable chiral sensitivity of second-order nonlinear optics. *ChemPhysChem.* 5:1301–1310.

13. Reeve, J. E., H. L. Anderson, and K. Clays. 2010. Dyes for biological second harmonic generation imaging. *Phys. Chem. Chem. Phys.* 12:13484–13498.

14. Oudar, J. L. and D. S. Chemla. 1977. Hyperpolarizabilities of the nitroanilines and their relations to the excited state dipole moment. *J. Chem. Phys.* 66:2664–2668.

15. Lacomb, R., O. Nadiarnykh, S. S. Townsend, and P. J. Campagnola. 2008. Phase matching considerations in second harmonic generation from tissues: Effects on emission directionality, conversion efficiency and observed morphology. *Opt. Commun.* 281:1823–1832.

16. Mertz, J. and L. Moreaux. 2001. Second-harmonic generation by focused excitation of inhomogeneously distributed scatterers. *Opt. Commun.* 196:325–330.

17. Gouy, L. G. 1890. Sur une propriete nouvelle des ondes lumineuses. *C. R. Acad. Sci. Paris* 110:1251.

18. Boyd, R. W. 1980. Intuitive explanation of the phase anomaly of focused light beams. *JOSA* 70:877.

19. Feng, S. and H. G. Winful. 2001. Physical origin of the Gouy phase shift. *Opt. Lett.* 26:485.

20. Moreaux, L., O. Sandre, S. Charpak, M. Blanchard-Desce, and J. Mertz. 2001. Coherent scattering in multi-harmonic light microscopy. *Biophys. J.* 80:1568–1574.

21. Moreaux, L., O. Sandre, M. Blanchard-Desce, and J. Mertz. 2000. Membrane imaging by simultaneous second-harmonic generation and two-photo microscopy. *Opt. Lett.* 25:320–322.

22. Stryland, E. W. V. and L. Chase. 1994. Two-photon absorption: Inorganic materials. In *Handbook of Laser Science and Technology, Supplement 2, Optical Materials*. M. Weber, editor. CRC Press, Boca Raton, FL.

23. Barad, Y., H. Eisenberg, M. Horowitz, and Y. Silberberg. 1997. Nonlinear scanning laser microscopy by third harmonic generation. *Appl. Phys. Lett.* 70:992.

24. Débarre, D., W. Supatto, and E. Beaurepaire. 2005. Structure sensitivity in third-harmonic generation microscopy. *Opt. Lett.* 30:2134–2136.

25. Boyd, R. W. 2008. *Nonlinear Optics*. Academic Press, London, UK.

26. Cheng, J.-X. and X. S. Xie. 2004. Coherent anti-Stokes Raman scattering microscopy: Instrumentation, theory, and applications. *J. Phys. Chem. B* 108:827–840.

27. Freudiger, C. W., W. Min, B. G. Saar, S. Lu, G. R. Holtom, C. He, J. C. Tsai, J. X. Kang, and X. S. Xie. 2008. Label-free biomedical imaging with high sensitivity by stimulated Raman scattering microscopy. *Science* 322:1857–1861.

28. Evans, C. L. and X. S. Xie. 2008. Coherent anti-Stokes Raman scattering microscopy: Chemical imaging for biology and medicine. *Annu. Rev. Anal. Chem. (Palo Alto Calif)* 1:883–909.

29. Puppels, G. J. 1999. Confocal Raman microscopy. In *Fluorescent and Luminescent Probes for Biological Activity*. W. T. Mason, editor. Academic Press, New York. p. 377.

30. Levenson, M. D. and S. S. Kano. 1988. *Introduction to Nonlinear Laser Spectroscopy*. Academic Press, San Diego.

31. Amos, W. 2003. How the confocal laser scanning microscope entered biological research. *Biol. Cell* 95:335–342.

32. Debarre, D., W. Supatto, A. M. Pena, A. Fabre, T. Tordjmann, L. Combettes, M. C. Schanne-Klein, and E. Beaurepaire. 2006. Imaging lipid bodies in cells and tissues using third-harmonic generation microscopy. *Nat. Methods* 3:47–53.

33. Sun, C.-K., S.-W. Chu, S.-Y. Chen, T.-H. Tsai, T.-M. Liu, C.-Y. Lin, and H.-J. Tsai. 2004. Higher harmonic generation microscopy for developmental biology. *J. Struct. Biol.* 147:19–30.

34. Jones, D. J., E. O. Potma, J.-X. Cheng, B. Burfeindt, Y. Pang, J. Ye, and X. S. Xie. 2002. Synchronization of two passively mode-locked, picosecond lasers within 20 fs for coherent anti-Stokes Raman scattering microscopy. *Rev. Sci. Instrum.* 73:2843.

35. Freudiger, C. W., W. Min, B. G. Saar, S. Lu, G. R. Holtom, C. He, J. C. Tsai, J. X. Kang, and X. S. Xie. 2008. Label-free biomedical imaging with high sensitivity by stimulated Raman scattering microscopy. *Science* 322:1857–1861.

36. Ait-Belkacem, D., A. Gasecka, F. Munhoz, S. Brustlein, and S. Brasselet. 2010. Influence of birefringence on polarization resolved nonlinear microscopy and collagen SHG structural protein. *Opt. Express* 18:14859.

37. Campagnola, P. 2011. Second harmonic generation imaging microscopy: Applications to diseases diagnostics. *Anal. Chem.* 83:3224–3231.

38. Nadiarnykh, O., R. B. LaComb, M. A. Brewer, and P. J. Campagnola. 2010. Alterations of the extracellular matrix in ovarian cancer studied by second harmonic generation imaging microscopy. *BMC Cancer* 10:94.

39. Chu, S.-W., S.-Y. Chen, G.-W. Chern, T.-H. Tsai, Y.-C. Chen, B.-L. Lin, and C.-K. Sun. 2004. Studies of (2)/(3) tensors in submicron-scaled bio-tissues by polarization harmonics optical microscopy. *Biophys. J.* 86:3914–3922.

40. Ajeti, V., O. Nadiarnykh, S. M. Ponik, P. J. Keely, K. W. Eliceiri, and P. J. Campagnola. 2011. Structural changes in mixed Col I/Col V collagen gels probed by SHG microscopy: Implications for probing stromal alterations in human breast cancer. *Biomed. Opt. Express* 2:2307–2316.

41. Nadiarnykh, O., S. Plotnikov, W. A. Mohler, I. Kalajzic, D. Redford-Badwal, and P. J. Campagnola. 2007. Second harmonic generation imaging microscopy studies of osteogenesis imperfecta. *J. Biomed. Opt.* 12:051805.

42. Tai, S.-P., T.-H. Tsai, W.-J. Lee, D.-B. Shieh, Y.-H. Liao, H.-Y. Huang, K. Zhang, H.-L. Liu, and C.-K. Sun. 2005. Optical biopsy of fixed human skin with backward-collected optical harmonics signals. *Opt. Express* 13:8231–8242.

43. Fung, D. T., J. B. Sereysky, J. Basta-Pljakic, D. M. Laudier, R. Huq, K. J. Jepsen, M. B. Schaffler, and E. L. Flatow. 2010. Second harmonic generation imaging and Fourier transform spectral analysis reveal damage in fatigue-loaded tendons. *Ann. Biomed. Eng.* 38:1741–1751.

44. Roth, S. and I. Freund. 1981. Optical second-harmonic scattering in rat-tail tendon. *Biopolymers* 20:1271–1290.

45. Nadiarnykh, O. and P. J. Campagnola. 2009. Retention of polarization signatures in SHG microscopy of scattering tissues through optical clearing. *Opt. Express* 17:5794.

46. Morishige, N., Y. Takagi, T. Chikama, A. Takahara, and T. Nishida. 2011. Three-dimensional analysis of collagen lamellae in the anterior stroma of the human cornea visualized by second harmonic generation imaging microscopy. *Invest. Ophthalmol. Vis. Sci.* 52:911–915.

47. Han, M., G. Giese, and J. F. Bille. 2005. Second harmonic generation imaging of collagen fibrils in cornea and sclera. *Opt. Express* 13:5791–5797.

48. Hsueh, C. M., W. Lo, W. L. Chen, V. A. Hovhannisyan, G. Y. Liu, S. S. Wang, H. Y. Tan, and C. Y. Dong. 2009. Structural characterization of edematous corneas by forward and backward second harmonic generation imaging. *Biophys. J.* 97:1198–1205.

49. Olivier, N., M. A. Luengo-Oroz, L. Duloquin, E. Faure, T. Savy, I. Veilleux, X. Solinas et al. 2010. Cell lineage reconstruction of early zebrafish embryos using label-free nonlinear microscopy. *Science* 329:967–971.

50. Komai, Y. and T. Ushikif. 1991. The three-dimensional organization of collagen fibrils in the human cornea and sclera. *Invest. Ophthalmol. Vis. Sci.* 32:2244.

51. Plotnikov, S. V., A. C. Millard, P. J. Campagnola, and W. A. Mohler. 2006. Characterization of the myosin-based source for second-harmonic generation from muscle sarcomeres. *Biophys. J.* 90:693–703.

52. Fu, Y., H. Wang, R. Shi, and J. X. Cheng. 2007. Second harmonic and sum frequency generation imaging of fibrous astroglial filaments in ex vivo spinal tissues. *Biophys. J.* 92:3251–3259.

53. Geisler, N., E. Kaufmann, and K. Weber. 1985. Antiparallel orientation of the two double-stranded coiled-coils in the tetrameric protofilament unit of intermediate filaments. *J. Mol. Biol.* 182:173–177.

54. Zipfel, W. R., R. M. Williams, R. Christie, A. Y. Nikitin, B. T. Hyman, and W. W. Webb. 2003. Live tissue intrinsic emission microscopy using multiphoton-excited native fluorescence and second harmonic generation. *Proc. Natl. Acad. Sci. USA* 100:7075–7080.

55. Carriles, R., D. N. Schafer, K. E. Sheetz, J. J. Field, R. Cisek, V. Barzda, A. W. Sylvester, and J. A. Squier. 2009. Invited review article: Imaging techniques for harmonic and multiphoton absorption fluorescence microscopy. *Rev. Sci. Instrum.* 80:081101.

56. Yuste, R. 2005. Fluorescence microscopy today. *Nat. Methods* 2:902–904.

57. Moreaux, L., O. Sandre, and J. Mertz. 2000. Membrane imaging by second-harmonic generation microscopy. *J. Opt. Soc. Am. B* 17:1685–1694.

58. Cheng, J. X., Y. K. Jia, G. Zheng, and X. S. Xie. 2002. Laser-scanning coherent anti-Stokes Raman scattering microscopy and applications to cell biology. *Biophys. J.* 83:502–509.

59. Pudney, P. D. A., M. Melot, P. J. Caspers, A. Van Der Pol, and G. J. Puppels. 2007. An *in vivo* confocal Raman study of the delivery of trans-retinol to the skin. *Appl. Spectrosc.* 61:804–811.

60. Wang, H. W., T. T. Le, and J. X. Cheng. 2008. Label-free imaging of arterial cells and extracellular matrix using a multimodal CARS microscope. *Opt. Commun.* 281:1813–1822.

61. Wurpel, G. W. H., J. M. Schins, and M. Mueller. 2002. Chemical specificity in three-dimensional imaging with multiplex coherent anti-Stokes Raman scattering microscopy. *Opt. Lett.* 27:1093–1095.

62. Nan, X., E. O. Potma, and X. S. Xie. 2006. Nonperturbative chemical imaging of organelle transport in living cells with coherent anti-Stokes Raman scattering microscopy. *Biophys. J.* 91:728–735.

63. Wang, H., Y. Fu, P. Zickmund, R. Shi, and J. X. Cheng. 2005. Coherent anti-Stokes Raman scattering imaging of axonal myelin in live spinal tissues. *Biophys. J.* 89:581–591.

II

Biomedical Imaging Using SHG

5

Francesco Vanzi
European Laboratory for Non-Linear Spectroscopy (LENS)

University of Florence

Leonardo Sacconi
European Laboratory for Non-Linear Spectroscopy (LENS)

National Institute of Optics—National Research Council (INO—CNR)

Riccardo Cicchi
National Institute of Optics—National Research Council (INO—CNR)

Chiara Stringari
University of California, Irvine

Francesco S. Pavone
European Laboratory for Non-Linear Spectroscopy (LENS)

University of Florence

Molecular Structure and Order with Second-Harmonic Generation Microscopy

5.1 Introduction

Advances in imaging technologies drive a constant progress in our capability of probing structures and their dynamics within cells and tissues. The application of nonlinear spectroscopy to optical microscopy (Denk et al., 1990; Helmchen and Denk, 2005; Zipfel et al., 2003b) has led to new perspectives both in basic research and in the potential development of very powerful noninvasive diagnostic tools. Some of these techniques permit optical probing of biological functions (Dombeck et al., 2004; Skala et al., 2007; Svoboda and Yasuda, 2006; Zipfel et al., 2003a), as well as monitoring molecular structure and dynamics *in vivo* (Nucciotti et al., 2010). In this chapter, we review the properties of second-harmonic generation (SHG) and its application for the characterization of biological samples in terms of degree of molecular order, structural organization, and dynamics. The coherent nature of second-harmonic generated light (Campagnola and Loew, 2003; Moreaux et al., 2001) makes this optical process intrinsically sensitive to the angular distribution of the emitting elements in the focal volume, allowing both high-contrast imaging of ordered versus disordered structures and quantitative analysis of molecular orientation (Moreaux et al., 2000; Pons et al., 2003; Sacconi et al., 2005). Applications of these principles range from voltage-sensitive membrane imaging via exogenous labeling (Dombeck et al., 2005; Jiang et al., 2007; Millard et al., 2003; Moreaux et al., 2003; Nuriya et al., 2006; Sacconi et al., 2006a, 2008) to probing order and structural organization in tissues rich in intrinsic second-harmonic emitters such as collagen (Brown et al., 2003; Cox et al., 2003; Freund and Deutsch, 1986; Jain et al., 2003; Stoller et al., 2002; Williams et al., 2005), myosin (Both et al., 2004; Campagnola et al., 2002; Plotnikov et al., 2008; Plotnikov, 2006; Vanzi et al., 2006), tubulin (Campagnola et al., 2002; Dombeck et al., 2003;

Kwan et al., 2008). We describe a method for interpreting SHG anisotropy in terms of molecular conformation of the emitting proteins within a living tissue. Due to the properties of nonlinear spectroscopy, this method empowers SHG microscopy with the unique advantage of noninvasively probing molecular structure and order *in vivo*.

The description of molecular structural dynamics occurring in a living cell and determining its biology requires a variety of techniques to encompass the spatial and temporal scales involved. In fact, while x-ray crystallography provides static structures with atomic resolution, complementary techniques are necessary to probe structural dynamics in the cell. A good example of use of complementary approaches to the study of structure–function relationship is represented by skeletal muscle. In fact, due to its structural organization, muscle tissue has long been a sample of choice for the development and application of novel biophysical methodologies. Knowledge of the atomic structures of myosin and actin, combined with information from cryo-EM (Piazzesi et al., 2007; Reedy, 2000), electron paramagnetic resonance (Thomas et al., 2009), x-ray diffraction (Huxley, 2004), fluorescence polarization (Corrie et al., 1999), birefringence (Irving, 1993; Peckham et al., 1994) has led to a structural description of the chemo-mechanical energy conversion in terms of the lever-arm hypothesis of the working stroke in the myosin motor (Holmes, 1997; Rayment et al., 1993). In this chapter, we illustrate the capability of SHG to probe myosin molecular conformation *in vivo*.

SHG is a nonlinear second-order optical process occurring in systems without a center of symmetry and which have a large molecular hyperpolarizability. This condition is easily fulfilled at the molecular level by a moiety in which an electron donor is connected to an electron acceptor by a π-conjugated system. Further, only systems with non-centrosymmetric packing and organization of the dipoles present large second-order susceptibility and can generate SHG. Examples of such symmetries are biological membranes, where only one leaflet is stained with a donor–acceptor molecule.

In proteins, the intrinsic SHG sources lie within the amide bonds of polypeptide chains and the peptide bond C–N between two amino acids can be considered as the elementary dipole (Conboy and Kriech, 2003; Mitchell et al., 2005). Hence, the capability of a protein to generate SHG depends on its three-dimensional structure and folding which determine the degree of alignment of peptide bonds. In secondary structures like helices the peptide bonds are very well ordered and, therefore, their harmonophoric units generate coherent second-harmonic signal leading to a constructive interference. The tertiary structure of the protein can also influence the SHG efficiency because the helices inside the protein may be aligned or oriented in random directions. SHG signal is experimentally observed only from proteins which have ordered structure: myosin which is mainly constituted by α-helices; collagen which is constituted by a triple-helix and microtubules.

The characterization of the SHG physical basis will be described in the following sections and will demonstrate the exquisite sensitivity of SHG to order and molecular organization. Further, employing a full reconstruction of the contributing elementary SHG emitters at atomic scale, we provide a molecular interpretation of the SHG measurements in terms of structural conformation and degree of organization of the proteins *in vivo*.

In Section 5.2, we describe the physical principle underlying the production of SHG through coherent summation. This description represents the theoretical basis for all subsequent applications of SHG microscopy to the study of structural order and molecular conformation *in vivo*. Most applications rely on the polarization properties of SHG. Section 5.3 provides a full mathematical framework for the analysis of SHG polarization anisotropy data in terms of the orientation distribution of HRS emitters within the focal volume. Section 5.4 shows how the dependence of sensitivity of SHG on molecular order is exploited in a simple geometry provided by the membrane lipid bilayer leading to high contrast membrane imaging. The sensitivity of SHG to molecular order also leads to intrinsic signal from specific tissues characterized by ordered lattices of proteins; the most prominent examples are described in Section 5.5; the source of HRS within proteins is then described in Section 5.6. With the concepts illustrated in Sections 5.2 and 5.3, and the molecular origin of HRS described in Section 5.6, Section 5.7 provides an explanation of the methods through which SHG polarization anisotropy can be used to probe molecular order and conformation within living tissues.

5.2 From Hyper Rayleigh Scattering to SHG

In the description of the interaction between light and matter, the optical properties of a molecule are determined by its dipole moment ($\vec{\mu}$) that can be described as

$$\vec{\mu} = \vec{\mu}_0 + \alpha\vec{E} + \tfrac{1}{2}\beta\vec{E}\vec{E} + \cdots \tag{5.1}$$

where $\vec{E}$ is the driving electro-magnetic field, $\vec{\mu}_0$ is the permanent molecular dipole, α is the molecular polarizability that describes linear absorption and scattering of light, and β is a tensor describing the first hyperpolarizability term, responsible for nonlinear scattering processes known as hyper-Rayleigh scattering (HRS). The condition for a nonzero value of first hyperpolarizability (β) in a molecule is the presence of an asymmetry of charge distribution, due to an electron donor and electron acceptor moieties. Such asymmetry perturbs the electron oscillations in the dipole, introducing frequencies different from that of the driving field. Figure 5.1 provides a simplified illustration of these principles.

Owing to the lack of energy absorption in scattering processes, HRS is characterized by phase and energy conservation, leading to the possibility of coherent summation of the scattered waves from all the irradiated molecules. To describe such coherent summation, we start from the behavior of a pair of HRS emitters and illustrate the strong dependence of detectable HRS on the relative orientation of the emitters. Figure 5.2a shows two parallel emitters located within a distance smaller than the optical wavelength. When the molecules are irradiated with an electro-magnetic field, due to the alignment of their donor–acceptor moieties, they emit in-phase HRS photons that will constructively interfere. If, on the other hand, the two molecules are oriented antiparallel, as in Figure 5.2b, the opposition of their donor–acceptor moieties produces out-of-phase HRS photons that will destructively interfere. Based on this simple example, it

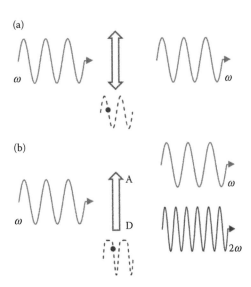

FIGURE 5.1 Basic principle of hyper-Rayleigh scattering. (a) The figure shows a symmetric dipole interacting with an incoming electric field oscillating at a frequency ω. The induced electron oscillation (shown below the dipole) follows the frequency and phase of the incident field, producing Rayleigh scattering at the same frequency ω. (b) The asymmetric nature of the dipole, highlighted by its donor (D) and acceptor (A) moieties, produces an asymmetry in the induced electron oscillation, introducing a component at frequency 2ω in the scattered light. This component is termed hyper-Rayleigh scattering (HRS). (Inspired from Mertz, J. P.S.B.R. Masters, ed. Applications of second-harmonic generation microsocpy. In *Handbook of Biomedical Nonlinear Optical Microscopy*, 2008. Oxford: Oxford University Press. By permission of Oxford University Press.)

(a)

(b)

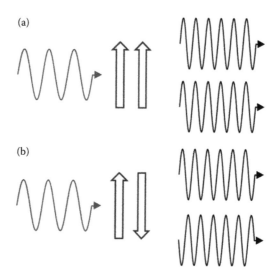

FIGURE 5.2 Dipole orientation and coherent summation. The figure shows examples of the HRS waves produced by two pairs of irradiated dipoles. The HRS wave generated by each dipole is represented on the right. (a) If the molecules are parallel, their HRS waves are in phase with the driving field and can interfere constructively. (b) If the molecules are antiparallel, their HRS waves have opposite phases and interfere destructively. (Inspired from Mertz, J. P.S.B.R. Masters, ed. Applications of second-harmonic generation microsocpy. In *Handbook of Biomedical Nonlinear Optical Microscopy*, 2008. Oxford: Oxford University Press. By permission of Oxford University Press.)

is clear that, considering a population of N molecules, the resulting signal is not simply the sum of the scattered powers from each molecule but rather the result of their overall interference, which is determined by the orientation distribution within the population. Therefore, the resulting HRS power strongly depends on the angular distribution of the population. In the extreme case of randomly oriented molecules, the HRS contributions have random phases, and the total HRS signal, produced incoherently, will scale as the number of radiating molecules N. On the other hand, if the molecules are well aligned, the phases of the individual HRS contributions are locked to the phase of the driving field. The coherent production of HRS occurring in the latter case is called second-harmonic generation (SHG) and its total power scales as N^2. A quantitative description of these processes is provided in the following section.

The dependence of signal on isotropic versus anisotropic HRS emitters distributions is the basis for high-contrast imaging of ordered structures. Moreover, the anisotropic distribution of the scatterers is reflected in coherent summation producing a polarized SHG signal. Quantitative measurements of SHG polarization anisotropy can, therefore, provide information on the scatterers distribution itself.

5.3 SHG Polarization Anisotropy

The general description of the polarization P in a medium is given by

$$P = \chi^{(1)}\vec{E} + \chi^{(2)}\vec{E}\vec{E} + \cdots \tag{5.2}$$

This equation is the bulk equivalent of Equation 5.1. The second-order susceptibility $\chi^{(2)}$ describes the relation between all the components of the polarization and the electric-field vectors for sum and difference frequencies generation. In the general case of three-wave mixing, the susceptibility tensor $\chi^{(2)}_{ijk}(\omega_1, \omega_2)$ is a third-rank tensor and contains 27 $(3 \times 3 \times 3)$ elements. In the specific case of SHG, in which two fields with the same frequency $(\omega_1 = \omega_2 = \omega)$ generate a third field with frequency 2ω, each component of the second-order polarization can be expressed as

$$P_i^{(2)}(2\omega) = \sum_{j,k} \chi_{ijk}^{(2)}(\omega,\omega)E_j(\omega_1)E_k(\omega_2) = \sum_{k,j} \chi_{ikj}^{(2)}(\omega,\omega)E_k(\omega_1)E_j(\omega_2)$$

$$= \sum_{j,k} \chi_{ijk}^{(2)}(\omega,\omega)E_j(\omega)E_k(\omega) \tag{5.3}$$

Therefore, the susceptibility tensor has the following symmetry:

$$\chi_{iik}^{(2)}(\omega,\omega) = \chi_{iki}^{(2)}(\omega,\omega) \tag{5.4}$$

With these symmetries, the number of independent elements of the tensor decreases to 18. Hence, the second-order induced polarization can be written as a function of the components of the tensor $\chi_{ijk}^{(2)}$ and of the electric field E_i as follows:

$$\begin{pmatrix} P_x^{(2)} \\ P_y^{(2)} \\ P_z^{(2)} \end{pmatrix} = \begin{pmatrix} \chi_{xxx}^{(2)} & \chi_{xyy}^{(2)} & \chi_{xzz}^{(2)} & \chi_{xyz}^{(2)} & \chi_{xxz}^{(2)} & \chi_{xxy}^{(2)} \\ \chi_{yxx}^{(2)} & \chi_{yyy}^{(2)} & \chi_{yzz}^{(2)} & \chi_{yyz}^{(2)} & \chi_{yxz}^{(2)} & \chi_{yxy}^{(2)} \\ \chi_{zxx}^{(2)} & \chi_{zyy}^{(2)} & \chi_{zzz}^{(2)} & \chi_{zyz}^{(2)} & \chi_{zxz}^{(2)} & \chi_{zxy}^{(2)} \end{pmatrix} \cdot \begin{pmatrix} E_x^2 \\ E_y^2 \\ E_z^2 \\ 2E_y E_z \\ 2E_x E_z \\ 2E_x E_y \end{pmatrix} \tag{5.5}$$

The susceptibility tensor $\chi_{ijk}^{(2)}$ can be expressed in the HRS molecule's system of coordinates (x', y', z') calculating each tensor component by the following rotation:

$$\chi_{ijk}^{(2)} = \sum_{i'j'k'} \cos\varphi_{ii'}\cos\varphi_{jj'}\cos\varphi_{kk'}\chi_{i'k'j'}^{(2)} \tag{5.6}$$

where $\varphi_{ii'}$ is the angle between the i and i' axes. Based on Equation 5.6, bulk susceptibility can be calculated from the HRS emitter hyperpolarizability $\beta_{ij'k'}$ as

$$\chi_{ijk}^{(2)} = \sum_{i'j'k'} \left\langle \cos\varphi_{ii'}\cos\varphi_{jj'}\cos\varphi_{kk'} \right\rangle \beta_{i'k'j'} \tag{5.7}$$

where the brackets indicate averaging over all the emitters. Considering HRS emitters with a single preferred axis of hyperpolarizability (as expected for push–pull resonance, see below) and defining the molecular system of coordinates with the y' axis coinciding with the hyperpolarizability axis, the only nonzero component of β is $\beta_{y'y'y'}$. As described in detail in the following sections, biologically relevant SHG-emitting samples are characterized by a distribution of HRS emitters with cylindrical symmetry. We define the laboratory system of coordinates (x, y, z) with the y-axis along the axis of cylindrical symmetry. Under the assumption that, within the cylindrical symmetry, the emitters are oriented at a fixed polar angle ϑ with respect to the symmetry axis (see Figure 5.3a), computation of the tensor components using Equation 5.7 produces the following nonzero components:

$$\chi_{yyy}^{(2)} = N\beta \cos^3 \vartheta$$

$$\chi_{yxx}^{(2)} = \chi_{xxy}^{(2)} = \chi_{yzz}^{(2)} = \chi_{zzy}^{(2)} = \frac{N}{2}\beta \cos\vartheta \sin^2 \vartheta \tag{5.8}$$

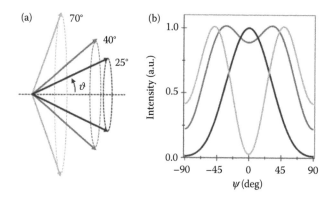

FIGURE 5.3 Dependence of SHG polarization anisotropy on polar angle in cylindrically symmetric sample. (a) Schematic representation of HRS distributed on the surface of a cone with an aperture angle ϑ. Different angles are coded in grayscale. (b) SHG polarization anisotropy (SPA) generated by the cylindrical distribution of HRS emitters in (a) with three different angles: $\vartheta = 25°$ (black), $\vartheta = 40°$ (dark gray), $\vartheta = 70°$ (light gray). The intensity of SHG is represented as a function of the angle ψ between the laser polarization and the cone axis (see panel a).

The second-order susceptibility tensor, therefore, can be written as

$$\chi^{(2)} = \begin{pmatrix} 0 & 0 & 0 & 0 & 0 & \chi_{xxy}^{(2)} \\ \chi_{yxx}^{(2)} & \chi_{yyy}^{(2)} & \chi_{yzz}^{(2)} & 0 & 0 & 0 \\ 0 & 0 & 0 & \chi_{zyz}^{(2)} & 0 & 0 \end{pmatrix} \qquad (5.9)$$

Considering an electric field propagating along the z-axis and linearly polarized at an angle ψ with respect to the y-axis (axis of cylindrical symmetry of the sample):

$$\vec{E} = E\sin\psi\,\hat{e}_x + E\cos\psi\,\hat{e}_y \qquad (5.10)$$

substituting Equation 5.9 into Equation 5.5, the second-order polarization can be written as

$$\vec{P}^{(2)} = 2E^2\sin\psi\cos\psi\,\chi_{yxx}^{(2)}\hat{e}_x + (E^2\sin^2\psi\chi_{yxx}^{(2)} + E^2\cos^2\psi\chi_{yyy}^{(2)})\hat{e}_y \qquad (5.11)$$

The intensity of SHG (I_{SHG}) is proportional to the square of the second-order polarization:

$$I_{SHG} \propto (\vec{P}^{(2)})^2 = E^4(\chi_{yxx}^{(2)})^2\left[\sin^2 2\psi + (\sin^2\psi + \frac{\chi_{yyy}^{(2)}}{\chi_{yxx}^{(2)}}\cos^2\psi)^2\right] \qquad (5.12)$$

The simple case illustrated in Figure 5.2a (extended to N molecules) can be described by setting the polar angle ϑ to zero so that

$$I_{SHG} \propto E^4 N^2 \beta^2 \cos^4\psi \qquad (5.13)$$

This equation provides a quantitative description of the constructive interference on the basis of SHG described in the previous section.

In general, Equation 5.12 provides the basis for using SHG measurements to assess the structural distribution of emitters in a sample. In fact, if I_{SHG} is measured as a function of the laser polarization angle ψ, the resulting SHG polarization anisotropy (SPA) data can be fitted with Equation 5.12 in the following form:

$$I_{SHG}(\psi) \propto \sin^2 2\psi + (\sin^2\psi + \gamma\cos^2\psi)^2 \tag{5.14}$$

with

$$\gamma \equiv \frac{\chi^{(2)}_{yyy}}{\chi^{(2)}_{yxx}} = \frac{N\beta\cos^3\vartheta}{(N/2)\beta\cos\vartheta\sin^2\vartheta} = \frac{2}{\tan^2\vartheta} \tag{5.15}$$

As an example, Figure 5.3b shows three different I_{SHG} profiles for samples characterized by different values of the angle ϑ. Therefore, SPA data can be used to access information on the structural distribution of HRS emitters in the sample.

5.4 Membrane SHG Imaging

For imaging applications, it is therefore highly desirable that the radiating molecules be organized rather than unorganized because much more signal is generated for the same number of molecules. As indicated above, the plasma membrane provides a plane for the ordered distribution of those HRSs which possess the physical property of embedding into the membrane bilayer itself. A molecular design strategy for obtaining this condition is shown in Figure 5.4: hydrocarbon (hydrophobic) side chains are grafted at one end of the dye and a polar head group at the other. The HRS capability, on the other hand, is conferred by a push–pull design with donor and acceptor moieties spanned by a uni-axial charge transfer path.

When perfused onto a lipid bilayer, the dye molecules insert themselves into the membrane with the same orientation: the hydrophobic side chains as far as possible from the aqueous surroundings, that is, directed into the membrane interior; and the polar head group pointing outward, into the aqueous surroundings (see Figure 5.5).

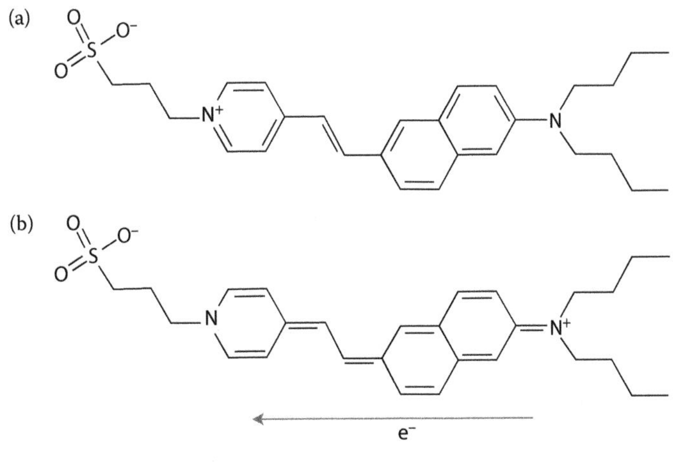

FIGURE 5.4 Chemical structure of the voltage-sensitive dye, di-4-ANEPPS. The dye is shown in two configurations (panels a and b) differing for charge distribution which is transferred intra-molecularly by a global shift of all the π-electron bonds from the donor to the acceptor ends of the molecule. The structure also shows the polar head and hydrophobic tail conferring to the molecule its ability to partition inside the membrane.

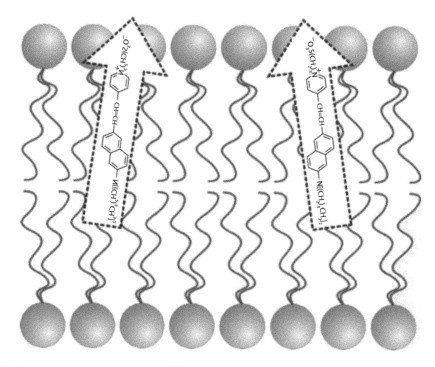

FIGURE 5.5   Insertion geometry of amphiphilic dye into the plasma membrane. The dye molecule is perfused extracellularly (top side) leading to insertion of the dye molecule into the outer lipid leaflet. Notice the alignment of the embedded molecules about the axes perpendicular to the membrane plane. (With kind permission from Springer Science+Business Media: *Cell Biochem Biophys*, Cell imaging and manipulation by nonlinear optical microscopy, 2006b, *45*, 289–302, Sacconi, L. et al.)

Significant energy is required to flip a molecule from one leaflet of the bilayer to the other, since such passage requires traversing of the polar head group through the nonpolar membrane interior. Such a process, although kinetically slow, does occur and is known as "flip-flop." Figure 5.6 shows two-photon fluorescence (TPF) and SHG images of a cell stained with a membrane HRS dye (RH 237). It is apparent from the TPF image that some dye molecules have been internalized into the cell cytoplasm (due to flip-flop, followed by diffusion of the dye inside the cell). The internalized dye molecules are randomly

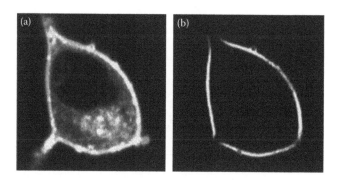

FIGURE 5.6   Membrane contrast of SHG signal in living cell. SY5Y cell labeled with RH237 membrane dye. (a) Two-photon fluorescence (TPF) image. (b) SHG image. (With kind permission from Springer Science+Business Media: *Cell Biochem Biophys*, Cell imaging and manipulation by nonlinear optical microscopy, 2006b, *45*, 289–302, Sacconi, L. et al.)

oriented and, therefore, they do not produce coherent SHG, resulting in a high-contrast SHG image, in which only the cell membranes are visible.

This mechanism of high-contrast imaging has been extensively used, in combination with voltage-sensitive dyes, to achieve improved sensitivity in the optical measurement of membrane potential (Stuart and Palmer, 2006; Wilt et al., 2009). In fact, a strong limitation of conventional membrane imaging techniques derives from the reduction in the measured dye response to membrane potential variations due to molecules that are in the focal volume but are not embedded in the plasma membrane, thus producing background. Variations in this background have the effect of making the observed signal difficult to quantify in terms of membrane potential. Because of the molecular alignment required for SHG, signal only emanates from properly oriented dye molecules in the membrane; thus, its effective membrane potential response is not significantly attenuated by background, allowing for a more quantitative relation between optical signal changes and membrane potential variations.

Beyond this fundamental background reduction, coherent summation also provides an order-based mechanism for membrane potential sensing. In fact, two mechanisms contribute to the voltage sensitivity: an electro-optic-induced alteration of the molecular hyperpolarizability and an electric-field-induced alteration of the degree of molecular alignment (Moreaux et al., 2003; Pons et al., 2003). The latter was experimentally characterized by SPA measurements showing that, with increasing electric field across the membrane, the embedded dipolar molecules reduce their angular dispersion, increasing the coherent summation.

The extreme sensitivity of SHG to dipole alignment in the membrane was demonstrated also by the disappearance of SHG signal upon fertilization in sea urchin eggs stained with di-8-ANEPPS (Millard et al., 2005): the production of micro-villi in the membrane, in fact, breaks the regular arrangement of the dye molecules in the plane of the membrane itself, leading to a dramatic reduction of SHG.

5.5 Endogenous SHG Imaging

As described above, the condition for a nonzero value of first hyperpolarizability in a molecule is the presence of an asymmetry of charge distribution spanned by a uni-axial charge transfer path.

Many biological molecules have this property, leading to the interesting prospective that biological samples may be rich of endogenous HRS emitters. In fact, the possibility of observing HRS in biomolecules was demonstrated more than 40 years ago, when SHG from amino acid crystals was detected (Rieckhoff, 1965). The ubiquitous distribution of amino acids in biological samples, therefore, insures the presence of HRS emitters in every cell and tissue. However, as illustrated above, a proper spatial organization of the HRS emitters is required for SHG, so that endogenous signal could arise, for example, in biological samples with semi-crystalline organization.

The first SHG biological imaging was reported by Freund et al. on connective tissue (Freund and Deutsch, 1986). In this tissue, in fact, the structural organization of collagen in fibrils and fibers disposes the HRS emitters in a lattice leading to coherent summation, that is, to SHG. Figure 5.7 shows SHG images of rat-tail tendon collagen taken at the dawn of SHG microscopy (panel a) and today (panel b).

Owing to the peculiarity of SHG (i.e., label-free micron-scale resolution in deep tissue), this imaging technique has tremendous potential for biomedical applications, ranging from morphological characterization of healthy and pathological connective tissue *in vivo* (Brown et al., 2003; Cicchi et al., 2007; Cicchi et al., 2010; Han et al., 2008; Wang et al., 2007) to quantitative measurement of fibril orientation within the pixel size (Stoller et al., 2002; Williams et al., 2005). Collagen type I is the most abundant protein in mammals, featuring a hierarchical structure ranging from the atomic and molecular scale to the macroscopic scale. It consists of tropo-collagen molecules of an approximate length of 280 nm and a diameter of 1.5 nm, staggered side-by-side to form fibrils. Fibrils can be considered the unitary structures constituting collagen. They are approximately 1 µm in length and 30 nm in diameter and they hierarchically organize themselves in bigger fibers or sheets in order to form collagen fiber bundles in skin dermis or lamellae in cornea.

(a)   (b)

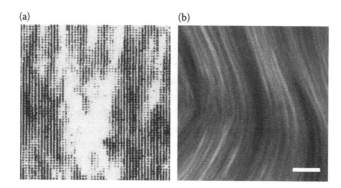

FIGURE 5.7 SHG from rat-tail tendon collagen. (a) First coherent second-harmonic image of rat-tail tendon. (Modified from Freund, I. and Deutsch, M. 1986. Second-harmonic microscopy of biological tissue. *Opt Lett 11*, 94. With permission of Optical Society of America.) (b) Contemporary image of rat-tail tendon collagen. (Modified from *Biophys J*, 88, Williams, R.M. Zipfel, W.R., and Webb, W.W., Interpreting second-harmonic generation images of collagen I fibrils, 1377, Copyright (2005), with permission from Elsevier.) Scale bar 10 μm. For comparison, the image size is 2.5×2.5 mm^2 and 56×56 μm^2 for panels a and b, respectively.

The sensitivity of SHG to molecular structural organization allows probing the molecular organization of collagen, as demonstrated by studies that monitored thermally induced conformational changes of collagen (Lin et al., 2005; Matteini et al., 2009; Sun et al., 2006; Theodossiou et al., 2002): generally it was observed that above 54°C the measured SHG intensity decreases in agreement with the thermal denaturation of collagen, which is known to occur at about 55°C. As illustrated above, additional information about molecular organization of collagen can be provided by SHG polarization anisotropy (SPA) measurements. For example, the loss of organization of the corneal collagen lattice induced by photo-thermal effects was investigated in porcine cornea treated with low-power laser irradiation or bulk heating. SPA measurements probe the changes in the structural anisotropy of fibrillar hierarchical levels, indicating that collagen thermal denaturation occurs as a sequence of three main events: interfibrillar misalignment (taking place at temperatures below the denaturation threshold and probably due to damage of the collagen supramolecular assembly); variation in the helix angle (taking place at temperatures above 60°C and ascribed to molecular unfolding); complete loss of intermolecular cross-links (taking place at higher temperatures and leading to complete loss of SHG signal).

The possibility to monitor collagen thermal modifications is an important issue in biomedical optics. In fact, several laser-based treatments (such as corneal thickening, vascular treatment, and skin rejuvenation) can cause collateral thermal damage. The response of collagen to heating has been studied using different methods (including differential scanning calorimetry, x-ray diffraction, NMR, and spectroscopy); however, SHG microscopy has the unique advantage to be performed *in vivo*.

Strong SHG signal was also detected in skeletal muscle (Both et al., 2004; Plotnikov, 2006). The signal arising from muscle tissue displays a striking alternation of bright and dark bands, similar to what was observed in transmission or polarization microscopy. Figure 5.8 shows an SHG image of a muscle fiber. The possibility of imaging sarcomeres by SHG was used to measure sarcomere length with 20 nm resolution in living cardiac myocytes (Boulesteix, 2004).

More generally, due to the advantage conferred by nonlinear microscopy, the possibility of imaging muscle *in vivo* (Brown et al., 2003; Llewellyn, 2008) with 3D capabilities holds great promise for the development of biomedical diagnostic tools for muscular pathologies involving alterations and/or loss of sarcomeric structure (Plotnikov et al., 2008; Ralston, 2008).

The organization of most muscle proteins in helical filaments and the distribution of such filaments in cylindrically symmetric, repetitive structures along the fiber clearly represent an ideal structural configuration to give rise to SHG, provided that one or more of the molecule constituents of muscle is

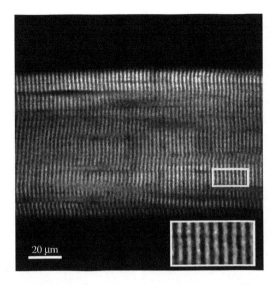

FIGURE 5.8 SHG image from skeletal muscle. SHG image of a single demembranated rabbit psoas fiber in rigor (sarcomere length = 2.4 μm). The bright bands in the image correspond to sarcomeric A bands. The inset shows, at 2.5× magnification, the area highlighted. (Modified from Nucciotti, V. et al. 2010. *Proc Natl Acad Sci USA 107*, 7763–7768.)

characterized by a high HRS. The possible molecular origin of SHG in the muscle has been studied in several types of preparations; the first interesting indication came from the colocalization of the bright SHG bands with the fluorescence from GFP-myosin heavy chain in nematode muscles (Campagnola et al., 2002). Using mouse myofibrils, Plotnikov et al. (2006) showed that SHG signal does not colocalize with either α-actinin (immunostained) or actin (labeled at the ends with rhodamine–phalloidin). Another measurement relevant for the identification of SHG source consists in the extraction of myosin: the extraction drastically reduced the SHG intensity of the bright bands, while not altering overall actin organization in the sarcomere as imaged by two-photon florescence of rhodamine–phalloidin (see Figure 5.9). These results suggested that the myosin filaments are necessary for the SHG signal but could not resolve whether myosin alone could produce SHG or, rather, the myosin/actin lattice would be the more likely source (Chu et al., 2004). This issue is addressed by the dependence of the width of the SHG bright band on the length of the sarcomeres (Plotnikov, 2006). In a range of lengths in which the degree of overlap between actin and myosin filaments changes linearly with the sarcomere length, the width of SHG bright band does not show significant changes.

This result strongly supports the hypothesis that SHG arises entirely from the myosin-containing band and does not require an overlap with actin filaments. On the other hand, myosin is the molecular motor undergoing the structural changes responsible for muscle contraction and SPA measurements can yield information on HRS orientation (via a geometrical parameter γ, as described above).

Therefore, the muscle tissue represents an excellent sample for testing the capabilities of SPA to monitor molecular conformation *in vivo*. Clearly, the possibility of using SHG for the measurement of myosin structural conformation relies on the location of the HRS emitters within the myosin molecule. In fact, this protein is constituted by a passive portion and a catalytically active actin-binding ATPase (S1). Figures 5.10a,b show a direct comparison of SPAs from muscle at rest and during contraction. In this experiment line-scan SPA measurement (Figure 5.10a) was needed due to the limited duration ($\approx$1 s) allowed for tetanic contraction (trace in Figure 5.10a). Figure 5.10b shows a clear difference in the SPA profiles between resting and active, yielding a dependence of γ on the physiological state: $\gamma_{rest} = 0.30 \pm 0.03$ (mean $\pm$ std, $n = 7$); $\gamma_{act} = 0.64 \pm 0.02$ ($n = 7$). Therefore, SPA results indicate

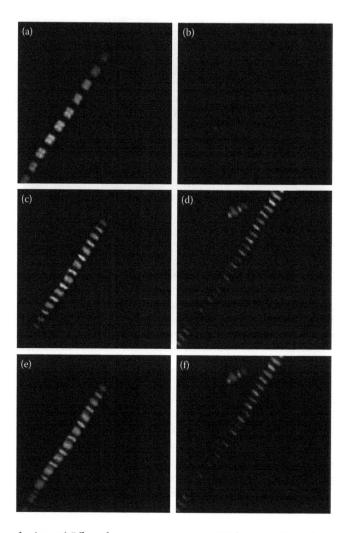

FIGURE 5.9 **(See color insert.)** Effect of myosin extraction on SHG from myofibrils. Panels a and b show myo-fibril SHG images before and after myosin extraction, respectively. Panels c and d show two-photon excitation images of actin labeled with rhodamine-phalloidin before and after myosin extraction, respectively. Panels e and f show a superposition of SHG image (in red) and TPE image (in green) before and after the extraction. The size of the images is 20×20 μm^2. (Modified from Vanzi, F. et al. 2006. *J Muscle Res Cell Motil 27*, 469–479. With permission.)

that the structural changes occurring in myosin S1 during force production can be monitored by SHG. During contraction, the fraction of myosin molecules which interact with actin to generate force can be modulated by changing sarcomere length and, thus, the degree of overlap between actin and myosin filaments. Figure 5.10c shows the dependence of γ on sarcomere length at rest (light gray circles) and at the plateau of isometric contraction (dark gray circles). In the sarcomere length range between 2.2 and 3.6 μm, the fraction of cross-bridges that can attach to actin decreases linearly, as indicated also by the tetanic isometric force (black squares and lines in Figure 5.10c). The γ_{act} data track remarkably the force behavior, confirming that SPA is sensitive to the fraction of attached myosin heads. As such fraction goes to zero at no overlap (sarcomere length > 3.6 μm), γ_{act} approaches γ_{rest}.

Force generation during active contraction occurs in the acto-myosin cross-bridges, which sample different biochemical/structural states during the myosin ATPase cycle. By depleting ATP from the muscle fiber, it is possible to induce a static and structurally homogenous state (the rigor state) in which

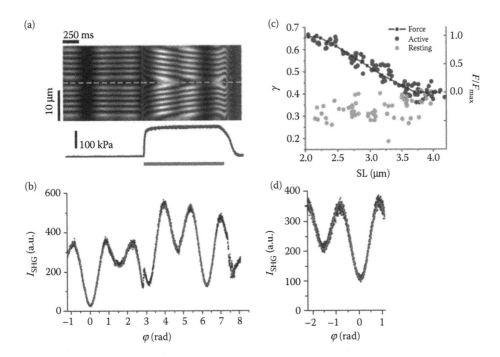

FIGURE 5.10 SHG polarization anisotropy (SPA) in intact muscle fibers. (a) Line scan measurement of SHG intensity at rest and during tetanic contraction. The dashed gray line shows the point of inversion of scanning direction. The trace shows the force recording and the dashed line indicates the period of stimulation. (b) SPA measurement: SHG intensity was measured from integration over each line of the scan. The best fit of Equation 5.14 to a portion of data covering a φ range of 2π for the resting (fitted $\gamma_{rest} = 0.351 \pm 0.005$) and the active state (fitted $\gamma_{act} = 0.664 \pm 0.005$) is plotted but not visible due to perfect overlap with the experimental data. Panels a and b share the same x-axis. (c) Measurements of γ versus sarcomere length (SL) in intact fibers at rest (light gray circles) and during contraction (dark gray circles). The black squares and line represent the normalized force (right axis). (d) SPA measurement in rigor (fitted $\gamma_{rig} = 0.691 \pm 0.004$). (Modified from Nucciotti, V. et al. 2010. *Proc Natl Acad Sci USA 107*, 7763–7768.)

all myosin heads are attached to actin in a well-characterized conformation, corresponding to the end of the power stroke. The measurement of SPA in rigor (Figure 5.10c, sarcomere length 2.2 μm) produces a value ($\gamma_{rig} = 0.74 \pm 0.02$, $n = 33$) statistically different ($p < 0.001$) from γ_{act}, further demonstrating that SPA monitors the structural conformation of the acto-myosin cross-bridge. The molecular modeling described in Section 5.7 will show how the SPA results can be used to infer information on molecular conformation.

Another fundamental issue in the successful application of this technique to muscle pertains to the spatial resolution. In this respect, a strong difference is observed between myofibrils and intact fibers. In detail, SHG imaging is characterized by a much higher spatial resolution in myofibrils than in fibers (see Figure 5.11).

The most likely explanation for this difference in resolution is that in myofibrils the diffraction-limited spot of the laser elicits SHG from a spatially coherent population of molecules, whereas in fibers myofibrils with different spatial phases are irradiated by the spot causing a loss of resolution. In myofibrils, a dark zone is visible in the middle of the bright bands (see Figures 5.11b,d). From sarcomere ultra-structure, it is known that the center of the myosin-based A-band is characterized by the juxtaposition of myosin filaments with opposing polarities, so that HRS emitters located within myosin molecule are oriented in opposing senses, leading to the destructive interference of HRS signal.

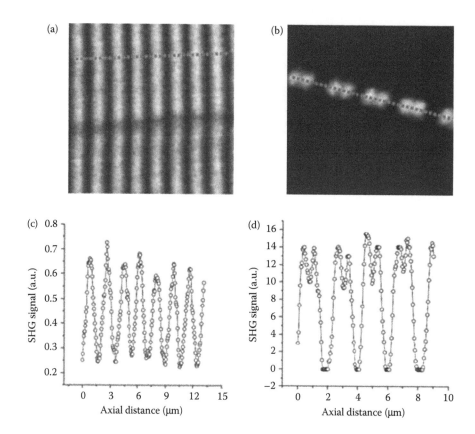

FIGURE 5.11 SHG spatial resolution. Endogenous SHG in single fiber from frog skeletal muscle (a) and in single myofibril from rabbit skeletal muscle (b) in rigor solution at sarcomere length of 2.2 μm. Image dimensions are 15×15 μm² in fiber and 9×9 μm² in myofibril. (c) SHG intensity profile along the axis of the fiber and of the myofibril (d). In the myofibril, the contours of the bands are better resolved and a dark zone is evident in the middle of the bright band. (Modified from Vanzi, F. et al. 2006. *J Muscle Res Cell Motil 27*, 469–479.)

An analogous effect of relative orientation of polarized polymers on detectable SHG was observed for microtubules. Endogenous SHG was observed in microtubules of the mitotic spindle (Campagnola et al., 2002; Dombeck et al., 2003) (see Figure 5.12a) and in neuronal axons (Dombeck et al., 2003; Kwan et al., 2008) (see Figure 5.12b).

The polarity of microtubules forms the basis for SHG. In fact, due to the geometrical symmetry of the microtubule itself, the HRS scatterers must be distributed with cylindrical symmetry and oriented with respect to the microtubule axis. When microtubules align both in direction and sense (as in axons and mitotic spindles), coherent summation takes place and SHG is detected. When, on the other hand, microtubules are disposed with opposing polarities (as in dendrites) destructive interference among HRS emitters cancels the signal out.

5.6 Source of the Endogenous SHG

Characterization of SPA data from both muscle and collagen has indicated that the average polar angle ϑ (see Equation 5.15) corresponds to the pitch angle of the helices in these samples (Bella et al., 1994; Isaac Freund, 1986; Tiaho et al., 2007), suggesting the HRS emitters of these proteins are located within the helical structure itself. In agreement with this observation, previous studies suggest that protein HRS lie within the amide groups (HN–CO) of polypeptide chains (Conboy and Kriech, 2003; Mitchell et al., 2005).

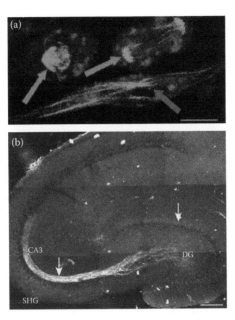

FIGURE 5.12 **(See color insert.)** SHG from microtubules. (a) SHG in RBL cells. SHG (green) arises from mitotic spindles (orange arrows) and from interphase microtubule ensembles (blue arrow). Scale bar: 10 μm. (b) SHG in hippocampal brain acute slice. SHG arises from the dense mossy fiber axon bundle between the dentate gyrus (DG) and CA3 rea of the hippocampus. Scale bar: 200 μm. (Modified from Dombeck, D.A. et al. 2003. *Proc Natl Acad Sci 100*, 7081–7086.)

Molecules with π-electron donors and acceptors exhibiting intramolecular charge transfer between the two groups show large nonlinear second-order optical susceptibility (Lalama and Garito, 1979). The amide HRS is due to the partial charge transfer in the peptide bond due to two resonance forms, as shown in Figure 5.13. As a consequence of this resonance, all peptide bonds are found to be almost planar.

SHG has been experimentally observed in polypeptide α-helix (Mitchell et al., 2005). An α-helix is a tightly coiled structure, which has an average of 3.6 amino acids per turn, pitch of 5.5 Å, and radius of 2.2 Å. An individual α-helix, therefore, is characterized by cylindrical symmetry with all HRS emitters tilted at a fixed polar angle with respect to the helical axis (the same geometry described in Section 5.3). The generation of SHG signal through coherent summation requires an anisotropic distribution of the scatterers. Proteins characterized by randomly oriented α-helices do not fulfill such anisotropy and are not expected to be good SHG emitters. On the other hand, proteins with a high degree of alignment of their α-helices should produce coherent summation. In other words, a first level of order (required for constructive interference) is achieved by organization of peptide bonds in a helical pattern; however, a second level of order is also necessary, consisting of substantial alignment of the helices themselves in the protein.

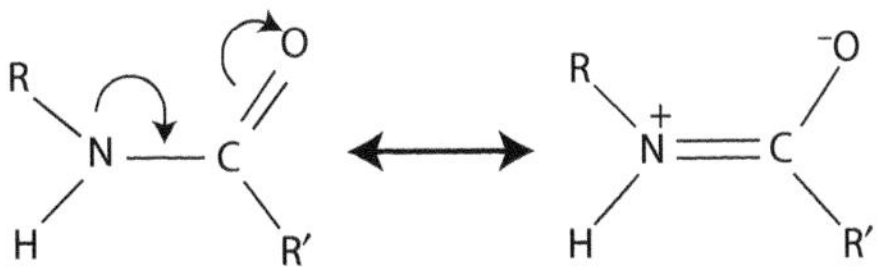

FIGURE 5.13 Resonance structure of the peptide bond. Delocalization of π-electrons allows charge transfer between the donor N and acceptor O.

Considering for example the two main constituents of muscle (i.e., myosin and actin), the α-helices present in G-actin display an orientational dispersion preventing SHG, while myosin is endowed with some extraordinarily long α-helices highly aligned, especially in the tail portion. Clearly, a single protein would produce too low an intensity of SHG to be detected. Thus, a third level of structural organization is required in which SHG-emitting proteins are arranged with a symmetry allowing further summation of the signal up to a detectable level. These considerations provide basis of interpretation for the experimental observation of SHG only in specific samples such as collagen and myosin in muscle.

The identification of the C–N peptide bond as the main HRS emitter in proteins allows applying the theory described in Section 5.3 for probing protein structural conformation through SPA measurements. In the case of proteins characterized by sequence repeats with amino acids containing methylene groups (e.g., proline), this additional element of resonance should be considered (Rocha-Mendoza et al., 2007). For example, in collagen (rich of the -ProHypGly- repeat), the second-order susceptibility arises mainly from peptide groups in the backbone (Deniset-Besseau et al., 2009), but also from the symmetric stretch of the methylene groups in the side chain. Analysis of collagen SPA data showed that the helical pitch angle estimated including methylene groups resonance more closely agrees (Su et al., 2011) with the known pitch angle of 45.3°. The analysis of large conformational changes in a protein (see the next section), on the other hand, can be satisfactorily conducted with the simplified assumption of all HRS emitters residing within the amide group.

5.7 Probing Protein Structural Conformation

Knowledge of the atomic structure of a protein allows placing all its HRS emitters in space so that the bulk second-order susceptible tensor ($\chi^{(2)}$) can be calculated using Equation 5.7. According to the previous section, we can assume that all C–N HRS emitters (Figure 5.14a) have the same nonlinear hyperpolarizability tensor, characterized by $\beta_{y'y'y'}$ as the only nonzero component in the $x'\, y'\, z'$ molecular reference system. Since all biological samples capable of SHG emission are characterized by a cylindrically symmetric distribution of their protein constituents, the theoretical framework developed in Section 5.3 can be used to experimentally extract structural information in terms of the factor γ from SPA data (see Equation 5.14). On the other hand, γ can also be calculated from the computed $\chi^{(2)}$ derived from an atomic model of the protein conformation (see Equations 5.7 and 5.15), so that one can calculate

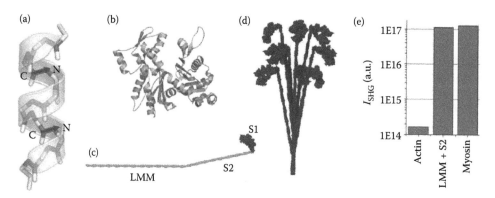

FIGURE 5.14 Structural modeling. (a) Example of location of two C-N HRSs in α-helix. (b) Atomic structure of an actin monomer in ribbon representation. (c) Atomic structure of myosin molecule with light meromyosin (LMM), S2 and the S1 globular head shown. (d) Model of the nine-myosin motif which repeats along the thick filament. The figure shows S1, S2, and only a small segment of LMM, with the fiber axis oriented vertically. (e) Computed SHG intensities. The relative contributions of actin and myosin were calculated modeling the acto-myosin array inside the excitation volume. The I_{SHG} ratio between full and S1-less myosin is 1.11.

the different values of γ associated with different protein conformations. A direct comparison of the experimentally measured γ with the theoretically computed ones allows determining which modeled protein structure is most representative of the conformation inside the tissue under investigation.

This approach can be used to interpret the muscle fibers SPA measurements shown in Figure 5.10 in terms of myosin conformation. The atomic structures of actin (Figure 5.14b) and myosin (Figure 5.14c), their polymeric organization in filaments, and the overall sarcomeric ultrastructure are known. In detail, the atomic-resolution structure of full-length myosin can be reconstructed using the atomic coordinates from the Proteins Data Bank: α-helix coiled-coil LMM and S2 (Blankenfeldt, 2006) and double-headed rigor S1 (Chen et al., 2002).

Further, based on the thick filament structure, full-length myosin molecules repeated with the proper axial periodicity and helical symmetry generate the quasi-helical 42.9-nm-long elementary unit containing nine myosin molecules (see Figure 5.14d). The structure of the actin filament, on the other hand, is published (Chen, 2002). In our calculations, a focal volume of ~1 fL contains 3.27×10^{6} actin molecules (372 C-N HRSs per molecule) and 5.28×10^{5} myosin molecules (4325 C–N HRSs per molecule). Equation 5.12 can be used to estimate the I_{SHG} from knowledge of hyperpolarizability tensor; the dependence of I_{SHG} on ψ can be accounted for by estimating an intensity indicator defined as $(\chi_{yxx}^{(2)})^{2} + (\chi_{yyy}^{(2)})^{2}$, which considers the contributions from both parallel and perpendicular incident polarizations.

Using this approach, myosin contribution results in three orders of magnitude larger than that of actin (see Figure 5.14e), in agreement with the experimental evidence of SHG as myosin based (see Section 5.5) and with structural distributions of HRS emitters within proteins (as described in Section 5.6).

Furthermore, the results shown in Figure 5.15 highlight SPA sensitivity to the conformation of myosin. The molecular origin of this sensitivity can be assessed by comparing the I_{SHG} from whole myosin and headless myosin. As shown in Figure 5.14e, the contributions from the coiled-coil portions (LMM and S2) quantitatively dominate myosin SHG. However, the small but not negligible (10% of the total) contribution from the motor head (S1) can provide SHG with fine sensitivity to the orientation of myosin motors.

The γ_{rest} and γ_{rig} values can be interpreted in structural terms by implementing the model with conformational flexibility within the myosin molecule. Figure 5.15 shows a direct comparison between γ values experimentally measured (gray bars) and computed for selected myosin conformations (dashed lines). The γ_{rig} value can be reproduced by the model only when the S1 heads are oriented about perpendicularly to the fiber axis. In particular, the rigor conformation described by cryo-EM (Chen, 2002) and

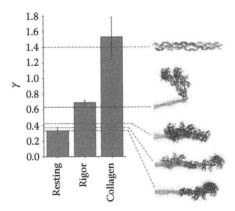

FIGURE 5.15 Structural interpretation of SPA data. Modeling of γ from muscle and collagen. The bars show experimental data (mean ± std): resting and rigor values were averaged over all recordings and sarcomere lengths; the collagen experimental value was taken from tendon (see text). The dashed lines show values computed from different myosin conformations and from collagen. The structures on the right show the terminal segment of double-headed myosin, with the fiber axis oriented horizontally.

x-ray diffraction (Reconditi et al., 2003) produces a modeled γ_{rig} in good agreement with the experimental value. The γ_{rest} value, on the other hand, can be reproduced by several conformations, all characterized by S1 heads oriented along the fiber axis.

The validity of this approach is also confirmed on a simple and static protein: collagen. From collagen atomic structure (Berisio et al., 2002), the value of γ can be computed, equal to 1.39. This value is in excellent agreement with the average of published experimental measurements of γ (Tiaho et al., 2007; Isaac Freund, 1986): 1.47 ± 0.10 (see Figure 5.15).

5.8 Conclusions

In this chapter, we have illustrated the physical origin of SHG from HRS and how coherent summation establishes the sensitivity of this signal to the order and geometrical orientation of the emitters. The use of this sensitivity through SHG polarization anisotropy measurements has been illustrated. Further, we describe the molecular origin of HRS within proteins and provide a full mathematical model for the interpretation of SPA data in terms of molecular order and protein conformation within the focal volume. The power of these techniques was illustrated with examples applied to collagen imaging and to the study of myosin conformation in skeletal muscle.

The capability of probing protein conformation and the geometrical distribution of proteins within ordered lattices in a tissue, coupled with the μm-scale resolution in deep tissue (allowed by the nonlinear nature of SHG), makes of SHG microscopy a unique technique for biomedical applications ranging from basic research to the development of novel diagnostic tools, as illustrated in other chapters of this book.

References

Bella, J., Eaton, M., Brodsky, B., and Berman, H.M. 1994. Crystal and molecular structure of a collagen-like peptide at 1.9 A resolution. *Science 266*, 7581.

Berisio, R., V.L., Mazzarella, L., and Zagari A. 2002. Crystal structure of the collagen triple helix model [(Pro-Pro-Gly)(10)](3). *Protein Sci, 11*, 262–270.

Blankenfeldt, W., Thoma, N.H., Wray, J.S., Gautel, M., and Schlichting, I. 2006. Crystal structures of human cardiac beta-myosin II S2-Delta provide insight into the functional role of the S2 subfragment. *Proc Natl Acad Sci USA 103*, 17713–17717.

Both, M., Vogel, M., Friedric, O., von Wegner, F., Kunsting, T. Fink, R. H. A., and Uttenweiler, D., 2004. Second harmonic imaging of intrinsic signals in muscle fibers in situ. *J Biomed Opt 9*, 882–892.

Boulesteix, T., Beaurepaire, E., Sauviat, M.P., and Schanne-Klein, M.C. 2004. Second harmonic microscopy of unstained living cardiac myocytes: Measurements of sarcomere length with 20 nm accuracy. *Opt Lett 29*, 2031–2033.

Brown, E. McKee, T., DiTomaso, E., Pluen, A., Seed, B., Boucher, Y., and Jain, R.K. 2003. Dynamic imaging of collagen and its modulation in tumors *in vivo* using second harmonic generation. *Nat Med 9*, 796–800.

Campagnola, P.J. and Loew, L.M. 2003. Second-harmonic imaging microscopy for visualizing biomolecular arrays in cells, tissues and organisms. *Nat Biotechnol 21*, 1356–1360.

Campagnola, P.J., Millard, A.C., Terasaki, M., Hoppe, P.E., Malone, C.J., and Mohler, W.A. 2002. Three-dimensional high-resolution second-harmonic generation imaging of endogenous structural proteins in biological tissues. *Biophys J 82*, 493–508.

Chen, L.F., Winkler, H., Reedy, M.K., Reedy, M.C., and Taylor, K.A. 2002. Molecular modeling of averaged rigor crossbridges from tomograms of insect flight muscle. *J Struct Biol 138*, 92–104.

Chu, S.W. S.Y.C., Chern, G.W., Tsai, T.H., Chen, Y.C., Lin, B.L. and Sun, C.K. 2004. Studies of chi(2)/chi(3) tensors in submicron-scaled bio-tissues by polarization harmonics optical microscopy. *Biophys J 86*, 3914–3922.

Cicchi, R., Kapsokalyvas, D., De Giorgi, V., Maio, V., Van Wiechen, A., Massi, D., Lotti, T., and Pavone, F.S. 2010. Scoring of collagen organization in healthy and diseased human dermis by multiphoton microscopy. *J Biophotons 3*, 34–43.

Cicchi, R., Massi, D., Sestini, S., Carli, P., De Giorgi, V., Lotti, T., and Pavone, F.S. 2007. Multidimensional non-linear laser imaging of basal cell carcinoma. *Opt Express 15*, 10135–10148.

Conboy, J.C. and Kriech, M.A. 2003. Measuring melittin binding to planar supported lipid bilayer by chiral second harmonic generation. *Anal Chim Acta 496*, 143–153.

Corrie, J.E., Brandmeier, B.D., Ferguson, R.E., Trentham, D.R., Kendrick-Jones, J., Hopkins, S.C., van der Heide, U.A., Goldman, Y.E., Sabido-David, C., Dale, R.E. et al. 1999. Dynamic measurement of myosin light-chain-domain tilt and twist in muscle contraction. *Nature 400*, 425–430.

Cox, G., Kable, E., Jones, A., Fraser, I.K., Manconi, F., and Gorrell, M.D. 2003. 3-Dimensional imaging of collagen using second harmonic generation. *J Struct Biol 141*, 53–62.

Deniset-Besseau, A., Duboisset, J., Benichou, E., Hache, F., Brevet, P.F., and Schanne-Klein, M.C. 2009. Measurement of the second-order hyperpolarizability of the collagen triple helix and determination of its physical origin. *J Phys Chem B 113*, 13437–13445.

Denk, W., Strickler, J.H., and Webb, W.W. 1990. Two-photon laser scanning fluorescence microscopy. *Science 248*, 73–76.

Dombeck, D.A., Blanchard-Desce, M., and Webb, W.W. 2004. Optical recording of action potentials with second-harmonic generation microscopy. *J Neurosci 24*, 999–1003.

Dombeck, D.A., Kasischke, K.A., Vishwasrao, H.D., Ingelsson, M., Hyman, B.T., and Webb, W.W. 2003. Uniform polarity microtubule assemblies imaged in native brain tissue by second-harmonic generation microscopy. *Proc Natl Acad Sci 100*, 7081–7086.

Dombeck, D.A., Sacconi, L., Blanchard-Desce, M., and Webb, W.W. 2005. Optical recording of fast neuronal membrane potential transients in acute mammalian brain slices by second-harmonic generation microscopy. *J Neurophysiol 94*, 3628–3636.

Freund, I. and Deutsch, M. 1986. Second-harmonic microscopy of biological tissue. *Opt Lett 11*, 94.

Han, X., Burke, R.M., Zettel, M.L., Tang, P., and Brown, E.B. 2008. Second harmonic properties of tumor collagen: Determining the structural relationship between reactive stroma and healthy stroma. *Opt Express 16*, 1846–1859.

Helmchen, F. and Denk, W. 2005. Deep tissue two-photon microscopy. *Nat Methods 2*, 932–940.

Holmes, K.C. 1997. The swinging lever-arm hypothesis of muscle contraction. *Curr Biol 7*, R112–118.

Huxley, H.E. 2004. Fifty years of muscle and the sliding filament hypothesis. *Eur J Biochem 271*, 1403–1415.

Irving, M. 1993. Birefringence changes associated with isometric contraction and rapid shortening steps in frog skeletal muscle fibres. *J Physiol 472*, 127–156.

Isaac Freund, M.D. and Aaron, S. 1986. Optical second-harmonic microscopy, crossed-beam summation, and small-angle scattering in rat-tail tendon. *Biophys J 50*, 693–712.

Jain, R.K., Brown, E., McKee, T., diTomaso, E., Pluen, A., Seed, B., and Boucher, Y. 2003. Dynamic imaging of collagen and its modulation in tumors *in vivo* using second-harmonic generation. *Nat Med 9*, 796–800.

Jiang, J., Eisenthal, K.B., and Yuste, R. 2007. Second harmonic generation in neurons: Electro-optic mechanism of membrane potential sensitivity. *Biophys J 93*, L26–L28.

Kriech, M.A. and Conboy, J.C. 2003. Label-free chiral detection of melittin binding to a membrane. *J Am Chem Soc 125*, 1148–1149.

Kwan, A.C., Dombeck, D.A., and Webb, W.W. 2008. Polarized microtubule arrays in apical dendrites and axons. *Proc Natl Acad Sci USA 105*, 11370–11375.

Lalama, S.J. and Garito, A.F. 1979. Origin of the nonlinear second-order optical susceptibilities of organic systems. *Phys Rev A 20*, 1179–1194.

Lin, S.J., Hsiao, C.Y., Sun, Y., Lo, W., Lin, W.C., Jan, G.J., Jee, S.H., and Dong, C.Y. 2005. Monitoring the thermally induced structural transitions of collagen by use of second-harmonic generation microscopy. *Opt Lett 30*, 622–624.

Llewellyn, M.E., Barretto, R.P.J., Delp, S.L., and Schnitzer, M.J. 2008. Minimally invasive high-speed imaging of sarcomere contractile dynamics in mice and humans. *Nature, 454,* 784–788.

Matteini, P., Ratto, F., Rossi, F., Cicchi, R., Stringari, C., Kapsokalyvas, D., Pavone, F.S., and Pini, R. 2009. Photothermally-induced disordered patterns of corneal collagen revealed by SHG imaging. *Opt Express 17,* 4868–4878.

Mertz, J. 2008. Applications of second-harmonic generation microsocpy. In *Handbook of Biomedical Nonlinear Optical Microscopy,* P.S. B.R. Masters, ed. Oxford: Oxford University Press.

Millard, A.C., Jin, L., Lewis, A., and Loew, L.M. 2003. Direct measurement of the voltage sensitivity of second-harmonic generation from a membrane dye in patch-clamped cells. *Opt Lett 28,* 1221–1223.

Millard, A.C., Terasaki, M., and Loew, L.M. 2005. Second harmonic imaging of exocytosis at fertilization. *Biophys J 88,* L46–L48.

Mitchell, S.A., McAloney, R.A., Moffatt, D., Mora-Diez, D.N., and Zgierski, M.Z. 2005. Second-harmonic generation optical activity of a polypeptide a-helix at the air/water interface. *J Chem Phys 122,* 114707.

Moreaux, L., Pons, T., Dambrin, V., Blanchard-Desce, M., and Mertz, J. 2003. Electro-optic response of second-harmonic generation membrane potential sensors. *Opt Lett 28,* 625–627.

Moreaux, L., Sandre, O., Charpak, S., Blanchard-Desce, M., and Mertz, J. 2001. Coherent scattering in multi-harmonic light microscopy. *Biophys J 80,* 1568–1574.

Moreaux, L., Sandre, O., and Mertz, J. 2000. Membrane imaging by second-harmonic generation microscopy. *J Opt Soc Am B 17,* 1685–1694.

Nucciotti, V., Stringari, C., Sacconi, L., Vanzi, F., Fusi, L., Linari, M., Piazzesi, G., Lombardi, V., and Pavone, F.S. 2010. Probing myosin structural conformation *in vivo* by second-harmonic generation microscopy. *Proc Natl Acad Sci USA 107,* 7763–7768.

Nuriya, M., Jiang, J., Nemet, B., Eisenthal, K.B., and Yuste, R. 2006. Imaging membrane potential in dendritic spines. *Proc Natl Acad Sci USA 103,* 786–790.

Peckham, M., Ferenczi, M.A., and Irving, M. 1994. A birefringence study of changes in myosin orientation during relaxation of skinned muscle fibers induced by photolytic ATP release. *Biophys J 67,* 1141–1148.

Piazzesi, G., Reconditi, M., Linari, M., Lucii, L., Bianco, P., Brunello, E., Decostre, V., Stewart, A., Gore, D.B., Irving, T.C. et al. 2007. Skeletal muscle performance determined by modulation of number of myosin motors rather than motor force or stroke size. *Cell 131,* 784–795.

Plotnikov, S.V., Millard, A.C., Campagnola, P.J., and Mohler, W.A. 2006. Characterization of the myosin-based source for second-harmonic generation from muscle sarcomeres. *Biophys J 90,* 693–703.

Plotnikov, S.V., Kenny, A.M., Walsh, S.J., Zubrowski, B., Joseph, C., Scranton, V.L., Kuchel, G.A., Dauser, D., Xu M., Pilbeam, C.C., Adams, D.J., Dougherty, R.P., Campagnola, P.J., and Mohler, W.A. 2008. Measurement of muscle disease by quantitative second-harmonic generation imaging. *J Biomed Opt 13,* 044018.

Pons, T., Moreaux, L., Mongin, O., Blanchard-Desce, M., and Mertz, J. 2003. Mechanisms of membrane potential sensing with second-harmonic generation microscopy. *J Biomed Opt 8,* 428–431.

Ralston, E., Swaim, B., Czapiga, M., Hwu, W.-L., Chien, Y.-H., Pittis, M.G., Bembi, B., Schwartz, O., Plotz, P., and Raben, N. 2008. Detection and imaging of non-contractile inclusions and sarcomeric anomalies in skeletal muscle by second harmonic generation combined with two-photon excited fluorescence. *J Struct Biol 162,* 500–508.

Rayment, I., Holden, H.M., Whittaker, M., Yohn, C.B., Lorenz, M., Holmes, K.C., and Milligan, R.A. 1993. Structure of the actin–myosin complex and its implications for muscle contraction. *Science 261,* 58–65.

Reconditi, M., Koubassova, N., Linari, M., Dobbie, I., Narayanan, T., Diat, O., Piazzesi, G., Lombardi, V., and Irving, M. 2003. The conformation of myosin head domains in rigor muscle determined by X-ray interference. *Biophys J 85,* 1098–1110.

Reedy, M.C. 2000. Visualizing myosin's power stroke in muscle contraction. *J Cell Sci 113*(Pt 20), 3551–3562.

Rieckhoff, K., and Peticolas, W.L. 1965. Optical second-harmonic generation in crystalline amino acids. *Science, 147*, 610–611.

Rocha-Mendoza, I., Yankelevich, D.R., Wang, M., Reiser, K.M., Frank, C.W., and Knoesen, A. 2007. Sum frequency vibrational spectroscopy: The molecular origins of the optical second-order nonlinearity of collagen. *Biophys J 93*, 4433–4444.

Sacconi, L., D'Amico, M., Vanzi, F., Biagiotti, T., Antolini, R., Olivotto, M., and Pavone, F.S. 2005. Second-harmonic generation sensitivity to transmembrane potential in normal and tumor cells. *J Biomed Opt 10*, 024014.

Sacconi, L., Dombeck, D.A., and Webb, W.W. 2006a. Overcoming photodamage in second-harmonic generation microscopy: Real-time optical recording of neuronal action potentials. *Proc Natl Acad Sci USA 103*, 3124–3129.

Sacconi, L., Mapelli, J., Gandolfi, D., Lotti, J., O'Connor, R.P., D'Angelo, E., and Pavone, F.S. 2008. Optical recording of electrical activity in intact neuronal networks with random access second-harmonic generation microscopy. *Opt Express 16*, 14910–14921.

Sacconi, L., Tolic-Norrelykke, I.M., D'Amico, M., Vanzi, F., Olivotto, M., Antolini, R., and Pavone, F.S. 2006b. Cell imaging and manipulation by nonlinear optical microscopy. *Cell Biochem Biophys 45*, 289–302.

Skala, M.C., Riching, K.M., Gendron-Fitzpatrick, A., Eickhoff, J., Eliceiri, K.W., White, J.G., and Ramanujam, N. 2007. *In vivo* multiphoton microscopy of NADH and FAD redox states, fluorescence lifetimes, and cellular morphology in precancerous epithelia. *Proc Natl Acad Sci USA 104*, 19494–19499.

Stoller, P., Reiser, K.M., Celliers, P.M., and Rubenchik, A.M. 2002. Polarization-modulated second harmonic generation in collagen. *Biophys J 82*, 3330–3342.

Stuart, G.J., and Palmer, L.M. 2006. Imaging membrane potential in dendrites and axons of single neurons. *Pflugers Arch 453*, 403–410.

Su, P.J., Chen, W.L., Chen, Y.F., and Dong, C.Y. 2011. Determination of collagen nanostructure from second-order susceptibility tensor analysis. *Biophys J 100*, 2053–2062.

Sun, Y., Chen, W.L., Lin, S.J., Jee, S.H., Chen, Y.F., Lin, L.C., So, P.T., and Dong, C.Y. 2006. Investigating mechanisms of collagen thermal denaturation by high resolution second-harmonic generation imaging. *Biophys J 91*, 2620–2625.

Svoboda, K., and Yasuda, R. 2006. Principles of two-photon excitation microscopy and its applications to neuroscience. *Neuron 50*, 823–839.

Theodossiou, T., Rapti, G.S., Hovhannisyan, V., Georgiou, E., Politopoulos, K., and Yova, D. 2002. Thermally induced irreversible conformational changes in collagen probed by optical second harmonic generation and laser-induced fluorescence. *Lasers Med Sci 17*, 34–41.

Thomas, D.D., Kast, D., and Korman, V.L. 2009. Site-directed spectroscopic probes of actomyosin structural dynamics. *Annu Rev Biophys 38*, 347–369.

Tiaho, F., Recher, G., and Rouede, D. 2007. Estimation of helical angles of myosin and collagen by second harmonic generation imaging microscopy. *Opt Express 15*, 12286.

Vanzi, F., Capitanio, M., Sacconi, L., Stringari, C., Cicchi, R., Canepari, M., Maffei, M., Piroddi, N., Poggesi, C., Nucciotti, V. et al. 2006. New techniques in linear and non-linear laser optics in muscle research. *J Muscle Res Cell Motil 27*, 469–479.

Wang, M., Reiser, K.M., and Knoesen, A. 2007. Spectral moment invariant analysis of disorder in polarization-modulated second-harmonic-generation images obtained from collagen assemblies. *J Opt Soc Am A Opt Image Sci Vis 24*, 3573–3586.

Williams, R.M. Zipfel, W.R., and Webb, W.W. 2005. Interpreting second-harmonic generation images of collagen I fibrils. *Biophys J 88*, 1377.

Wilt, B.A., Burns, L.D., Wei Ho, E.T., Ghosh, K.K., Mukamel, E.A., and Schnitzer, M.J. 2009. Advances in light microscopy for neuroscience. *Annu Rev Neurosci 32*, 435–506.

Zipfel, W.R., Williams, R.M., Christie, R., Nikitin, A.Y., Hyman, B.T., and Webb, W.W. (2003a). Live tissue intrinsic emission microscopy using multiphoton-excited native fluorescence and second harmonic generation. *Proc Natl Acad Sci USA 100*, 7075–7080.

Zipfel, W.R., Williams, R.M., and Webb, W.W. (2003b). Nonlinear magic: Multiphoton microscopy in the biosciences. *Nat Biotechnol 21*, 1369–1377.

6

3D SHG Imaging and Analysis of Fibrillar Collagen Organization

Paul J. Campagnola
*University of
Wisconsin—Madison*

6.1 Introduction

Second harmonic generation (SHG) imaging microscopy has great potential for visualization of disease states where there is change in collagen structure. This is because SHG, as a second-order nonlinear optical process, requires a noncentrosymmetric environment, and tissue alterations modify the overall symmetry of the collagen architecture. Specifically, SHG is an exquisitely sensitive probe of the fibrillar structure in tissues as it directly visualizes the supramolecular assembly, over the size scale of collagen fibrils to fibers, that is, from ~50 nm to a few microns. While there are approximately 20 isoforms of collagen, type I collagen (also known as col I) is the most abundant one and in fact is the most abundant protein in the body. It is either the primary component or at least a component of the structure in the matrix in diverse tissues such as tendon [1–3], skin [4,5], cornea [6,7], blood vessels [8], and bone and also in internal organs such as lung [9], liver [10], and kidney [11]. Given this range of tissues, it is possible that SHG could be used to image collagen changes in a wide range of pathologies. For example, many connective tissue disorders including osteogenesis imperfecta (OI) and scleroderma are characterized by abnormal collagen assembly and SHG may reveal differences in the morphology of diseased fibers not possible by other optical methods [12]. In addition, it is becoming increasingly documented that extracellular matrix (ECM) changes occur in most cancers. Similarly, fibrosis, that is, an increase in collagen, secretion is associated with several diseases and is further associated with poor prognoses.

The ability to probe ECM structure in diverse tissues gives the SHG imaging modality a great potential as a clinical diagnostic tool. In terms of human health, the greatest impact may be in early cancer detection. SHG has already shown early promise in imaging cancer since malignant tumors often have abnormal assembly of collagen relative to normal tissue [4,13–17]. This has now been shown in both animal models as well as human tissues *ex vivo* and *in vivo*. Besides simple visualization of fiber morphology, an additional enabling property for diagnostic imaging arises from the coherent nature of the SHG process. This is manifested in the initial directionality of the emission, where the morphology observed in the forward and backward channels is reflective of the fibril size distribution as well as the order of the packing. This is of particular importance as the fibril size and distribution may be different in healthy and diseased tissues, and we have shown this to be the case for the oim murine model for OI [12,18] and more recently in ovarian cancer [14].

The results from our lab and those of other labs now suggest that SHG has the potential to be developed into a clinical tool to analyze *ex vivo* biopsies or to perform *in vivo* imaging through endoscopes. A large remaining challenge is how to quantify and standardize 3D SHG image data for these diagnostic purposes. In this chapter, we describe our efforts to provide a general approach for measuring and modeling 3D SHG data to differentiate normal and diseased tissues based on SHG creation physics as well as photon propagation. We begin by describing some other analysis methods and discuss their limitations. Next, we describe experimental and modeling methods, present a theoretical treatment of SHG process in tissues, and then show examples of the general approach for OI and ovarian cancer, and finally conclude with some perspectives.

6.1.1 Limitations of Existing SHG Analysis Techniques

6.1.1.1 Comparison with Histological Analysis

A primary motivation for pursuing SHG for biomedical imaging lies in the ability to provide more quantitative/less subjective analysis than possible by classical histology. SHG microscopy has several advantages over standard histological scoring procedures for diagnostic imaging. First, SHG microscopy acquires 3D image sets through tissues of several hundred microns of thickness and can obtain more data than possible by histologic sections. In the latter, tissues are fixed, sliced in cross section into thin slides (~5–10 microns in thickness), and stained. The process can lead to artifacts. Perhaps more importantly, the interpretation depends highly on the skill of the pathologist. Comparisons between histology and SHG have been reported for several tissues, mostly for the purposes of visual assessment. More quantitative analyses of the collagen from histological sections have been reported for fibrosis. The first demonstration of SHG microscopy for scoring fibroses was done in a mouse model by Schanne-Klein [19]. For quantification of the collagen changes, they used a thresholding process for image segmentation to identify individual fibers. Through this process, they were able to successfully discriminate normal versus fibrotic tissues. A somewhat more sophisticated segmentation approach was taken by Yu and coworkers for imaging liver fibrosis in a rat model [20]. They used Otsu segmentation to score the amount of resulting collagen. The results of their automated approach were compared to pathology analysis and revealed a good correlation between these approaches, especially in areas of low collagen coverage. These results are promising but the specificity and generality of the analysis requires further study.

6.1.1.2 Quantification of Fiber Alignment in Cancer

It has been suggested that changes in collagen alignment in the ECM during cancer progression can be a useful metric. This is because there is an increase in collagen deposition, or desmoplasia, in many epithelial cancers. To determine if changes in collagen can be an early diagnostic of breast cancer, and, further, if SHG is sensitive to these alterations, Keely and coworkers measured the alignment of collagen fibers in murine tumor models of over a range of disease progression. In these efforts, they characterized three "tumor-associated collagen signatures (TACS)," which are reproducible during defined stages of tumor progression [16,21]. These signatures (1, 2, and 3) are characterized by (i) the presence of dense collagen localized around small tumors during early disease; (ii) collagen fibers that are parallel to the tumor boundary for *in situ* carcinoma; and (iii) collagen fibers that are normal to the tumor boundary for invasive disease. This approach relies on alignment and the results may be tissue specific, hence the generalization to other cancers requires further investigation.

6.1.1.3 Signal Processing Approaches to SHG Microscopy of Musculo-Skeletal Disorders

There is considerable interest in signal processing methods, for example, Fourier transforms, wavelet transforms, and texture analysis that can analyze and classify whole images without examining individual features. In healthy muscle fibers, bands of sarcomeres composed of actomyosin complexes are straight and evenly spaced (~2–3 microns), with each band lying nominally orthogonal to the axis of contraction. Damaged cells display a range of visible deviations from this norm. To measure changes that accompany muscle diseases, Plotnikov et al. [22] applied the Helmholtz equation for wave number to calculate the local striation spacing and angle of orientation with respect to the long axis of the myofiber (90° for ideal case). They tested this approach by comparing controls to three models of muscular disorders of varying severity: disuse-induced atrophy, mild and severe hereditary muscular dystrophy, and sarcopenia of aging. Analyzing these images for all the models, we found a consistently negative correlation between severity of the disorder and the mean sarcomere length. While quantitative, the scheme relies on the periodicity and alignment of sarcomeres. However, this regularity does not exist in most tissues and more general approaches are still needed.

6.1.1.4 Need for a More General Approach

The analysis approaches described earlier have all demonstrated potential for discriminating normal and diseased tissues. However, they rely on fiber morphologies and may not be applicable for all cases, especially if the morphologies are too complex or irregular to quantify. As an alternative, our lab developed a general approach to quantify 3D (up to several hundred microns) SHG imaging data. In a tissue imaging experiment, the measured SHG signal is composed of a convolution of the initially emitted SHG photons as well as the subsequent scattering of these photons at λ_{SHG}. The initial SHG emission from tissues has a distribution of emitted forward and backward components, whose ratio we denote F_{SHG}/B_{SHG}, which depends on the regularity of the fibril/fiber assembly. Additionally, more ordered tissues will give rise to brighter SHG due to higher photon conversion efficiency. Collectively, we refer to the SHG emitted directionality and the emitted intensity as the creation attributes. Following generation of SHG in tissue, the generated photons will propagate based on the scattering coefficient and scattering anisotropy at λ_{SHG}. The scattering coefficient μ_s is a measure of density, where it is the inverse of the average distance a photon will propagate before undergoing a scattering collision and changing direction. For most tissues, these scattering lengths are typically ~20–50 microns in the visible/near infrared region of the spectrum. The scattering anisotropy, g, is related to the directionality of the scattering, and varies from 0 to 1, where higher values correspond to greater organization. While scattering limits the achievable depth in any microscopy experiment, there is useful information in the scattering as well. Like the SHG creation attributes, the scattering properties also depend on tissue structure, and can differ between normal and diseased states.

The description of tissue scattering in terms of bulk optical properties is well developed and, as will be explained in a later section, we incorporate previous treatments into our approach. However, no complete, rigorous description of the SHG creation in tissues has been previously given. This is an essential step for this imaging modality to become a useful clinical diagnostic tool for monitoring disease severity and progression. Our overall premise is that different fibril/fiber size and packing will be different in disease states resulting in SHG properties. Examination of the collective findings of the literature has not yielded a unified relationship between SHG directionality, fibril morphology, and fibril size, suggesting that fibril size considerations alone are insufficient for complete SHG image interpretation. To solve this problem, we presented a new heuristic model based on phase-matching consideration, which provides a mathematical framework leading to the necessary insight to enable a thorough understanding of the relationship between fibril size, and assembly to the SHG response [23]. This will be described in detail in Section 6.4.

6.2 Methods

6.2.1 SHG Imaging System

The SHG instrument consists of a laser scanning unit (Fluoview 300; Olympus) mounted on an upright microscope stand (BX61, Olympus), coupled to a mode-locked titanium sapphire femtosecond laser (Mira; Coherent). All the SHG imaging described here was performed with an excitation wavelength of 890 nm with an average power of ~20 mW on the specimen using a water immersion 40 × 0.8 NA objective. This wavelength and numerical aperture (NA) result in lateral and axial resolutions of approximately 0.7 and 2.5 microns, respectively. SHG images were obtained using circularly polarized excitation as it probes all fiber orientations equally. The desired polarization at the focus was achieved as previously described [24]. The microscope simultaneously collects both the forward (*F*) and backward (*B*) components of the SHG intensity using identical calibrated detectors (7421 GaAsP photon counting modules; Hamamatsu). The relative efficiencies of the two detection paths are calibrated using fluorescence imaging of either small beads or dye-slides with fluorophores emitting near the SHG wavelength. The SHG signal (445 nm) was isolated with a 20 nm wide bandpass filter (Semrock, Rochester, NY) in both channels.

6.2.2 3D SHG Imaging Measurements

The measured depth dependence of the forward–backward intensity ratio (*F/B*) of the SHG signal is one of the means to characterize structural changes in the ECM between normal and diseased tissues. This axial response arises from a convolution between the initial SHG directional emission ratio (which we denote F_{SHG}/B_{SHG}) and subsequent SHG propagation through the tissue, which is based on μ_s and g at λ_{SHG} (445 nm). The F_{SHG}/B_{SHG} is highly dependent upon the fibril diameter, the packing density, and regularity relative to the size-scale of the SHG wavelength [23]. The bulk optical properties are related to density (primarily μ_s) and organization (primarily g) of the fibrillar assembly. The measured SHG directional (*F/B* ratios) values were determined by integration of the intensity in each optical section every few microns of depth using ImageJ software (http://rsb.info.nih.gov/ij/).

The measured attenuation of the forward SHG signal, that is, rate of intensity decrease with increasing depth into tissue, is also used to characterize structural changes in the ECM. The attenuation results from a convolution of the SHG creation attributes (the F_{SHG}/B_{SHG} emission directionality and relative SHG intensity), the primary filter effect (loss of laser intensity due to scattering) and secondary filter effect (loss of SHG signal). The relative SHG brightness or conversion efficiency and the primary filter effects have the largest impact. Since biological tissues have intrinsic heterogeneity in concentration, we have found a normalized approach necessary to account for local variability in SHG intensities in the same tissue (different fields) and to make relative comparisons between tissues [18]. To this end, the data of each optical series for each tissue are self-normalized to the optical section with the average maximum intensity. The normalized forward attenuation data were taken concurrently with the *F/B* data.

6.2.3 Measurement of Bulk Optical Parameters

The measured depth-dependent forward to backward ratios (*F/B*) and attenuation of SHG intensities are determined by both SHG creation attributes as well as the subsequent photon propagation dynamics. The latter is governed by the bulk optical coefficients, including scattering coefficient (μ_s), absorption coefficient (μ_a), scattering anisotropy (g), and index of refraction of the tissue at the fundamental and SHG wavelengths. We determined these at the laser and approximate SHG wavelengths (890 and 457 nm, respectively) for the OI and ovarian cancer examples described later. The diffuse reflected and transmitted intensities were measured by placing the specimen (~100 micron thickness) between a three- and a two-port dual integrating sphere setup. This setup yields the absorption coefficient, μ_a, and the reduced scattering coefficient, μ'_s, where

$$\mu'_s = \mu_s(1 - g) \tag{6.1}$$

The refractive indices necessary for the extraction of the scattering and absorption coefficients [25] were obtained using the method of Li [26], where the specimen is placed on a cylindrical lens and the critical angle for total internal reflection is measured. To experimentally determine the anisotropy factor, g, the Henyey–Greenstein function was fitted to experimental data following a similar technique as demonstrated by Marchesini et al. [27]. We recorded the angular scattering profile by rotating a photon detector with a slit-aperture about a fixed central specimen. The intensity of the scattered light from the tissues was measured from 5 to 45 degrees and the normalized values were fit to the following expression:

$$p(\cos \theta) = (1 - g^2)/(1 + g^2 - 2g\cos \theta)^{3/2} \tag{6.2}$$

Utilizing the diffuse reflectance, transmittance, index of refraction, and anisotropy, g, we performed a multilayer inverse Monte Carlo simulation [28,29] and calculated the absorption coefficient μ_a and scattering coefficient μ_s.

6.2.4 Monte Carlo Simulations

Monte Carlo simulations based on photon diffusion using the bulk optical parameters were performed to analyze the measured depth-dependent F/B ratio and forward intensity attenuation in terms of decoupling the contributing factors to the total response, which cannot be achieved directly. This allows the isolation of the most sensitive factors that can discriminate normal and diseased tissue. Our approach is based on the MCML framework of Wang and Jacques [30], where we added the necessary modifications to simulate the 3D SHG response [18]. The Monte Carlo technique is a stochastic approach that utilizes probability distribution functions to perform a three-dimensional random walk to estimate the transport equation [30] given by

$$\frac{dJ(r,s)}{ds} = -\alpha_t J(r,s) + \frac{\alpha_s}{4\pi} \int_{4\pi} p(s,s')J(r,s')d\omega \tag{6.3}$$

where $p(s,s')$ is the phase function of a scattered photon from direction s' into s, ds is the incremental path length, and $d\omega$ is the incremental solid angle about direction s. If the scattering is symmetric about the optical axis, the phase function can be written as the form of Equation 6.3. The radiance $J(r,s)$ relates to the observable quantity, intensity I, through the relation

$$I = \int_{4\pi} J(r,s)\, d\omega \tag{6.4}$$

The six principle operations that influence an individual photons trajectory are the launch of the laser, excitation pathway generation, absorption, scattering, elimination, and detection. As the MCML framework is well documented, we only present our modifications to the basic approach required to simulate the SHG directional and attenuation responses as a function of focal depth. In terms of creation attributes, these are based upon an assumed initial F_{SHG}/B_{SHG} creation ratio and scattering cross-sectional window σ, which represents the relative value of the $\chi^{(2)}$ susceptibility tensor [31]. We also account for the primary and secondary filter effects on the laser and SHG signal, respectively. A flowchart for the simulations is shown in Figure 6.1. First, to simulate optical

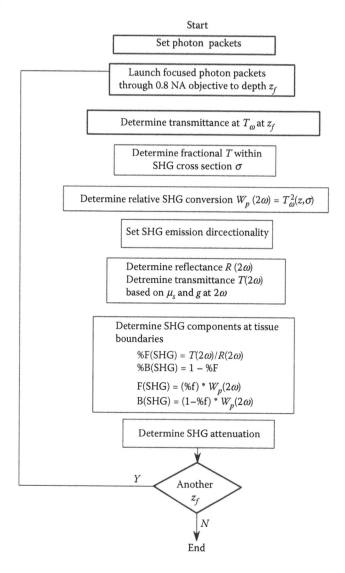

FIGURE 6.1 Flowchart of the algorithm used in the Monte Carlo simulations of the axial dependences of the measured SHG directionality and attenuation. (Reprinted from *Biophys. J.* 94, Lacomb, R., O. Nadiarnykh, and P. J. Campagnola, Quantitative SHG imaging of the diseased state osteogenesis imperfecta: Experiment and simulation, 4504–4514, Copyright (2008), with permission from Elsevier.)

sectioning, the incident photons are focused by an objective lens to an axial spot within the tissue at depth z_f, where the beam has a width or $1/e^2$ radius of ω_{beam} using the experimental NA = 0.8. The transmission of the laser, T_ω, that reaches the focal depth z_f is then determined based upon bulk optical parameters at the fundamental wavelength, and the cone formed by the experimental NA. This then accounts for the primary filter effect on laser intensity and determines the density of ballistic photons that arrive at the focus. To estimate the relative SHG brightness, that is, $\chi^{(2)}$ values of the normal and diseased tissues, or different tissues, we next define a two-dimensional scattering cross-sectional window $\sigma = 2\pi r_s$, where r_s is the radius in which fundamental photons are converted to second harmonic photons. This then yields the initial SHG intensity $T_\omega^2(z,\sigma)$, the initial weight of which we designate $W_p(2\omega)$. We next define the SHG emission directionality in terms of the forward to backward creation ratio (F_{SHG}/B_{SHG}) based solely on initial emission before scattering. The secondary filter effect is then determined by the bulk optical parameters at the SHG wavelength by simulating the transmission, $T_{2\omega}$, and reflection, $R_{2\omega}$. By running the trajectories of 50,000 photons, the detected forward (F) and backward (B) components as well as the attenuation are then simulated. This approach allows for the decomposition of the SHG creation and propagation dynamics, where experimentally the sources of the photons, that is, from direct emission or arising from multiple scattering, are indistinguishable.

6.3 SHG Phasematching in Tissues

6.3.1 Introduction to Phasematching

Under the right conditions, that is, angle of incidence, and polarization or artificially induced periodic grating, uniaxial birefringent crystals such as KDP and BBO can display ideal phase matching for SHG, that is, $\Delta k = k_{2\omega} - 2k_\omega = 0$, where $k_{2\omega}$ is the wave vector for SHG photon and k_ω is the wave vector for the incident photon. Accordingly, this condition is also characterized by infinite coherence length, $L_c = 2\pi/\Delta k$, and is 100% forward propagating. However, the physical situation in tissues is different and there is always some extent of phase mismatch, that is, $\Delta k \neq 0$, due to the underlying polycrystalline nature of most collagenous tissues [32]. As a consequence of this structure, these materials contribute axial momentum to the lattice, altering the ideal phasematching conditions described earlier. Specifically, the fibrillar packing density and randomness alter the conservation of momentum establishing a quasicoherent process, which results in an emission directionality where the relative shares of the forward and backward components depend on the extent of the mismatch. In general, strict phasematching conditions are not applicable in this case, as there are no type I (or angle-tuned) phasematching conditions, and the minimum mismatch for the forward SHG (F_{SHG}) is governed by the dispersion between the laser and SHG wavelengths, ($n_{2\omega} \neq n_\omega$). Additionally, as will be shown later, backward emitted SHG (B_{SHG}) requires strong axial momentum contributions from the media if the SHG-created photons are to travel in the opposite direction of the incident photons and still conserve momentum. The SHG conversion efficiency (related to $\chi^{(2)}$) is also then determined by the phase mismatch, by axial momentum contributions from the media, as well as decreased by the randomness inherent to biological tissues.

Previously, using antenna theory, Mertz predicted that spatial inhomogeneities (axially periodic and spherically localized distributions) are capable of contributing such momentum to the phasematching condition and, under appropriate conditions, can account for the creation of backward emitted SHG [33]. This poses a question as to how tissues, in practice, contribute sufficient axial momentum to create B_{SHG}. Through our model, we demonstrated that only through quasiphasematching (QPM) can appreciable B_{SHG} be produced by intensity buildup along multiple fibrils. Moreover, this treatment explains why backward SHG creation is not appreciably observed in dye-labeled membrane case due to lack of distributive amplification along its very thin axial extent (~4 nm). As Mertz's treatment only considered a single scattering cluster (i.e., individual dye molecules in a membrane) and neglected dispersion and randomness, this theory does not include the factors necessary for consideration of the SHG directionality and conversion efficiency in tissues. This situation differs from the limiting membrane case, as axially adjacent fibrils are packed sufficiently close (on the order of the coherence length) to interact and contribute to the overall SHG response.

6.3.2 Heuristic Model of Phasematching in Collagenous Tissues

To describe SHG creation in fibrillar collagen, we build upon and expand Mertz's formalism, by defining relaxed phasematching conditions. We introduce the concept of a domain of SHG-producing structures, which can be a collection of smaller fibrils, larger fibrils, or fibers packed together in the axial direction on the size scale of less than λ_{SHG}. This generalizes previous efforts based only on fibril size and without consideration for packing [2,6]. Our model includes contributions from QPM and additional phase mismatch due to dispersion and randomness, all of which lead to the creation of nonideally phasematched SHG in both the forward and backward directions. QPM allows the buildup of SHG intensity between anisotropic domains (here either fibers or assembly of small fibrils) without the need for strict phase-matching conditions, with maximum effect when the domain size and spacing are on the order of the coherence length of radiation. For example, QPM theory has been utilized to describe the buildup of SHG in ferroelectric crystals [34], and has been used in the design and fabrication of efficient backward SHG-producing periodically poled crystals [35]. We note that while periodically poled crystals utilize periodic structures designed to maximize conversion efficiencies with single values of Δk, similar, although less efficient, effects are present in tissues, where the collagen has been described as a nematic liquid crystal. Therefore, it is instructive to associate the high QPM conversion efficiency characteristic of periodic poled crystals with a completely periodic (hypothetical) collagenous structure, and on the other extreme, low QPM conversion efficiency consistent with a totally random structure. The physiological case will lie somewhere in between these limits where the conversion efficiency and emitted directionality are dependent upon the fibrillar diameter, interfibrillar spacing, and randomness of the tissue assembly.

To explicitly examine the impact of the relaxed phasematching conditions including axial contributions from the media, and a mismatch term Δk arising from dispersion and randomness, we begin by first considering the simple case of the propagation of a plane wave moving through a nonlinear media in the direction of its $\bar{k}$ vector. Even though this is not a strictly accurate description of the actual case, which involves focused excitation (NA = 0.8), we argue that general inferences may be obtained for the case where both the focused beam length (depth of focus) and the coherence length of the incident laser are longer than the coherence length of SHG radiation. In our experiment, the coherence length of the Ti:sapphire laser is ~30 microns, whereas the maximum forward coherence length of collagenous tissues based on dispersion is ~7 microns.

By following the coupled wave treatment of Munn [36] utilizing the slowly varying field approximation, the distributed amplification of the second harmonic within a homogeneous region or domain is given by the following equation:

$$\frac{dE(2\omega)}{dz} = -\frac{i\omega}{n_{2\omega}c} d_{\text{eff}} E^2(\omega) \exp(i\Delta kz) \tag{6.5}$$

where the effective hyperpolarizability coefficient $d_{\text{eff}}(2\omega)$ is proportional to the second-order bulk susceptibility χ^2, $n_{2\omega}$ is the index of refraction for the second harmonic wavelength, c is the speed of light, and $\Delta k = k_{2\omega} - 2k_\omega$ is the magnitude of wave vector mismatch between the incident and second harmonic waves.

Assuming propagation in the z direction and neglecting walk-off [37], the total second harmonic radiation at length L (domain length) is the vector sum of all the intermediate constituents (taking into consideration their respective phases) from lengths $0 < z < L$, and, utilizing the boundary condition $[E(2\omega, z = 0) = 0]$, can be expressed by

$$E(2\omega, z = L) = \kappa E^2(\omega) \left[\frac{1 - \exp(i(\Delta k))L}{-i\Delta k} \right] \exp(i(2\omega t - k_{2\omega}L)) \tag{6.6}$$

with the coupling coefficient $\kappa = \omega d_{\text{eff}}/nc$. When the phase matching term $\Delta k = k_{2\omega} - 2k_\omega$ is equal to zero, the maximum conversion efficiency is achieved.

For nonzero Δk's, the second harmonic amplitude $E_{2\omega}(z)$ at a position L along the propagation direction is given by

$$E_{2\omega} = \kappa E_\omega^2 \frac{\sin(\Delta k L/2)}{\Delta k L/2} \tag{6.7}$$

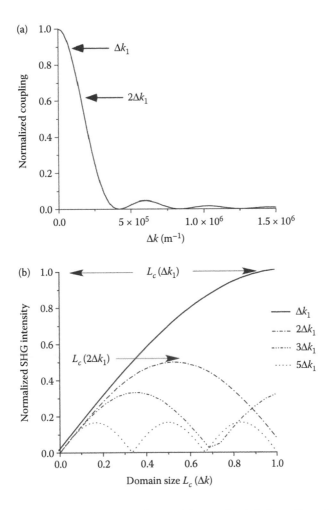

FIGURE 6.2 Effects of phase mismatch on SHG efficiency. (a) Normalized SHG coupling as function of Δk for values of Δk_1 and $2\Delta k_1$. The increased phase mismatch of the wave vector results in relative decreased SHG conversion efficiency. (b) Normalized SHG buildup as functions of Δk_1 and domain size. The conversion oscillates as $\sin(\Delta k L/2)$ and domain length, L, reaching its first maximum at the respective coherence length. The respective coherence lengths for the two largest ($m = 1$ and 2) are denoted by arrows. (From Lacomb, R. et al. 2008. *Opt. Commun.* 281:1823–1832. With permission.)

A plot illustrating the normalized conversion as a function of Δk is given in Figure 6.2a. In biological tissues, the minimum phase mismatch is governed by dispersion and is denoted by Δk_1. Two sample values (Δk_1 and $2\Delta k_1$, where Δk_1 is the wave vector mismatch due to change in index of refraction between the incident and second harmonic frequency) are shown as examples to demonstrate the decrease in intensity for increasing values of Δk for relatively small mismatches. Unlike birefringent crystals with $\Delta k = 0$ at specific incident angles of excitation for phase matching between the ordinary and extraordinary waves [38], axial momentum contributions from the media (which are nonsingular due to randomness) and dispersion result in a large distribution of Δk values, where the overall SHG intensity includes contributions from this entire assembly. It will be shown later that the dominant Δk values contributing to the overall SHG conversion efficiency are determined by both the Δk distribution and domain size.

Here, we present the more general case of an SHG conversion within a domain, which we define as a local source or distributed collection of SHG radiators, which in tissues can be either single fibrils/fibers or smaller fibrils packed closely together, while taking into account the regularity of the packing in the axial direction. For a given effective phasematching condition (Δk), the incremental conversion amplitude along the propagation direction is approximated by Equation 6.7 assuming a homogeneous domain of length L. Owing to the randomness associated with biological tissues, we cannot measure actual Δk values as they will not be single valued. However, as an example, we can utilize multiples of the mismatch due to dispersion (Δk_1) to investigate trends associated with higher mismatch. Figure 6.2b illustrates the normalized buildup of the second harmonic with a phase mismatch of Δk_1 (for a $\Delta n = 0.02$, which is approximately the case for collagen over the visible–NIR wavelength range) over the course of one normalized coherence length $L_{cl} = 2\pi/\Delta k_1$. Analogous curves are also given for the distributive buildup of higher mismatched conditions given by $m\Delta k$ (m is an integer) for multiples of Δk_1. We note that maximal SHG conversion efficiency will occur for small Δk values and interaction lengths L (or domains) on the order of L_c. For illustrative purposes, the respective coherence lengths for $m = 1$ and 2 are depicted parallel to the x axis. We see that for each phasematching condition, the conversion scale is proportional to $\sin(m\Delta k\,L/2)$ and is therefore domain length dependent, reaching its first maximum at the respective L_c, which is normalized by L_{cl} in Figure 6.2b. If the propagation lengths exceed the respective coherence lengths, the amplitude oscillates sinusoidally (as depicted by the curves associated with larger Δk values). This suggests that for domains on the order of the coherence length, fields supported by relatively small Δk values (i.e., regular structures) will dominate, while fields associated with larger Δk values will be characterized by less-efficient SHG conversion at their corresponding maximum value. In sum, we associate large Δk values with lower SHG conversion efficiency. Next, we will continue this analysis by showing that large Δk values support backward SHG emission through relaxed phasematching conditions.

6.3.3 Relaxed Phasematching Conditions and SHG Directionality

Here, we consider respective phasematching conditions for forward and backward SHG and how these relate to fibrillar domains in collagenous tissues. As pointed out by Mertz, backward emission arises when the SHG-producing assembly provides axial momentum, K, which alters the direction of the created photon [33]. We stress here that this is specific to the SHG creation step and is unrelated to subsequent multiple scattering of the generated signal in tissue. Owing to the fibrillar hierarchy of collagen (often described as polycrystalline in nature) and measured dispersion ($\Delta n = n(2\omega) - n(\omega) = 0.02$), we assume that such Δk values will exist that the coherence length of the created SHG is on the order of the interfibrillar spacing, thus allowing for the possibility of QPM. This then results in the following relaxed phase conditions:

$$\Delta k_f = K_f - (k_{2\omega} - 2k_\omega) \tag{6.8}$$

and

$$\Delta k_b = K_b - (k_{2\omega} + 2k_\omega) \tag{6.9}$$

where Δk_f and Δk_b are the phase mismatches for the forward and backward SHG creation, respectively, and K_b and K_f are the respective axial momentum contributions to the backward and forward SHG creation. These equations are identical to those given by Canalias [35] used to describe periodically poled crystals. However, here, we do not associate K with a single grating wave vector but rather an assembly of values (due to inherent randomness of collagenous tissues) provided by the medium. Backward SHG creation implies that in terms of magnitude $K_b > K_f$ and therefore $\Delta k_b > \Delta k_f$, resulting in a distribution of "lower" efficiency SHG components making up the overall B_{SHG}. Consequently, shorter coherence lengths are associated with this component. These equations can be modified to account for focused initial radiation by replacing k_ω by ξk_ω, where ξ is the effective reduction in axial propagation vector due to the Gouy phase shift [37]. The description that follows later holds for the case where the axial spread of the focused spot is comparable or larger to $1/\Delta k_1$, that is, the material coherence length based on dispersion. This condition is valid for image acquisition performed at medium NA.

Forward SHG will be dominated by phasematching with smaller Δk_f values for domains on the order of $L_c = 2\pi/\Delta k_f$, corresponding to small K_f values. Although the overall forward SHG signal is a summation of all the Δk terms, the lower Δk terms will dominate due to their relatively high conversion efficiency (see Figure 6.2b). Thus, we associate forward SHG primarily with Δk_1 (i.e., the maximum coherence length). By contrast, backward SHG is entirely dependent upon axial momentum provided by the lattice to redirect the created wave. Therefore, for significant (on the order of F_{SHG}) backward SHG intensity, the domain size should be less than the coherence length of the forward field (corresponding to the linear region of Figure 6.2b), while the interfibrillar spacing must be on the order of the coherence length associated with the backward field. Thus, phasematching conditions support the association of B_{SHG} with relatively larger Δk values. We note that randomness increases the distribution of available K values contributed by the medium and therefore the distribution of both Δk_f and Δk_b, which effects the overall distribution of SHG creation. Thus, we cannot specify the coherence lengths, as they are not single valued but state that the F_{SHG} has an upper bound limited by the material dispersion and as a consequence of Equation 6.8 and 6.9, it is characterized by longer L_c than that for B_{SHG}.

For the purposes of comparative computational analysis, we will associate F_{SHG} and $\Delta k_f = \Delta k_1$ and B_{SHG} with larger multiples of Δk_1 values (assigned values $\Delta k_b = m\Delta k_1$, where m is an integer and pertains to effective mismatch within a single domain). Doing so, one can predict the $\%F_{SHG}$ as a function of normalized domain (normalized to $L_{c1} = 2\pi/\Delta k_1$) by dividing the SHG intensity of Δk_1 over the sum of itself and the respective $m\Delta k_1$ term (calculated from Figure 6.2b). Figure 6.3 shows the resulting $\% F_{SHG}$ for $m = 2$–4. Utilizing superposition of these curves, the calculation shows that domains with values close to Lc_1 will support predominantly forward emission, while shorter domains will produce essentially even distributions. The emission directionality is highly sensitive to both the domain size and magnitude of Δk, where larger Δk results, in steeper transitions to higher F_{SHG}. Based upon arguments made earlier, we also attribute increased randomness with higher Δk values (i.e., shorter L_c) and lower conversion efficiency within a single domain. This analysis demonstrates that both domain size and randomness play an integral part in SHG emission directionality and that considerations based solely on fibril size do not form a complete description of the process. We will demonstrate this explicitly for the OI disease model in Section 6.4.1.2.

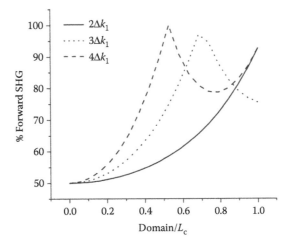

FIGURE 6.3 Calculated $\%F_{SHG}$ as a function of normalized domain size for several phasematching conditions (multiples of Δk_1). This calculation shows that domains with values close to L_{c1} will support predominantly forward emission, while smaller domains will produce essentially 50–50% forward and backward distributions. (From Lacomb, R. et al. 2008. *Opt. Commun.* 281:1823–1832. With permission.)

It is also illustrative to relate the normalized domain used in Figure 6.2b to λ_{SHG} with experimental observations that fibrils on the order of $\lambda_{SHG}/10$ produce creation ratios of $F_{SHG}/B_{SHG} \sim 1$. We point out that based on Equation 6.8, the forward emission coherence length is affected by both K and Δk (which is at least Δk_1). To justify a forward coherence length on this order, we note that the maximum coherence length as limited by dispersion is on the order of 7 μm and any axial momentum contribution from the medium acts to decrease this value, potentially by an order of magnitude. Thus, by normalizing the domain to the SHG coherence length, we observe that for domains on the order of $\lambda_{SHG}/10$ (normalized to 0.1 in Figure 6.2b), the %F_{SHG} is close to 50%, as suggested by other work [2,6] and increases to approximately 100% F_{SHG} for domains on the order of λ_{SHG} (normalized to 1 in Figure 6.2b).

6.3.4 Fiber Morphology

It has been suggested in other work that differences in forward and backward detected morphology are most likely to be manifested in the observation of smaller features. Specifically, several reports [2,24] have shown the existence of segmented appearing fibrils in the backward channel, where these same features appear to be continuous in the forward geometry. We can now explain this phenomenon in terms of the difference in forward and backward coherence lengths associated with the respective relaxed phasematching considerations and intensity amplification due to QPM. As an example, we show the forward and backward collected images from *Valonia* cellulose in the left and right panels of Figure 6.4. These specimens are approximately 30 microns in thickness, and the MFP is ~130 microns [24]. Thus, the contrast in the backward channel arises predominantly from direct quasicoherent emission and will not contain a significant multiply scattered contribution. We note that the fibrils observed in the forward channel are long and continuous, whereas these frequently have a segmented appearance in the backward channel. This can now be interpreted by our assignment of F_{SHG} and B_{SHG} with relatively small and large Δk values, respectively, which assigns a shorter coherence length to the latter. This result predicts that if the fibril packing in the axial direction is on the order of the backward coherence length, destructive interference occurs for the backward signal. By contrast, F_{SHG} is characterized by a relatively longer coherence length, and such destructive interference between fibrils does not contribute to the forward contrast. In other word, the fibrils are not physically segmented but their appearance as such in the backward channel arises from the quasicoherence of SHG in tissue.

This can also explain the findings of Williams et al. [2], where they showed that the morphology in mature rat tail tendon fibrils displayed similar features in the forward and backward collection geometries, whereas immature fibrils had a similar, segmented appearance. We suggest this arose because the immature tendons have effectively smaller domains and interfibril spacing. We further suggest the

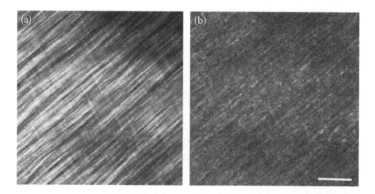

FIGURE 6.4 Forward (a) and backward (b) SHG images of *Valonia* cellulose. Segmented features appear in the backward channel due to destructive interference. Scale bar = 20 microns. (From Lacomb, R. et al. 2008. *Opt. Commun.* 281:1823–1832. With permission.)

fibrils and interfibril spacing of the mature tendons were sufficiently large to produce predominantly forward SHG and that the similar appearance in the backward channel may have been due to multiple scattering of the initial F_{SHG}. This description can also be utilized to explain the observation of bright B_{SHG} from sclera by Han [6]. The hollow fibrils in this tissue are on the order of 300 nm and *a priori* would be expected to produce predominantly F_{SHG}. However, they reported $F_{SHG}/B_{SHG} \sim 1$. This result is predicted by our current theory if one associates the domain length with the shell sidewall (presumably much thinner than the diameter) rather than the fibril diameter.

6.4 Results of 3D Imaging and Analysis

6.4.1 SHG Imaging of the Murine Model of Osteogenesis Imperfecta

Having discussed our experimental methods, our simulation framework, and our heuristic model of SHG production in tissues, we present the combined analysis for the murine oim model of osteogenesis imperfecta. This is a heritable disease of humans characterized by recurrent bone fractures, stunted growth, defective teeth, and other symptoms from abnormal tissues composed of type I collagen. OI results from mutations within the *Col1A1* or *Col1A2* genes that affect the primary structure of the collagen chain and induce changes in the secondary structure of the collagen trimers, which incorporate the mutant chains. The ultimate outcome is collagen fibrils that are either abnormally organized, small, or both.

6.4.1.1 Determination of the Bulk Optical Parameters for Oim and Wild Type Skin

While the most dramatic clinical presentation of OI is in bone, we chose to examine skin. This tissue is the most feasible for a virtual optical biopsy (i.e., backward collection geometry), and certainly be more accessible than bone for *ex vivo* analysis. Representative single optical sections for the oim and wild type (WT) skin are shown in Figure 6.5. While these tissues are distinct by visual inspection, where the images suggest that the fibrils are less ordered for the diseased skin, for future clinical diagnosis, a quantitative description is required. The less dense packing in the oim case suggests that the scattering properties may be different. Here, we measured μ_s' and g (effective) values for the oim and WT skin at both the fundamental and SHG wavelengths.

This scattering anisotropy, g, is a measure of the directionality of photon scattering, varies from 0 to 1, and is typically in the range of ~0.6–0.95 for most connective tissues. The upper limit corresponds to highly forward-directed scattering and is characteristic of very highly ordered tissues such as tendon, whereas in the other limit, brain is random and has low anisotropy. Thus, g can be used as a measure of the organization of the tissue. The resulting g values for 900 and 457 nm are shown in Table 6.1. At the SHG wavelength, fits of the experimental data to Equation 6.2 yield respective values for the oim and WT

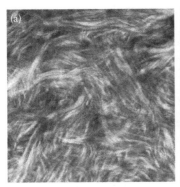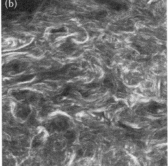

FIGURE 6.5   SHG single optical sections of murine WT (a) and oim dermis (b). Scale bar = 25 microns. (Reprinted from *Biophys. J.* 94, Lacomb, R., O. Nadiarnykh, and P. J. Campagnola, Quantitative SHG imaging of the diseased state osteogenesis imperfecta: Experiment and simulation, 4504–4514, Copyright (2008), with permission from Elsevier.)

TABLE 6.1 Bulk Optical Parameters for Oim and WT Skin Measured at the Fundamental and SHG Wavelengths

	Oim 457 nm	Oim 900 nm	WT 457 nm	WT 900 nm
μ_s (cm^{-1})	177 ± 17	130 ± 13	302 ± 45	106 ± 19
μ_a (cm^{-1})	1.5 ± 0.1	1.9 ± 0.1	1.8 ± 0.4	1.1 ± 0.3
g	0.65 ± 0.04	0.80 ± 0.02	0.80 ± 0.02	0.83 ± 0.01
μ_s' (cm^{-1})	57 ± 9	26 ± 3	61 ± 10	18 ± 3

Source: Reprinted from *Biophys. J.* 94, Lacomb, R., O. Nadiarnykh, and P. J. Campagnola, Quantitative SHG imaging of the diseased state osteogenesis imperfecta: Experiment and simulation, 4504–4514, Copyright (2008), with permission from Elsevier.

of 0.65 ± 0.04 and 0.80 ± 0.02. These values at 457 nm are statistically significantly different ($p = 0.03$), whereas at the fundamental wavelength, they are statistically similar ($p = 0.17$). We also point out that the measured g values are to be considered effective, as the 100-micron-thick biopsies include both dermal and adipose layers and can support 1–2 scattering events; thus, the actual values might be somewhat higher. This is because the convolution of single scattering with additional scattering events results in a broader distribution of measured angles and lower g. However, as we have not observed significant differences in extracted values for tissues 100–200 microns in thickness, we believe the extracted g values are attributed to a dominant contribution from the dermis. Moreover, the measurements were made to provide a comparison between the tissues, and as they were performed in a self-consistent manner, this does not affect the subsequent analysis.

The dual integrating sphere setup was used to extract μ_s' and μ_a. The analysis was achieved by using a multilayer inverse Monte Carlo simulation that also considered the scattering within the adipose layer underlying the dermis. Values for the adipose were measured separately in the absence of the dermis and verified by comparison with published values [39]. By this analysis, we find reduced scattering coefficients (shown in Table 6.1) that are in the same range as those determined in human skin by Tuchin [40]. We observe insignificant absorption at these wavelengths, which are on the red side above the type I collagen autofluorescence band. The μ_a values are to be considered upper bounds, as these measurements were near the noise floor. Thus, the mean free path (MFP) is effectively the inverse of scattering coefficient.

Using separately measured g, values for μ_s are determined by Equation 6.1. We observe that the oim tissue is statistically less scattering ($p = 0.009$) and more isotropic ($p = 0.03$) than the WT at the SHG wavelength and we interpret these results to be indicative that the matrix of the former is less densely packed. This is consistent with the smaller, shorter fibrils observed for the oim skin in Figure 6.5. We suggest that this approach relating morphological disorder based on bulk optical parameters in conjunction with the SHG image data is further consistent with the clinical presentation of a weakened matrix for the oim tissue. We note that these values at 900 nm were not statistically different, indicating that a descriptive metric of tissues must consist of more information than single wavelength measurements of the bulk optical parameters. In the next two sections, we show how a combination of 3D SHG imaging, the bulk optical parameters, and Monte Carlo simulations provides such a metric.

6.4.1.2 Forward/Backward SHG versus Depth: Experiment and Simulation

Here, we use the measured forward–backward intensity ratio of the SHG signal as part of the overall metric to differentiate oim and WT skin. This measurement arises from the SHG directional emission creation ratio (F_{SHG}/B_{SHG}) and the secondary filter effects on the subsequent SHG propagation through the tissue. Through simulation, we will extract the respective values of F_{SHG}/B_{SHG}, which as described previously are characteristic of the tissue but cannot be directly measured.

The experimentally measured F/B (composed of both components) versus depth plots for oim and WT skin are shown in Figure 6.6. These data result from 14 mice (7 each), with 5–10 3D stacks acquired from separate regions of the dermis. We observe that at all depths the oim is more forward directed, which we

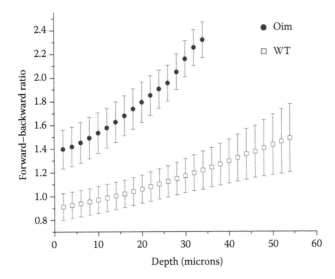

FIGURE 6.6 Ratio of experimentally measured forward and backward collected SHG as a function of depth into oim and WT skin. These photon propagation data are consistent with a multiple scattering process. (Reprinted from *Biophys. J.* 94, Lacomb, R., O. Nadiarnykh, and P. J. Campagnola, Quantitative SHG imaging of the diseased state osteogenesis imperfecta: Experiment and simulation, 4504–4514, Copyright (2008), with permission from Elsevier.)

can attribute to having a smaller scattering coefficient than the WT at the SHG wavelength, that is, forward directed photons have a higher probability of remaining forward directed in the less scattering tissues. This is consistent with the bulk scattering measurements shown in Table 6.1, where these values at 450 nm were ~300 and 175 cm^{-1} for the WT and oim, respectively. To validate the distinction between the oim and WT F/B data, we have performed a t-test at every depth and obtained p values in the range of 0.06–0.10, showing that they are statistically distinct at the 10% level. We note that while the dermis is <50–60 microns thick in murine skin, the tissue biopsies were in the range of 100–200 microns in total thickness, being composed of the epidermis, dermis, and adipose layers. While only the dermis (and only from the collagen component) provides SHG contrast, the entire thickness represents a scattering medium.

We also observe that for both tissues the F/B increases with increasing depth into the tissue. This result is consistent in the framework of photon diffusion theory, where at least one MFP is required between the location of the emitted photon and the forward boundary of the specimen for efficient multiple scattering to occur [30]. Thus, at increasing focal depths, the probability of multiple scattering events decreases as the forward pathlength to the tissue boundary shortens, and subsequently the F/B ratio must increase.

Owing to the fibrillar morphology in these tissues, the SHG has an initial emission directionality composed of both forward and backward components, and we incorporate this factor into the simulations through the creation ratio F_{SHG}/B_{SHG}. The trends of how this emission ratio is determined by fibril size and packing into SH producing domains were described in Section 6.3 on phasematching. Representative simulated curves of the depth-dependent F/B assuming 100% and 50% forward for oim and WT skin are shown in Figure 6.7a. The simulations used a four-layer model comprising the dermis and adipose layers. If we first only consider the 100% forward emission curves, we observe that these simulations reproduce the overall experimental trend in that the oim is characterized by a larger F/B than the WT at all depths because of the lower scattering coefficient. Despite providing qualitative distinction between the tissues, we note that this simulation largely overestimates the magnitude of the ratio (approximately by a few fold) for each tissue. This result indicates that it is inappropriate to assume that all SHG photons are emitted in the forward direction, and additionally that bulk optical parameters alone constitute an insufficient description. The case of 100% forward emission in tissue is not physically

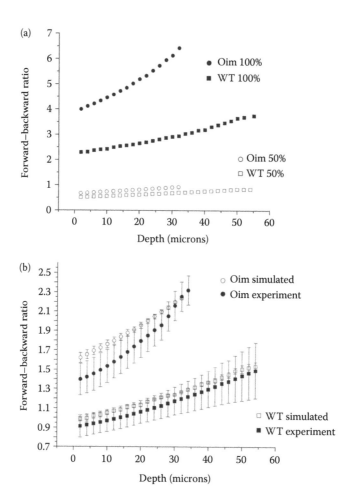

FIGURE 6.7 Comparison of experimental and simulated forward/backward response. (a) Representative Monte Carlo simulations of the measured depth-dependent directionality (*F/B*) for the oim and WT skin with SHG creation emission directions of F_{SHG} = 100% and 50%. (b) Comparison of the Monte Carlo simulations of the oim and WT *F/B* assuming 77.5% and 72.5% F_{SHG} creation emission for the oim and WT, respectively, with the experimental data. The standard error in the simulations results from the standard errors in the bulk optical parameters shown in Table 6.1 at the SHG wavelength. Chi-squared tests for both the WT and oim indicate that the respective experimental and simulated results are not significantly different. (Reprinted from *Biophys. J.* 94, Lacomb, R., O. Nadiarnykh, and P. J. Campagnola, Quantitative SHG imaging of the diseased state osteogenesis imperfecta: Experiment and simulation, 4504–4514, Copyright (2008), with permission from Elsevier.)

realizable as there will also be at least a small backward component due to the phase mismatch from dispersion. In the current case, the skin fibrils are $\sim\lambda_{SHG}/5$ in diameter (70 and 100 nm for the oim and WT, respectively) and, based on our phasematching model, are predicted to produce significant backward SHG (but always smaller than the forward component). To estimate this value, we ran simulations varying the ratio of the initial emission directionality from 40% to 100% (F_{SHG}) at 2.5% increments. We use these simulations to fit to the initial directionality by squaring and summing the residuals between the simulations and experimental data to calculate the R^2 parameter. Taking the minimum of R^2 then yields %F_{SHG} of 77.5% and 72.5%, for the oim and WT, respectively, where the uncertainty in each case is approximately ±3%, determined by the shallowness of this function around the minimum. The corresponding Monte Carlo simulations with standard error generated from the measured bulk optical parameters are shown with the experimental SHG data in Figure 6.7b. The chi-squared test results in values of 0.13 and 0.29 for the WT and oim, respectively, indicating that, for both tissues at the $\alpha = 0.05$ level, the experimental and simulated results are not significantly different. This good agreement demonstrates that the combination of 3D imaging, measuring the bulk optical parameters, and then performing simulations combining these aspects provides a robust method of modeling the directional response and extracting out the emission directionality. As described earlier, this cannot be directly measured in tissues of thickness >1 MFP, and further cannot be measured reliably in thin sections due to the unevenness of the tissue slicing.

We now apply the theoretical findings of Section 6.3 to comparing the SHG response in the oim and normal tissues. Using SEM imaging, we measured the average fibril diameters for these tissues and found average values of 70 and 100 nm for the oim and WT, respectively. Based solely on size considerations, one might expect that the F_{SHG}/B_{SHG} from the WT should be larger than that of the oim skin. To explain the observed similar creation ratios, we must also remember that the F_{SHG}/B_{SHG} is a function of Δk as demonstrated in Figure 6.4. To utilize these figures as a descriptive aid, we must consider the effective domain size $D = n\bar{L}/\lambda_{SHG}$ (i.e., normalized to λ_{SHG}), where ($\bar{L}$) is the average fibril diameter. Using these domain sizes and assuming $\Delta k_f = \Delta k_1$, we can then estimate effective Δk_b values that produce 75% F_{SHG} for both oim and WT skin. This results in ~20% higher value for oim over that of WT (with effective values of $6\Delta k_1$ and $5\Delta k_1$, respectively). By this description, one can make the connection that larger Δk_b values are associated with a higher degree of randomness in the collagen matrix, as is evidenced in the oim SHG image relative to the WT (Figure 6.5). The increased randomness of the oim tissue decreases the QPM contribution to the overall SHG; thus, the emission is more forward directed (although with lower conversion efficiency) compared to the more regularly packed fibrils of the same size. Thus, the fact that the same F_{SHG}/B_{SHG} occurs for the WT and oim is a coincidence that arises from offsetting contributions from the larger fibril size in the WT and increased randomness in the oim skin. This example further shows that a treatment based solely on fibril size is insufficient to describe the emission direction. In this case, the bulk optical parameters were significantly different, resulting in different measured F/B responses for the normal and diseased skin.

6.4.1.3 Depth-Dependent Attenuation of Forward SHG Intensity: Experiment and Simulation

The next part of our integrated metric for differentiating normal and diseased tissues is measurement and simulation of the depth-dependent attenuation of the forward SHG intensity. This axial response arises from the relative $\chi^{(2)}$ values, the square of the primary filter effect, SHG creation directionality, and secondary filter effects governing subsequent propagation. The forward attenuation data were taken concurrently with the F/B data (Figure 6.6) and the resulting averaged data with standard errors are shown in Figure 6.8 for the oim and WT dermis. As the absolute SHG intensity of the diseased skin is less than that in WT, the data are normalized to each other by using the maxima in each image stack. We observe that this method of measuring the SHG attenuation provides clear separation between the WT and oim skin in terms of the attenuation. Interestingly, despite being characterized by a similar and smaller μ_s at the fundamental and SHG wavelengths, respectively, the oim skin displays a more rapid decrease in intensity with increasing depth than the WT, demonstrating the inadequacies of the use of bulk optical parameters alone as a quantitative description.

In conventional plane wave scattering experiments, the attenuation can be estimated by fitting the response to an exponential decay. This is not possible for the SHG case as the attenuation results from a compounded mechanism composed of the wavelength-dependent bulk optical effects (distinct at the fundamental and second harmonic frequencies), SHG conversion efficiency (large effect, determined by simulation), and SHG creation directionality (small effect, determined by simulation), all of which culminate to produce the measured response. As a consequence, the initial intensity of the SHG at a given depth is linked to the laser intensity at that point (having been decreased by scattering with increasing depth)

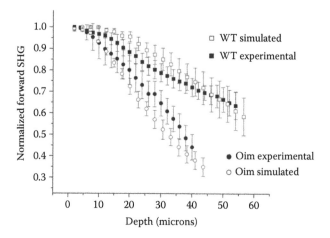

FIGURE 6.8 Comparison of the experimental forward SHG attenuation data with Monte Carlo simulations (with associated standard errors) based on the bulk optical parameters at both the fundamental and SHG wavelengths (Table 6.1). The creation directionality was taken from Figure 6.7b, and relative SHG conversion efficiency of 2.54-fold larger for the WT was used. As absolute magnitude of the SHG intensity from the oim is smaller than that of the WT, the data are normalized to their respective maximum and also to the maximum in each series to account for local variability in the tissues. Chi-squared tests for both the WT and oim indicate that the respective experimental and simulated results are not significantly different. (Reprinted from *Biophys. J.* 94, Lacomb, R., O. Nadiarnykh, and P. J. Campagnola, Quantitative SHG imaging of the diseased state osteogenesis imperfecta: Experiment and simulation, 4504–4514, Copyright (2008), with permission from Elsevier.)

and dominates the observed response. The measured intensity is further determined by the extent of the remaining tissue the photons must travel through to be collected (secondary filter effects); however, in all tissues, we have examined this is a small effect in this wavelength range and tissue thickness.

We must also consider that the relative SHG intensity from the oim skin (using the same laser excitation power) is weaker than that from the WT. Thus, the normalized SHG intensity from the oim will decay faster relative to the WT due to fewer initially generated photons at subsequent depths. A similar mechanism was proposed by Welch et al. [41] for fluorescence measurements in tissue, where they introduced the idea of "weighted photons" that accounted for local absorption coefficients and fluorescence quantum yields. We draw upon this idea to compare the SHG signal propagation in these different SHG-producing tissues. Rather than an absorption coefficient, the SHG intensity is determined by the second-order nonlinear susceptibility $\chi^{(2)}$. While we do not determine absolute $\chi^{(2)}$ coefficients, as the apparent efficiency will be convolved with scattering for tissues of greater thickness than 1 MFP, we can estimate the relative conversion efficiencies for the WT and oim tissues based on SHG intensity measurements. We cannot measure these values directly in skin as it is not possible to slice specimens of insufficient thickness (<30 microns) such that the initial SHG intensities can be measured in the absence of scattering. As an alternative, we performed these measurements in thin oim and WT bone slices (6 microns), where multiple scattering will be insignificant. This approach assumes similar changes in the collagen between skin and bone in the diseased state. Additionally, bone cryosections are fairly uniform, whereas analogous sections of skin can display substantial cutting nonuniformities. These measurements reveal that the WT was ~2.54 ± 0.22 ($p = 0.04$)-fold brighter than the oim bone [12]. We previously reported that the collagen concentration in these tissues was similar (based on quantitative Sirius Red staining); thus, the observed intensity differences can be ascribed to the difference in $\chi^{(2)}$. As $\chi^{(2)}$ is the spatially averaged macroscopic analog of the molecular hyperpolarizability, β, it is expected to have a lower value in the more disordered tissue, even if β is the same between the tissues. It is likely, however, that the β will also be different due to changes in helical structure in the diseased tissue.

To decouple the forward SHG attenuation data provided in Figure 6.8 into relative conversion efficiency, primary and secondary filter effects, we utilize the Monte Carlo framework used for the directionality. Mathematically, we account for the differences in $\chi^{(2)}$ by the use of a smaller scattering cross section σ in the simulation (see flowchart in Figure 6.1). Using the relative SHG conversion efficiency ratio of 2.5:1 for WT versus oim, the resulting simulations (with standard error from the measured bulk parameters) are illustrated in Figure 6.8 along with the experimental data. We observe that for both the WT and oim, most of the depth points are characterized by overlap of the experimental and corresponding simulated data. The chi-squared test results in values of 0.07 and 0.28 for the WT and oim, respectively, indicating that for both tissues, at the $\alpha = 0.05$ level, the experimental and simulated results are not significantly different.

This simulation demonstrates how the bulk optical parameters and relative $\chi^{(2)}$ values strongly affect the measured attenuation of the forward SHG for each tissue. Moreover, the use of simulations enables us to isolate the relative effects of the contributing factors of the measured signal and establishes the sensitivity of the various factors. Specifically, this approach would also allow us to determine relative SHG conversion efficiencies between different tissues once the respective bulk optical parameters are known at the fundamental and SHG wavelengths. This would be accomplished by running simulations varying the relative conversion efficiency and then comparing the results to the experimental data to achieve the best fit (in analogy with the directional data).

6.4.2 Ovarian Cancer

Here, we present the same analysis on human ovarian cancer as was shown for osteogenesis imperfecta above. Representative optical SHG sections for normal and malignant ovarian tissues are given in Figure 6.9. The large difference in fiber morphology and collagen packing in malignant tissues suggests that the scattering coefficient and scattering anisotropy may be different than that of the normal tissue. The bulk optical parameters (μ_s, μ_a, and g) for the normal and malignant tissues at the laser (890 nm) and approximate SHG (457 nm) wavelengths are given in Table 6.2. The malignant tissues are more highly scattering at the SHG wavelength compared to the normal ones (267 vs. 172 cm^{-1}), where this difference was significant ($p = 0.008$). We interpret the higher scattering of the cancer to be indicative that the matrix is more densely packed than the normal tissue. We point out that these measurements were on fixed tissues, where the procedure will affect the absolute values of the scattering coefficient. However, previous EM studies have shown that fixation does not significantly alter the fibrillar structure and only results in a slight reduction in volume (~20%) [42]. We have also compared SHG images of fully hydrated and fixed specimens of tendon and skin, and found that the fibrillar morphology was similar

Normal   Cancer

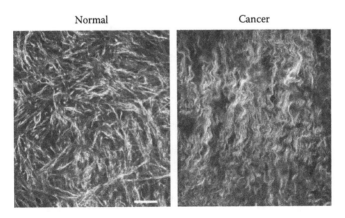

FIGURE 6.9 SHG single optical sections from normal and malignant human ovarian tissues. Scale bar = 25 microns. (Reproduced in part from Nadiarnykh, O. et al. 2010. *BMC Cancer* 10:94.)

TABLE 6.2   Measured Bulk Optical Parameters at the SHG and Fundamental Wavelengths with Standard Deviations

	Cancer $n = 3$ 457 nm	Cancer 890 nm	Normal $n = 5$ 457 nm	Normal 890 nm
μ_s (cm^{-1})	267 ± 19	195 ± 26	172 ± 39	161 ± 43
μ_a (cm^{-1})	6.3 ± 3.1	6.6 ± 1.5	7.3 ± 2.4	5.6 ± 1.2
g	0.82 ± 0.02	0.86 ± 0.02	0.83 ± 0.01	0.94 ± 0.01
μ_s' (cm^{-1})	46.2 ± 2.0	29 ± 3.3	29.6 ± 6.5	10.3 ± 3.8

with comparable associated shrinking (unpublished data). We further note that in this analysis, we are making comparisons between two tissues and both of which would be similarly affected by the fixation procedure.

While not all the respective μ_s and g values for the normal and malignant tissues were different at the laser and SHG wavelengths, we first note that the reduced scattering coefficient μ_s' was statistically different at both wavelengths (890 nm, $p = 0.0004$; 457 nm, $p = 0.02$). Additionally, the spectral dependence or spectral slope [43] of μ_s' was different where the malignant and normal tissues were characterized by a 1.5- and 3-fold respective increase in μ_s' between the laser wavelength and SHG wavelengths. This flatter spectral slope for the more ordered malignant tissue (exemplified by highly periodic helical fibrils compared to the more randomly appearing normal tissue) is predicted by the recent theoretical treatment by Backman [44]. We have also seen this results in measurements on tendon, which has similar regularity [45]. Thus, the spectral dependence of μ_s' provides one piece of quantitative evidence of the change in tissue structure that occurs during ovarian cancer. These bulk optical parameters are also incorporated in Monte Carlo simulations of the depth-dependent SHG directionality and attenuation later.

6.4.2.1 SHG Directional Measurements and Simulations

The averaged experimentally measured F/B versus depth plots for normal ($n = 5$) and malignant ovaries ($n = 3$) are shown in Figure 6.10. At all depths below the surface epithelium, the SHG from the normal tissues is more forward directed than from the cancers, which is consistent with the lower scattering coefficient of the former (172 vs. 267 cm^{-1}), such that photons that are initially forward directed have

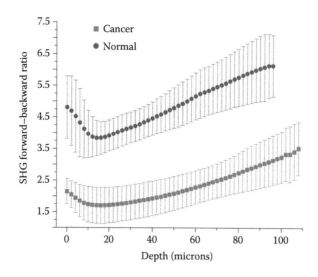

FIGURE 6.10   Averaged measured forward/backward SHG intensities as a function of depth for normal (circles) and malignant (squares) ovaries. (Reproduced from Nadiarnykh, O. et al. 2010. *BMC Cancer* 10:94.)

a higher probability of continuing to propagate in this direction. To validate the distinction between these tissues, *t*-tests were performed at 10 micron depth intervals, and the differences were statistically significant ($p < 0.01$ in all cases).

As in the OI case, we utilized Monte Carlo simulations of these plots using the measured bulk optical parameters as inputs to decouple the initial emission directionality (i.e., F_{SHG}/B_{SHG}) from the SHG propagation (based on μ_s and g). Representative simulations for the normal and malignant biopsies are shown in Figures 6.11a and 6.11b, respectively. We then fit to the initial directionality by squaring and summing the residuals between the series of simulations and the experimental data. Taking the minimum of the R^2 function yielded $\%F_{SHG}$ of 93% and 77%, for the normal and cancer, respectively, where the uncertainty in each case is approximately ±3%. The corresponding Monte Carlo simulation generated from the best fit for each tissue type (open squares = normal and open circles = cancer) is overlapped with the experimental SHG data in Figure 6.11c. The chi-squared test between the experimental and simulated data resulted in values of 0.40 and 0.30 for the normal and cancer, respectively, indicating that, for both tissues at the $\alpha = 0.05$ level, the data and corresponding simulation are not significantly different. This good fit between the simulated and measured data thus gives us confidence in the extracted $\%F_{SHG}$ values, as was the case for the OI tissues.

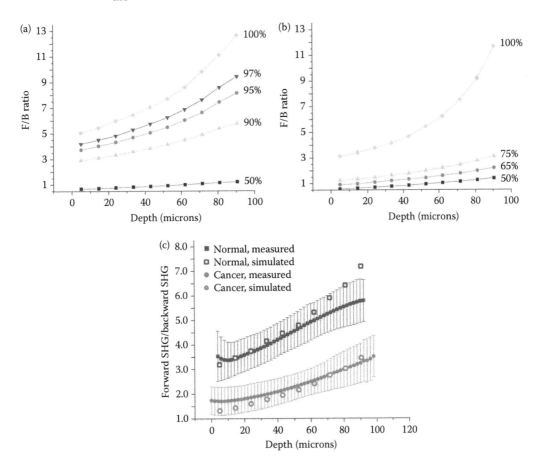

FIGURE 6.11 Monte Carlo Simulations of the measured *F/B* response, where (a) and (b) show the results for normal (a) and malignant (b) ovaries using the bulk optical parameters in Table 6.1 over a range of initial emission distributions. The best-fit simulation to the data in each case is overlapped with the experimental data in (c), where the $\%F_{SHG}$ was determined to be 77% and 93% for the malignant and normal tissues, respectively. (Reproduced from Nadiarnykh, O. et al. 2010. *BMC Cancer* 10:94.)

The SHG emission directionalities between the tissues are significantly different and can be interpreted by the difference in fibril assembly. Using a space-filling analysis of TEM images, we found the packing of the fibrils in the malignant tissue to be more regular relative to the normal (10% vs. 15% interfibrillar space, respectively). Based on our mathematical model of SHG in fibrillar tissues [23], the fibril assembly of the malignant tissue, that is, regularly packed fibrils on the order of the coherence length, would give rise to efficient backward emitted SHG [23]. In contrast, the more random assembly in the normal would result in more predominantly forward initial emission directionality (i.e., higher $\%F_{SHG}$), as was extracted from the simulation of the data.

6.4.2.2 SHG Attenuation Measurements and Simulations

The averaged normalized forward attenuation data with standard errors are shown in Figure 6.12a for the normal ($n = 5$) and malignant ($n = 3$) tissues. Unlike the *F/B* response, the SHG attenuation provides no clear separation between the tissues. To understand this effect, we need to consider all the factors that give rise to the measured attenuation. As described earlier, it is not possible to directly determine relative $\chi^{(2)}$ values in intact tissues as the measured signal is convolved with scattering when the tissues are thicker than one MFP or ~50 microns. However, this can be achieved using much thinner H&E histological sections (~5 microns). We note that the eosin staining does not contribute to the observed SHG. Measurement of the relative SHG intensities from these sections yields a factor of 3.9 ± 0.1 ($p < 0.005$) increased brightness for the cancer.

We now use all our measured factors as inputs into Monte Carlo simulations of the SHG attenuation. The simulated data for the normal (open squares) and malignant tumors (open circles) based on the bulk optical parameters (see Table 6.2) and the relative $\chi^{(2)}$ values are shown in Figure 6.12b. Like the experimental data, the simulations are highly similar for these two tissues. However, the rate of decay of the simulated SHG intensity is somewhat greater for the experimental data, where the differences are most pronounced at the bottom of the slice. While an exact match was not obtained, this approach still allows us to understand the similarity in the measured data for these tissues. It arises from the offsetting parameters of the increased conversion efficiency ($\chi^{(2)}$) and larger μ_s for the cancers, as these separately would result in slower and faster normalized attenuations, respectively.

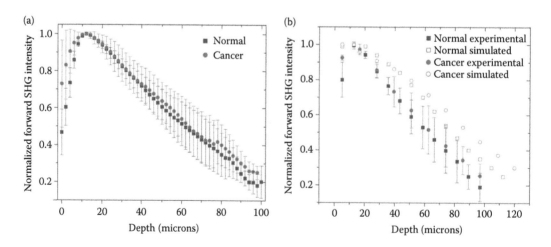

FIGURE 6.12   Forward SHG attenuation data and simulations for normal and malignant ovarian biopsies. (a) Shows the experimental data for normal (squares) and cancer (circles). (b) Shows the experimental data (closed circles and squares) every 10 microns and the simulations (open circles and squares) based on the measured bulk optical parameters in Table 6.1 and the relative $\chi^{(2)}$ values that were determined from the histological sections. (Reproduced from Nadiarnykh, O. et al. 2010. *BMC Cancer* 10:94.)

6.5 Discussion

The enabling aspect of the SHG contrast results from the quasicoherence of the process as well as the intrinsic symmetry constraints, yielding sensitivity to morphological and physical properties that may in general be different for normal and diseased tissues. For example, the axial directionality response (Figures 6.6 and 6.10 for the oim and ovary work, respectively) arises, in part, from the initial emission directionality, which is directly related to the fibrillar assembly of the tissue in terms of fibril size as well as organization. The measured F/B versus depth is also governed by the secondary filter effects on the generated signal, which are statistically different for the oim and WT. The axial dependence of the SHG intensity (Figures 6.8 and 6.12) provides an additional piece of the metric for tissue characterization as it is governed by SHG conversion efficiency, as well as the primary and secondary filter effects during subsequent propagation. The conversion efficiency is directly related to the organization of the tissue, such that at the same collagen concentration, uniformly aligned fibrils will yield a larger second-order response, $\chi^{(2)}$ than a more random assembly. We stress it is not possible to generally associate more randomness with diseased states. While the collagen in the OI case was more random and produced weaker SHG (~threefold), the ECMs in the malignant ovaries were more organized, of higher density, and had fourfold brighter SHG intensity

We also point out that the extent or regularity in the order, or in the other limit, the randomness, is not directly reflected in the scattering coefficient, which is essentially a measure of density. While the scattering anisotropy, g, is related to the order, the SHG conversion efficiency is of more direct relevance due to the inherent need for nonrandom assembly to satisfy the second-order asymmetry constraint. However, the simulations still require bulk optical parameters at both the fundamental and second harmonic wavelengths. Thus, we submit that when taken together, the SHG signatures (initial emission direction, conversion efficiency, and subsequent propagation) more directly and completely reflect the tissue organization than possible by consideration of the bulk optical parameters or SHG properties alone. Lastly, we note that the Monte Carlo simulations are essential in decoupling the factors that give rise to the 3D SHG data. This not only provides physical insight into tissue structure but may also identify the most sensitive factors in discriminating normal and diseased tissues and thus have diagnostic/prognostic value.

While we utilized this analysis for the murine model of OI and human ovarian cancer, we submit that the approach is a general means to analyze tissue structures. For example, we used this same combination of experiments (imaging and measurement of bulk optical properties) simulations to investigate the mechanism of optical clearing in skeletal muscle and tendon (see Chapter 8). In those efforts, we showed consistency with the extent of clearing and the reduction of scattering in thick tissue slices. While experimentally and computationally intensive, our integrated approach should be applicable for comparative analysis between any type of tissues that are composed of collagen. As a result, a wide range of pathologies such as cancer, connective tissue disorders, fibrosis, skin damage, and pathologies of the cornea could be analyzed. In contrast, signal processing schemes to date rely on changes in fiber alignment to provide quantitative discrimination between tissues and it remains to be seen how generalizable more advanced approaches like texture analysis with feature selection will be.

Through our exploratory studies derived from a basic science perspective, we have identified a collection of physical/structural properties of the ECM that change in both the OI model and in human ovarian cancer. While these were *ex vivo* studies with no opportunity for follow-up, the methods could be used to monitor the status of disease and/response to treatment. For example, for human OI patients, we foresee the method as being especially useful in monitoring the status of individual patients relative to their initial screen, where patients would already have a genetic profile. Thus, imaging several areas of skin would provide a reference point for future screenings. Additionally, this approach may permit monitoring the efficacy of treatment. For example, the effect of treatment with bis-phosphonates has been typically performed by bulk bone density and mineralization measurements [46]. Perhaps more insight into the action of such drugs can be gained by analyzing the fibrillar structure of the matrix at high resolution and monitoring patients over time intervals. Performing this through skin biopsies (or ultimately through a true virtual biopsy using only backward detection) would be much less invasive than performing unpleasant bone biopsies. Analogously, women at high risk for cancer could be monitored every several years through a combination of a laparascope and endoscope.

We have stressed that the overall approach is generally in terms of analyzing the ECM in diseased tissues; there remain challenges in making the scheme practical. However, we believe these can be overcome. Currently, we measure the bulk optical properties through integrating spheres and rotating goniometers. There are not simple measurements and can obviously only be performed on *ex vivo* tissues. However, recently, Jacques presented a method using reflectance confocal microscopy to measure local optical properties [47]. The Jacque and Campagnola labs used this approach in classifying normal and oim skin tissues. This method could be ultimately implemented *in vivo* through a microendoscope, given the rapid advances in fiber optics and miniaturized scanning being developed in several labs. Our method of measuring the SHG directionality also cannot be implemented directly *in vivo*. However, we suggest both the *F/B* and attenuations method can be performed in the backward collection geometry and would yield all the same data as shown here in the forward case. For attenuation, this is straightforward, although the signal levels would be smaller. For the directional measurements, devices we envision that a multiple probe scheme could be constructed where the same information could be extracted. Recently, Brown showed how the *F/B* could be extracted using a single objective [48]. We note that the directional measurements could be applied to *ex vivo* biopsies of more accessible epithelial tissues such as skin and breast, where ECM remodeling also accompanies carcinogenesis.

References

1. Stoller, P., K. M. Reiser, P. M. Celliers, and A. M. Rubinchik. 2002. Polarization-modulated second harmonic generation in collagen. *Biophys. J.* 82:3330–3342.
2. Williams, R. M., W. R. Zipfel, and W. W. Webb. 2005. Interpreting second-harmonic generation images of collagen I fibrils. *Biophys. J.* 88:1377–1386.
3. Theodossiou, T. A., C. Thrasivoulou, C. Ekwobi, and D. L. Becker. 2006. Second harmonic generation confocal microscopy of collagen type I from rat tendon cryosections. *Biophys. J.* 91:4665–4677.
4. Lin, S. J., S. H. Jee, C. J. Kuo, R. J. Wu, W. C. Lin, J. S. Chen, Y. H. Liao, C. J. Hsu, T. F. Tsai, Y. F. Chen, and C. Y. Dong. 2006. Discrimination of basal cell carcinoma from normal dermal stroma by quantitative multiphoton imaging. *Opt. Lett.* 31:2756–2758.
5. Tai, S.-P., T.-H. Tsai, W.-J. Lee, D.-B. Shieh, Y.-H. Liao, H.-Y. Huang, K. Zhang, H.-L. Liu, and C.-K. Sun. 2005. Optical biopsy of fixed human skin with backward-collected optical harmonics signals. *Opt. Express* 13:8231–8242.
6. Han, M., G. Giese, and J. F. Bille. 2005. Second harmonic generation imaging of collagen fibrils in cornea and sclera. *Opt. Express* 13:5791–5797.
7. Yeh, A. T., N. Nassif, A. Zoumi, and B. J. Tromberg. 2002. Selective corneal imaging using combined second-harmonic generation and two-photon excited fluorescence. *Opt. Lett.* 27:2082–2084.
8. Zoumi, A., X. Lu, G. S. Kassab, and B. J. Tromberg. 2004. Imaging coronary artery microstructure using second-harmonic and two-photon fluorescence microscopy. *Biophys. J.* 87:2778–2786.
9. Rothstein, E. C., M. Nauman, S. Chesnick, and R. S. Balaban. 2006. Multi-photon excitation microscopy in intact animals. *J. Microsc.* 222:58–64.
10. Sun, W., S. Chang, D. C. Tai, N. Tan, G. Xiao, H. Tang, and H. Yu. 2008. Nonlinear optical microscopy: Use of second harmonic generation and two-photon microscopy for automated quantitative liver fibrosis studies. *J. Biomed. Opt.* 13:064010.
11. Pena, A. M., A. Fabre, D. Debarre, J. Marchal-Somme, B. Crestani, J. L. Martin, E. Beaurepaire, and M. C. Schanne-Klein. 2007. Three-dimensional investigation and scoring of extracellular matrix remodeling during lung fibrosis using multiphoton microscopy. *Microsc. Res. Tech.* 70:162–170.

12. Nadiarnykh, O., S. Plotnikov, W. A. Mohler, I. Kalajzic, D. Redford-Badwal, and P. J. Campagnola. 2007. Second harmonic generation imaging microscopy studies of osteogenesis imperfecta. *J. Biomed. Opt.* 12:051805.

13. Brown, E., T. McKee, E. diTomaso, A. Pluen, B. Seed, Y. Boucher, and R. K. Jain. 2003. Dynamic imaging of collagen and its modulation in tumors *in vivo* using second-harmonic generation. *Nat. Med.* 9:796–800.

14. Nadiarnykh, O., R. B. Lacomb, M. A. Brewer, and P. J. Campagnola. 2010. Alterations of the extracellular matrix in ovarian cancer studied by second harmonic generation imaging microscopy. *BMC Cancer* 10:94.

15. Sahai, E., J. Wyckoff, U. Philippar, J. E. Segall, F. Gertler, and J. Condeelis. 2005. Simultaneous imaging of GFP, CFP and collagen in tumors *in vivo* using multiphoton microscopy. *BMC Biotechnol.* 5:14.

16. Provenzano, P. P., D. R. Inman, K. W. Eliceiri, J. G. Knittel, L. Yan, C. T. Rueden, J. G. White, and P. J. Keely. 2008. Collagen density promotes mammary tumor initiation and progression. *BMC Med.* 6:11.

17. Chen, S. Y., S. U. Chen, H. Y. Wu, W. J. Lee, Y. H. Liao, and C. K. Sun. 2010. *In vivo* virtual biopsy of human skin by using noninvasive higher harmonic generation microscopy. *IEEE J. Sel. Top. Quant.* 16:478–492.

18. Lacomb, R., O. Nadiarnykh, and P. J. Campagnola. 2008. Quantitative SHG imaging of the diseased state osteogenesis imperfecta: Experiment and simulation. *Biophys. J.* 94:4504–4514.

19. Strupler, M., A. M. Pena, M. Hernest, P. L. Tharaux, J. L. Martin, E. Beaurepaire, and M. C. Schanne-Klein. 2007. Second harmonic imaging and scoring of collagen in fibrotic tissues. *Opt. Express* 15:4054–4065.

20. Sun, W. X., S. Chang, D. C. S. Tai, N. Tan, G. F. Xiao, H. H. Tang, and H. Yu. 2008. Nonlinear optical microscopy: Use of second harmonic generation and two-photon microscopy for automated quantitative liver fibrosis studies. *J. Biomed. Opt.* 13:064010.

21. Provenzano, P. P., K. W. Eliceiri, J. M. Campbell, D. R. Inman, J. G. White, and P. J. Keely. 2006. Collagen reorganization at the tumor-stromal interface facilitates local invasion. *BMC Med.* 4:38.

22. Plotnikov, S. V., A. Kenny, S. Walsh, B. Zubrowski, C. Joseph, V. L. Scranton, G. A. Kuchel et al. 2008. Measurement of muscle disease by quantitative second-harmonic generation imaging. *J. Biomed. Opt.* 13:044018.

23. Lacomb, R., O. Nadiarnykh, S. S. Townsend, and P. J. Campagnola. 2008. Phase Matching considerations in second harmonic generation from tissues: Effects on emission directionality, conversion efficiency and observed morphology. *Opt. Commun.* 281:1823–1832.

24. Nadiarnykh, O., R. B. LaComb, P. J. Campagnola, and W. A. Mohler. 2007. Coherent and incoherent SHG in fibrillar cellulose matrices. *Opt. Express* 15:3348–3360.

25. Reichman, J. 1973. Determination of absorption and scattering coefficients for nonhomogeneous media. 1: Theory. *Appl. Opt.* 12:1811–1815.

26. Li, H., and S. Xie. 1996. Measurement method of the refractive index of biotissue by total internal reflection. *Appl. Opt.* 35:1793–1795.

27. Marchesini, R., A. Bertoni, S. Andreola, E. Melloni, and A. E. Sichirollo. 1989. Extinction and absorption coefficients and scattering phase functions of human tissues in vitro. *Appl. Opt.* 28: 2318–2324.

28. Chen, C., J. Q. Lu, H. F. Ding, K. M. Jacobs, Y. Du, and X. H. Hu. 2006. A primary method for determination of optical parameters of turbid samples and application to intralipid between 550 and 1630 nm. *Opt. Express* 14:7420–7435.

29. Palmer, G. M., and N. Ramanujam. 2006. Monte Carlo-based inverse model for calculating tissue optical properties. Part I: Theory and validation on synthetic phantoms. *Appl. Opt.* 45:1062–1071.

30. Wang, L., S. L. Jacques, and L. Zheng. 1995. MCML—Monte Carlo modeling of light transport in multi-layered tissues. *Comput. Methods Programs Biomed.* 47:131–146.

31. Moreaux, L., O. Sandre, and J. Mertz. 2000. Membrane imaging by second-harmonic generation microscopy. *J. Opt. Soc. Am. B* 17:1685–1694.

32. Prockop, D. J., and A. Fertala. 1998. The collagen fibril: The almost crystalline structure. *J. Struct. Biol.* 122:111–118.

33. Mertz, J., and L. Moreaux. 2001. Second-harmonic generation by focused excitation of inhomogeneously distributed scatterers. *Opt. Commun.* 196:325–330.

34. Baldwin, G. C. 1969. *An Introduction to Nonlinear Optics*. Plenum Publishing Press, New York, NY.

35. Canalias, C., V. Pasiskevicius, M. Fokine, and F. Laurell. 2005. Backward quasi-phase-matched second-harmonic generation in submicrometer periodically poled flux-grown KTiOPO4. *Appl. Phys. Lett.* 86, article no. 181105.

36. Munn, R.W., and C. N. Ironside. 1993. *Nonlinear Optical Materials*. Blackie Academic and Professionals, London, UK.

37. Moreaux, L., O. Sandre, S. Charpak, M. Blanchard-Desce, and J. Mertz. 2001. Coherent scattering in multi-harmonic light microscopy. *Biophys. J.* 80:1568–1574.

38. Yariv, A. 1989. *Quantum Electronics*. Wiley, New York.

39. Bashkatov, A. N., E. A. Genina, V. I. Kochubey, and V. V. Tuchin. 2005. Optical properties of the subcutaneous adipose tissue in the spectral range 400–2500 nm. *Opt. Spectrosc.* 99:836–842.

40. Bashkatov, A. N., E. A. Genina, V. I. Kochubey, and V. V. Tuchin. 2005. Optical properties of human skin, subcutaneous and mucous tissues in the wavelength range from 400 to 2000 nm. *J. Phys. D: Appl. Phys.* 38:2543–2555.

41. Welch, A. J., C. Gardner, R. Richards-Kortum, E. Chan, G. Criswell, J. Pfefer, and S. Warren. 1997. Propagation of fluorescent light. *Lasers Surg. Med.* 21:166–178.

42. Meek, K. M. 1981. The use of glutaraldehyde and tannic acid to preserve reconstituted collagen for electron microscopy. *Histochemistry* 73:115–120.

43. Liu, Y., R. E. Brand, V. Turzhitsky, Y. L. Kim, H. K. Roy, N. Hasabou, C. Sturgis, D. Shah, C. Hall, and V. Backman. 2007. Optical markers in duodenal mucosa predict the presence of pancreatic cancer. *Clin. Cancer Res.* 13:4392–4399.

44. Rogers, J. D., I. R. Capoglu, and V. Backman. 2009. Nonscalar elastic light scattering from continuous random media in the Born approximation. *Opt. Lett.* 34:1891–1893.

45. LaComb, R., O. Nadiarnykh, S. Carey, and P. J. Campagnola. 2008. Quantitative SHG imaging and modeling of the optical clearing mechanism in striated muscle and tendon. *J. Biomed. Opt.* 13:021108.

46. McCarthy, E. A., C. L. Raggio, M. D. Hossack, E. A. Miller, S. Jain, A. L. Boskey, and N. P. Camacho. 2002. Alendronate treatment for infants with osteogenesis imperfecta: Demonstration of efficacy in a mouse model. *Pediatr. Res.* 52:660–670.

47. Samatham, R., S. L. Jacques, and P. J. Campagnola. 2008. Optical properties of mutant vs wildtype mouse skin measured by reflectance-mode confocal scanning laser microscopy (rCSLM). *J. Biomed. Opt.* 13:041309.

48. Han, X. X., and E. Brown. 2010. Measurement of the ratio of forward-propagating to back-propagating second harmonic signal using a single objective. *Opt. Express* 18:10538–10550.

7

Second-Harmonic Generation Imaging of Microtubules

Alex C. Kwan
University of California, Berkeley

Cornell University

Microtubules are a well-studied class of cytoskeleton and serve numerous vital functions for the cell. They are the mechanical supports for cellular compartments and the roadways for active intracellular transport. A single microtubule consists of 13 parallel protofilaments, each a linear chain of the tubulin proteins connected end to end. The protofilaments are attracted to each other lengthwise; so, each microtubule is a 25-μm-diameter rod with a hollow core. Tubulin, the basic repeating unit, is a heterodimer of α- and β-tubulins. One end of the tubulin dimer has an accessible, bound guanosine-5'-triphosphate (GTP). This GTP can be hydrolyzed readily so that the end can bond with another tubulin, and therefore, this is the fast-growing end, or the "plus end." Since tubulins are joined end to end to form protofilaments, each microtubule is structurally asymmetric. This polarity is particularly important for active transport because it dictates the travel direction of molecular motors, which carry cargoes from the soma to distal cellular compartments (Hirokawa and Takemura, 2005).

The structural polarity of tubulin also manifests itself as hyperpolarizability. As a result, microtubules can participate in second-order nonlinear optical processes such as second-harmonic generation (SHG). Presumably, because the hyperpolarizability of tubulin is small, SHG imaging at single-microtubule resolution is not yet possible. However, if the neighboring microtubules have similar polarity, such that the SHG amplitudes coherently add, then the SHG intensity increases as the square of the number of scattering tubulins. SHG signals can be reliably observed from a single axon, that is, ~50 microtubules within the focal volume.

SHG from microtubules has been imaged in a variety of preparations (Figures 7.1 and 7.2), including mitotic spindles in *Caenorhabditis elegans* embryos (Campagnola et al., 2002, Mohler et al., 2003), sea urchin embryos (Mohler et al., 2003), mouse oocytes and embryos (Hsieh et al., 2008a), zebra fish embryos (Chu et al., 2003, Chen et al., 2006, Hsieh et al., 2008b, Olivier et al., 2010), and cultured rat

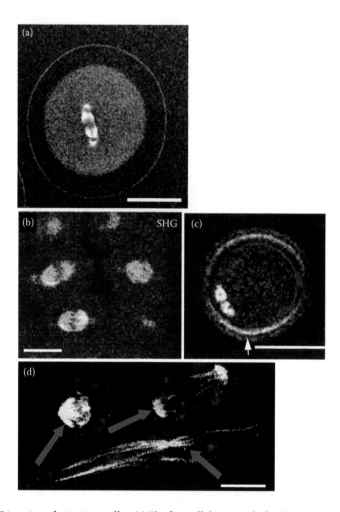

FIGURE 7.1 SHG imaging of mitotic spindles. (a) The first cell division of a fertilized sea urchin egg. (Reprinted from *Methods*, 29, Mohler, W., Millard, A. C., and Campagnola, P. J. Second harmonic generation imaging of endogenous structural proteins. 97–109. Copyright 2003, with permission from Elsevier.) (b) Multiple dividing sister cells in a zebra fish embryo. (Reprinted from Olivier, N. et al., 2010. Cell lineage reconstruction of early zebrafish embryos using label-free nonlinear microscopy. *Science*, 329, 967–971. Copyright 2010, with permission from the American Association for the Advancement of Science.) (c) An *in vitro*, cultured mouse oocyte. The arrowhead indicates the zona pellucida, a boundary region that also generated SHG signal. (Reprinted from Hsieh, C. S. et al., 2008a. Higher harmonic generation microscopy of *in vitro* cultured mammal oocytes and embryos. *Optics Express*, 16, 11574–11588. Copyright 2008, with permission from the Optical Society of America.) (d) Dividing rat basophilic leukemia cells in culture. The light gray arrow indicates the mitotic spindles and dark gray arrow indicates interphase microtubule ensembles. (Reprinted from Dombeck, D. A. et al., 2003. Uniform polarity microtubule assemblies imaged in native brain tissue by second-harmonic generation microscopy. *Proceedings of the National Academy of Sciences of the United States of America*, 100, 7081–7086. Copyright 2003, with permission from the National Academy of Sciences, USA.) Scale bar = (a) 45 μm, (b) 20 μm, (c) 60 μm, and (d) 10 μm.

basophilic leukemia cells (Dombeck et al., 2003). In the brain, SHG from microtubule bundles within axons, nascent neurites, and mature apical dendrites have been reported in rat primary neuronal culture (Dombeck et al., 2003, Kwan et al., 2008, Psilodimitrakopoulos et al., 2009), mouse primary neuronal culture (Stoothoff et al., 2008), *Aplysia* neuronal culture (Mertz, 2004), rat acute hippocampal slice (Dombeck et al., 2003), and mouse acute hippocampal and neocortical slices (Kwan et al., 2008, 2009,

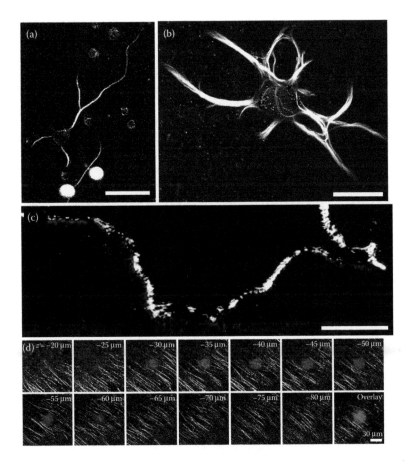

FIGURE 7.2 SHG imaging of microtubule bundles. (a) Single neurites of a neuron from a 3-day-old dissociated rat hippocampal culture. The bright solid circles are fluorescent beads used for intensity calibration. (Reprinted from Kwan, A. C. et al., 2008. Polarized microtubule arrays in apical dendrites and axons. *Proceedings of the National Academy of Sciences of the United States of America*, 105, 11370–11375. Copyright 2008, with permission from the National Academy of Sciences, USA.) (b) Dense microtubule networks within the neurites of an *Aplysia* neuron imaged after 3 days in culture. The neuron has a large nucleus so that the centrally located microtubules may be within adhesion areas that are underneath the cell body. (Reprinted from *Current Opinion in Neurobiology*, 14, Mertz, J. Nonlinear microscopy: New techniques and applications. 610–616. Copyright 2004, with permission from Elsevier.) (c) Cilia lining the walls of the aquaductus cerebri in rat brain stem slices. (Reprinted from Dombeck, D. A. et al., 2003. Uniform polarity microtubule assemblies imaged in native brain tissue by second-harmonic generation microscopy. *Proceedings of the National Academy of Sciences of the United States of America*, 100, 7081–7086. Copyright 2003, with permission from the National Academy of Sciences, USA.) (d) The apical dendrites of pyramidal neurons in an acute hippocampal slice of a transgenic Alzheimer's disease mouse model. The spherical object in the center of the image is a senile plaque, the hallmark pathological lesion for the disease. (Reprinted from Kwan, A. C. et al., 2009. Optical visualization of Alzheimer's pathology via multiphoton-excited intrinsic fluorescence and second harmonic generation. *Optics Express*, 17, 3679–3689. Copyright 2009, with permission from the Optical Society of America.) Scale bar = 30 μm, (a); 50 μm, (b); 100 μm, (c); and 30 μm, (d).

Barnes et al., 2010). SHG from microtubule bundles in the cilia of the rat brain stem (Dombeck et al., 2003), axoneme of sea urchin (Odin et al., 2009), and astroglial filaments of the spinal cord (Fu et al., 2007) have also been described.

In addition to the biophysical argument based on structural polarity, there are two main lines of experimental evidence confirming microtubules as a source of SHG. First, there is a clear correspondence of SHG images to microtubule structures with distinct morphology such as the mitotic spindles. In less obvious situations, such as in neurons, simultaneous imaging of fluorescently labeled tubulin (Campagnola et al., 2002) or post hoc immunohistochemistry of microtubule-associated proteins (Dombeck et al., 2003, Kwan et al., 2008) revealed a tight relationship between the microtubule distribution and SHG intensity. It is possible that microtubule-associated proteins may contribute to the observed signal, but this is unlikely to be the primary source because SHG is present in many cell types, animal species, and brain and body regions that do not share the same composition of microtubule-associated proteins. Second, the application of microtubule-depolymerizing drugs such as nocodazole (Dombeck et al., 2003, Kwan et al., 2008, Barnes et al., 2010) and colchicine (Stoothoff et al., 2008) significantly reduces the SHG intensity, whereas actin-specific agents such as cytochalasin D has no effect.

7.1 Why Study Microtubules with SHG Imaging?

SHG imaging is an optical technique; so, it is noninvasive and can provide information at high spatial and temporal resolution. Moreover, when applied to microtubules, SHG imaging has numerous unique, useful features. Here, we will discuss these features and also some drawbacks with an emphasis on comparing SHG imaging to alternative methods for visualizing microtubules.

7.1.1 SHG Intensity Reflects Microtubule Polarity

One unique property of SHG is coherent summation, where signal amplitudes from scattering units are summed. Phase differences can enhance or reduce SHG intensity depending on whether the inference is constructive or destructive. SHG signal is largest when the scatterers are oriented similarly, that is, from tubulins that have the same orientation. As a result, SHG is sensitive to the microtubule orientation within the focal volume and therefore can be used as a probe of polarity with submicron spatial resolution. However, the sign of the polarity, whether it is plus- or minus-end pointing, is not known.

The SHG intensity dependence on microtubule polarity can be best illustrated by comparing the fluorescence and SHG images of a mitotic spindle (Figure 7.3). Green fluorescent protein (GFP)-labeled tubulin proteins imaged with two-photon-excited fluorescence microscopy revealed a spatial pattern that directly relates to the tubulin concentration. The fluorescence image showed the characteristic shape of a mitotic spindle, consisting of two spindle poles with microtubules forming asters, most of which met at the equator, where duplicated chromosomes were to be separated. In contrast, SHG intensity depended on both the tubulin concentration and the microtubule polarity; so, it displayed a different

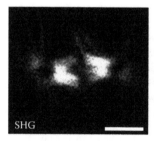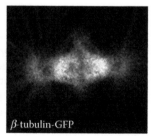

FIGURE 7.3 SHG and two-photon-excited fluorescence images of a *C. elegans* embryo during first mitosis. The microtubules in the mitotic spindle were fluorescent because the cell expresses *β*-tubulin fused with GFP. SHG signals were detected in a subset of microtubules; it was absent in the spindle equator, distal ends of the astral microtubules, and the center of the spindle poles. This profile of SHG intensity is consistent with coherent summation and the quadratic dependence on microtubule density. Scale bar = 10 μm. (Reprinted from *Biophysical Journal*, 82, Campagnola, P. J. et al., Three-dimensional high-resolution second-harmonic generation imaging of endogenous structural proteins in biological tissues. 493–508. Copyright 2002, with permission from Elsevier.)

pattern. The strongest SHG signals came from the area between the poles and the equator, where nearly parallel microtubules had the same orientation. At the equator, overlapping microtubules from the two poles had opposite orientations, leading to destructive interference and no SHG signal. Away from the spindle poles, the radially projecting astral microtubules were just weakly detected. SHG intensity has a quadratic dependence on the number of scattering units; therefore, it falls quickly for low tubulin concentration, unlike fluorescence that varies linearly with the fluorophore concentration.

An optical method for finding polarized microtubules is desirable because the conventional approach, the hook method, has severe limitations. To use the hook method, cells are lysed and permeated so that, when exogenously added tubulin would adhere to the existing microtubules, it forms hook-like appendages (Heidemann and McIntosh, 1980). Following fixation and using transmission electron microscopy, the chirality of individual hooks can be measured in a cross-sectional view of the microtubules. The number of clockwise and counterclockwise hooks corresponds to the number of plus-end microtubules pointing out of and into the cross section. Therefore, unlike SHG intensity that reports only the polarity magnitude, the hook method can be used to determine the magnitude and sign of the polarity. The hook method has been applied to characterize microtubule polarity in axons and dendrites of primary cultured neurons (Baas et al., 1988, 1989), excised nerves (Burton and Paige, 1981, Heidemann et al., 1981), and thinly sectioned brain tissues (Burton, 1988, Rakic et al., 1996). One drawback for the hook method is the large fraction of microtubules with ambiguously oriented hooks. Exogenous tubulin must be able to enter the lysed cells so that the hook method is also limited to thin and fixed specimens.

The recent discovery and characterization of plus-end-tracking proteins (Schuyler and Pellman, 2001) have added a new tool for characterizing the microtubule polarity. As their name suggests, these proteins are exclusively bounded to the fast-growing plus ends. By fluorescently tagging the plus-end-tracking proteins, it is possible to optically track their motion and locate the plus ends of growing microtubules (Stepanova et al., 2003). The large arsenal of genetic approaches means that this method can be easily applied to *in vitro* as well as *in vivo* preparations (Rolls et al., 2007, Stone et al., 2008). Moreover, the sign and magnitude of the polarity can be estimated. One potential problem for using fluorescently labeled plus-end-tracking protein is that many microtubules are stabilized by capping proteins and are less dynamic *in vivo*; therefore, polarity estimates may be skewed by mostly observing the dynamic fraction.

7.1.2 Endogenous Signal in Scattering Tissues and Whole Embryos

A key advantage for SHG microscopy is its applicability to scattering tissues (Figure 7.4). Similar to two-photon-excited fluorescence microscopy, the excitation volume is confined by the quadratic dependence of emission on the laser intensity; so, nondescanned detection can be used to collect all scattered photons as the signal and enables an imaging depth of >500 μm in scattering tissues (Denk et al., 1990, Helmchen and Denk, 2005). The ability to image deeply in the scattering tissue distinguishes SHG microscopy from other optical imaging techniques that have been heavily used to characterize the microtubules, such as fluorescence–speckle microscopy (Waterman-Storer et al., 1998) and dark-field microscopy (Horio and Hotani, 1986). Furthermore, the microtubule is an intrinsic SHG source that does not require staining with dyes. This simplifies the experimental procedure and protects delicate specimens. Maintaining the native state is critical for certain applications such as time-lapse imaging of embryos, where the developmental process can be easily stunned.

One restrictive requirement for SHG imaging of microtubules is the need for detecting the forward-directed emission, which places a practical limit on imaging depth. SHG intensity from microtubules is mostly forward directed and the forward-over-backward intensity ratio is ~5–100 depending on the spatial distribution of tubulins (Kwan et al., 2008). Since the backward emission from microtubules is small, two approaches have been used. One, the forward emission can be directly imaged with a photomultiplier tube behind the condenser lens. Two, it is possible to use the epipathway to detect back-scattered forward emission, in addition to the direct backward emission (Legare et al., 2007).

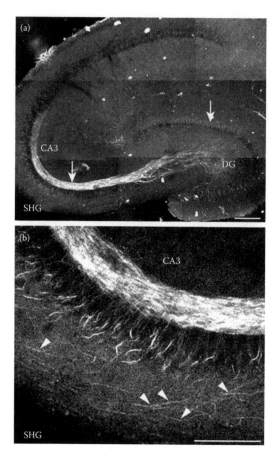

FIGURE 7.4 SHG imaging of microtubules in scattering tissues. (a) SHG from a transverse hippocampal slice, typically 250–400 µm thick and prepared from 14- to 20-day-old rat pups. The neurons were kept healthy by perfusing with oxygenated artificial cerebrospinal fluid. SHG signals were clearly seen from the mossy fibers, an axonal bundle that connects the dentate gyrus (DG) to the CA3 region of the hippocampus. (b) In a magnified view, SHG signals were also observed from thin processes, putative single axons, in the CA3 region. Scale bar = (a) 200 µm and (b) 100 µm. (Reprinted from Dombeck, D. A. et al., 2003. Uniform polarity microtubule assemblies imaged in native brain tissue by second-harmonic generation microscopy. *Proceedings of the National Academy of Sciences of the United States of America*, 100, 7081–7086. Copyright 2003, with permission from the National Academy of Sciences, USA.)

Furthermore, by measuring SHG signal from different angular acceptance angles, the fraction of the direct and scattered fractions may be separated (Han and Brown, 2010). In practice, detecting backscattered signal is more suitable for strong SHG sources such as collagen; so, all reported imaging studies of microtubules have focused on collecting the forward emission. The need to detect forward emissions limits the utility of SHG imaging of microtubules in en bloc tissues or intact animals.

Interestingly, fixatives such as paraformaldehyde abolish the SHG signal from microtubules. This is likely because protein cross-linking due to fixation has altered the hyperpolarizability or the spatial arrangement of tubulins. Moreover, in acute brain slices where the tissue health deteriorates over the course of several hours, the SHG signal from microtubules also gradually decreases. These anecdotal observations suggest that SHG is possible only in intact microtubules within their native environment. SHG intensity could potentially be used to assess the structural and functional integrity of microtubules *in situ* (Barnes et al., 2010).

7.1.3 Combining with Other Imaging Modalities

An SHG microscope has nearly identical requirements as a two-photon-excited fluorescence microscope; so, the two modalities can be easily applied together on the same specimen. For studying microtubules, it is often useful to visualize the surrounding cellular environment. For example, Thy1-GFP mice have sparsely labeled neurites (Feng et al., 2000), which enable the identification of neurite types that contain polarized microtubules (Kwan et al., 2008). Another example is using fluorescently tagged histones to label the chromosomes, providing more clues for classifying the different phases of cell division (Olivier et al., 2010). SHG imaging of microtubules can also be combined with third-harmonic generation imaging. The triple detection of two-photon-excited fluorescence, third-harmonic generation, and SHG have been used to visualize the various aspects of cellular architectures during embryonic development (Chu et al., 2003, Chen et al., 2006, Olivier et al., 2010). Mechanical manipulation can be applied via laser dissection of microtubules *in situ* (Tolic-Norrelykke et al., 2004). Complementary modalities such as coherent anti-Stokes Raman scattering can also be simultaneously used, for example, to visualize the myelin fibers (Fu et al., 2008).

Separating the emissions of SHG from other imaging modalities is straightforward. The SHG signal from microtubules is more strongly excited by the low end of the tuning range of a typical Ti:sapphire laser (Figure 7.5a). This is opposite to some common fluorophores such as GFP that are optimally excited by higher wavelengths. Moreover, SHG emission is narrowband, centering at half the excitation wavelength with a bandwidth of ~$1/\sqrt{2}$ of the excitation bandwidth. For a Ti:sapphire laser, the SHG emission bandwidth is ~10 nm. The background broadband fluorescence signal can be reduced by using narrowband emission filters (Figure 7.5b). By tuning the excitation wavelength, the SHG emission can be shifted to avoid the peak wavelength of the fluorescence emission.

7.1.4 Drawbacks

Despite evidence that tubulin is the basic unit responsible for SHG, the factors that affect molecular hyperpolarizability remain unclear (Gualtieri et al., 2008). As a result, it is difficult to interpret the observed SHG intensity quantitatively in terms of the properties of the microtubules. The main question is how do structural and conformational changes, for example, via posttranslational modification of tubulin or addition of microtubule-associated proteins, influence the SHG intensity? For structural changes, numerical simulation can predict how differences in the number density, intermicrotubule distances, and polarity can affect the SHG intensity. For molecular and conformational changes, which are more relevant to many biological questions, the answer is less obvious and requires correlated imaging and biochemical analysis. One study has shown that the expression of tau, a microtubule-binding protein, can lead to an increase in SHG intensity (Stoothoff et al., 2008). More studies will be required before SHG intensity can be a quantitative probe for characterizing *in situ* modifications to microtubules.

Compared to other methods for measuring microtubule polarity, SHG imaging is sensitive to the magnitude but not to the sign. This drawback may be remedied by comparing the phase of SHG amplitude from microtubules with a reference to extract the sign of polarity (Kemnitz et al., 1986). The weak SHG signal from microtubules would make such calibration a challenging task. Another consequence of weak intensity is the signal contamination from other endogenous SHG sources. For example, muscles and skin contain collagen, which generates an SHG signal that would overwhelm the signal from microtubules. Therefore, SHG imaging of microtubules is possible only when other dominant SHG sources are absent.

Despite the drawbacks, SHG microscopy is uniquely capable of identifying polarized microtubule ensembles and is applicable to native, scattering tissues. As a result, SHG imaging of microtubules has found a niche role in several fields of study. The following sections highlight two specific applications.

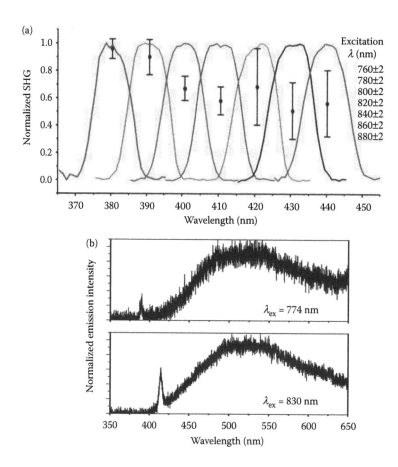

FIGURE 7.5 Emission spectra from microtubules and other subcellular structures. (a) Normalized emission intensities from the mossy fibers of a rat hippocampal slice. The emission spectra always center at half of the excitation wavelength and are stronger for short wavelengths. The mean normalized SHG intensity excited with constant excitation power at different wavelengths is plotted with dots and error bars. (Reprinted from Dombeck, D. A. et al., 2003. Uniform polarity microtubule assemblies imaged in native brain tissue by second harmonic generation microscopy. *Proceedings of the National Academy of Sciences of the United States of America*, 100, 7081–7086. Copyright 2003, with permission from the National Academy of Sciences, USA.) (b) The emission spectra obtained from a senile plaque in a brain slice of an Alzheimer's disease mouse model. The molecular origin of this SHG signal is not known. When the excitation wavelength was changed from 774 to 830 nm, the SHG-related spectra shifted accordingly, distinguishing the SHG signal from the broadband autofluorescence. (Reprinted from Kwan, A. C. et al., 2009. Optical visualization of Alzheimer's pathology via multiphoton-excited intrinsic fluorescence and second harmonic generation. *Optics Express*, 17, 3679–3689. Copyright 2009, with permission from the Optical Society of America.)

7.2 Application 1: Mapping the Distribution of Polarized Microtubule Bundles in Native Brain Tissues

Many proteins and messenger RNAs are synthesized in the cell body and are then actively transported to other compartments. Neuron is an extreme example with many distal sites along the long axon and branching dendrites. Cargoes are carried by molecular motors, such as kinesin and dynein, which move along microtubules with preference toward the plus- or minus-end direction. As a result, the polarity of microtubule bundles within axons and dendrites has implications for sorting cargo traffic and limiting the speed and throughput of transport.

Using the hook method, the microtubule polarity of neurites has been measured in excised nerves (Burton and Paige, 1981, Heidemann et al., 1981) and cultured neurons (Baas et al., 1988, 1989). These studies showed that early in neuronal development, all neurites have uniform polarity microtubules. However, as the neurites mature and their identities as axons and dendrites are established, the microtubule organization also evolves. The axons contain uniform polarity, plus-end-distal microtubules, whereas dendrites contain mixed-polarity microtubules with equal numbers of plus- and minus-end-distal microtubules. The distinct difference between axons and dendrites is maintained in the mature neuronal cultures. There are exceptions to this *in vitro* developmental rule, for example, when the axon is injured (Baas et al., 1987).

Can the *in vitro* microtubule organization principles be extrapolated to understand neurites *in vivo*? SHG intensity is sensitive to microtubule polarity, providing a tool for measuring in native brain tissues that are too thick for the hook method. An early study showed that in acute hippocampal brain slices of young rats, postnatal 14- to 20-day-old, axons generated SHG signal, agreeing with earlier observations that axons contain uniform polarity microtubules (Dombeck et al., 2003). In a follow-up study in mice over a wide range of ages, it was found that mature apical dendrites of pyramidal neurons, the principal excitatory cell in the hippocampus and in the neocortex, also generate SHG (Figure 7.6). This result implies that, in addition to axons, certain classes of dendrites can also contain polarized microtubule arrays (Kwan et al., 2008). Recently, a separate study that tracked the movement of fluorescently labeled microtubule plus-end-tracking proteins in *Drosophila* neurons found that their proximal dendrites contain polarized, minus-end-distal microtubules (Stone et al., 2008). Taken together, these new results demonstrate that uniform polarity, plus-end-distal microtubules in axons may be a universal rule, whereas microtubule organization in dendrites can differ for proximal versus distal compartments and for *in vivo* versus *in vitro* conditions.

Mapping the distribution of polarized microtubule ensembles using SHG imaging is applicable to native brain tissues so it can be combined with electrophysiology and molecular manipulations. For example, the dependence of long-term potentiation on microtubule-based axonal transport was investigated in the acute brain slice, where SHG intensity was used to monitor the pharmacological destruction of the microtubule network (Barnes et al., 2010). Another application area is the study of neurodegenerative disease. One hypothesis for the pathological mechanism in Alzheimer's disease is the disruption of axonal transport. SHG imaging has been applied to study the structure of polarized microtubules near senile plaques in brain slices from Alzheimer's disease mouse models (Kwan et al., 2009) and to investigate how overexpression of tau, a protein genetically linked to the disease, can affect the SHG intensity (Stoothoff et al., 2008). It is expected that SHG imaging will be a useful tool for assessing the structural and functional integrity of microtubule networks within neurons.

7.3 Application 2: Endogenous Time Stamp and Marker for Cell Division for Whole-Embryo Developmental Studies

A central problem in developmental biology is embryogenesis. From the beginning of a single cell, the fertilized egg goes through successive series of cell divisions. The number of cells grows exponentially until the organism takes shape. To unravel this process, it is essential to track cell differentiation and movement with high spatiotemporal specificity. An ideal tool should be able to observe subcellular details at a temporal resolution finer than the duration of each phase of cell division, over a time span that covers the entire embryogenesis. Moreover, because later cell division cycles involve a large number of cells, an automated process of segmentation and annotation is a necessity.

The multiple forms of nonlinear optical signals, including SHG, two-photon-excited fluorescence, and third-harmonic generation, could excite different endogenous molecules to illuminate various aspects of the tissue (Zipfel et al., 2003). An early study used a Cr:forsterite laser source in the ~1200–1350 nm range to observe mitosis inside live zebra fish embryos (Chu et al., 2003). Long excitation wavelengths reduced photodamage and enabled time-lapse recording for over 20 h (Chen et al., 2006). For the development

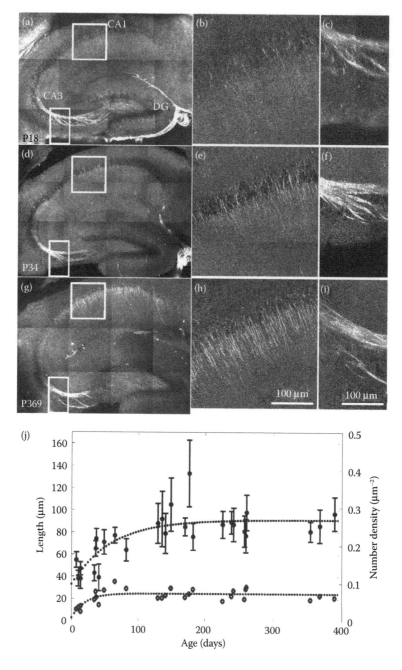

FIGURE 7.6 Mapping the age-dependent distribution of polarized microtubules in native brain tissues. (a through c) In an acute hippocampal brain slice from a 13-day-old mouse, SHG signals were observed in the axons of mossy fibers, but not in the neuropil of the CA1 stratum radiatum region that primarily consisted of apical dendrites of pyramidal neurons. (d through f, g through i) Same as (a) through (c) but for 34-day- and 369-day-old mice, where SHG signals appeared in the CA1 stratum radiatum in an age-dependent manner, demonstrating that mature apical dendrites contain polarized microtubule bundles. (j) Quantification of the length and density of microtubule bundles in the CA1 stratum radiatum as a function of age. (Reprinted from Kwan, A. C. et al., 2008. Polarized microtubule arrays in apical dendrites and axons. *Proceedings of the National Academy of Sciences of the United States of America*, 105, 11370–11375. Copyright 2008, with permission from the National Academy of Sciences, USA.)

of the zebra fish nervous system, the recording period allows imaging of nerve fibers from the initial bud stage, starting from the neural plate to the eventual formation of the neural tube. These proof-of-concept studies are an important step toward phenotype screening, where the nervous system development can be compared between normal and mutant zebra fish whose genes have been manipulated (Hsieh et al., 2008b).

A recent study has pushed the technology further by quantifying and automatically annotating the SHG intensity from microtubules as a marker for cell division (Olivier et al., 2010). During each cell division, the mitotic spindle transiently appears to separate the duplicated chromosomes; so, the SHG intensity rises and falls. This change in SHG intensity can be fitted to provide a time stamp for each cell division within the embryo (Figure 7.7). Combining with third-harmonic imaging of the cell boundaries,

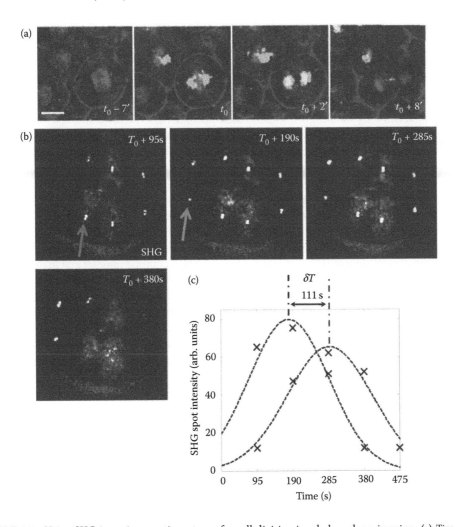

FIGURE 7.7 Using SHG intensity as a time stamp for cell division in whole-embryo imaging. (a) Time-lapse sequence of SHG images from a zebra fish embryo during cell division. Continuous imaging did not prevent normal embryonic development. Scale bar = 20 μm. (b) Large field-of-view images contain multiple dividing sister cells. The spot intensity of SHG was used to mark the time and location of cell division. (c) Quantification of the time-lapse SHG intensity revealed that two sister cells, indicated by arrowheads in (b), were divided at different times during the same division cycle. (Reprinted from Olivier, N. et al., 2010. Cell lineage reconstruction of early zebrafish embryos using label-free nonlinear microscopy. *Science*, 329, 967–971. Copyright 2010, with permissions from the American Association for the Advancement of Science.)

Olivier et al. developed algorithms to automatically trace the entire cell differentiation process through the successive cleavage stages in live zebra fish embryos. The experiments followed the embryos up to the 10th cell division cycle, where there are ~1000 cells within an embryo. The detailed cell lineage tree and location atlas should be invaluable for understanding how internal and external factors combine to influence each cell, thereby orchestrating the complex steps of embryogenesis.

7.4 Technical Notes for SHG Imaging of Microtubules

Since the SHG intensity from microtubules is weak, a number of practical details should be considered for improving the signal-to-noise ratio of the observed signal. To optimize detection, photon counting should be used instead of simple integration of the photocurrent of the photomultiplier tubes. Photon counting will exclude the dark current of the photomultiplier tube when calculating pixel intensity, and therefore accumulates less noise for long image acquisition. For example, an SHG image of microtubules in acute mouse brain slice requires an acquisition time of ~20 s (Figure 7.6). In addition, narrowband emission filters should exactly match the narrowband SHG emission; so, contaminating signals such as broadband autofluorescence are minimized (Figure 7.5b). Similarly, a compromise should be made between using short excitation wavelengths, which generate higher SHG intensity from microtubules (Figure 7.5a), versus long wavelengths, which reduce photodamage. Theoretically, SHG is a scattering process that involves a virtual state; so, photodamage is nonexistent. However, in practice, photodamage can result from exciting other endogenous photon-absorbing molecules. The mechanisms for photodamage are often not known and should be calibrated for the specimen under study (Sacconi et al., 2006).

Fluorescence can overwhelm the weak SHG signal from microtubules; so, it is essential to check for the signatures of SHG in every experiment. A few of the following factors will be sufficient to verify that SHG is responsible for the measured intensity: a squared dependence on excitation intensity, dependence on excitation polarization, tuning the excitation wavelength so that the emission wavelength shifts outside the narrowband emission filter range, and angular dependence of the SHG emission. So far, it has been demonstrated by numerous studies that the SHG signals in mitotic spindles and in neurons originate from microtubules, but for other specimens, the identity of the SHG source needs to be verified.

The SHG emission has angular dependence. For microtubule bundles, the forward emission can be ~5–100 times stronger than the backward emission. Efficient collection of the forward emission requires using a high-numerical aperture objective lens in the forward pathway that matches the field of view of the objective lens used to focus the excitation beam. Moreover, collecting in the forward pathway from thin samples such as cultured neurons is easier and should be attempted first. For imaging acute brain tissue, there is a compromise between thick slices, where the forward emission would be more scattered, versus thin slices, where the microtubule integrity may be disrupted. SHG signals from microtubules are abolished from fixed or dying neurons.

The SHG intensity is also dependent on the polarization of the excitation laser source relative to the molecular orientation of the tubulins. The optimal condition for SHG occurs when the laser polarization is parallel to the long axis of the microtubules (Figure 7.8). This alignment is possible over the entire field of view when it is anatomically known that all microtubules orient similarly, for example, in the apical dendrites in area CA1 of acute hippocampal slices (Figure 7.6h). Alternately, using circular polarization or linear polarization with a constantly rotating sample will ensure that all microtubules within the field of view are equally excited.

Finally, from the SHG signal of microtubules, a couple of studies have measured a hyperpolarizability angle that matches well with the physical dimensions of tubulin (Psilodimitrakopoulos et al., 2009, Odin et al., 2009), implicating tubulin as the basic unit that scatters and participates in SHG. For a given spatial concentration of tubulins, it is possible to calculate the expected intensity and angular distribution of the SHG signal (Moreaux et al., 2000, Williams et al., 2005). For a bundle of microtubules, numerical simulation using typical values for axon and dendrite diameters and intermicrotubule spacing had been carried out (Figure 7.9). The simulation results showed that in contrast to membrane

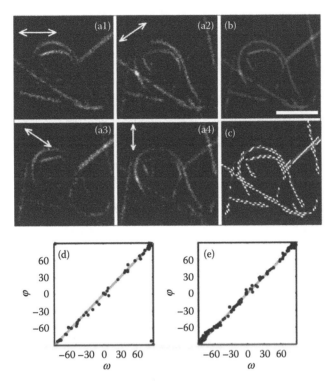

FIGURE 7.8 Polarization dependence of SHG intensity from microtubules. (a1 through a4) Fresh axonemes were prepared from sea urchin sperms. SHG images were collected using linearly polarized excitation at four different angles of polarization. (b) The mean image. Scale bar = 10 μm. (c) From the intensity recorded at each of the polarization angles, a direction map for hyperpolarizability is computed, which matches well with the physical orientation of the axonemes. (d, e) Quantification of the correlation between hyperpolarizability direction, φ, and physical orientation, ω, for images obtained with four or six different polarization angles. (Reprinted from Odin, C. et al., 2009. Second harmonic microscopy of axonemes. *Optics Express*, 17, 9235–9240. Copyright 2009, with permissions from the Optical Society of America.)

SHG imaging, where the forward emission is maximal at two lobes that are ~25° away from the forward axis, the emission from microtubules is almost entirely along the forward axis. Moreover, the calculated forward-over-backward emission intensity ratio for microtubule bundles is significantly higher than those calculated for collagen, another well-characterized biological SHG source. This agrees with the observation that it is easier to image collagen in the epipathway and to suggest a possible method for separating different sources of SHG signals. For example, forward-to-backward emission intensity ratios have been used to estimate the overall polarity of the microtubule arrays within axons and apical dendrites (Kwan et al., 2008).

7.5 Summary

SHG imaging of microtubules has been applied to a large variety of preparations, particularly for visualizing mitotic spindles and microtubule bundles in neurons. The SHG signal offers unique advantages: the method is sensitive to polarized microtubule ensembles and is applicable to scattering tissues. The development of SHG imaging of microtubules has so far focused on a few biological applications, including mapping polarized microtubules in neuronal compartments and marking time of cell division during embryonic development. Because of its unique capabilities, SHG imaging of microtubules should continue to thrive in areas where no comparable technique exists.

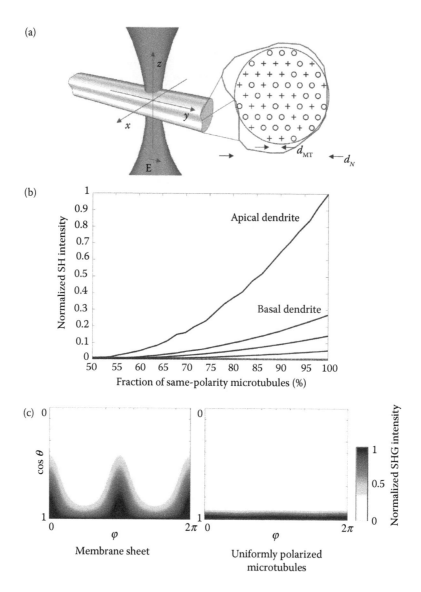

FIGURE 7.9 Numerical simulation of SHG intensity expected from a microtubule bundle. (a) Schematic of the numerical simulation. The excitation beam, focused with high-numerical aperture objective, is linearly polarized at the same direction as the microtubule long axis. The parameters including the number of microtubules, intermicrotubule spacing, and microtubule polarity were varied to calculate the effect on SHG intensity. (b) SHG intensity has a quadratic dependence on the microtubule polarity as a result of coherent summation. With typical intermicrotubule distances and neurite diameters, the forward-emitted intensities were strongest in the order apical dendrite > basal dendrite > axon > distal dendrite. (The last two neurite types were plotted but were not labeled.) The parameters used for the simulation were $d_{MT} = 64$ nm, $d_N = 3$ μm for apical dendrite, $d_{MT} = 64$ nm, $d_N = 1$ μm for basal dendrite, $d_{MT} = 64$ nm, $d_N = 0.5$ μm for distal dendrite, and $d_{MT} = 22$ nm, and $d_N = 0.17$ μm for axon. (c) The forward-emitting angular distribution of SHG intensity for scatterers arranged in two spatial configurations: on a sheet such as a membrane, or in an array of tubular filaments such as in a neurite as depicted in (a). (Reprinted from Kwan A. C. et al., 2008. Polarized microtubule arrays in apical dendrites and axons. *Proceedings of the National Academy of Sciences of the United States of America*, 105, 11370–11375. Copyright 2008, with permission from the National Academy of Sciences, USA.)

References

Baas, P. W., Black, M. M., and Banker, G. A. 1989. Changes in microtubule polarity orientation during the development of hippocampal-neurons in culture. *Journal of Cell Biology,* 109, 3085–3094.

Baas, P. W., Deitch, J. S., Black, M. M., and Banker, G. A. 1988. Polarity orientation of microtubules in hippocampal-neurons—Uniformity in the axon and nonuniformity in the dendrite. *Proceedings of the National Academy of Sciences of the United States of America,* 85, 8335–8339.

Baas, P. W., White, L. A., and Heidemann, S. R. 1987. Microtubule polarity reversal accompanies regrowth of amputated neurites. *Proceedings of the National Academy of Sciences of the United States of America,* 84, 5272–5276.

Barnes, S. J., Opitz, T., Merkens, M., Kelly, T., Von Der Brelie, C., Krueppel, R., and Beck, H. 2010. Stable mossy fiber long-term potentiation requires calcium influx at the granule cell soma, protein synthesis, and microtubule-dependent axonal transport. *Journal of Neuroscience,* 30, 12996–13004.

Burton, P. R. 1988. Dendrites of mitral cell neurons contain microtubules of opposite polarity. *Brain Research,* 473, 107–115.

Burton, P. R. and Paige, J. L. 1981. Polarity of axoplasmic microtubules in the olfactory nerve of the frog. *Proceedings of the National Academy of Sciences of the United States of America-Biological Sciences,* 78, 3269–3273.

Campagnola, P. J., Millard, A. C., Terasaki, M., Hoppe, P. E., Malone, C. J., and Mohler, W. A. 2002. Three-dimensional high-resolution second-harmonic generation imaging of endogenous structural proteins in biological tissues. *Biophysical Journal,* 82, 493–508.

Chen, S. Y., Hsieh, C. S., Chu, S. W., Lin, C. Y., Ko, C. Y., Chen, Y. C., Tsai, H. J., Hu, C. H., and Sun, C. K. 2006. Noninvasive harmonics optical microscopy for long-term observation of embryonic nervous system development *in vivo. Journal of Biomedical Optics,* 11, 8.

Chu, S. W., Chen, S. Y., Tsai, T. H., Liu, T. M., Lin, C. Y., Tsai, H. J., and Sun, C. K. 2003. *In vivo* developmental biology study using noninvasive multi-harmonic generation microscopy. *Optics Express,* 11, 3093–3099.

Denk, W., Strickler, J. H., and Webb, W. W. 1990. 2-Photon laser scanning fluorescence microscopy. *Science,* 248, 73–76.

Dombeck, D. A., Kasischke, K. A., Vishwasrao, H. D., Ingelsson, M., Hyman, B. T., and Webb, W. W. 2003. Uniform polarity microtubule assemblies imaged in native brain tissue by second-harmonic generation microscopy. *Proceedings of the National Academy of Sciences of the United States of America,* 100, 7081–7086.

Feng, G. P., Mellor, R. H., Bernstein, M., Keller-Peck, C., Nguyen, Q. T., Wallace, M., Nerbonne, J. M., Lichtman, J. W., and Sanes, J. R. 2000. Imaging neuronal subsets in transgenic mice expressing multiple spectral variants of GFP. *Neuron,* 28, 41–51.

Fu, Y., Huff, T. B., Wang, H. W., Wang, H. F., and Cheng, J. X. 2008. *Ex vivo* and *in vivo* imaging of myelin fibers in mouse brain by coherent anti-Stokes Raman scattering microscopy. *Optics Express,* 16, 19396–19409.

Fu, Y., Wang, H. F., Shi, R. Y., and Cheng, J. X. 2007. Second harmonic and sum frequency generation imaging of fibrous astroglial filaments in *ex vivo* spinal tissues. *Biophysical Journal,* 92, 3251–3259.

Gualtieri, E. J., Haupert, L. M., and Simpson, G. J. 2008. Interpreting nonlinear optics of biopolymer assemblies: Finding a hook. *Chemical Physics Letters,* 465, 167–174.

Han, X. X. and Brown, E. 2010. Measurement of the ratio of forward-propagating to back-propagating second harmonic signal using a single objective. *Optics Express,* 18, 10538–10550.

Heidemann, S. R., Landers, J. M., and Hamborg, M. A. 1981. Polarity orientation of axonal microtubules. *Journal of Cell Biology,* 91, 661–665.

Heidemann, S. R. and Mcintosh, J. R. 1980. Visualization of the structural polarity of microtubules. *Nature,* 286, 517–519.

Helmchen, F. and Denk, W. 2005. Deep tissue two-photon microscopy. *Nature Methods,* 2, 932–940.

Hirokawa, N. and Takemura, R. 2005. Molecular motors and mechanisms of directional transport in neurons. *Nature Reviews Neuroscience,* 6, 201–214.

Horio, T. and Hotani, H. 1986. Visualization of the dynamic instability of individual microtubules by dark-field microscopy. *Nature,* 321, 605–607.

Hsieh, C. S., Chen, S. U., Lee, Y. W., Yang, Y. S., and Sun, C. K. 2008a. Higher harmonic generation microscopy of *in vitro* cultured mammal oocytes and embryos. *Optics Express,* 16, 11574–11588.

Hsieh, C. S., Ko, C. Y., Chen, S. Y., Liu, T. M., Wu, J. S., Hu, C. H., and Sun, C. K. 2008b. *In vivo* long-term continuous observation of gene expression in zebrafish embryo nerve systems by using harmonic generation microscopy and morphant technology. *Journal of Biomedical Optics,* 13, 7.

Kemnitz, K., Bhattacharyya, K., Hicks, J. M., Pinto, G. R., Eisenthal, K. B., and Heinz, T. F. 1986. The phase of 2nd-harmonic light generated at an interface and its relation to absolute molecular-orientation. *Chemical Physics Letters,* 131, 285–290.

Kwan, A. C., Dombeck, D. A., and Webb, W. W. 2008. Polarized microtubule arrays in apical dendrites and axons. *Proceedings of the National Academy of Sciences of the United States of America,* 105, 11370–11375.

Kwan, A. C., Duff, K., Gouras, G. K., and Webb, W. W. 2009. Optical visualization of Alzheimer's pathology via multiphoton-excited intrinsic fluorescence and second harmonic generation. *Optics Express,* 17, 3679–3689.

Legare, F., Pfeffer, C., and Olsen, B. R. 2007. The role of backscattering in SHG tissue imaging. *Biophysical Journal,* 93, 1312–1320.

Mertz, J. 2004. Nonlinear microscopy: New techniques and applications. *Current Opinion in Neurobiology,* 14, 610–616.

Mohler, W., Millard, A. C., and Campagnola, P. J. 2003. Second harmonic generation imaging of endogenous structural proteins. *Methods,* 29, 97–109.

Moreaux, L., Sandre, O., and Mertz, J. 2000. Membrane imaging by second-harmonic generation microscopy. *Journal of the Optical Society of America B-Optical Physics,* 17, 1685–1694.

Odin, C., Heichette, C., Chretien, D., and Le Grand, Y. 2009. Second harmonic microscopy of axonemes. *Optics Express,* 17, 9235–9240.

Olivier, N., Luengo-Oroz, M. A., Duloquin, L., Faure, E., Savy, T., Veilleux, I. et al. 2010. Cell lineage reconstruction of early zebrafish embryos using label-free nonlinear microscopy. *Science,* 329, 967–971.

Psilodimitrakopoulos, S., Petegnief, V., Soria, G., Amat-Roldan, I., Artigas, D., Planas, A. M., and Loza-Alvarez, P. 2009. Estimation of the effective orientation of the SHG source in primary cortical neurons. *Optics Express,* 17, 14418–14425.

Rakic, P., Knyiharcsillik, E., and Csillik, B. 1996. Polarity of microtubule assemblies during neuronal cell migration. *Proceedings of the National Academy of Sciences of the United States of America,* 93, 9218–9222.

Rolls, M. M., Satoh, D., Clyne, P. J., Henner, A. L., Uemura, T., and Doe, C. Q. 2007. Polarity and intracellular compartmentalization of Drosophila neurons. *Neural Development,* 2, 14.

Sacconi, L., Dombeck, D. A., and Webb, W. W. 2006. Overcoming photodamage in second-harmonic generation microscopy: Real-time optical recording of neuronal action potentials. *Proceedings of the National Academy of Sciences of the United States of America,* 103, 3124–3129.

Schuyler, S. C. and Pellman, D. 2001. Microtubule "plus-end-tracking proteins": The end is just the beginning. *Cell,* 105, 421–424.

Stepanova, T., Slemmer, J., Hoogenraad, C. C., Lansbergen, G., Dortland, B., De Zeeuw, C. I., Grosveld, F., van Cappellen, G., Akhmanova, A., and Galjart, N. 2003. Visualization of microtubule growth in cultured neurons via the use of EB3–GFP (end-binding protein 3–green fluorescent protein). *Journal of Neuroscience,* 23, 2655–2664.

Stone, M. C., Roegiers, F., and Rolls, M. M. 2008. Microtubules have opposite orientation in axons and dendrites of Drosophila neurons. *Molecular Biology of the Cell,* 19, 4122–4129.

Stoothoff, W. H., Bacskai, B. J., and Hyman, B. T. 2008. Monitoring tau–tubulin interactions utilizing second harmonic generation in living neurons. *Journal of Biomedical Optics,* 13, 9.

Tolic-Norrelykke, I. M., Sacconi, L., Thon, G., and Pavone, F. S. 2004. Positioning and elongation of the fission yeast spindle by microtubule-based pushing. *Current Biology,* 14, 1181–1186.

Waterman-Storer, C. M., Desai, A., Bulinski, J. C., and Salmon, E. D. 1998. Fluorescent speckle microscopy, a method to visualize the dynamics of protein assemblies in living cells. *Current Biology,* 8, 1227–1230.

Williams, R. M., Zipfel, W. R., and Webb, W. W. 2005. Interpreting second-harmonic generation images of collagen I fibrils. *Biophysical Journal,* 88, 1377–1386.

Zipfel, W. R., Williams, R. M., Christie, R., Nikitin, A. Y., Hyman, B. T., and Webb, W. W. 2003. Live tissue intrinsic emission microscopy using multiphoton-excited native fluorescence and second harmonic generation. *Proceedings of the National Academy of Sciences of the United States of America,* 100, 7075–7080.

SHG and Optical Clearing

Oleg Nadiarnykh
VU University

Paul J. Campagnola
*University of
Wisconsin—Madison*

8.1 Introduction

There has been ever-increasing interest in the development of high-resolution optical imaging of the tissue structure for biological and biomedical applications, both *in vitro* and *in vivo*. A special emphasis is being made on minimally and noninvasive clinical imaging systems for diagnostics and monitoring of cancers and other disorders. The most recent additions to the clinical tools are optical diffusion tomography methods based on the measurement of scattering, absorption, and fluorescence from the tissue over a depth of a few centimeters [1]. The main drawback of these methods is their little specificity due to low resolution (~1 mm) that is inadequate for analysis of the microscopic tissue structure, for example, remodeled collagen fibril/fiber assembly in the vicinity of tumors [2,3], disrupted myosin fibers and sarcomeric patterns in muscle disorders [4,5], and various localized dynamic or static inhomogeneities associated with disease states. Much better resolution is available through optical coherence tomography (OCT, 1–2 μm lateral, 3–10 μm axial) and nonlinear optical (NLO) modalities: multiphoton (MPM) and second-harmonic generation (SHG) microscopy (~0.5 μm). Although these techniques rely on near-infrared (IR) wavelengths (700–1000 nm), imaging depths in tissues are still limited to ~0.5–1 mm at best due to high multiple scattering of the excitation and even stronger scattering of emission signals (shorter wavelengths), thus restricting imageable penetration depths for *in vivo* applications. We have shown that the primary filter effect dominates for NLO processes in this spectral region [6]. We note that OCT is carried out at 1300 nm, enabling increased depth, although still lacking true cellular resolution. The biological window ends at approximately 1300 nm, where increased absorption from water overtones becomes significant and the limiting factor.

One of the promising solutions for this intrinsic depth limitation is the optical clearing method, where a high refractive index hyperosmotic agent is added to the tissue to increase its transparency and to improve signal detection. The most accepted mechanism is via refractive index matching that reduces

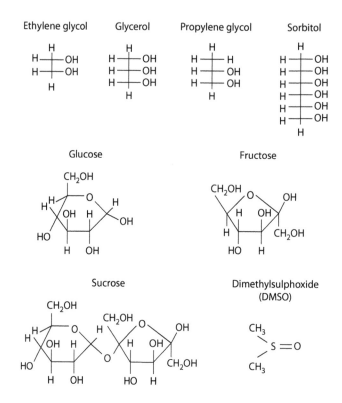

FIGURE 8.1 Chemical structure representations of the most common optical clearing agents.

scattering. Optical clearing for tissues was first proposed by Tuchin et al. [7], and quite encouraging results have been reported by several research groups since then. For example, immersion of a sample of murine tail tendon in a test tube with 50% glycerol solution gives a rather dramatic demonstration of optical clearing. In fact, the sample turns literally transparent, and locating it in glycerol becomes somewhat challenging. Yet, upon transfer to PBS (phosphate-buffered saline), the sample regains its native white appearance [8].

Successful optical clearing *in vitro* has been reported for diverse tissues such as skin (epidermis and dermis) [9–11], cerebral membrane (*dura mater*) [12], blood [13,14], tendon [6], gastrointestinal tissue [15,16], and muscle [17], where imaging was carried out with bright-field, OCT, MPM, and SHG. More recently, a clearing effect was observed *in vivo* in human [10,18,19] and rat skin [20], human and rabbit sclera [21], and human *dura mater* [22] in some cases within as few as 10 min after injection or topical application of optical clearing agents (OCAs). The clearing agents investigated so far include glycerol, glycerol–water solutions, sugars and sugar alcohols, propylene glycol, polyethylene glycol, dimethylsulfoxide (the most typical clearing agents are shown in Figure 8.1), and even sunscreen creams and other pharmaceutical products. This chapter offers an overview of the current understanding of optical clearing mechanism, its capabilities, applicability, and toxicity considerations, with special attention given to SHG imaging.

8.2 Physical Background

Consider a typical soft tissue on a microscopic scale. It is formed by a hydrated network of fibers that hosts groups of cells. The individual tissue components have quite different refractive indices. In other words, tissues consist of scatterers with high refractive index such as fibers and cell organelles, distributed in a medium of lower background index consisting of interstitial fluid and cytoplasm. A photon of light traveling through the tissue encounters a continuous structure with local spatial variations of refractive index as well as step changes of index if the interfaces are sharper in terms of refractive index change.

The index of refraction of the scattering medium n_s can be described as the sum of the average background index n_0 and the mean index variation, $<\Delta n>$:

$$n_s = n_s + <\Delta n> \tag{8.1}$$

As explained by Tuchin [23], the average background index n_0 is merely the weighted average of indices of the cytoplasm ($n_{cp} = 1.367$), and the interstitial fluid ($n_{is} = 1.355$) taken at their respective volume fractions f using the law of Dale and Gladstone

$$n_0 = f_{cp}n_{cp} + (1 - f_{cp})n_{is} \tag{8.2}$$

where we arrive at $n_0 = 1.362$ if we assume about 60% of the total fluid is retained in the intracellular compartment. To estimate the mean index variation within the tissue, we consider the average weighted differences between the respective pairs of refractive indexes (fibers n_{fib} and interstitial liquid n_{is}, nucleus n_{nucl} and cytoplasm n_{cp}, organelles n_{org} and cytoplasm) as follows:

$$<\Delta n> = f_{fib}(n_{fib} - n_{is}) + f_{nucl}(n_{nucl} - n_{cp}) + f_{org}(n_{org} - n_{cp}) \tag{8.3}$$

In fibrous tissues, collagen has a considerably higher refractive index from 1.411 in the human cornea to 1.47 in tendon, depending on the degree of hydration. Taking into account the different contents of collagen in various tissue types (from 3% in nonmuscular internal organs to 70% of fat-free dermis), Equation 8.3 estimates the mean index variation $<\Delta n>$ to range between 0.04 and 0.09.

The effect of the refractive indices mismatch on tissue scattering properties; hence, optical transmittance and reflectance can be examined applying Mie theory, as described by Tuchin [23]. Strictly speaking, Mie theory holds for spherical particles, but its results are still useful for cumulative properties of the tissue composed of irregularly shaped obstacles. For a simplified monodisperse model of dielectric spheres, the reduced scattering coefficient μ_s is given by

$$\mu_s' = 3.28\pi a^2 \rho_s \left(\frac{2\pi a}{\lambda}\right)^{0.37} (m - 1)^{2.09} \tag{8.4}$$

where a is the sphere radius, ρ_s is the sphere volume density, and $m = n_s/n_0$ is the ratio of the refractive index between the scattering particle and the background. This equation shows that μ_s' is a steep function of the magnitude of this mismatch, and it can be reduced considerably if index variation is lowered, and approaches zero for the index-matched case ($n_s = n_0$).

Index matching has been argued as the main mechanism of optical clearing [7,24]. However, due to complexity of interactions between OCAs and tissue components, other mechanisms have been offered, including reversible collagen dissociation [25], tissue dehydration by OCAs [7,26,27], and two studies even claimed no correlation between OCA's refractive index, osmolarity, and optical clearing potential [28,29].

For SHG, in particular, additional consideration of any changes in local packing of molecular sources is important [24]. Since SHG arises from the polarization induced over the noncentrosymmetrically arranged dipoles, the emission is coherent and extremely sensitive to local molecular organization below the optical resolution, where the relevant structure size scales from $\sim \lambda_{SHG}/10$ to λ_{SHG}. Owing to second-order nonsymmetry constraints, SHG is not possible in centrosymmetric environment, whereas the SHG conversion efficiency increases for well-ordered nonsymmetric structures. The efficiency is given by

$$E_{2\omega} \propto E_\omega^2 \frac{\sin(\Delta k L/2)}{\Delta k L/2}$$

where L is the length of the harmonophore assembly in the axial direction and Δk is the phase mismatch, which increases for more random assemblies [30]. As a result, the relative alignment of molecular sources, fibrils, and fibers is reflected in the magnitude of the second-order nonlinear susceptibility tensor $\chi^{(2)}$ and hence in the experimentally observed SHG intensity. Moreover, the SHG directionality and polarization properties contain additional subresolution information. As a consequence of the imperfect phase matching in tissues, the SHG signal has a distribution of emitted forward and backward components, whose ratio we denote as F_{SHG}/B_{SHG}. The extent of mismatch (and resulting directionality) depends on the regularity of the fibril/fiber assembly, where structures that are ordered on the size of λ_{SHG} in the axial direction will give rise to predominantly forward SHG, whereas smaller structures will be less forward directed. Therefore, SHG imaging is a valuable tool to investigate the mechanism of optical clearing in collagen and striated muscle due to structural changes that occur in this process.

Below, we will examine the currently available experimental evidence for the mechanism of optical clearing. Throughout this treatment, the interpretation of the experimental data will be enhanced with theoretical simulations.

8.3 Monte Carlo Simulations

In a tissue imaging experiment, the measured SHG signal will be composed of a convolution of the initially emitted directionality (F_{SHG}/B_{SHG}; described above) as well as the subsequent scattering of these photons at λ_{SHG}. The scattering coefficient, μ_s, is a measure of density, where it is the inverse of the distance a photon will propagate before undergoing a scattering collision and changing direction. For most tissues, these scattering lengths are typically ~20–50 μm in the visible/near-IR region of the spectrum. The scattering anisotropy, g, is related to the directionality of the scattering, and varies from 0 to 1, where higher values correspond to more ordered structural organization. Both the SHG creation and propagation properties undergo changes in the clearing process, as fibril/fiber assembly will be effectively altered by OCA, which will change the $\chi^{(2)}$ tensor and consequently the scattering parameters. The sources of forward- and backward-propagating photons (i.e., from direct emission or from multiple scattering) cannot be determined experimentally, and a Monte Carlo simulation approach is used to decouple these processes. Similarly, the experimentally measured dependence of SHG intensity as a function of depth (i.e., the attenuation) into the tissue cannot be fitted with the single exponential as it represents convolution of two depth-dependent "filter" effects. In general, the combination of three-dimensional (3D) SHG imaging with Monte Carlo simulations of experimental data is a powerful tool for revealing additional subresolution information, as discussed in detail in Chapters 1 and 6.

In the case of optical clearing, the simulations give us insight into the clearing mechanism. Our Monte Carlo approach is based on the framework of Jacques and Wang [31], where we modified this to include all the factors necessary to model a 3D SHG imaging. Monte Carlo simulations are based on photon diffusion using the following bulk optical parameters as inputs: scattering coefficient μ_s, absorption coefficient μ_a, anisotropy of scattering g, and refractive index n. Since all these parameters are wavelength dependent, they must be measured at both the fundamental and SHG wavelengths. Then, simulation algorithm is used to trace the fate of every photon that can be scattered multiple times, and it can be either transmitted or absorbed with certain probability. A flowchart of the simulation algorithm is shown in Figure 8.2 [32].

The laser excitation is attenuated by scattering before it arrives at the focal plane (primary filter). Then, SHG signal travels through the scattering medium until it exits the sample (secondary filter). Obviously, the relative contributions of these two filters vary with depth. Experimental data alone does not allow one to decouple the two processes, but Monte Carlo simulations provide that missing part of the puzzle. To determine these contributions, first, optical sectioning is simulated at the excitation numerical aperture by calculating the fraction of incident laser photons that arrives at the focal point at a given depth and the resulting SHG efficiency is calculated based on the square of the remaining intensity. The secondary filter effects are then modeled by calculating the propagation losses governed

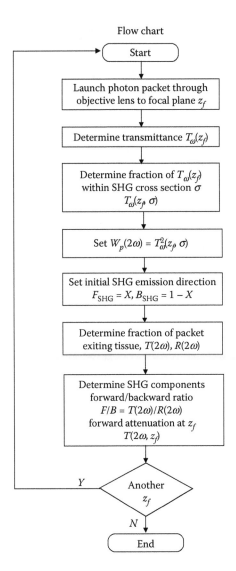

FIGURE 8.2 Flowchart of the algorithm used in the Monte Carlo simulations of the axial dependences of the measured SHG directionality and attenuation. (Reprinted from *Biophys. J.*, 94, Lacomb, R., O. Nadiarnykh, and P. J. Campagnola. Quantitative SHG imaging of the diseased state *Osteogenesis imperfecta*: Experiment and simulation, 4104–4104, Copyright 2008, with permission from Elsevier.)

by the bulk optical properties at the SHG wavelength. Analogously, these simulations can be done for bulk spectral measurements [31]. In the following discussion of optical clearing in muscle and tendon simulation, the results will be addressed along with the experimental data.

8.4 Mechanism of Optical Clearing in Muscle

Optical clearing with glycerol significantly improves the imaging depth in striated muscle as shown in Figure 8.3, where the *xz* projections are reconstructed from optical slices imaged with SHG, and muscle cells are visible as deep as 500 μm [6]. In this work, *ex vivo* muscle slices were treated for several hours with 25%, 50%, and 75% glycerol solutions resulting in the axial attenuation profiles shown in Figure 8.3b in comparison with untreated (control) muscle. As a side effect, glycerol penetration results

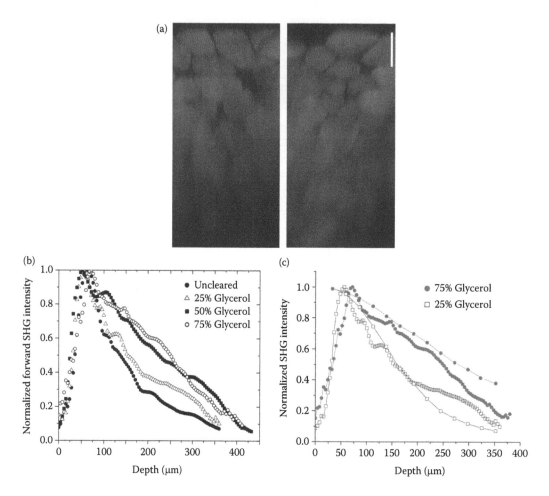

FIGURE 8.3 3-D SHG imaging of cleared striated muscle tissue. (a) *x–z* projections of the SHG image stacks for 25% (left) and 75% (right) glycerol treatments. Scale = 50 μm. (b) The normalized axial attenuation of the forward SHG intensity for the control and three glycerol treatments. All the glycerol treatments result in greater imaging depth relative to the control. (c) Comparison of the simulations and experimental data for the 75% and 25% glycerol treatment, considering a relative SHG creation efficiency of twofold greater for the former treatment. (Reproduced from LaComb, R. et al. 2008. *J. Biomed. Opt.* 13:021108.)

in swelling of the muscle cells, where the higher concentrations result in thicker tissues for the same exposure time. This is due to the hyperosmotic effect of the glycerol, which replaces the lower refractive index cytoplasm. Thus, swelling must be taken into account when directly comparing the data because the physical thickness affects the subsequent measured SHG signal attenuation and directionality. Owing to decreased scattering and absorption losses, the 50% and 75% glycerol treatments result in similar axial attenuation profiles, suggesting that 50% glycerol already provides sufficient clearing effect. The measured bulk optical parameters for the muscle are listed in Table 8.1, where the data for SHG wavelength are approximated to the closest available argon laser line of 457 nm. The values for μ_s' and μ_a were measured by a dual integrating sphere setup. We took the *g* values (0.95) from the literature as reliable values are difficult to obtain by goniometry when *g* is >0.9 and the light is essentially forward scattered, as is the case for highly organized tissues such as the muscle and tendon. At both wavelengths, scattering is at least an order of magnitude stronger than absorption. More notably, 25% glycerol reduces scattering by a factor of 2 over the control tissue whereas 50% and 75% glycerol solution deliver an order

TABLE 8.1   Bulk Optical Parameters for Murine Striated Muscle at the Fundamental and SHG Wavelengths

λ	457 nm				890 nm			
% Glycerol	Control	25%	50%	75%	Control	25%	50%	75%
g	0.96	0.96	0.96	0.96	0.96	0.96	0.96	0.96
μ_s (cm^{-1})	509 ± 225	229 ± 71	71 ± 25	28 ± 2.4	285 ± 113	138 ± 45	29 ± 5	26 ± 6
μ_a (cm^{-1})	9.1 ± 3.1	8.8 ± 1.4	3.8 ± 1.5	3.3 ± 1.2	5.9 ± 1.5	2.9 ± 0.5	1.9 ± 0.5	1.4 ± 0.2

Source: Reproduced from LaComb, R. et al. 2008. *J. Biomed. Opt.* 13:021108.

of magnitude improvement in tissue transparency. The absorption coefficients also decrease possibly due to replacement of cytoplasmic proteins, but absorption contribution is already relatively negligible in the control sample (~10 cm^{-1}). Even though the reduced scattering coefficient is twice higher at the SHG wavelength than the near-IR excitation (890 nm), it is shown in Ref. [6] that the axial attenuation response is largely dominated by the primary filter effect (since SHG signal scales as the square of the excitation intensity) rather than scattering and absorption of SHG photons. Figure 8.3c shows the comparison between the Monte Carlo simulations using the measured optical parameters and the experimental data for the 25% and 75% treatments. In both cases, good agreement is achieved. This analysis shows that the extent of clearing can be predicted in an SHG imaging experiment if the bulk optical properties before and after clearing are known.

The analysis of the experimental and simulated curves of forward-to-backward (*F/B*) ratio of SHG intensities as a function of depth offers additional insight into the optical clearing mechanism. The detected backward SHG consists of direct quasi-coherent emission and a multiple scattered incoherent component in proportions that depend on local packing of dipoles in SHG-producing domain and bulk optical properties, respectively [30]. Optical clearing can affect both these attributes. In the muscle, the whole myofibril (~1–3 μm) is the SHG-producing element, whereas myofibrils stacked together create an extended SHG-producing domain that is larger than λ_{SHG}. Consequently, the initial SHG emission in the muscle is nearly forward directed. This is demonstrated by the simulations plotted in Figure 8.4b. Uncleared muscle is one of the most scattering of tissues. If we compare the

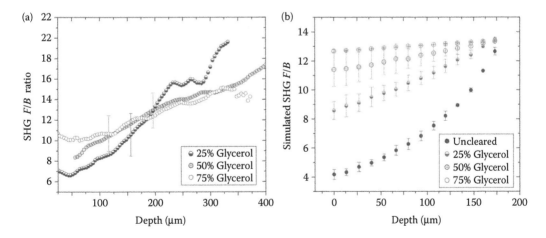

FIGURE 8.4   Axial dependence of the forward-to-backward SHG intensity (*F/B*) ratio measured in the muscle sample cleared in 25%, 50%, and 75% glycerol–PBS solution. The data for the experimental control (uncleared) sample were in the noise floor, and a meaningful *F/B* ratio could not be ascertained. (a) Experimental data. (b) Monte Carlo simulation results assuming 100% forward creation directionality. (Reproduced from LaComb, R. et al. 2008. *J. Biomed. Opt.* 13:021108.)

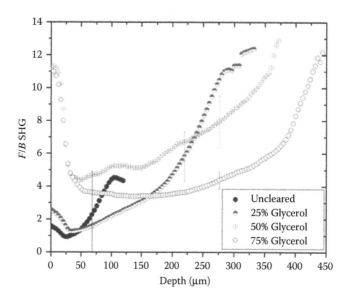

FIGURE 8.5 Axial dependence of the *F/B* SHG intensity ratio measured in the muscle sample cleared in 25%, 50%, and 75% glycerol–PBS solution. (Reproduced from LaComb, R. et al. 2008. *J. Biomed. Opt.* 13:021108.)

simulations to the experiments, we find a good agreement with the overall trends. However, there is a discrepancy between the experiments and simulations for the 25% treatment. This arises because a significant fraction of the backward-propagating photons is scattered into a larger radius and is missed by a microscope objective or elsewhere in the collection path. Since scattering is reduced in samples treated with 25% glycerol, *F/B* ratio can be detected up to the depth of 200 μm (Figure 8.5). The tissues treated with 50% and 75% glycerol solution exhibit improved quality of the backward-detected signal, and *F/B* ratio increases steadily over 400 μm until the remaining thickness falls below one mean-free path of SHG photon. If appreciable initial SHG emission occurred in the backward direction, the *F/B* ratio would be experimentally observable at all depths until the exit from the tissue (500 μm). Therefore, optical clearing has no significant effect on initial SHG directionality in striated muscle. We note that deep in the tissue, both the primary and secondary filters are operative in measuring the backward component accurately (although the measured attenuation is still governed by the square of the primary filter).

While the bulk of the striated muscle is composed of acto-myosin complexes, every muscle cell is wrapped around by a 1–2 μm thick layer of collagenous endomysium. The refractive index mismatch at these interfaces between collagen ($n = 1.47$) and muscle cells ($n = 1.38$) gives rise to strong scattering (see Mie equation 8.4 above). These interfaces are responsible for the oscillations in the attenuation data in Figure 8.3 and as shown previously [17]. We can use the weighted average law of Gladstone and Dale to determine the effective intracellular refractive indices for 25%, 50%, and 75% glycerol treatments, where they become 1.40, 1.43, and 1.45, respectively. Then, it follows from Equation 8.4 that μ_s at SHG wavelength will be decreased by factors of ~2, 6, and 25, respectively, which is in agreement with the data in Table 8.1. The simulation of experimental data also revealed considerably different SHG conversion efficiencies in the cleared and uncleared tissues (e.g., twofold greater efficiency for 25% glycerol, than that for 75% glycerol). This can be explained by 50–100% swelling of the tissue during clearing that causes increased spacing of the myofibrils. Harmonophores from adjacent myofibrils can no longer interact coherently; so, the size of SHG-producing domains is reduced together with SHG field build up. At the same time, the domains do not shorten sufficiently or do not become randomized to support appreciable direct SHG backward emission.

8.5 Optical Clearing in Collagenous Tissues

Analogous experiments coupled with Monte Carlo simulations showed even more pronounced reduction of scattering in murine tendon *ex vivo*. The treatment with 50% glycerol decreased the scattering coefficient μ_s from 510 to 27 cm^{-1} at SHG wavelength, and from 399 to 3 cm^{-1} at 890 nm excitation, Table 8.2 [24]. The reduction of scattering is achieved by replacing interfibril water with index-matching glycerol solution, which increases the spacing between the fibrils and results in swelling of the tissue. Interestingly, in human skin samples treated with high glycerol concentration (75%), electron microscopy revealed that diameters of collagen fibers become significantly smaller [19]. This particular finding helps us to understand better the *F/B* SHG curves for the tendon treated with 25%, 50%, and 75% glycerol [6]. In the previous section, we discussed muscle, where there was no difference in the attenuation observed between 50% and 75% glycerol solutions, as the resulting bulk optical properties were also similar. In contrast, in tendon samples, *F/B* ratio drops from 5 to 3.5 when glycerol concentration is increased from 50% to 75%, causing the individual fibrils to dehydrate and decrease in size. This suggests that SHG-producing domains can now give rise to a significant share of SHG emitted backward directly. Consequently, collagen will be a much better target for *in vivo* imaging with SHG, where only the backward-propagating signal can be detected. Recall that the emission in muscle was highly forward directed even in the cleared case.

It is important to further determine if there are any detrimental effects that accompany optical clearing. One of the controversial subjects of this quest is to understand the deterioration or complete disappearance of backward-detected SHG signal from collagenous tissues treated with high-concentration glycerol solution, as there have been conflicting arguments made by different research groups. Initially, Yeh et al. [25] showed decreased SHG signal from RAFT (real architecture for three-dimensional tissue) models and skin, and suggested the mechanism of reversible collagen dissociation, where clearing agents reversibly affect higher-order collagen structures. Specifically, the interaction between collagen and optical clearing agents screens noncovalent attractive forces, inhibiting collagen self-assembly in the solution and destabilizing higher-order collagen structures at microscopic and ultrastructural levels. In this eventuality, the asymmetry of dipoles arrangement would be broken and no SHG would be produced. Yeh et al. [33] reported a correlation between collagen solubility and optical clearing, where they studied a *self-assembly* of solubilized collagen from murine tail tendon incubated for 24 h with highly concentrated clearing agent.

In a similar study, both dimethyl sulfoxide (DMSO) and glycerol were shown to exhibit a weak concentration-dependent inhibition of the self-assembly of collagen molecules into fibrils *in vitro* [34]. Previously, the efficiency of collagen fibrillogenesis was shown to decrease with the length of molecular chain of clearing agents (e.g., as shown in Figure 8.1) [35,36]. Most recently, Yeh and coworkers [37] published additional experimental and theoretical arguments to support the adverse effects of sugars and sugar alcohols on higher-order collagen structures. The interactions between the clearing agents with collagen are described as destabilizing, but are nonreactive and reversible upon agent removal. These interactions were investigated using a combination of optical spectroscopy, integrating sphere measurements of bulk optical parameters, and molecular dynamics simulations. The simulations revealed a theoretical correlation between the rate of optical clearing and the affinity to form hydrogen bond bridges between a clearing agent and collagen molecules. The study suggested that such a bridge formation disturbs the collagen hydration layer and facilitates water replacement by an OCA. According to

TABLE 8.2   Bulk Optical Parameters for Murine Tendon at the Fundamental and SHG Wavelengths

λ	457 nm		890 nm	
% Glycerol	Control	50%	Control	50%
g	0.96	0.96	0.96	0.96
μ_s (cm^{-1})	510 ± 300	27 ± 7.1	399 ± 44	3.1 ± 2.1
μ_a (cm^{-1})	5.0 ± 1.7	2.1 ± 1.1	2.2 ± 0.3	1.5 ± 0.7

Source: Reproduced from LaComb, R. et al. 2008. *J. Biomed. Opt.* 13:021108.

simulation results, the position of hydroxyl (OH) groups on clearing agent molecules imposes steric constraints on surface bridge formation on collagen triple helix (see Figure 8.6). This finding was validated by experimentally observing the kinetics of skin dehydration by 1,2- and 1,3-propanediol. These two agents have very close molecular weights, refractive indices, and osmolarity, but the 1,3-isomer was found to be more effective after the first hour of clearing. Meanwhile, the steady state after 24 h of clearing showed no difference between the two agents. Additionally, formalin fixation was shown to reduce the efficiency of optical clearing by glycerol in rodent skin and collagen gel [38]. However, since clearing is ultimately achieved in fixed tissues (albeit at much longer exposure times), Hirshburg et al. [37] suggested that the destabilization happens at the molecular level and not at the fiber scale.

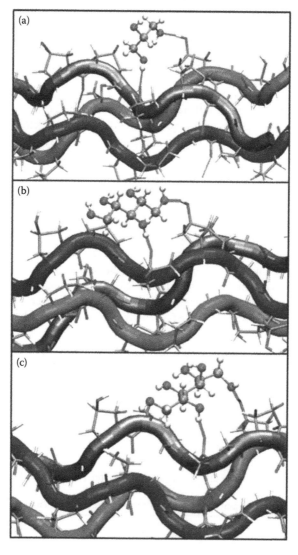

FIGURE 8.6 Typical hydrogen bond bridges in alcohols. The bridge of −OH groups between (a) one and three carbon positions (type II) in glycerol, (b) one and three carbon positions (type II) in xylitol, and (c) one and five carbon positions (type IV) in sorbitol. Higher bridge types, as in (c), span further across the collagen surface and can potentially disrupt collagen–collagen and collagen–water interactions better than lower bridge types (Hirshburg, 2010). (Reproduced with permission from Hirshburg, J. M. et al. 2010. *J. Biomed. Opt.* 15.)

However, in a separate quantitative SHG imaging with both forward- and backward-detected signals from native murine tendon [6], we observed only decreased SHG intensity and threefold tissue swelling, but no adverse effects on fibril morphology in both forward and backward channels. Swelling of the tendon over a course of 5-h clearing with 50% glycerol is shown in x–z projections and in a plot of cross-sectional area increase with time (Figure 8.7). It must be noted that in such experiments, a sample is restricted between a glass slide and a coverslip; so, the swelling is limited and will be different *in vivo* depending on tissue mechanical properties as well. The decrease of SHG intensity in individual optical sections is caused by increase in interfibrillar spacing and thus, lower concentration of SHG-producing dipoles. Interestingly, the overall SHG intensity integrated over the entire cross section of the swollen tissue is in fact two times higher than that in the control sample. This suggests the retention of morphological structure through the clearing process, while the overall higher SHG intensity is consistent with the reduction of scattering. We stress that optical clearing would not be of such an interest for clinical applications were it not reversible. Fortunately, upon washing in PBS, the tendon loses glycerol and returns to its original size with no signs of altered morphology (Figure 8.8).

Both optical clearing mechanisms of collagenous structures as described above seem to be coexisting. In fact, partial disruption (or screening) of collagen helices by sugars and sugar alcohols may well facilitate dehydration and index matching; so, all these processes contribute to the observed optical clearing. However, the collagen dissociation hypothesis suggested earlier would require complete disappearance of fibrillar morphology observed by SHG imaging. The experimental observation of SHG in

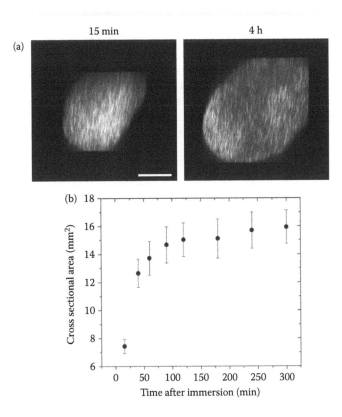

FIGURE 8.7 3-D SHG imaging of 50% glycerol-treated tendon. (a) x–z projections of the SHG image stacks for 15 min (left) and 5 h (right) after immersion. Scale = 50 μm. Significant swelling is observed. (b) The resulting cross-sectional area over a 5 h time course. The area increases by approximately twofold. (Reproduced from LaComb, R. et al. 2008. *J. Biomed. Opt.* 13:021108.)

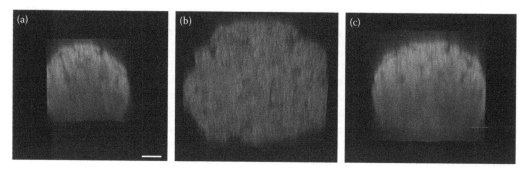

FIGURE 8.8   Optical clearing in the tendon with 50% glycerol is reversible. The 3-D renderings are for the SHG of (a) uncleared tendon, (b) cleared tendon, and (c) tendon following washing in PBS. Scale bar = 20 µm. (Reproduced from LaComb, R. et al. 2008. *J. Biomed. Opt.* 13:021108.)

cleared collagenous tissues proves that the dipoles are still arranged noncentrosymmetrically in optically cleared collagen triple helices. Here, we note that no similar arguments have been raised in regard to the mechanism of optical clearing in the muscle although the collagenous layers of perimysium present in the muscle are subject to the same considerations as described above. However, the clearing in the muscle is not likely to be reversible as the cell membranes are permeabilized in the glycerol–cytoplasm exchange.

8.6 Optical Clearing in Skin

Owing to their regularity and highly organized assembly, striated muscle and tendon were useful examples in which we can examine optical clearing in depth. We now turn our attention to the more challenging imaging environment of the skin. Skin is a far more complex collagenous tissue where optical clearing could offer immediate benefits, as it is readily accessible in clinics. In particular, the dermis is of significant interest due to the wide range of skin pathologies. This structure is a 1–5 mm thick layer sandwiched between the epidermis on top and subcutaneous tissue below and is composed of collagen, elastin, and fibroblasts.

The clearing of the human skin *ex vivo* with DMSO has been shown to introduce morphological changes to the collagen matrix on a submicrometer scale in addition to refractive index matching [39]. Obviously, the reduction of scattering at the price of decomposition of higher-order collagen structures limits the use of DMSO only to *ex vivo* imaging. Glycerol does not exhibit such toxicity, especially at moderate concentrations, since the clearing is reversible as discussed above, and might be a viable agent for *in vivo* imaging.

One of the most recent studies of *in vivo* optical clearing was carried out on the skin, where 20%, 30%, and 75% glycerol was injected in the dermis of rat dorsal skin, followed by recording of reflectance spectra *in vivo*, although SHG images were taken from excised samples [19]. A simple reflectance measurement only 10 min after the injection showed stronger reduction of scattering consistent with increasing concentration of glycerol (see Figure 8.9). For lower concentrations, reflectance recovered and exceeded its initial level within 30 min time frame, and samples swelled due to the hyperosmotic effect. However, in the case of 75% glycerol, reflectance remained very low and the samples became thinner instead of swelling, consistent with almost a 30% decrease in average fibril diameter as measured with electron microscopy (from 109.34 ± 20 to 79.47 ± 13 nm). The thinning of fibrils can be explained by two-step dehydration: first the extracellular matrix is dehydrated, creating a gradient of liquid between the fibers and their surroundings, and then the fibers are dehydrated as well. Neither dissociation of collagen nor any deterioration of fibers was observed in SHG images, and the signal intensity remained strong with all three concentrations. In addition to refractive index matching, more regular packing of collagen fibrils is thought to facilitate optical clearing in the skin [19].

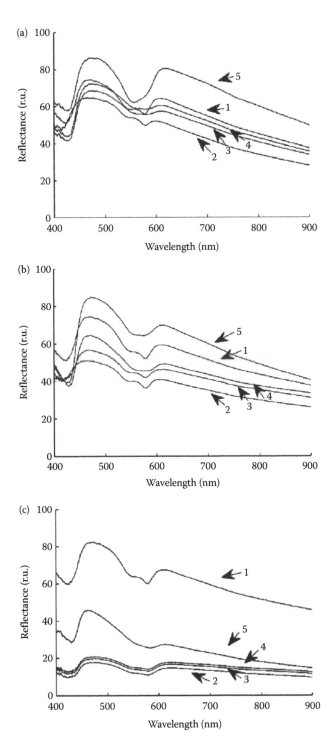

FIGURE 8.9 Typical reflectance spectrum of the skin at 0 min, 5 min, 10 min, and 30 min after injection of (a) 20% glycerol, (b) 30% glycerol, and (c) 75% glycerol, with (1) original spectrum, (2) 0 min spectrum, (3) 5 min spectrum, (4) 10 min spectrum, and (5) 30 min spectrum. (Wen, X. et al. *In vivo* skin optical clearing by glycerol solutions: Mechanism. *J. Biophoton*. 2010. 3:44–52. Copyright Wiley-VCH Verlag GmbH & Co. KGaA. Reproduced with permission.)

The toxicity of glycerol in rat skin *in vivo* was assessed by Mao et al. [29]. While 75% glycerol caused local skin edema (a condition of abnormal accumulation of liquid under the skin) in 24 h, followed by tissue suppuration and necrosis, 30% glycerol cleared the skin without an injury. Thus, it is important to balance between concentrations of OCAs strong enough for sufficient clearing, yet gentle on tissues *in vivo*.

8.7 Optical Clearing in Kidney Tissue

As a final example, we look at optical clearing of the kidney investigated by MPM fluorescence microscopy. The kidney tissue is composed of vast vasculatures, renal corpuscules, and tubules, all covered by fibrous capsules. These numerous interfaces of different refractive indices result in strongly scattering environment, where image contrast deteriorates with depth due to signal attenuation and spherical aberrations. The latter results from yet another refractive index mismatch that we address for the first time in this chapter—between the immersion fluid and the sample [40]. Indeed, although the sample is homogenized in terms of the effective refractive index by the clearing agent, we now have a strong mismatch if a water immersion objective is used. This can be improved with oil immersion objective as demonstrated by Young et al. [41], who investigated clearing capabilities of several glycerol-based agents with refractive indices ranging from 1.3386 to 1.5297. These were prepared by the addition of 2% of 1,4-diazabicyclo [2,2,2] octane (DABCO) and either PBS or benzyl alcohol in various proportions. Additionally, kidneys were intravenously labeled with 0.2 μm fluorescent microspheres distributed through the volume of the tissue to assess the axial resolution with depth. As expected, the penetration depth increases with refractive index of the OCA used due to better index matching within the sample and optimal reduction of scattering. This effect is clearly demonstrated in Figure 8.10 by *XY* images taken at 25, 75, and 125 μm deep, and the reconstructed *XZ* projection, as imaged with 60× NA1.4 oil

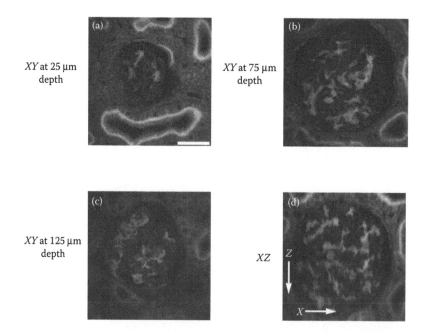

FIGURE 8.10 Two-photon microscopy of the kidney tissue labeled with Hoechst, lens culinaris, agglutinin–fluorescein and phalloidin–rhodamine, and mounted in the media with refractive index 1.51. The image volume is collected with Olympus 60 × NA 1.4 oil immersion objective. The pixel dimensions are 0.345 × 0.35 μm. Scale bar = 30 μm. (Young, P. A. et al. The effects of spherical aberration on multiphoton fluorescence excitation microscopy. *J. Microsc.* 2011. 242: 157–165. Copyright Wiley-VCH verlag GmbH & co. KGaA Reproduced with permission.)

immersion objective. The secondary filter effect discussed earlier is evident here: the shortest emission wavelengths are attenuated faster than the longer ones. There is far more red signal (560–650 nm) at a depth of 125 μm (Figure 8.10c) as opposed to the depth of 25 μm (Figure 8.10a), where blue emission (380–480 nm) dominates. Similarly, this can be seen from *XZ* cross section, where the fraction of red signal increases with depth (Figure 8.10d).

The quantitative analysis of microsphere point-spread functions with respect to depth showed clear difference between the oil and water immersion objectives. Although there is no difference in the achievable depth of imaging, the depth-induced degradation in axial resolution can be eliminated if oil immersion fluid ($n = 1.515$) is matched with a high refractive index OCA. For images taken with water immersion objective (60× N.A. 1.2), a kidney sample cleared by OCA with refractive index of 1.34 had axial resolution just above 1 μm at 40 μm deep into the tissue (Figure 8.11a). Let us compare that

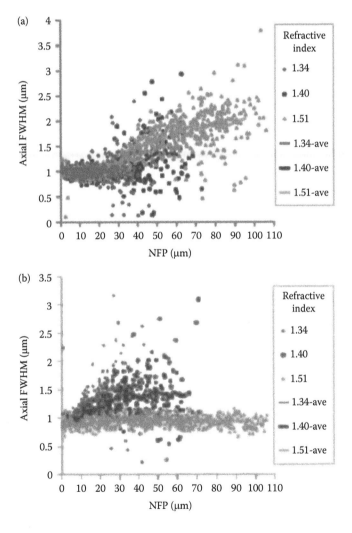

FIGURE 8.11 **(See color insert.)** Axial resolution as a function of depth into kidney tissue measured from two-photon fluorescence of microspheres collected with (a) Olympus 60× NA 1.2 water immersion objective and (b) Olympus 60× NA 1.4 oil immersion ($n = 1.515$) objective. The samples were cleared with agents of different refractive indices as shown in the right panels. (Young, P. A. et al. The effects of spherical aberration on multiphoton fluorescence excitation microscopy. 2011. *J. Microsc.* 242: 157–165. Copyright Wiley-VCH Verlag GmbH & Co. KGaA. Reproduced with permission.)

to OCA with refractive index of 1.51 and axial resolution of 1.5 and 2.0 μm at depths of 40 and 80 μm, respectively. A striking contrast is seen with oil immersion objective where the sample cleared with an OCA with refractive index of 1.51 showed virtually no change in axial resolution over the same depth (Figure 8.11b). By matching all the refractive indices, Young et al. [42] demonstrated the improvement of both the imaging depth and axial resolution in kidney tissue over the entire working distance (150 μm) of a 60× NA1.4 oil immersion objective.

8.8 Alternative Investigation of Optical Clearing: Artificial Environment

As discussed earlier in Section 8.6, biological tissues have sophisticated structures with multiple components that complicate our understanding of optical clearing mechanisms by presenting additional possibilities of nontrivial interactions between the OCAs and biomolecules. These effects can be excluded from investigation by using tissue-simulating phantoms. One of the suitable choices is Intralipid, a brand name for the fat emulsion that is the aqueous suspension of lipid droplets safe for human use. The scattering particles in Intralipid are soybean oil spheres encapsulated by lecithin. Not only can bulk scattering properties of Intralipid be predicted by Mie theory for spherical scatterers, but the measured scattering by vesicles of Intralipid is also conveniently close to that caused by lipid bilayer membrane structures of cells. Moreover, diffusion of an OCA is not an issue with a phantom, while it can present additional problem as in the skin tissue.

In a recent study, quantitative changes of Intralipid scattering properties upon optical clearing were investigated by direct observation, theoretical calculation with Mie theory, and spectral measurement

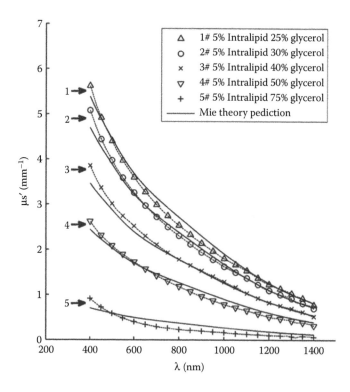

FIGURE 8.12 Mie theory predictions and measurements of the reduced scattering coefficient of 5% intralipid with 25%, 30%, 40%, 50%, and 75% glycerol. (From Wen, X. et al. 2009. Controlling the scattering of intralipid by using optical clearing agents. *Phys. Med. Biol.* 54:6917–6930.)

[43]. The measured dependence of reduced scattering coefficient on wavelength and corresponding theoretical fits is plotted in Figure 8.12. The experimental trends are well approximated by the Mie equation, and, as expected, the reduced scattering coefficient decreases with increased glycerol concentrations. In the case of Intralipid, a very high correlation of 0.97 was found between the reduced scattering coefficient and the OCA refractive index, clearly showing that refractive index matching is the mechanism of optical clearing in the absence of diffusion through tissue and molecular interactions. Consequently, scattering properties of tissue phantoms can be precisely controlled to simulate biological tissues.

8.9 Retention of SHG Polarization Signatures through Optical Clearing

The polarization response exhibited by SHG is one of the exclusive benefits of this optical modality. It is governed by the local geometrical distribution of SHG-producing dipoles that result in a respective tensor of second-order nonlinear susceptibilities [44,45]. It has been shown that dependence of SHG intensity on the polarization of the fundamental wave is related to the pitch angle of the protein helix [46], whereas anisotropy of SHG signal indirectly measures regularity of dipole assembly [47]. Thus, polarization measurement is yet another way SHG imaging can probe submicron structures with resolution beyond any other optical method. For example, whenever protein assembly is altered in disease state, the SHG polarization response will change accordingly. However, the polarization responses in tissues become scrambled due to multiple scattering since each photon collision introduces depolarization, both in fundamental and SHG waves [48]. In fact, at a depth of just a few mean-free paths, additional information related to polarization can be lost completely, as shown in Figure 8.13, which plots the dependence of forward SHG intensity on the angle between collagen fibril and excitation polarization. The data are taken from murine tail tendon at two depths of 5 and 45 μm, the latter showing a complete loss of the typical collagen response [46], since the fundamental wave gets highly depolarized by the time the photons reach the focal volume. The second panel shows the polarization profiles from depths of 10 and 100 μm measured in the tendon cleared with 50% glycerol. Owing to reduction of scattering coefficient by two orders of magnitude, the fundamental retains its polarization and the profile measured deep in the tissue essentially overlaps with the one from the top. However, minor deviations from uncleared sample are evident here as the minimum becomes sharper, whereas both maxima appear wider. This might partially support the argument for alteration of higher-order collagen structures by glycerol.

Similar trends occur in SHG from striated muscle, where we observed that the input polarization is again completely randomized at the depth of 100 μm, although the exact depth threshold has not been experimentally verified. The treatment with 50% glycerol retains the polarization response over the first 100 μm of depth, although considerable depolarization takes place at 180 μm, where amplitude of the curve decreases by about 40% (data not shown).

One way the effects of optical clearing on SHG polarization state can be assessed is by measurement of the SHG signal anisotropy defined by

$$\beta = (I_{par} - I_{orth})/(I_{par} + 2I_{orth}) \tag{8.5}$$

where I_{par} and I_{orth} are components of SHG signal parallel and orthogonal to the polarization of the excitation laser. These are successively measured with the respectively oriented Glan polarizer in the detection path. The signal anisotropy β varies from 1 (all dipoles are aligned with the laser polarization) through 0 (isotropic state, circular polarization) to −0.5 (completely out-of-phase polarization response). When β is measured from samples of any thickness beyond a single isolated fiber, there will be some depolarization present due to scattering. In fact, forward SHG signal that originated at the top of 60 μm tendon is almost depolarized with $\beta < 0.2$, and is completely randomized deeper in the sample (Figure 8.14, squares). Both primary and secondary filter effects are responsible for the loss of

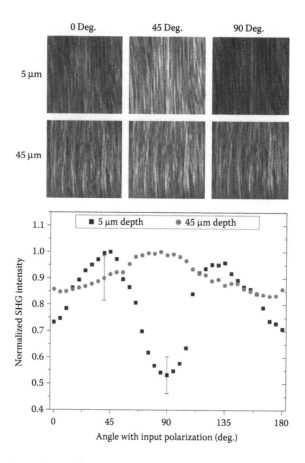

FIGURE 8.13 Angular dependence of the forward SHG intensity from control tendon at depths of 5 and 45 µm. The representative optical sections at several polarization angles are shown in the top panels. The representative error bars (standard error) are given for one laser polarization for each depth. (Adapted from Nadiarnykh, O. and P. J. Campagnola. 2009. *Opt. Express* 17:5794–5806.)

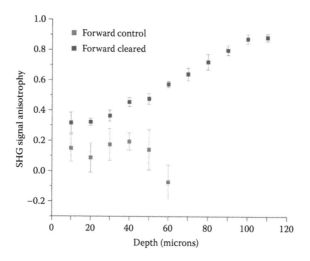

FIGURE 8.14 Forward SHG anisotropy β measured for forward SHG from control (black squares) and cleared (gray squares) tendon. (Adapted from Nadiarnykh, O. and P. J. Campagnola. 2009. *Opt. Express* 17:5794–5806.)

polarization. This is the reason why the tendon cleared in 50% glycerol exhibits a completely different trend of β versus depth, where forward signal anisotropy increases through 110 μm total thickness (doubled due to swelling), and changes from 0.3 at the top of the sample to 0.85 at the last optical plane at the bottom (Figure 8.14, circles).

The anisotropy increases with the decrease in the propagation distance to the exit from the sample. Still, due to residual scattering at both wavelengths, the anisotropy does not reach a value of 1.0 as one would expect for the highly ordered collagen fibrils in the tendon. Again, similar results have been found in the muscle, although the anisotropy values are inherently lower.

Overall, polarization dependencies are largely preserved in both the tendon and striated muscle cleared with 50% glycerol and consistent with the reduction of scattering that we measured by integrating spheres [6].

8.10 Final Remarks

The potential safety issues with the use of clearing agents *in vivo* must be thoroughly investigated, including considerations for the OCA concentrations and exposure times. Optical clearing causes swelling in the striated muscle and tendon, although there is no significant disruption of tissue structure, and the process is reversible in the case of tendon and 50% glycerol. Still, there remains a concern over possible alteration of higher-order protein structures by OCA's. At the same time, shrinking has been observed in glycerol-treated gastrointestinal tissue due to dehydration of the intercellular space [8]. Additionally, all the *ex vivo* samples can be easily cleaned with a uniformly distributed agent over a few hours of incubation. When used *in vivo*, both topically or injected into the tissue, the clearing agent will be subject to further diffusion from the region of interest. Hence, even if toxicity is not a problem, in the clinic, we will not have the luxury of exposure for several hours for complete replacement of interstitial water with an OCA; hence, there is no guarantee of uniform reduction of scattering. Even if optical clearing is limited to *ex vivo* tissues, it still offers advantages over traditional histological staining methods and complements them with subresolution information through a much less time- and labor-consuming approach.

Regarding SHG specifically, the *F/B* ratio of SHG intensity is highly dependent on the extent of optical clearing, due to the effects of both scattered SHG component and of directly emitted backward SHG signal. Fibril separation is also increased compared to that of the native tissue. Most likely the *F/B* metric will not be consistent for diagnostic purposes, although visualization deeper into the tissue has been shown to improve. On the contrary, SHG polarization measurements greatly benefit from optical clearing, since polarization signatures are randomized only in two to three scattering events, but are retained through at least 100–200 μm of cleared tissue, as demonstrated in the muscle and tendon. Finally, the need remains to fully understand the mechanism of optical clearing during complex interactions of OCAs and tissues to make the process predictable and controllable.

References

1. Tuchin, V. V. 2007. Tissue optics: Light scattering methods and instruments for medical diagnosis. SPIE, Bellubgham, WA, USA.
2. Theret, N., O. Musso, B. Turlin, D. Lotrian, P. Bioulac-Sage, J. P. Campion, K. Boudjema, and B. Clement. 2001. Increased extracellular matrix remodeling is associated with tumor progression in human hepatocellular carcinomas. *Hepatology* 34:82–88.
3. Nadiarnykh, O., R. B. Lacomb, M. A. Brewer, and P. J. Campagnola. 2010. Alterations of the extracellular matrix in ovarian cancer studied by second harmonic generation imaging microscopy. *BMC Cancer* 10:94.
4. Plotnikov, S. V., A. Kenny, S. Walsh, B. Zubrowski, C. Joseph, V. L. Scranton, G. A. Kuchel, D. Dauser, M. Xu, C. Pilbeam, D. Adams, R. Dougherty, P. J. Campagnola, and W. A. Mohler. 2008. Measurement of muscle disease by quantitative second-harmonic generation imaging. *J. Biomed. Opt.* 13:044018.

5. Plotnikov, S. V., A. C. Millard, P. J. Campagnola, and W. A. Mohler. 2006. Characterization of the myosin-based source for second-harmonic generation from muscle sarcomeres. *Biophys. J.* 90:693–703.
6. LaComb, R., O. Nadiarnykh, S. Carey, and P. J. Campagnola. 2008. Quantitative SHG imaging and modeling of the optical clearing mechanism in striated muscle and tendon. *J. Biomed. Opt.* 13:021108.
7. Tuchin, V. V., I. L., Maksimova, D. A., Zimnyakov, I. L., Kon, A. H., Mavlyutov, and A. A. Mishin. 1997. Light propagation in tissues with controlled optical properties. *J. Biomed. Opt.* 2:401–417.
8. Wang, R. K. K., X. Q. Xu, Y. H. He, and J. B. Elder. 2003. Investigation of optical clearing of gastric tissue immersed with hyperosmotic agents. *IEEE J. Select. Top. Quantum Electron.* 9:234–242.
9. Khan, M. H., B. Choi, S. Chess, K. M. Kelly, J. McCullough, and J. S. Nelson. 2004. Optical clearing of *in vivo* human skin: Implications for light-based diagnostic imaging and therapeutics. *Lasers Surg. Med.* 34:83–85.
10. Khan, M. H., S. Chess, B. Choi, K. M. Kelly, J. S. Nelson. 2004. Can topically applied optical clearing agents increase the epidermal damage threshold and enhance therapeutic efficacy? *Lasers Surg. Med.* 35(2):93–95.
11. Yeh, A. T. and J. Hirshburg. 2006. Molecular interactions of exogenous chemical agents with collagen—Implications for tissue optical clearing. *J. Biomed. Opt.* 11:014003.
12. Bashkatov, A. N., E. A. Genina, Y. P. Sinichkin, V. I. Kochubey, N. A. Lakodina, and V. V. Tuchin, 2003. Glucose and mannitol diffusion in human *dura mater. Biophys. J.* 85:3310–3318.
13. Tuchin, V. V., X. Q. Xu, and R. K. K Wang. 2002. Dynamic optical coherence tomography in studies of optical clearing, sedimentation, and aggregation of immersed blood. *Appl. Opt.* 41:258–271.
14. Xu, X. Q., L. F. Yu, and Z. P. Chen. 2008. Optical clearing of flowing blood using dextrans with spectral domain optical coherence tomography. *J. Biomed. Opt.* 13:021107.
15. Wang, R. K. K., X. Q. Xu, Y. H. He, and J. B. Elder. 2003. Investigation of optical clearing of gastric tissue immersed with hyperosmotic agents. *IEEE J. Sel. Top. Quant.* 9:234–242.
16. Xu, X. Q. and R. K. K. Wang. 2004. Synergistic effect of hyperosmotic agents of dimethyl sulfoxide and glycerol on optical clearing of gastric tissue studied with near infrared spectroscopy. *Phys. Med. Biol.* 49:457–468.
17. Plotnikov, S., V. Juneja, A. B. Isaacson, W. A. Mohler, and P. J. Campagnola. 2006. Optical clearing for improved contrast in second harmonic generation imaging of skeletal muscle. *Biophys. J.* 90:328–339.
18. Tuchin, V. V., A. A. Gavrilova, G. B. Pravdin, A. N. Bashkatov, E. V. Migacheva, D. Tabatadze, J. Childs, I. Yaroslavsky, and G. Altshuler. 2006. Controlling spectral properties of skin with optical clearing. *Lasers Surg. Med.* 2:2.
19. Wen, X., Z. Mao, Z. Han, V. V. Tuchin, and D. Zhu. 2010. *In vivo* skin optical clearing by glycerol solutions: Mechanism. *J. Biophoton.* 3:44–52.
20. Galanzha, E. I., V. V. Tuchin, A. V. Solovieva, T. V. Stepanova, Q. Luo, and H. Cheng. 2003. Skin backreflectance and microvascular system functioning at the action of osmotic agents. *J. Phys. D: Appl. Phys.* 36:1739–1746.
21. Tuchin, V. V., X. Q. Xu, and R. K. Wang. 2002. Dynamic optical coherence tomography in studies of optical clearing, sedimentation, and aggregation of immersed blood. *Appl. Opt.* 41:258–271.
22. Bashkatov, A. N., E. A. Genina, Y. P. Sinichkin, V. I. Kochubey, N. A. Lakodina, and V. V. Tuchin. 2003. Glucose and mannitol diffusion in human *dura mater. Biophys. J.* 85:3310–3318.
23. Tuchin, V. V. 2005. Optical clearing of tissues and blood using the immersion method. *J. Phys. D-Appl. Phys.* 38:2497–2518.
24. LaComb, R., O. Nadiarnykh, S. Carey, and P. J. Campagnola. 2008. Quantitative second harmonic generation imaging and modeling of the optical clearing mechanism in striated muscle and tendon. *J. Biomed. Opt.* 13:021108.
25. Yeh, A. T., B. Choi, J. S. Nelson, and B. J. Tromberg. 2003. Reversible dissociation of collagen in tissues. *J. Invest. Dermatol.* 121:1332–1335.
26. Oliveira, L., A. Lage, C. M. Pais, and V. V. Tuchin. 2009. Optical characterization and composition of abdominal wall muscle from rat. *Opt. Lasers Eng.* 47:667–672.

27. Rylander, C. G., O. F. Stumpp, T. E. Milner, N. J. Kemp, J. M. Mendenhall, K. R. Diller, and A. J. Welch. 2006. Dehydration mechanism of optical clearing in tissue. *J. Biomed. Opt.* 11.

28. Choi, B., L. Tsu, E. Chen, T. S. Ishak, S. M. Iskandar, S. Chess, and J. S. Nelson. 2005. Determination of chemical agent optical clearing potential using *in vitro* human skin. *Lasers Surg. Med.* 36:72–75.

29. Mao, Z., Z. Han, X. Wen, Q. Luo, and D. Zhu. 2008. Influence of glycerol with different concentrations on skin optical clearing and morphological changes *in vivo*. In 2008 *SPIE Proceedings*, Vol. 7278, art. 72781T.

30. Lacomb, R., O. Nadiarnykh, S. S. Townsend, and P. J. Campagnola. 2008. Phase matching considerations in second harmonic generation from tissues: Effects on emission directionality, conversion efficiency and observed morphology. *Opt. Commun.* 281:1823–1832.

31. Wang, L., S. L. Jacques, and L. Zheng. 1995. MCML—Monte Carlo modeling of light transport in multi-layered tissues. *Comput. Methods Programs Biomed.* 47:131–146.

32. Lacomb, R., O. Nadiarnykh, and P. J. Campagnola. 2008. Quantitative SHG imaging of the diseased state *osteogenesis imperfecta*: Experiment and simulation. *Biophys. J.* 94:4104–4104.

33. Yeh, A. and J. Hirshburg. 2006. Molecular interactions of exogenous chemical agents with collagen—Implications for tissue optical clearing. *J. Biomed. Opt.* 11:014003-014001–014003-014006.

34. Bui, A. K., R. A. McClure, J. Chang, C. Stoianovici, J. Hirshburg, A. T. Yeh, and B. Choi. 2009. Revisiting optical clearing with dimethyl sulfoxide (DMSO). *Lasers Surg. Med.* 41:142–148.

35. Hayashi, T. 1972. Factors affecting interactions of collagen molecules as observed by *in-vitro* fibril formation. 1. Effects of small molecules, especially saccharides. *J. Biochem-Tokyo* 72:749.

36. Kuznetsova, N., S. L. Chi, and S. Leikin. 1998. Sugars and polyols inhibit fibrillogenesis of type I collagen by disrupting hydrogen-bonded water bridges between the helices. *Biochemistry* 37:11888–11895.

37. Hirshburg, J. M., K. M. Ravikumar, W. Hwang, and A. T. Yeh. 2010. Molecular basis for optical clearing of collagenous tissues. *J. Biomed. Opt.* 15:055002.

38. Yeh, A. T., B. Choi, J. S. Nelson, and B. J. Tromberg. 2003. Reversible dissociation of collagen in tissues. *J. Invest. Dermatol.* 121:1332–1335.

39. Zimmerley, M., R. A. McClure, B. Choi, and E. O. Potma. 2009. Following dimethyl sulfoxide skin optical clearing dynamics with quantitative nonlinear multimodal microscopy. *Appl. Opt.* 48:D79–D87.

40. Muriello, P. A. and K. W. Dunn. 2008. Improving signal levels in intravital multiphoton microscopy using an objective correction collar. *Opt. Commun.* 281:1806–1812.

41. Young, P. A., S. G. Clendenon, J. M. Byars, R. S. Decca, and K. W. Dunn. 2011. The effects of spherical aberration on multiphoton fluorescence excitation microscopy. *J. Microsc.* 242:157–165.

42. Young, P. A., S. G. Clendenon, J. M. Byars, and K. W. Dunn. 2011. The effects of refractive index heterogeneity within kidney tissue on multiphoton fluorescence excitation microscopy. *J. Microsc.* 242:148–156.

43. Wen, X., V. V. Tuchin, Q. Luo, and D. Zhu. 2009. Controlling the scattering of intralipid by using optical clearing agents. *Phys. Med. Biol.* 54:6917–6930.

44. Sun, S.-W., C.-Y. Wang, C. C. Yang, Y.-W. Kiang, I.-J. Hsu, and C. W. Lin. 2001. Polarization gating in ultrafast-optics imaging of skeletal muscle tissues. *Opt. Lett.* 26:432–434.

45. Chu, S. W., S. Y. Chen, G. W. Chern, T. H. Tsai, Y. C. Chen, B. L. Lin, and C. K. Sun. 2004. Studies of chi(2)/chi(3) tensors in submicron-scaled bio-tissues by polarization harmonics optical microscopy. *Biophys. J.* 86:3914–3922.

46. Plotnikov, S. V., A. C. Millard, P. J. Campagnola, and W. A. Mohler. 2006. Characterization of the myosin-based source for second-harmonic generation from muscle sarcomeres. *Biophys. J.* 90:693–703.

47. Campagnola, P. J., A. C. Millard, M. Terasaki, P. E. Hoppe, C. J. Malone, and W. A. Mohler. 2002. 3-Dimensional high-resolution second harmonic generation imaging of endogenous structural proteins in biological tissues. *J. Opt. Soc. Am. A* 82:493–508.

48. Nadiarnykh, O. and P. J. Campagnola. 2009. Retention of polarization signatures in SHG microscopy of scattering tissues through optical clearing. *Opt. Express* 17:5794–5806.

9

Holographic SHG Imaging

Etienne Shaffer
École Polytechnique

Christian
Depeursinge
École Polytechnique

9.1 Introduction

Second-harmonic generation (SHG) microscopy has mostly been developed alongside of incoherent non-linear microscopy (multiphoton excitation fluorescence), and has largely contributed to the emergence of coherent nonlinear microscopy (e.g., higher harmonic generation or coherent anti-Stokes Raman scattering). One particularity of coherent nonlinear microscopy is that the detected signals originate from instantaneous interaction of incident electromagnetic radiation with the specimen, making ultrafast imaging possible. This is because these nonlinear interactions are of a scattering nature and, as such, do not involve absorption of light by the specimen. As a consequence, the light source for nonlinear microscopy does not need to be limited to a narrow absorption band, as it is the case in fluorescence, and can be, oppositely, selected to match a spectral window of transmission for the specimen, thus avoiding photo-damage.

Another advantage of SHG—and, more generally, of coherent nonlinear microscopy—lies in its coherent nature, which makes possible not only polarization-based measurements, but also retrieval of both the SHG amplitude and phase with a proper phase-sensitive imaging technique such as holographic interferometry.

The principle of holographic SHG imaging is to record an intensity-only hologram formed from the interference of the second harmonic generated in the specimen with an externally generated second-harmonic reference wave. An interesting consequence of the coherent nature of SHG is that the second-harmonic signals generated in different media by radiation originating from the same light source are mutually coherent. Because of it, both the amplitude and the phase of the second harmonic generated by the specimen may be encoded in the form of a fringe pattern and retrieved by either optical or numerical reconstruction of the hologram. Either way, hologram reconstruction produces 3D images of the SHG intensity contrast (holographic SHG intensity images), but numerical reconstruction also enables quantitative SHG phase contrast (holographic SHG phase images).

9.1.1 The Short History of Holographic SHG Imaging

To the best of our knowledge, holographic SHG imaging was first reported by Ye Pu and Demetri Psaltis (2006). At that time, Pu, a postdoctoral fellow, was working with Psaltis, director of the DARPA Center for Optofluidic Integration at the California Institute of Technology, and the two of them were looking for means to monitor highly dynamic systems. Monitoring highly dynamic systems poses two big challenges. First, it requires high-speed image acquisition, which can only be provided by nonscanning imaging techniques, and second, it needs a 3D imaging scheme that does not suffer from heavy background scattering. It happened that digital holography of second-harmonic signals fulfilled those needs.

In the first years of holographic SHG imaging, only two groups worldwide were active in the field. The only other group, based in Switzerland, was led by Christian Depeursinge, an adjunct professor at the École Polytechnique Fédérale de Lausanne (EPFL), and specialized in digital holographic microscopy. Etienne Shaffer, at the time a doctoral student working in Depeursinge's group, had started investigating possible application of digital holographic microscopy to nonlinearly generated signals, such as SHG. That year, a strange turn of fate had Pu and Psaltis moving their lab to EPFL, thus concentrating all world research on holographic SHG microscopy in one single Swiss institution.

In 2008, Pu et al. detailed the holographic SHG microscopy technique more deeply and investigated its benefits over direct SHG imaging in terms of signal-to-noise ratio (Pu et al., 2008). Then, Psaltis' team dedicated a lot of efforts in the development of highly efficient nanoprobes for SHG imaging and Chia-Lung Hsieh, a doctoral student working in Psaltis' group, demonstrated 3D imaging of these markers by holographic SHG microscopy (Hsieh et al., 2009). A year later, Hsieh reported successful antibody conjugation for specific labeling (Hsieh et al., 2010b), as well as its use for phase conjugation imaging (Hsieh et al., 2010a,c).

During that time, Depeursinge's team devoted its efforts to exploitation of the SHG phase signal. Shaffer et al. reported the first representation and interpretation of the SHG phase and, as a proof of concept, demonstrated that it could reveal the polarization component of the focused laser illumination responsible for generation of second harmonic at a glass/air interface (Shaffer et al., 2009). One year later, they established the relation between the detected SHG phase, the medium refractive index and the axial position at which second harmonic is generated. This SHG phase relation made possible nanometer-scale, 3D tracking of SHG-emitting nanoparticles (Shaffer et al., 2010a). At the same time, SHG phase was also proposed as a contrast agent for imaging of label-free biological sample (Shaffer et al., 2010b), revealing phase-matching conditions.

Earlier that same year, Omid Masihzadeh, a doctoral student working with Randy Bartels, associate professor at the Colorado State University, reported on the label-free holographic SHG intensity images of biological specimen (Masihzadeh et al., 2010).

Second-harmonic interferometry has been around for some time (Chang et al., 1965), but its use for imaging purposes, as in the so-called interferometric SHG microscopy (Yazdanfar et al., 2004), is somewhat newer. Even more recent is holographic SHG imaging; its nonscanning counterpart recently made possible by the development of both ultrafast lasers and very sensitive digital sensors. While holographic SHG imaging has only started gathering interests and enthusiasm, it has already paved the way to other nonlinear holographic imaging techniques (Xu et al., 2010b). In the years to come, and as ultrafast lasers and digital sensors technology will develop, holographic SHG imaging is expected to become of even greater interest in the scientific community.

9.1.2 Chapter Overview

In this chapter, we aboard holographic SHG imaging by first introducing the reader to the basic principles of holography. We make the distinction between classical and digital holography and discuss the opportunities and related drawbacks the digital era has brought to the field. This section is very important since we address holographic SHG imaging only through digital and not classical media.

Then, we explain how a holographic SHG setup can be implemented and SHG holograms be digitally recorded. Discussing the possible setup implementations, we insist on some key components, most especially light sources and detectors, to explain both why holographic SHG has only recently been made possible and why it appears more and more appealing. Addressing the core of the technique, we describe the numerical reconstruction process of holograms that yield both amplitude and phase of the object wavefront, before commenting further on the types of image contrast accessible with holographic SHG imaging.

At last, we review a few fields of application where holographic SHG imaging has been reported. Notably, we describe its application to imaging of biological structures. Also, we explain how retrieval of 3D images from a single hologram is made possible by the extended depth of field peculiar to digital holography and, more particularly, how this is of interest for real-time tracking of nanoprobes. Finally, we also comment on phase conjugation imaging based on holographic SHG characterization of turbid media.

9.2 Principle of Holography

9.2.1 Introduction

Most light detectors, human eye included, are directly sensitive only to the intensity; they neither perceive nor record the phase or the polarization state. While this is of minor importance when working with incoherent, unpolarized light sources, it is of greater consequences with polarized and/or coherent light sources, since any phase- or polarization-related information goes missing when light is recorded by a detector sensitive to intensity only.

Holography is an interferometric method by which the amplitude and phase of light can be recorded by an intensity-sensitive detector and successively retrieved. The name holography originates from the Greek *holos*, meaning whole, and *grafe*, meaning writing or drawing. Holography differs from standard interferometry by being an imaging technique. Its invention is attributed to the Hungarian-British physicist Dennis Gabor in 1948 (Gabor, 1948) and earned him the Nobel Prize in Physics in 1971.

9.2.2 Coherence

As holography requires spatially and temporally coherent light sources, we will introduce the reader to the notion of coherence. Coherence describes the correlation properties of the physical quantities of waves. An electromagnetic wave is said to be coherent when there is a fixed phase relationship between its electric field values at different locations or at different times.

The temporal coherence is a measure of the degree of monochromaticity of a wave. The narrower the spectral bandwidth of a wave, the higher its temporal coherence will be. Temporal coherence is measured in time units, and the coherence time is the time interval within which the phase of the wave is, on average, predictable. Sometimes, however, it is more convenient to quantify the temporal coherence in distance units. After all, temporal coherence is often measured with a Michelson interferometer in which the length of one arm is varied. In such cases, the coherence length simply

TABLE 9.1   Coherence Lengths and FTL Pulse Duration of Gaussian Pulses, for Different Pulse Bandwidths

Bandwidth (nm)	10	12	15	20	30	40	60	80	100
Coherence length (μm)	28	24	19	14	9	7	5	4	3
FTL pulse duration (fs)	94	78	63	47	31	23	16	12	9

corresponds to the distance a wave travels in a time interval corresponding to the coherence time. A light source having a Gaussian spectrum centered at wavelength λ_0 and a bandwidth $\Delta\lambda$ has a coherence length given by

$$L_C = \frac{2\ln 2}{\pi} \cdot \frac{\lambda_0^2}{\Delta\lambda}. \tag{9.1}$$

As an indicator, helium–neon lasers will have high temporal coherence (a few tens of centimeters), while femtosecond laser will have much smaller temporal coherence (a few tens of micrometers at most). Coherence lengths of ultrafast lasers, for several pulse bandwidth, can be found in Table 9.1. Interestingly, Fourier-transform-limited (FTL) ultrafast lasers have a coherence length that actually corresponds to the spatial extension of the pulse, given by the product of its duration with the speed of light. The corollary is that non-FTL ultrafast lasers have a coherence length that is smaller than the spatial extension of their pulses.

The spatial coherence is a measure of how predictable, on average, is the phase of a wave at different spatial coordinates. More precisely, the spatial coherence is the time-independent cross-correlation between two points in a wave. Lasers all have relatively good spatial coherence properties resulting from stimulated emission of light. In opposition, incandescent light bulbs have very low spatial coherence properties. A high spatial coherence source can however be made from an incandescent light bulb by diffraction through a very small aperture, as in Young's double-slit experiment of 1803.

Because holography is based on interferometric principles, a light source having relatively good coherence properties is needed to record holograms. Similarly, the polarization state of the light source, and how predictable it is, on average, also impact the recording of a hologram.

9.2.3 Hologram Recording

In holography, the phase and amplitude of the light diffracted by an object is encoded in an intensity-only image, called hologram, by means of interference with a reference wave (Figure 9.1). Using standard nomenclature, the light diffracted by the object is referred to as the object wave and is designated by o. Similarly, the reference wave is designated by r and is generally a plane or spherical wave. Of course, both o and r are complex functions of space and time, but only their time-independent behavior (wavefront) is of interest here. Indeed, the oscillation frequency of visible electromagnetic waves reaches almost to the petaHertz range, that is, orders of magnitude faster than the integration time of any detector. Therefore, only the time-averaged wavefront can be detected. Mathematically, the intensity I resulting from interference of o with r at the detector plane can be expressed by

$$I = (o \quad r)\begin{pmatrix} 1 & g_{o,r} \\ g_{o,r} & 1 \end{pmatrix}\begin{pmatrix} o \\ r \end{pmatrix}^*, \tag{9.2}$$

where the star symbol (*) denotes the complex conjugate. For interference to occur, o and r must have some mutual (temporal, spatial, and polarization) coherence properties, expressed here by the normalized mutual coherence function $g_{o,r}$.

If the object and reference waves have no mutual coherence ($g_{o,r} = 0$), then $I = |o|^2 + |r|^2$ is simply the sum of the respective intensity of o and r. In such case, no holographic information exists. Oppositely, if

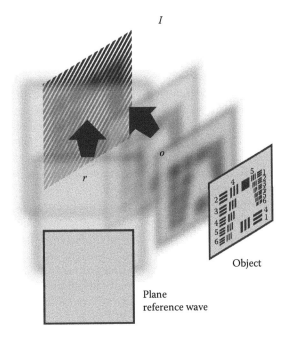

FIGURE 9.1 Optical recording of a hologram in an off-axis configuration: a plane reference wave *r* interferes with an object wave *o* to produce an intensity-only hologram *I*. This off-axis configuration takes its name from the nonzero angle subtended by the direction of *r* with respect to that of *o*.

the object and reference waves exhibit perfect mutual coherence ($g_{o,r} = 1$), then $I = |o|^2 + or^* + ro^* + |r|^2$ provides an ideal support for the holographic information, contained in the cross-terms that are referred to as imaging terms. These terms produce an interference pattern, in which the successive dark and bright fringes can be regarded as contour lines expressing the phase difference between the object and the reference waves.

Classically, a photo-sensitive plate is used to record the hologram. After exposure, the plate is photochemically developed and only then is the hologram revealed as a plate of transmissivity proportional to *I*. This process is, in itself, quite complicated. Indeed, exposure has to be rigorously controlled, even more than with film photography, to ensure working in the linear regime of the photo-sensitive medium. As if this is not enough, all development steps, requiring multiple chemical baths, also have to be precisely controlled. Nevertheless, this art has been mastered both for scientific and artistic purposes by the holographists of the time and has helped establish the foundations for the later digital holography.

It is not the purpose of this section to go into any more details concerning recording of holograms in classical holography. Interested readers should refer to dedicated books (Collier et al., 1971; Hariharan, 1996).

9.2.4 Hologram Reconstruction

The hologram is a diffraction grating: its dark and bright fringes form a structure on which light can diffract to restitute all the information about the amplitude and the phase of the object wavefront. This is how the information contained in the hologram is classically revealed.

To understand how this works, let us assume that the photo-chemically developed hologram plate has a transmissivity corresponding exactly to *I*, and is illuminated by a wave *u*. After expliciting *I* according to Equation 9.2, the transmitted wave *ψ* can be expressed as

$$\psi = uI = u|o|^2 + ug_{o,r}or^* + ug_{o,r}ro^* + u|r|^2. \tag{9.3}$$

In the interesting case where the hologram is illuminated by a replica of the reference wave used for its recording ($u = r$), the hologram would diffract part of that light in a way that mimics the object wave. This is illustrated in Figure 9.2a. By replacing u with r in Equation 9.3, one sees that the second term of ψ produces a replica of the object wave, weighted by the product of mutual coherence function with the intensity of the reference wave. Consequently, an observer looking at the hologram along the direction indicated by the viewpoint arrow would see an image with the apparent depth and parallax properties identical to that of the real object. This image is called the virtual image, since placing a detector (or screen) at the position of the image would not reveal it.

Another interesting case is that for which the hologram is illuminated by a wave replicating the conjugate of the reference wave used for recording ($u = r^*$). In this case, the hologram would diffract part of the light in the direction of the object, as illustrated in Figure 9.2b. That diffracted light would mimic the complex conjugate of the object wave, as indicates the third term of Equation 9.3, in which u would be replaced by r^*. This means that light diffracted from the hologram would form a real image of the object, exactly at the position it was located for the recording of the hologram.

Sometimes, the two cross-terms forming the images we have just described are called terms of ± 1 order of diffraction. With such nomenclature, the first and last terms of Equation 9.3, that do not carry holographic image information, are consequently referred to as the zero-order terms.

9.2.5 Elimination of Zero-Order Terms and Twin Image

From a holographic imaging point of view, the zero-order terms are a nuisance. In the worst possible case, where o and r are collinear, the zero-order terms are completely superimposed on both imaging terms and thus deteriorate the image quality.

A very simple solution to diminish the influence of the zero-order terms is to record the hologram using a plane reference wave of intensity much stronger than that of the object wave. This way, oo^* is negligible compared to the imaging terms, and rr^*, because it is uniform, much less disturbs the image.

But even then, the two imaging terms will overlap, one being in sharp focus, and its twin image being out of focus. This problem was identified long ago, and many methods were developed to get rid of the noise introduced in the images by the zero-order and twin image terms. The most widely used method simply consists in using a reference wave subtending a nonzero angle with the object wave, as in Leith and Upatnieks (1962). In this so-called off-axis configuration, the angle between the two waves modulates the two imaging terms in such a way that illuminating the hologram with a reconstruction wave u will see the two imaging terms diffract in different directions. For large off-axis angles and/or for large object-to-hologram distances (vs. the object dimensions), complete spatial separation of the imaging and zero-order terms can be achieved. Exact retrieval of the object wave, that is, unaltered by the presence of the twin image nor the zero-order terms, then becomes possible.

However, this method requires highly coherent light sources, such as lasers. To understand why it is so, let us consider a plane object wave propagating along the optical axis, perpendicular to the detector's surface, and a plane off-axis reference wave, as illustrated in Figure 9.1. At different points on the detector where I is measured, the reference wave will have traveled a different optical path length. In opposition, the object wave would have traveled the same optical path length, regardless of where it hits the detector. Therefore, for interference to occur over the entire detector, the coherence length of the source should be larger than the optical path length difference of the reference wave at all points on the detector. If not, then interference will appear only a region smaller than the size of the detector. This point will be further discussed in Section 9.4.2, for the case of digital holography.

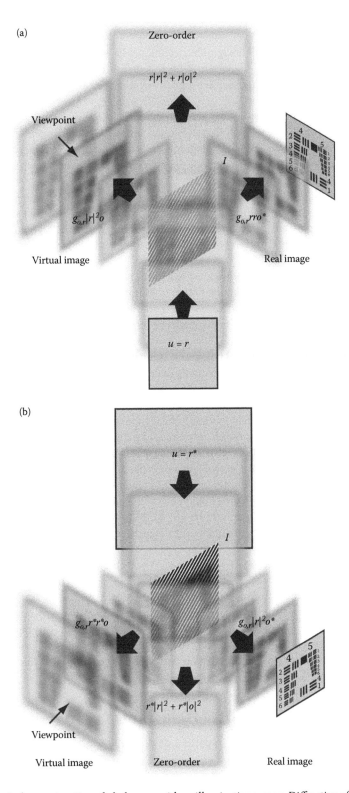

FIGURE 9.2 Optical reconstruction of a hologram with an illumination wave u. Diffraction of u on the hologram produces real and virtual images of the object. (a) Reconstruction with $u = -r$ and (b) reconstruction with $u = r^*$.

9.2.6 Going Digital

By digital holography, we refer to all holographic techniques in which the image reconstruction is performed numerically, from a digital hologram. With this definition, we exclude the field of computer-generated holograms, in which holograms are digitally synthesized and exported to physical form for optical image reconstruction.

Digital hologram recording, and subsequent numerical image reconstruction, is attributed to Goodman and Lawrence (1967) and was first demonstrated in 1967. Their work was motivated by the desire to obtain an electronic detection of holograms at the earliest possible opportunity, that is, when the signal-to-noise ratio is the highest, to provide the high sensitivity needed to detect very weak objects of small angular subtense.

Because digital holography differs from its classical counterpart on both hologram recording and reconstruction steps, it is no surprise that it offers unique advantages and drawbacks related to these two key steps of the holographic process. Compared to classical holography, digital holography offers many advantages for hologram recording. For one, digital format makes possible stable, long-term storage of holograms.

Much more importantly though, the time for hologram recording and reconstruction has dropped by a few orders of magnitude. Where it took hours, or at least minutes, to correctly develop and optically reconstruct a photographic plate, transferring a frame from the sensor to the computer, which makes the numerical reconstructions, is now a matter of milliseconds. Thanks to the advent of personal computers, made more accessible in the 1980s, and to the emergence of integrated solid-state digital sensors, live, video-rate reconstruction of holograms is now easily achievable. With these technological advances, digital holography has seen an increasing interest that fueled the development of the field and really allowed it to take off.

The ease of changing the exposure time of the sensor should not be neglected, as it gives, in combination with live histogram display, the tools to set the right exposure level. This is rather important as best results are obtained for highest possible intensities that do not result in overexposure (saturation) of some pixels. Overexposure should be avoided, because it introduces nonlinearities in the signal by adding harmonics into the hologram spectrum. It is considered as a noise source which can fortunately be avoided by appropriate exposure adjustment, which is made easy by the digital technology.

The one big challenge intrinsic to digital holography is the adequate separation of diffraction orders (imaging and zero-order terms). Digital sensors are, to this day, essentially 2D detectors that cannot record volume holograms. All information about the object wavefront therefore has to be recorded on this 2D surface which possesses relatively large pixel pitches that limit the maximum off-axis angle still resulting in appropriately sampled interference fringes. Overall, these characteristics of digital sensors limit the separation power of diffraction orders in digital holography. Fortunately, judicious choices in the optical configuration may still allow separation of diffractions orders by off-axis scheme. Other dedicated methods, such as phase-shifting interferometry, were also developed to overcome the problem.

Overall, the digital era, by making holography capable of high-speed imaging, has made it very appealing for a wide range of new applications requiring fast frame rates. In principle, it also made holography capable of truly exploiting the ultrafast nature of nonlinear light-matter interactions, which is of great interest for SHG imaging.

9.3 Recording of Digital SHG Holograms

As proposed by Goodman, digital holography involved recording of holograms on an orthicon, that is, a video camera tube similar in operation to a cathode ray tube. Such pixel-scanning recording devices convert light intensity to an electrical signal that can in turn be digitalized. For example, Goodman's orthicon converted the optical hologram to a 256×256 pixels digital hologram with intensity values quantized to eight gray levels (3-bits) only.

The invention in the late 1960s of digital detectors such as charge-coupled devices (CCDs) and active-pixel sensors (also known as CMOS sensors) provided the ideal solution for digital recording of optical holograms. These two-dimensional devices directly convert light intensity to electric charges that can be quantized by a built-in analog-to-digital converter. More importantly, they can be manufactured at relatively low cost and they easily provide video or better frame rate. Today, most of the work carried in digital holography is based either on CCD or CMOS sensors.

Despite completely new technology that sees digital sensors replacing photographic plates, schemes for recording of holograms in digital holography are very similar, if not identical, to those in classical holography. In turn, recording of SHG holograms is very much alike recording of digital holograms in bright-field digital holography, with the main differences being attributed not to the digital recording medium, but to all the required apparatus to generate the second-harmonic object and reference waves.

9.3.1 Setting up a Holographic SHG Microscope

All digital holographic microscopes, no matter how they are physically implemented, comprise the same functional groups of elements. In this section, we present implementations of transmitted light (Figure 9.3a) and reflected light (Figure 9.3b) off-axis holographic SHG microscopes.

It should however be noted that to this day, we are unaware of any reflected light holographic SHG microscope and only the transmitted light types have yet been reported. In this sense, the description of

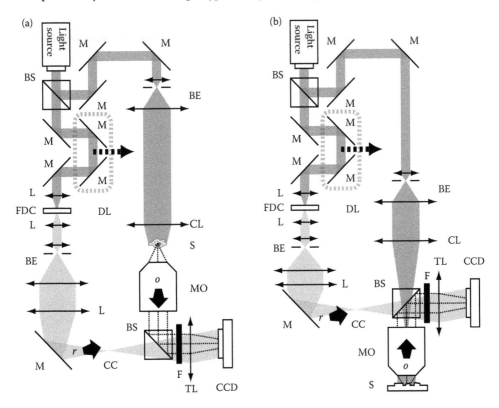

FIGURE 9.3 Schematics of typical holographic microscopes in a Fresnel configuration. (a) Transmission holographic SHG microscope and (b) reflection holographic SHG microscope. BS, beam splitter; L, lens; M, mirror; BE, beam expander; FDC, frequency doubler crystal; CL, condenser lens; MO, ∞-corrected microscope objective; TL, tube lens; *o*, object wave; *r*, reference wave; S, specimen; F, filter, and CCD digital sensor. Beam envelopes are represented by uniform gray shades (dark for fundamental and pale for the second harmonic), while image formation is represented by dashed rays.

the reflection setup we provide here should be viewed as a setup proposal and not as the description of a state-of-the-art microscope for backscattered second-harmonic signal.

The two implementations of Figure 9.3 are based on Mach–Zehnder interferometers which, because they are of very simple design and do not have folded optical path like Michelson's, are very convenient for generation and shaping of the reference and object waves. In fact, they are up to now the only type of interferometer reported for holographic SHG microscopes. In the setups of Figure 9.3, a pair of beam splitters are used to separate the laser beam into the reference and object arms, and to recombine them in the infinity space of the microscope objective.

A variable optical delay line (DL) in the reference arm makes possible to match very precisely the optical path of both arms, so that o and r may interfere on the digital sensor (CCD). Coherence lengths of femtosecond lasers are typically in the order of tens of micrometers so temporal coherence overlap has to be very precisely adjusted.

The second-harmonic reference wave is generated by focusing the beam in a frequency doubler crystal (FDC). It is then collimated and expanded by pair of lenses with different focal lengths. A pinhole may be introduced in the focal plane of the beam expander (BE) to clean the beam of any high-frequency content. A similar beam expander is located in the object arm to match the beam size to the clear aperture of the condenser lens (CL).

The condenser lens very loosely focuses the beam on the specimen. Since holographic SHG microscopy is a bright-field technique, the illumination must cover the entire field of view. This constitutes a major difference with scanning SHG microscopes, where the illumination is focused on a very tight spot. There are two very simple ways to ensure that the illumination is large enough. One is to focus the illumination at some distance below or above the specimen. The other is to use a condenser lens of long focal length and/or small diameter that will make a diffraction limited spot matching the field of view. This last method is generally preferable, since it makes the specimen lie in the waist of the beam, where the wavefront is more or less planar. This becomes even more important when investigating the second-harmonic response of the specimen with regard to the incident polarization. Although not illustrated in Figure 9.3, a Köhler illumination could also be used, since, after all, holographic SHG microscopy is based on a bright-field illumination.

To record holograms, it is preferable to have the reference wave (instead of the object wave) subtending the off-axis angle with the optical axis. This way, the object wave can propagate along the optical axis and hit the sensor perpendicular to its surface. Such arrangement is not only easier to align, but also makes the image plane parallel to the sensor surface. In the proposed implementations, an infinity-corrected objective is used and its tube lens serves both for forming the image of the object and to collimate the reference wave. Laterally moving the center of curvature (CC) of the reference wave gives it an off-axis angle upon collimation by the tube lens. One advantage is that the recombining beam splitter is located in the infinity space of the objective, where the diffraction orders propagate parallelly.

Finally, for both reflected-light and transmitted-light implementations, the digital sensor (CCD) is positioned at some distance before the system image plane, to record out-of-focus SHG holograms in the Fresnel configuration.

9.3.2 Key Elements of a Holographic SHG Microscope

This section discusses the key elements needed in a typical holographic SHG microscope. These elements include the laser source and the detector, essential to any digital holographic setup, but also elements specific to holographic SHG imaging like, for example, optical delay lines and frequency doubling crystals.

9.3.2.1 Ultrafast Laser Source

The most important element of a holographic SHG microscope is, without any doubt, its laser source. Putting it this way is probably an understatement, since nonlinear microscopy, in general, exists only thanks to development of ultrafast lasers: without ultrashort pulse lasers, there would be no nonlinear microscopy.

Nevertheless, the development of ultrafast laser sources is probably more important to holographic SHG than it is to other nonlinear microscopy techniques. The many reasons for this can all be summarized in one: holography is a nonscanning technique. This means that to produce an image, the light source must illuminate the entire field of view, as it is the case with bright-field microscopy. Therefore, to deliver the same peak power density to the specimen that of a confocal scanning microscope, the laser source for holography has to be much more powerful, in terms of mean power. Very high-power lasers would make possible to illuminate a large region of the specimen, thus providing a large field of view, while keeping the local peak power densities similar to those used in scanning microscopy, to generate comparable signals.

Provided very fast digital cameras, the maximum frame rate becomes limited by the quantity of second-harmonic generated photons that produce the hologram contrast. Therefore, the more powerful the laser source, the larger the possible field of view and the higher the maximum frame rate. At least, until photo-induced damage becomes an issue.

Photo-induced specimen damage can be a problem in nonlinear microscopy. The weak signals of interest are an incentive to increase the laser power. There is however a limit to the energy one can deliver onto the specimen without damaging it. That limit varies from specimen to specimen and depends on the wavelength of the laser. The specimen damage threshold is related to the total absorbed energy per unit volume per unit of time. As typical duration of ultrafast laser pulses are much smaller than the characteristic time of bio-physical phenomena, for example, heat dissipation, this limit can be reported to a total absorbed energy per unit volume per pulse, assuming that the time lapse between two pulses is large enough to allow the specimen to return to its equilibrium state. As the absorption depth is generally limited by the specimen thickness or extinction coefficient, these parameters can be reported in terms of energy per surface. For a given pulse duration, the striking conclusion to this is that delivering more energy over a proportionally larger surface maintains both the SHG signal per unit surface and the risk of photo-induced damage constant.

For the same pulse duration and energy per pulse, the SHG intensity scales linearly with the repetition rate. The higher the repetition rate, the stronger the SHG signal. When selecting a laser for holographic SHG microscopy, the thumb rule for the repetition rate thus is: the higher the better, as long as the time between two successive pulses is long enough to allow return of the specimen to equilibrium state. As an indicator, 80 MHz repetition rate lasers generally do not pose a problem. If also specimen-safe, a 2.0 GHz laser with comparable pulse duration and energy per pulse would yield a 25 times enhancement of the SHG signal.

The amplitude of the second harmonic generated varies with the square of the amplitude of the instantaneous electric field induced by each laser pulse. For given energy per pulse, the shortest the pulse duration, the higher the instantaneous electric fields and, in turn, the strongest the generated SHG. Of course, as the pulse duration becomes very short, its spectrum broadens and its temporal coherence declines—see Table 9.1. The consequence of this is a much reduced area on the detector where interference occurs between the object and the reference wave. This effect, called beam walk-off, could pose a serious problem for holography. Fortunately, different physical implementations of the holographic interferometer have already been proposed to overcome the beam walk-off problem (Maznev et al., 1998; Ansari et al., 2001). With such implementations, it is expected that the large spectrum broadening that comes with much reduced pulse duration would not pose a problem to the holographic scheme.

9.3.2.2 Detector

Being a nonscanning imaging technique, holographic SHG requires a full 2D digital sensor, typically a CCD or CMOS camera. The nature of holography and of SHG imposes some requirements on the specifications of such sensors. When looking for a digital sensor, specifications like frame rate, exposure time, and gain range (and their increments) immediately come to mind. However, apart maybe for fluorescence imaging that requires very long exposure times, these specifications are not so much related to the imaging technique used, but rather to the imaged subject. When looking for a sensor for digital holography, one must consider some other important specifications, like pixel pitch and total number of pixels.

One of the major problems of digital imaging in general is the shot noise. Shot noise is a type of noise that occurs when the finite number of particles that carry energy is small enough to give rise to detectable statistical fluctuations in a measurement, according to Poisson statistical distribution. Shot noise is independent of the quality of the electronic of a sensor and is only dependent on the number of photo-generated charge carriers. To reduce its influence, one should work with the highest possible number of photo-generated charge carriers. Critical for digital imaging is the (full) well capacity, that is, the number of charge carriers a pixel can contain before it saturates.

As the collected SHG intensity is rather weak, there is an interest in having a camera multiplying electrons before the readout is made. Electron multiplying CCD (EMCCD) and, possibly, intensified CCD cameras, especially suited for any low-light-level application, thus appear ideal for holographic SHG.

For some specific applications, for example, fluorescence or nonlinear imaging, the operating temperature of the sensor is important. When working with very weak signals, such as SHG, the thermally generated dark noise can become quite important, compared to the intensity of the signal of interest. Ideally, the detector would be cooled down to low temperature, in order to reduce thermal noise. A rule of thumb states that the dark current reduces by half for every 9 K of temperature drop.

The bit depth is actually not very relevant for holography. We recall that Goodman's experiment was carried out with a very low 3 bits depth, giving only eight quantized gray levels (Goodman and Lawrence, 1967). In fact, Mills and Yamaguchi reported evidence that 4-bit quantization is enough to provide a satisfactory visual image, and that not much difference could be seen for holograms recorded with 6- or 8-bit quantization (Mills and Yamaguchi, 2005). Therefore, an 8-bit sensor is already enough for digital holography.

More important is the total number of pixels which sets the speed of hologram reconstruction. As most of the computation involves Fourier transforms, it is preferable (faster) to work with images having dimensions N_x and N_y that are integer power of 2, for example, 512×512 pixels, although nothing prevents the two dimensions to be different: the image can very well be rectangular. In any cases, the computing time for digital reconstruction scales with the total number of pixels $N = N_x N_y$, so that holograms with very high number of pixels might not be reconstructed live. In such cases, a stack of holograms can still be recorded live at high speed, and reconstructed afterwards.

Finally, pixel pitch is a delicate specification. It must be large enough to have a high full well capacity, but small enough to sample appropriately the hologram fringes. Because it is important that modulations from both the off-axis angle (or the curvature mismatch, in the case of an in-line configuration) and from the diffraction on the specimen produce interference fringes that can be sampled, the pixel pitch imposes some limitations on the optical design of a digital holographic microscope. A tradeoff therefore has to be found between separation power of undesired terms and noise level in the hologram. We have found that 6.0–6.5 μm pixels, with full well capacity of 16–18 k photons, provide quite reasonable noise levels and would not recommend going below that size limit. On the other hand, with typical second-harmonic wavelengths in the visible spectrum (for instance 400 or 532 nm), pixel sizes larger than 10–12 μm would not provide a good off-axis separation power, forcing one to either use advanced scheme to retrieve the imaging term of interest or, ultimately, phase-shifting holography.

It would appear from the above discussion that scientific cameras designed for fluorescence imaging are also good choices for holographic SHG imaging, assuming pixels are reasonably sized. One must however make sure that the camera maximum frame rate and exposure time range are compatible with high-speed image acquisition, one of the strengths of holographic SHG imaging.

9.3.2.3 Microscope Objective

Requirements of holographic SHG imaging in terms of microscope objective are basically the same as for scanning SHG microscopy. For one, the objective has to withstand high laser power, since most practical implementations of holographic SHG microscopes involve the ultrafast laser source going through the objective. For a reflected-light microscope, the objective will serve to carry the excitation beam to the specimen, while, for a transmitted-light microscope, it will collect the fundamental laser illumination along with the second-harmonic signal. It is therefore essential that the objective is capable of withstanding high power without damaging.

Another important criterion in the selection of a microscope objective is its transmission efficiency. If the objective serves to deliver the ultrafast laser pulses to the specimen, one must also make sure that the objective has a good transmittance at the fundamental wavelength. Also, while the transmittance of all microscope objectives is very good in the visible spectrum, it abruptly drops, for some models and manufacturers, close to or below 400 nm. Of course, this is only relevant for fundamental laser wavelengths in the red, near-infrared spectrum that generate second harmonic in the extreme blue or near-UV wavelengths.

For example, a Leica Hi-Plan 63×, 0.75 NA objective was used in Shaffer et al. (2010b), while a Zeiss Epiplan 50×, 0.5 NA objective was used in Masihzadeh et al. (2010).

9.3.2.4 SHG Reference Crystal

A very important aspect of holographic SHG imaging is the generation of the second-harmonic reference wave. It is one thing to generate second harmonic in the specimen, but one must also produce a clean reference wave at the same frequency for holograms to be possibly recorded. This is typically achieved by inserting a frequency doubler crystal in the reference arm.

Most of the time, frequency doubling is made in a beta-barium borate (BBO) crystal (Pu et al., 2008; Shaffer et al., 2009), but use of potassium dihydrogen phosphate (KDP) crystals has also been reported (Masihzadeh et al., 2010). Such crystals generally support phase matching of type *I*, in which two photons having an ordinary polarization with respect to the crystal combine to form one SHG photon having an extraordinary polarization. In opposition, type *II* phase matching requires the two initial photons to have a perpendicular polarization. Actually, there now exist many solutions for generation of the second-harmonic reference wave and it is most likely that many more will become available in the future.

9.3.2.5 Optical Delay Lines

The ultrafast lasers used for holographic SHG imaging generally have very short temporal coherence length. For interference to occur, the optical paths of the object and the reference arms must be precisely matched to this coherence length value. As an indicator, a 250 fs pulse has a coherence length close to 75 μm. Therefore, precise adjustable optical delay lines are necessary to match the optical paths.

A very simple implementation of such delay lines can be realized with a mirror pair on a translation stage, as in Pu et al. (2008) and Masihzadeh et al. (2010) and illustrated in Figure 9.3.

9.3.2.6 Optical Configurations for Recording of Digital Holograms

There exist many optical configurations for hologram recording, and almost every one has its own related recipe for hologram reconstruction. In fact, there might be as many configurations as there are holographists and slight variations on a common theme are frequent. Here, we distinguish between three different configurations for hologram recording, classified by the reconstruction process required to reconstruct an in-focus image from the hologram. We briefly describe their main differences, but go no further. See Kreis (2005) for more details.

9.3.2.7 Fourier Configuration

In this configuration, the sensor records a hologram of the optical Fourier transform of the object. This can be achieved either by placing the digital sensor in the Fourier plane of an imaging lens, or by placing it very far from the object, in a lensless configuration. In the Fourier configuration, the hologram reconstruction process is rather simple and only consists of a Fourier transform of the hologram to retrieve the in-focus object wavefront. No numerical field propagation is required.

The most famous example of such implementation is the lensless Fourier configuration used by Goodman and Lawrence (1967).

9.3.2.8 In-Focus Configuration

In this configuration, the sensor is placed in the image plane of an optical imaging system. It records the hologram of the in-focus object wave interfering with the reference wave. Processing of the hologram, for example, by off-axis filtering in the Fourier domain or by phase-shifting holography, directly leads to retrieval of the in-focus object wavefront. Again, no numerical field propagation is required.

9.3.2.9 Fresnel Configuration

In this configuration, the sensor is located at some position between an image plane and a Fourier plane, along the optical axis. It records the hologram of the out-of-focus object wave, often called a Fresnel zone plate, interfering with the reference wave. One characteristic of this configuration is that hologram processing retrieves the out-of-focus object wavefront as it was in the recording plane. Numerical field propagation is therefore essential to end up with an in-focus image, and, as we will see, offers some interesting advantages.

9.4 Reconstruction of Digital SHG Holograms

In this section, we address the numerical reconstruction process of digital holograms in general and, more particularly, of holograms recorded from second-harmonic signals.

9.4.1 Numerical Field Propagation

In Goodman's experiment of 1967, the optical setup was in a Fourier transform configuration, meaning that the digital sensor recorded the intensity of the Fourier transform of the image subject. In this simple case, the image reconstruction consists of a numerical Fourier transform of the recorded frame.

A very important progress in digital holography is the report of numerical field propagation by Kronrod et al. (1972). In their experiment, a photographic plate located some finite distance away from an image plane was successively exposed, developed, optically magnified and, finally, digitized by means of a input–output device. But because they were working in a Fresnel configuration, their image reconstruction algorithm had to cope with numerical field propagation to produce an in-focus image. Numerical field propagation is based on diffraction algorithms should be seen as the numerical counterparts of physical light propagation.

9.4.1.1 Numerical Focusing and Extended Depth of Field

The ability to numerically simulate light propagation has had a huge impact in digital holography and really contributed to its development. For one, it removed all constraints for keeping the specimen and the digital sensor exactly in conjugate planes. Numerical focusing made holographists completely forget about small focus errors. In a Fresnel configuration, the hologram never records the in-focus object wave, but thanks to numerical field propagation always returns the in-focus image. Focusing problems are quite common in high-magnification microscopy. Numerical focusing is thus very useful in this field, especially for very long-term experiments, for example, lasting several hours, where the focus may change due to mechanical relaxation. It is also very promising for microscopic investigation of specimens in very rough environment, for example, a moving vehicle, satellite, or a random positioning machine (Pache et al., 2010).

Numerical field propagation does not only serve to compensate focus errors, but it is also a way to achieve an extended depth of field for the complex field retrieved from a single hologram can be reconstructed at different distances to bring in focus different sections of a specimen which height exceeds the depth of field of a given imaging system (Ferraro et al., 2005; Colomb et al., 2010). Forty times increases in depth of field have been reported by numerical field propagation in digital holography (Colomb et al., 2010).

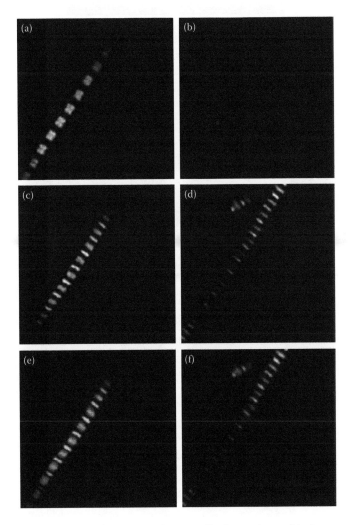

FIGURE 5.9 Effect of myosin extraction on SHG from myofibrils. Panels a and b show myofibril SHG images before and after myosin extraction, respectively. Panels c and d show two-photon excitation images of actin labeled with rhodamine-phalloidin before and after myosin extraction, respectively. Panels e and f show a superposition of SHG image (in red) and TPE image (in green) before and after the extraction. The size of the images is $20 \times 20 \, \mu m^2$. (Modified from Vanzi, F. et al. 2006. *J Muscle Res Cell Motil 27*, 469–479. With permission.)

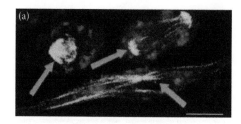

FIGURE 5.12 SHG from microtubules. (a) SHG in RBL cells. SHG (green) arises from mitotic spindles (orange arrows) and from interphase microtubule ensembles (blue arrow). Scale bar: 10 μm. (Modified from Dombeck, D.A. et al. 2003. *Proc Natl Acad Sci 100*, 7081–7086.)

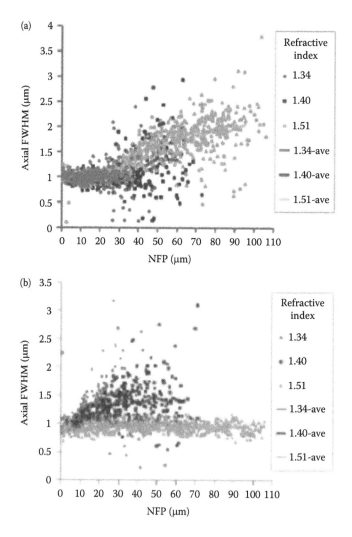

FIGURE 8.11 Axial resolution as a function of depth into kidney tissue measured from two-photon fluorescence of microspheres collected with (a) Olympus 60× NA 1.2 water immersion objective and (b) Olympus 60× NA 1.4 oil immersion ($n = 1.515$) objective. The samples were cleared with agents of different refractive indices as shown in the right panels. (Young, P. A. et al. The effects of spherical aberration on multiphoton fluorescence excitation microscopy. 2011. *J. Microsc.* 242: 157–165. Copyright Wiley-VCH Verlag GmbH & Co. KGaA. Reproduced with permission.)

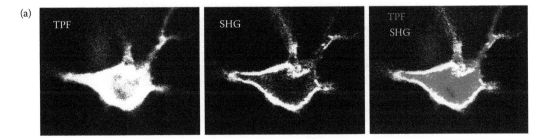

FIGURE 10.2 Characterization of SHG responses to membrane potential. (a) Comparison of two-photon fluorescence (TPF) and SHG signals. TPF and SHG signals were collected simultaneously from a neuron intracellularly stained with FM4-64 through the patch pipette. While TPF signals were obtained from FM4-64 dyes distributed both at the plasma membrane and cytoplasm, SHG signals were only collected from the plasma membrane.

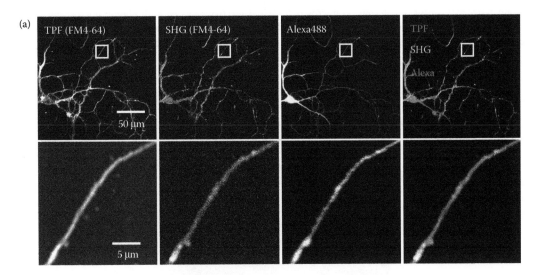

FIGURE 10.4 Membrane potential measurements at axons. (a) Visualization of axons by SHG. Primary cultured dissociated neurons were extracellularly stained with FM4-64 and loaded with Alexa 488 intracellularly. The whole neuronal morphologies including fine processes of axons can be clearly visualized by SHG. (Adapted from Nuriya, M. and M. Yasui. 2010. *J Biomed Opt* 15:020503. With permission.)

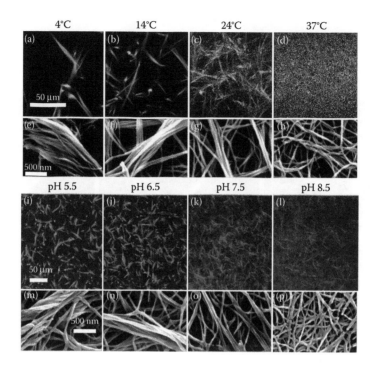

FIGURE 11.1 Simultaneously collected SHG (blue signal) and TPF (green signal) images of 4 mg/mL acellular collagen hydrogels polymerized at various temperatures (a–d) and pH values (i–l), with corresponding SEM images (e–h, m–p). The scale bars are indicated in the figure. (Reprinted from *Biophys J.*, 92, Raub, C. B. et al., Noninvasive assessment of collagen gel microstructure and mechanics using multiphoton microscopy, 2212–2222, Copyright 2007; *Biophys J.*, 94, Raub, C. B. et al., Image correlation spectroscopy of multiphoton images correlates with collagen mechanical properties, 2361–2373, Copyright 2008, with permission from Elsevier.)

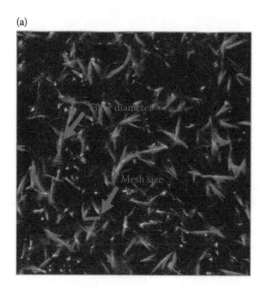

FIGURE 11.8 (a) Combined SHG + TPF (SHG, blue signal; TPF, green signal) image of acellular collagen hydrogel showing examples of measurements of collagen fiber diameter and network mesh size. (Reprinted from *Biophys J.*, 94, Raub, C. B. et al., Image correlation spectroscopy of multiphoton images correlates with collagen mechanical properties, 2361–2373, Copyright 2008, with permission from Elsevier.)

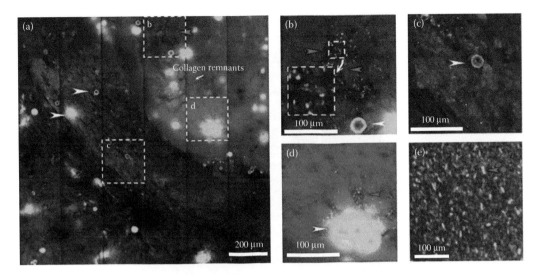

FIGURE 12.4 (a) Large-area, multiphoton autofluorescence (green) and backward SHG (blue) images of the *ex vivo* human cornea that have been infected with *Acanthamoeba castellinii* and *Pseudomonas aeruginosa*. Intrinsic autofluorescence shows the presence of *Acanthamoeba* cysts (yellow arrowhead) and *Pseudomonas bacteria* (red arrowhead) while backward SHG signal shows the degradation of corneal collagen. Increase in autofluorescence was also observed in infected cornea. (b), (c), (d) Magnified images from the selected regions of interests in (a). The presence of pathogens can be clearly visualized from autofluorescence. (e) Multiphoton autofluorescence image of isolated *Pseudomonas aeruginosa*. (Adapted from Tan, H. Y. et al. 2007. *J Biomed Opt* 12:024013.)

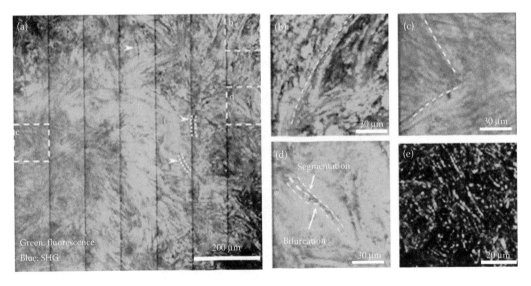

FIGURE 12.5 (a) Large-area, multiphoton autofluorescence (green) and backward SHG (blue) images of the *ex vivo* human cornea that have been infected with *Alternaria* sp. (b), (c), and (d) represent the magnified areas b, c, and d in (a). Parallel distributed fluorescence in the background and residual SHG-generating collagen remnants were found in (b) and (c). Possible fungal hyphae with characteristic morphology of bifurcation and segmentation (white arrows) can be visualized both in the large-scale (a) and detailed image (d). (e) Multiphoton autofluorescence image of isolated *Alternaria* sp. (Adapted from Tan, H. Y. et al. 2007. *J Biomed Opt* 12:024013.)

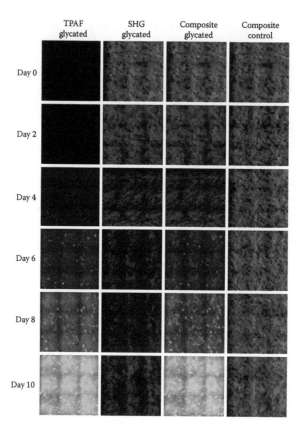

FIGURE 12.10 Time-coursed multiphoton autofluorescence (green) and SHG (red) images of glycated and control (nonglycated) samples of bovine cornea. Frame size is 620×620 mm². (Adapted from Tseng J. Y. et al. 2011. *Biomed Opt Express* 2(2):218–230.)

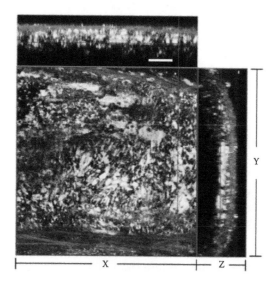

FIGURE 13.4 Atherosclerotic lesion in a ligated mouse carotid artery. Elastin autofluorescence is green, lipid droplets labeled with Nile Red is yellowish green, and collagen SHG is red. Scale bar: 40 μm. (Reproduced from Yu, W. M. et al. 2007. *J Biomed Opt* 12. With the permission of Lippincott Williams & Wilkins.)

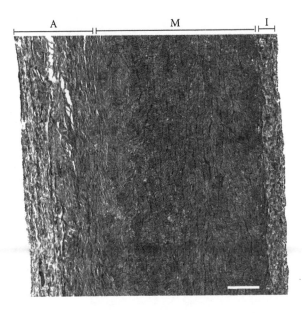

FIGURE 13.5 Structure of the porcine arterial wall. I, intima; M, media; A, adventitia. 25× magnification, scale bar = 400 μm.

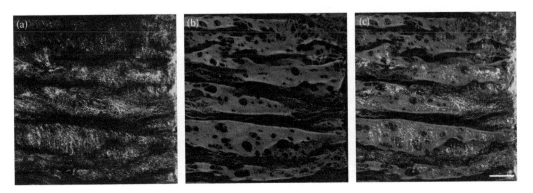

FIGURE 13.6 Porcine renal artery wall. *En face* view of 3D reconstructed artery. (a) Collagen SHG. (b) Elastin autofluorescence. (c) Merged image. Scale bar = 50 μm.

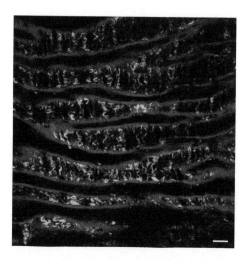

FIGURE 13.7 Collagen and elastin microstructure of the porcine carotid artery. Two-photon image 24 mm below the surface of the porcine carotid intima. Elastin autofluorescence is red; collagen SHG is green. A wavy sheet of elastin with circular holes throughout envelops the luminal surface of the artery. Individual, bunched-up collagen fibrils are radially arranged within the inner folds of the elastin lamellae. Scale bar = 20 μm.

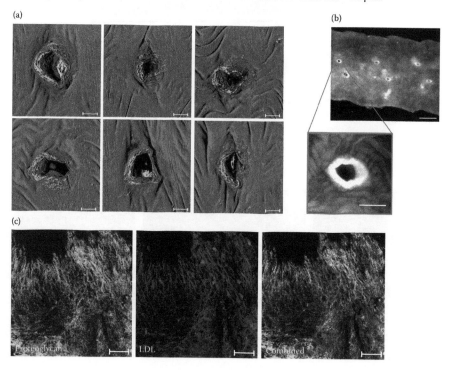

FIGURE 13.8 LDL binds to collagen- and proteoglycan-rich, elastin-poor aortic branch points. (a) Three-dimensional surface reconstructions of atherosclerosis-susceptible intervertebral branch points along the mouse thoracic aorta revealing with a ring of exposed collagen immediately surrounding the ostia. Elastin autofluorescence is red; collagen SHG is green. Scale bar = 50 μm. (b) LDL binding along the mouse aorta detected by epifluorescence. Alexa 647-tagged LDL bound extensively to a circular region around intervertebral branch points. Scale bar = 500 μm. Zoom scale bar = 100 μm. (c) Colocalization analysis of LDL binding at porcine coronary arterial branch points. Two-photon image of LDL (red) colocalization with immunolabeled proteoglycans (green). (Adapted from Kwon, G. P. et al. 2008. *Circulation* 117:2919–2927. With permission from Wolters Kluwer.)

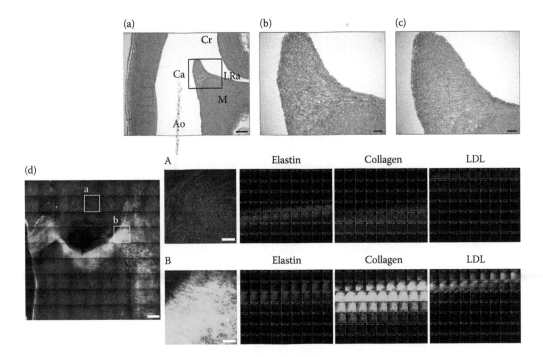

FIGURE 13.9 LDL binds to the atherosclerosis-susceptible renal artery ostial diverter at the aortic entrance to the renal artery. (a, b, c) Histological analysis. Sagital sections of renal diverter stained with Masson's trichrome stain (a, b) for collagen (blue) and smooth muscle (red), and Movat stain (green) for proteoglycans (c). The caudal aspect (Ca) of the renal ostium (a) is thickened, angulated, and more rigid than the cranial aspect (Cr). Both the intimal (rectangle) and medial (*M*) layers of the caudal renal ostium are thickened, forming a flow diverter at the entrance to the renal artery. The thickened intimal layer on the caudal side of the renal ostium forms an elongated cap (rectangle in a). A zone of proteoglycan enrichment (c, green) just below the surface of the caudal ostium surrounds a densely collagenous core (b, blue) in the intimal "fibrous cap"-like structure. (a) Scale bar = 500 μm; (b and c) Scale bars = 10 μm. (d) LDL binding to renal diverter microstructural components assessed by two-photon multimodal microscopy. (d) Tiled (9 × 9) two-photon maximum projection of z-series images through 180 μm of the luminal surface of the aortic entrance to the left renal artery. Merged image. Collagen SHG is green, elastin autofluorescence is red, and Alexa 647-LDL is blue. LDL binds eccentrically to the collagen-enriched, elastin-poor, atherosclerosis-susceptible, caudal side of the renal ostium. The tiles (A) and (B) outlined in the cranial and caudal regions of the renal ostium in (d) are shown to the right in (A) and (B), respectively. Scale bar = 1 mm. (Right) Z-series images at 0–124 μm from the luminal surface of the aortic wall, for tiles (A) and (B), respectively. Elastin autofluorescence is red, collagen SHG is green, and Alexa 647-LDL fluorescence is white. Collagen is enriched in the caudal compared to the cranial ostium. LDL binding is also greatly enhanced in the caudal compared to the cranial ostium. LDL binding to the surface occurs prior to the appearance of collagen at both the cranial and caudal sites, but overlaps to a considerable extent with collagen in the caudal ostium. Scale bars = 250 μm. (Reprinted from *Atherosclerosis* 211, Neufeld, E. B. et al. The renal artery ostium flow diverter: Structure and potential role in atherosclerosis, 153–158, Copyright (2010), with permission from Elsevier Ireland LTD.)

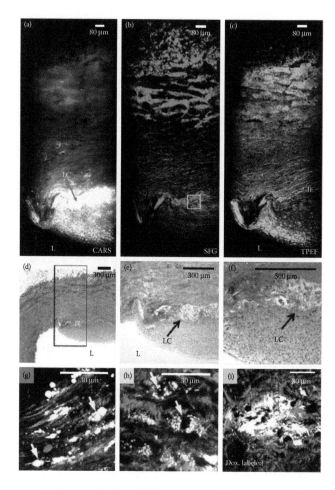

FIGURE 13.10 Representative images of a Type IV lesion inspected by CARS (a), SFG (b), and TPEF (c). The lipid core (LC) was identified by CARS. (d and e) Masson's trichrome staining. Black rectangle: the corresponding area in (a) through (c). (f) H&E staining. (g and h) Colocalized NLO images corresponding to the red and yellow squares in (a), respectively. (i) Doxorubicin-labeled (red) image, around lipid core, colocalized with CARS and SFG signals. (Reproduced from Wang, H. W. et al. 2009. *Arterioscl Throm Vas* 29:1342–1348. With the permission of SPIE.)

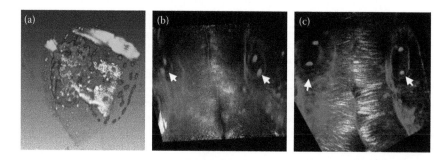

FIGURE 14.4 *In vivo* 3D images of (a) a zebrafish heart and (b)–(c) a zebrafish brain. The 3D images are reconstructed from stacks of HGM images versus depth. (a) The 3D structures of the cardiac muscles (SHG; green), cardiac cells (2PF; red), and red blood cells (THG; yellow); and (b) the 3D structures of the neural tube (THG; purple), otic vesicles (arrows; THG; purple), and the nerve fibers (SHG; green) can be observed with a submicron resolution. Image size: 240 × 240 μm².

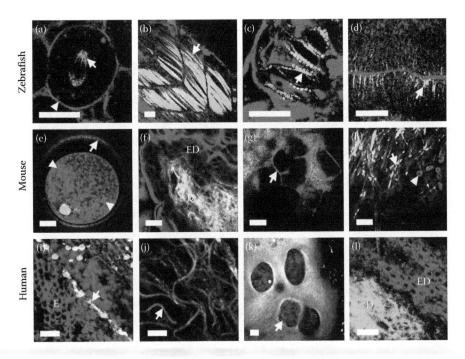

Zebrafish

Mouse

Human

FIGURE 14.6 See main text for figure caption.

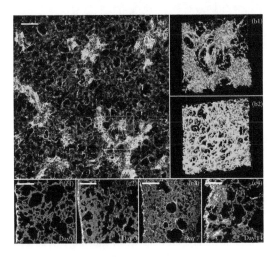

FIGURE 15.2 SHG/2PEF imaging of lung fibrosis in bleomycin-treated mice. SHG (green: collagen) and 2PEF (red: elastin, macrophages) signals excited with 50 mW excitation power at 860 nm. Yellow color: SHG/2PEF colocalization. (a) Fresh unlabeled fibrotic lung (day 14), at 42 μm under the pleura, evidencing the heterogeneous fibrosis distribution; (b) 3D reconstruction of the SHG signal from the underlined area in (a), within the tissue (b1) and in the pleurae (b2); (c) unstained histological sections embedded in paraffin, from (c1) control, (c2) day 3, (c3) day 7, and (c4) day 14 bleomycin lungs. Bleomycin treatment induces (i) inflammation, which decreases after day 7, as evidenced by the 2PEF signal, and (ii) fibrosis, evidenced by the SHG signal detected as early as day 3. Scale bar: 100 μm. (From Pena, A.-M. et al. Three-dimensional investigation and scoring of extracellular matrix remodeling during lung fibrosis using multiphoton microscopy. *Microsc. Res. Tech.* 2007. 70:162–170. Copyright Wiley-VCH Verlag GmbH & Co. KGaA. Reproduced with permission.)

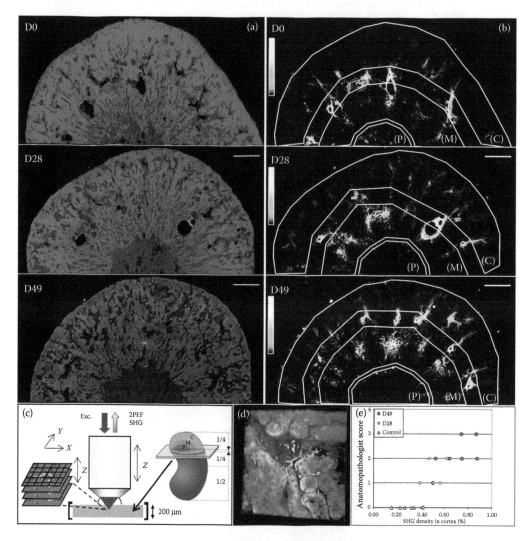

FIGURE 15.3 Multiphoton images of fibrotic murine kidney tissue. (a) SHG/2PEF and (b) segmented SHG images of coronal renal section from a control, 28-days, and 49-days Angiotensin II infused mouse showing renal papilla (P), medulla (M), outer cortex (C), and arcuate arteries along the boundary between cortex and medulla. SHG (green color) reveals collagen fibers in the tubular interstitium and in the arterial adventitia, and 2PEF (red color) underlines tubules. These 4.8×2.4 mm^2 images are obtained by stitching 10×20 laser scanned images (dimension: 270×270 μm^2, pixel size: 0.8×0.8 μm^2) acquired sequentially by moving the kidney sample with a motorized microscope stage. Scale bar: 500 μm. The white lines underline segmentation used for SHG scoring (excluding the artery region and the borders). (c) Scheme of the laser scanning multiphoton microscope showing epidetection of z-stacks of SHG/2PEF images and kidney coronal slicing for multiphoton microscopy. (d) 3D reconstruction showing interstitial fibrosis ($270 \times 270 \times 40$ μm^3 with $0.4 \times 0.4 \times 0.5$ μm^3 voxel size). (e) Automated SHG scoring of cortical fibrosis plotted as a function of a semi-quantitative estimate of interstitial damage for control and hypertensive mice. Both scores are correlated ($\tau = 0.68$, $p < 0.01$). (From Strupler, M. et al. 2008. *J. Biomed. Optics* 13:054041. With permission of SPIE.)

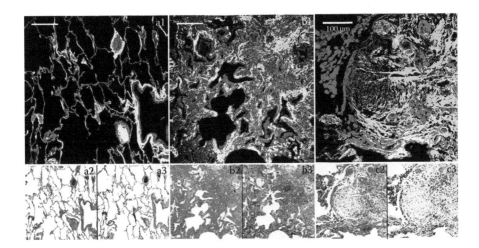

FIGURE 15.5 Normal and fibrotic human lung samples. (Top) SHG/2PEF images and (bottom) transmitted-light images of serial histological sections (scale bar: 200 μm). (Top) Unstained section, with SHG in green and 2PEF in red; (bottom) HPS staining, and Masson's trichrome staining. (a) Control lung showing a bronchovascular axis and surrounding alveolar spaces with thin alveolar walls, as revealed by HPS staining (a2). Masson's trichrome staining (a3) highlights fibrosis in green, which is seen mainly around the vessels and the bronchovascular axis, and correlates to the SHG collagen distribution (a1, green and yellow). (b) Fibrotic lung (idiopathic pulmonary fibrosis) showing marked architectural change with fibrosis and distorted alveolar spaces. (c) Fibroblastic focus composed by myo-fibroblasts (c2), and surrounded by fibrosis (c3, Masson's trichrome). SHG/2PEF image (c1) highlights the collagen distribution mainly in the periphery of the focus, with thinner fibers in its center, and the central accumulation of fibroblasts. (From Pena, A.-M. et al. Three-dimensional investigation and scoring of extracellular matrix remodeling during lung fibrosis using multiphoton microscopy. *Microsc. Res. Tech.* 2007. 70:162–170. Copyright Wiley-VCH Verlag GmbH & Co. KGaA. Reproduced with permission.)

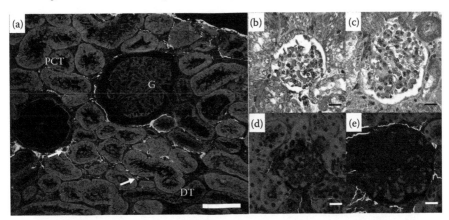

FIGURE 15.6 SHG microscopy versus histological staining of kidney cortex. (a) Multiphoton image of an unstained human renal implant biopsy. SHG (green color) reveals collagen fibers in the Bowman capsule and the tubular interstitium. Endogenous 2PEF (red color) underlines tubules, glomeruli, and arterioles. PCT, convoluted proximal tubules; DT, distal tubules; G, glomerulus; white arrows, arterioles. (b,c) histological versus (d,e) multiphoton images (same colors as for a) of serial renal sections of control mouse (b,d) and hypertensive mouse infused with AngII for 28 days (c,e). Note that hypertension promotes the accumulation of nonfibrillar ECM within the glomerular tuft stained with Masson's trichrome (c) (SHG silent in (e)) and the accumulation of fibrillar collagen in the Bowman capsule and the surrounding interstitium, similarly to the observation in the human biopsy in (a). Scale bars: 100 μm (a), 20 μm (b,c,d,e). (Simplified from Strupler, M. et al. 2008. *J. Biomed. Optics* 13:054041. With permission of SPIE.)

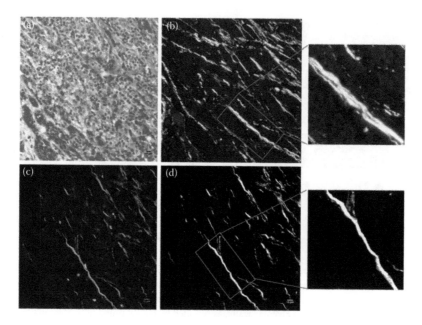

FIGURE 16.4 Collagen in stroma imaged by both polarized picrosirius and SHG images all taken at 40× magnification from Col1a1^{tm1Jae}/+ PyVT (+/+) palpable mammary mouse tissue. (a) H/E, (b) polarized picrosirius red, (c) multiphoton and second harmonic composite, (d) second harmonic. Zoomed in (150%) regions demonstrate differences in collagen fiber detection between techniques. Scale bars = 10 μm.

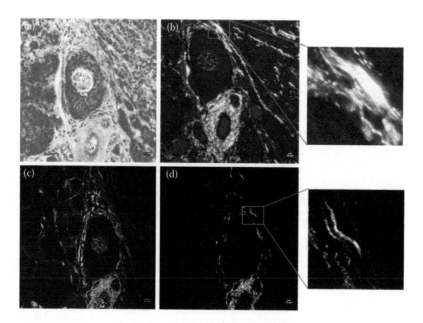

FIGURE 16.5 Collagen in early dysplasia imaged by both polarized picrosirius and SHG images all taken at 40× magnification from Col1a1^{tm1Jae}/+ PyVT (+/+) palpable mammary mouse tissue. (a) H/E, (b) polarized picrosirius red, (c) multiphoton and second-harmonic composite, (d) second harmonic. Zoomed in (150%) regions demonstrate differences in collagen fiber detection between techniques. Scale bars = 10 μm.

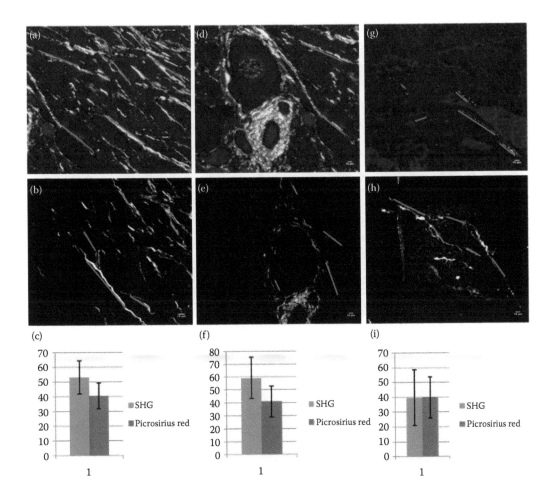

FIGURE 16.6 Highlighted fiber orientation relative to the horizontal was measured in Gimp, a freely available imaging processing tool. The graph shows average angle orientation with error bars indicating (+/−) standard deviations of the 5 angle measurements per sample of both the SHG and Picrosirius techniques. Scale bars = 10 μm.

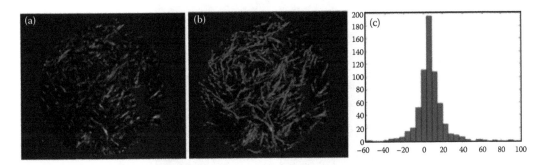

FIGURE 16.11 Use of CurveAlign to quantify collagen orientation in a human breast cancer specimen. An SHG image of a histopathological section (a) was analyzed by CurveAlign (b) and exported as a histogram (c).

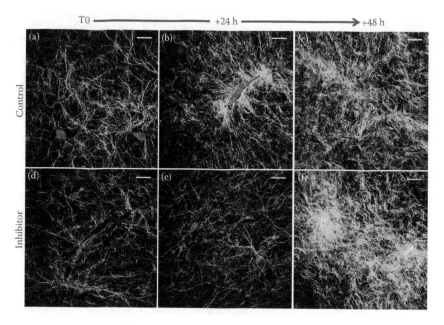

FIGURE 18.3 See main text for figure caption.

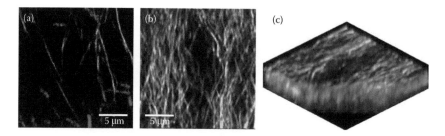

FIGURE 18.5 See main text for figure caption.

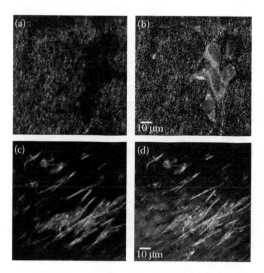

FIGURE 18.9 See main text for figure caption.

9.4.1.2 Fresnel Approximation

Let us define (x,y) and (ξ,η), respectively, as the lateral coordinates in the hologram plane $(z = 0)$ and in the reconstruction plane located at $z = d$, as illustrated in Figure 9.4. In the Fresnel approximation, it is shown in Goodman (1968) that, given the wavefront $\psi_0(\xi,\eta) = u(\xi,\eta,z = 0)I(\xi,\eta,z = 0)$ in the hologram plane, the reconstructed wave in an arbitrary plane $z = d$ can be expressed as

$$\psi_d(\xi,\eta) = \frac{\exp(ikd)}{i\lambda d} \iint \psi_0(\xi,\eta) \exp\left\{ \frac{i\pi}{\lambda d}\left[(\xi - x)^2 + (\eta - y)^2 \right] \right\} dx\,dy, \tag{9.4}$$

where λ is the wavelength. A review article by Schnars and Juptner specifically treats two of the possible implementations of the above equation for numerical field propagation (Schnars and Juptner, 2002). Here, let us summarize these two implementations.

9.4.1.3 Convolution Approach to Numerical Field Propagation

Equation 9.4 can be viewed as a convolution of the form

$$(f \otimes g)(\xi,\eta) \overset{\text{def}}{=} \int\limits_{-\infty}^{\infty}\int f(x,y)\, g(\xi - x, \eta - y)\, dx\,dy \tag{9.5}$$

between ψ_0 and a complex exponential kernel function of x and y. Using the equivalence between the convolution in the space domain and the multiplication in the frequency domain, one can write ψ_d as

$$
\begin{aligned}
\psi_d &= \frac{\exp(ikd)}{i\lambda d}\left\{ \psi_0 \otimes \exp\left[\frac{i\pi}{\lambda d}(x^2 + y^2) \right] \right\} \\
&= \frac{\exp(ikd)}{i\lambda d}\, \mathrm{F}^{-1}\left\{ \mathrm{F}\{\psi_0\} \cdot \mathrm{F}\left\{ \exp\left[\frac{i\pi}{\lambda d}(x^2 + y^2) \right] \right\} \right\} \\
&= \frac{\exp(ikd)}{i\lambda d}\, \mathrm{F}^{-1}\left\{ \mathrm{F}\{\psi_0\} \cdot \exp\left[i\pi\lambda d\left(k_x^2 + k_y^2 \right) \right] \right\}
\end{aligned}
\tag{9.6}
$$

where (F) and (F^{-1}), respectively, denote the direct and inverse Fourier transform operator.

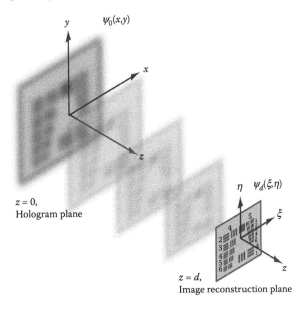

FIGURE 9.4 Hologram reconstruction. Convention for variables.

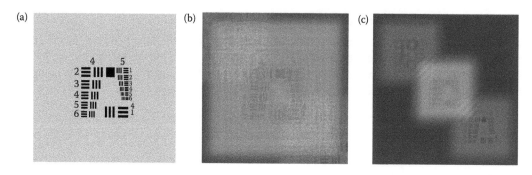

FIGURE 9.5 Amplitude of an object wave and of its reconstructions from an off-axis hologram, with different approaches for numerical field propagation. (a) Initial object, (b) convolution, and (c) single Fourier transform.

The convolution formulation of the propagation, despite the fact of being a little bit more time consuming in computing, has several advantages over the FFT version. Indeed, the fields in the hologram and image plane have the same sampling step. This means that the size of the propagated image is at the same scale as the one in the hologram plane. This is illustrated in Figure 9.5b, where the size (in pixel) of the reconstructed image has the same size, independently of the reconstruction distance. Probably for that reason alone, it is the most widely used implementation.

9.4.1.4 Single 2D Fourier Transform Numerical Field Propagation

Equation 9.4 can be developed to a single, two-dimensional Fourier transform:

$$\psi_d = \frac{\exp(ikd)}{i\lambda d} \exp\left[\frac{i\pi}{\lambda d}\left(\xi^2 + \eta^2\right)\right] F\left\{\psi_0 \exp\left[\frac{i\pi}{\lambda d}\left(x^2 + y^2\right)\right]\right\}, \qquad (9.7)$$

where $\xi/\lambda d$ and $\eta/\lambda d$ are the spatial frequencies. Equation 9.7 can be straightforwardly discretized and computer implemented, as, for example, in Schnars and Juptner (2002). As it consists of only one Fourier transform, it is very fast. In this implementation of the Fresnel propagation, however, the sampling step in the reconstruction plane varies with both λ and d. This leads to reconstructions like the one in Figure 9.5c, where the size (in pixels) of the imaging terms varies with the reconstruction distance. The advantage of this implementation is to allow spatial separation of the zero-order and diffraction terms, just as it is the case in classical off-axis holography.

9.4.2 Elimination of Zero-Order Terms and Twin Image

In Section 9.2.5, we discussed the elimination of the zero-order and twin image terms in classical holography. We have mentioned that appropriate balancing of the object and reference wave intensities could help reduce the importance of these undesired terms, and we have also described a method to completely remove them that is based on an off-axis optical configuration, in which an angle between the object and the reference waves spatially separates the reconstructed terms to avoid overlap.

With digital holography, there exist two schools of thoughts for elimination of the undesired zero-order and twin image terms. Spatial frequency filtering in an off-axis optical configuration is 1. The other relies on phase-shifting algorithms.

9.4.2.1 Off-Axis Holography

The off-axis holography was initially proposed to spatially separate the zero-order and different imaging terms upon hologram reconstruction in classical holography (Leith and Upatnieks, 1962). It therefore naturally appeared as a good solution to the same problem in digital holography.

However, there are some fundamental differences between classical and digital holography. Among other things, the pixels of digital sensors are much larger and fewer than the grains of photographic plates and consequently impose an upper limit on the range of off-axis angle resulting in interference fringes that can be appropriately sampled. Assuming square pixels of pitch $\Delta x = \Delta y$, the maximum spatial frequency that can be sampled by the sensor is $(2\Delta x)^{-1}$ pairs per unit distance. For two plane waves of wavelength λ, this corresponds to an off-axis angle θ_{MAX} of

$$\theta_{MAX} = \sin^{-1}\left(\frac{\lambda}{2\Delta x}\right). \tag{9.8}$$

Yet, because the object wave generally would have a nonzero frequency bandwidth, unlike the plane wave of this example, the off-axis angle would need to be somewhat smaller than θ_{MAX}.

In off-axis digital holography, there are two ways to eliminate zero-order and twin image terms: one operates in the spatial domain, the other in the spectral domain. To get a good understanding of these methods, let us consider the case of an off-axis configuration, where an object wave propagates along the optical axis z and a plane reference wave subtends an off-axis angle θ with that object wave, as illustrated in Figure 9.6a. A detector, located in the xy plane, records the hologram resulting from the interference pattern of o with r. The intensity of such hologram is given by Equation 9.2 and consists of two imaging terms ($g_{o,r}or^*$ and $g_{o,r}ro^*$) and two zero-order terms (oo^* and rr^*).

The first method is to spatially separate all terms by numerical field propagation, with the single 2D Fourier transform implementation. For a given wavelength and pixel pitch, judicious selection of reconstruction distance and off-axis angle should make possible complete separation of all terms. Unfortunately, by limiting the usable off-axis angle, the pixel pitch also affects the spatial separation power of off-axis configuration. For that reason, there might still remain a partial overlap between imaging zero-order terms, as it is the case in Figures 9.5c.

The alternative method is based on spatial frequency filtering. The 2D Fourier transform of the hologram comprises four terms and resembles the sketch of Figure 9.6b. If r is a plane wave, then the spectrum associated to rr^* is a Dirac delta, located at the origin of the (ω_x, ω_y) plane. The spectrum associated with oo^*, a term expressing the autocorrelation of o, is also centered at the origin of the Fourier plane and, if o has a bandwidth B, has a bandwidth $2B$ that is twice as large. Finally, because r is a plane wave, the two imaging terms have a bandwidth B and are modulated by the carrier spatial frequency of the interference fringes. The off-axis angle θ and the azimuth angle ϕ, respectively, determine the norm and the angle of the carrier frequency, in polar coordinates. Mathematically, the Fourier transform of the hologram is

$$\begin{aligned} F\{I\}(k_x, k_y) &= \hat{I}(k_x, k_y) \\ &= \hat{o} \otimes \hat{o}^*(k_x, k_y) + \hat{o} \otimes \hat{r}^*(k_x + k_{0,x}, k_y + k_{0,y}) \\ &\quad + R^2 \delta(k_x, k_y) + \hat{r} \otimes \hat{o}^*(k_x - k_{0,x}, k_y - k_{0,y}), \end{aligned} \tag{9.9}$$

with

$$k_{0,x} = \frac{2\pi}{\lambda}\sin\theta\cos\phi; \quad k_{0,y} = \frac{2\pi}{\lambda}\sin\theta\sin\phi. \tag{9.10}$$

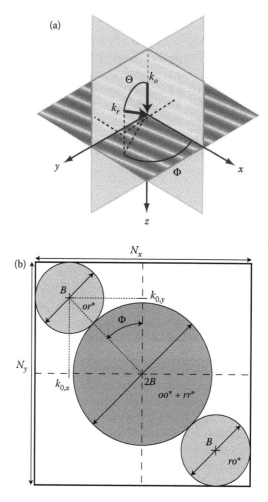

FIGURE 9.6 Digital holography in an off-axis configuration. The off-axis angle θ determines the carrier frequency modulation, and hence the separation of the zero-order and image terms, in the spatial frequency domain. (a) Angle definition and (b) Fourier spectrum.

Very often, the azimuth angle is chosen as $\pi/4$, $3\pi/4$, and so on, so that the carrier frequency modulates the signal equally in the k_x and k_y direction. It does not have to be so, but it is more elegant and more effective for spectral separation of zero-order terms. The ideal off-axis angle is that for which complete separation of the imaging and zero-order terms is achieved. However, this only occurs when the object wave as a diffraction-limited discrete bandwidth B that satisfies

$$|B| = 2\left(\frac{NA}{n\lambda}\right)\left(\frac{N_x \Delta x}{M}\right) \leq 2\left(\frac{N_x}{2 + 3\sqrt{2}}\right), \tag{9.11}$$

where NA and M are the numerical aperture and magnification of the microscope objective and where, the sake of simplicity, we have assumed the hologram to be square: N_x thus represents the number of pixels of the hologram in each direction.

In Equation 9.11, $(NA/n\lambda)$ represents the physical cutoff frequency of the microscope objective. When complete separation is possible, a simple binary mask can be designed to eliminate all spatial frequency content but that of the desired imaging term. If complete separation is not possible, one can still use this

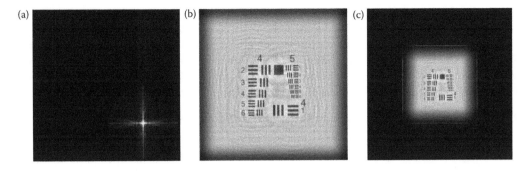

FIGURE 9.7 Elimination of zero-order terms and twin image by spatial frequency filtering, in an off-axis configuration. (a) Masked Fourier spectrum, (b) convolution, and (c) single Fourier transform.

method, but at the expense of introducing some noise in the reconstructed images. To this day, other Fourier-based methods have been proposed to eliminate the zero-order and the twin image terms, even when their respective spectrum overlaps. These methods include nonlinear filtering (Pavillon et al., 2009) and reconstruction using Gabor wavelet transform algorithms (Weng et al., 2008), to cite only two.

Once the zero-order and twin image terms have been filtered out, the retrieved imaging term still has to be demodulated. Here, demodulation is the process of extracting the phase of the initial object wavefront by removing the linear phase gradient introduced in r by the off-axis configuration. One must recall that the imaging term is made of the product of o with r^* (or o^* with r) and that its phase is therefore the sum of the phases of o and r^* (or o^* with r). For an optical configuration in which the reference wave has a wavevector k_0 subtending an off-axis angle θ with the optical axis, a large tilt corresponding to xy projection of k_0 will be introduced in the retrieved phase. It is common practice to demodulate this phase contribution, as detailed in Section 9.4.3.

Results of off-axis spatial frequency filtering and subsequent demodulation are presented in Figure 9.7.

9.4.2.2 Phase-Shifting Holography

Phase-shifting holography was proposed by Yamaguchi and Zhang (1997) and relies on combinations of multiple holograms to eliminate the zero-order and twin image terms.

In its original form, it required that four different holograms are recorded, each with exactly a $\pi/2$ phase shift compared to the previous. Assuming that I_0, $I_{\pi/2}$, I_π, and $I_{\pi/2}$ are four such holograms, then the complex object field ψ_0 can be retrieved by the following combination:

$$\psi_0 = (I_{3\pi/2} - I_{\pi/2}) + i(I_0 - I_\pi) = Ae^{i\varphi}, \tag{9.12}$$

where the amplitude A and phase φ of ψ_0 are, respectively:

$$\varphi = \tan^{-1}\left(\frac{I_{3\pi/2} - I_{\pi/2}}{I_0 - I_\pi}\right) \tag{9.13a}$$

$$A = \sqrt{(I_{3\pi/2} - I_{\pi/2})^2 + (I_0 - I_\pi)^2}. \tag{9.13b}$$

The phase shifts are generally induced by translating an optical component, for example, a mirror, with a piezoelectric transducer. Other schemes have been proposed and include rotating a wave plate, using a diffraction grating or an acousto-optic modulator. No matter how they are generated, the phase shifts have to be controlled with high precision, since the algorithm is based on the hypothesis that each hologram is in phase quadrature with the previous.

Today, there exist many other phase-shifting algorithms to retrieve the amplitude and phase of the object wavefront. Some require as few as three holograms when the phase shifts between the holograms can be very precisely controlled, others require more holograms but no a priori knowledge of the phase shifts between them. See Kreis (2005) for more information.

9.4.2.3 Comparison between Phase-Shifting and Off-Axis Methods

The major advantage of off-axis separation of zero-order and twin image terms is that it can be performed from a single hologram. This makes off-axis digital holography more suitable than its phase-shifting counterpart for real-time imaging. It also allows off-axis digital holography to excel in conditions where phase-shifting digital holography is simply impossible to use: for example, in high vibration environments, such as moving vehicles like a space satellite, as well as for specimens moving or changing shape at very high speed.

On the other side, phase-shifting holography uses the entire spatial bandwidth of the digital sensor and thus produces sharper images, whereas off-axis holography, because it oversamples the optical frequency content, does not look as sharp. This does not mean that off-axis holography suffers from a loss of resolution. Most off-axis holographic microscopy setup have, anyway, diffraction-limited resolution, determined by the microscope objective's numerical aperture. Rather, it means that this diffraction-limited resolution corresponds to many pixels in the image, hence the loss of sharpness.

9.4.3 Aberration Correction

The term optical aberration describes the nonideal behavior of imaging system that leads to distortion of the produced image. To set ideas in context, the optical aberrations we are referring to are of the monochromatic type, since digital holography generally uses monochromatic light sources. Every optical element in a setup is a potential source of monochromatic aberrations and each can have a unique effect, or signature, on the wavefront. Obviously, the more optical elements there are, the more complicated to characterize may become the resulting optical aberrations.

Digital holography is a very powerful imaging technique, for it retrieves both the amplitude and the phase of the object wavefront. As such, it is particularly adapted to characterize and correct for optical aberrations. Indeed, the phase retrieved by digital holography is nothing but the simple addition of these contributions:

1. The absolute phase shift introduced by the specimen
2. The tilt aberration due to the off-axis geometry
3. The curvature mismatch between o and r
4. Any other aberration induced by the optical elements of the setup

Methods exist to identify nonspecimen-related contributions to the retrieved phase, and to compensate them. In this section, we briefly present two different methods to numerically correct for optical aberrations in a digital holographic setup, namely the numerical parametric lens and the reference hologram correction.

9.4.3.1 Numerical Parametric Lens

This method is based on the assumption that any optical aberration can be decomposed in a series of linearly independent coefficients in a given polynomial base. This polynomial base can be, for instance, the Zernicke polynomials or the standard polynomials of x and y, the lateral coordinates in the hologram plane. For the sake of simplicity, only the latter will be treated here.

Let us suppose that the phase change introduced in the wavefront by the sum of all optical aberrations can be expressed as

$$\Gamma(x, y) = \frac{2\pi}{\lambda} \sum_{\alpha=0}^{N_x} \sum_{\beta=0}^{N_y} C_{\alpha,\beta} \, x^{\alpha} \, y^{\beta}. \tag{9.14}$$

Then, $C_{\alpha,\beta}$ defines a set of aberration coefficients, relating to the linearly independent polynomials of x and y and each coefficient corresponds to a unique aberration type. For example, coefficient $C_{0,0}$ defines a uniform phase offset, or piston aberration, while $C_{1,0}$ and $C_{0,1}$ respectively describe tilts along x and y. In an off-axis holographic microscope, some tilt aberrations are expected due to the off-axis angle between the reference and the object waves that introduces a linear phase gradient

$$\Gamma_{TILT}(x, y) = \frac{2\pi}{\lambda}(C_{1,0} \, x + C_{0,1} \, y) = k_{0,x} \, x + k_{0,y} \, y, \tag{9.15}$$

where the lateral coordinates x and y are expressed in physical distance units. Correcting for this linear phase gradient demodulates the carrier frequency of an off-axis hologram, thus shifting the hologram spectrum so that one initially modulated imaging term becomes centered in the spatial frequency space. This correction can be easily implemented and automated, since $k_{0,x}$ and $k_{0,y}$ can be deduced from the carrier frequency of the hologram (Equation 9.10).

As for coefficients $C_{2,0}$ and $C_{0,2}$, they can, in a parabolic approximation, compensate for possible curvature mismatch between the object and the reference waves. If such a mismatch exists, the interference fringes will no longer be straight lines, but rather will have some hyperbolic shape. This can also be observed by a broadening of the carrier frequency in the hologram spectrum.

In summary, all these aberrations are easily corrected by numerical parametric lenses and other, higher-order aberrations can be corrected similarly. It is however generally rare that a microscope presents a significant higher-than-second-order aberrations. Finally, an automated procedure for aberration correction with numerical parametric lenses can be found in Colomb et al. (2006a).

9.4.3.2 Reference Hologram Correction

This method is based on a two-step recording process. In one step, a reference hologram is recorded without the presence of the specimen and one of its imaging term is reconstructed to quantitatively characterize the influence of optical aberrations on the retrieved phase. In another step, the desired hologram is recorded with the specimen, and one of its associated imaging term is reconstructed. The method of reference hologram correction then simply consists in dividing the wavefront reconstructed in the second step by the one reconstructed from the reference hologram.

Mathematically, let us define the object waves $o_1 = A_1 e^{i\varphi_1}$ and $o_0 = A_0 e^{i\varphi_0}$, respectively, with and without the presence of the specimen. Reconstructing the reference hologram provides the imaging term $\psi_0 = g_{o,r} o_0 r$. Similarly, imaging term $\psi_1 = g_{o,r} o_1 r$ is retrieved from reconstruction of the second hologram. Dividing ψ_1 by ψ_0 yields:

$$\frac{\psi_1}{\psi_0} = \frac{g_{o,r} o_1 r}{g_{o,r} o_0 r} = \frac{o_1}{o_0} = \frac{A_1}{A_0} e^{i(\varphi_1 - \varphi_0)}. \tag{9.16}$$

In principle, A_0/A_1 compensates for nonuniformities in the illumination profile, while $\varphi_1 - \varphi_0$ corrects the phase aberrations.

While reference hologram may seem to complicate the recording process, it needs to be performed only once for a given microscope. It is therefore not much time consuming, and does not limit the technique in any way. Thanks to its simplicity and its excellence at compensating all aberrations, this method is widely used in digital holography. Unfortunately, it is not compatible with background-free techniques such as holographic SHG imaging, in which the object wave contains no signal if the specimen is removed. The reference hologram could possibly be recorded by replacing the specimen with a frequency doubler crystal. While this could compensate for most of the setup aberrations, it is however very likely that such a nonlinear crystal will introduce other aberrations that will reverberate in the reference hologram-corrected reconstructions.

See Colomb et al. (2006b) and Miccio et al. (2007) for more details on this technique.

9.5 Image Contrasts in Digital Holographic SHG Imaging

Holography differs from most optical imaging techniques, in the sense that it does not produce images of light intensity contrasts. Instead, holography encodes in an intensity contrast hologram both the amplitude and the phase of an object wavefront. With classical holography, this leads to a 3D appearance of the reconstructed images. In digital holography, amplitude and phase can be simultaneously reconstructed in separate images of different contrast (Cuche et al., 1999).

Holographic SHG imaging also makes possible retrieval of both second-harmonic amplitude and phase and may thus provide similar image contrast. There is altogether one important difference between holographic SHG imaging and bright-field digital holography: SHG images are background-free, meaning that there is a signal (amplitude and phase) only where SHG occurs. In opposition, the object wave in bright-field digital holography generally has intensity all over the field of view, even in regions where there is no object. As a result, part of the light incident on the specimen is not diffracted and reaches the detector unaltered. Yet not diffracted by the specimen, this light still interferes with the reference wave and thus provides a support for the carrier frequency of the hologram fringes. This is not the case with holographic SHG imaging and, as we have already seen, it has dramatic consequences, notably making use of reference hologram correction impossible.

Here, we show how holography makes possible the retrieval of amplitude-, intensity-, and phase-contrast images. For the following discussion, let us suppose that hologram reconstruction yields the wave $\psi = Ae^{i\varphi}$.

9.5.1 Amplitude Images

The amplitude contrast image is obtained by assigning $\psi = |A|$ to a colormap. The amplitude contrast image is proportional to the electric field distribution. An interesting observation is that, for a given dynamic range, amplitude contrast provides a better sampling of weak signals than intensity contrast.

To illustrate this, let us consider an Airy disk amplitude object imaged with an ideal 8-bit detector that fully exploits its given bit depth. Now, let us compare an image of the object amplitude retrieved with a holographic method to one obtained by simple intensity-based imaging. For this comparison, we suppose that the image is recorded in focus for both methods and that, since the detector is ideal, its dynamic range scales from 0 and 255 in both cases.

Even then, weak signals are sampled differently by the two methods. On one hand, Figure 9.8a illustrates the square root (i.e., the amplitude) of the 8-bit intensity image recorded by our ideal detector. On the other, Figure 9.8b illustrates the amplitude of the complex wavefront retrieved from the 8-bit digital hologram. Looking at the profiles plotted in Figure 9.8c, one sees that both methods provide a similar sampling of the central spot, but that holography provides a better sampling of the diffraction rings. Indeed, more rings are visible in the holographic image. Furthermore, discrete intensity levels are seen for the profile obtained with intensity-only imaging that are not observed on the profile obtained with the holographic method, even though the hologram was discretized to the same bit depth. This is due to the modulation introduced in the hologram by the interference fringes and to the image reconstruction process. We note that it would have been even more advantageous, with such an ideal detector, to record the hologram out of focus so that the intensity of the central region would spread over a larger surface, therefore, reducing the dynamic range of the object wave.

In summary, because holography retrieves the amplitude of the object wave, it is more suited than intensity-based imaging for detection of weak signals.

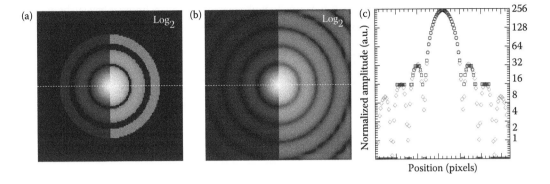

FIGURE 9.8 (a,b): Amplitude-contrast images of an Airy diffraction pattern recorded respectively with intensity-based imaging and with holographic methods. The simulation supposed that in both cases, the image was recorded in focus and that the detector returned a 8-bit depth image. Images are horizontally separated in two: the left part has a linear grayscale colormap, the right part has a binary logarithmic grayscale colormap. (c) Comparison of the amplitude profiles along the dashed lines in previous images. Diamonds ($\diamond$) are from the holographic image and squares ($\square$) are from the direct intensity image.

9.5.2 Intensity Images

While digital holography intrinsically retrieves the amplitude and phase of a complex wave ψ, an intensity image I_o of the object wave can still be obtained from ψ, provided that the mutual coherence function $g_{o,r}$ and the intensity of the reference wave I_r are known. According to Equation 9.3, I_o is

$$I_o = \frac{\psi^2}{g_{o,r}^2 I_r} \tag{9.17}$$

and quantitatively compares to the intensity of the object wave that any intensity-based detector would have recorded.

Obtaining I_r is very easy. One simply has to record an intensity image of the reference wave by blocking the object wave. As the reference wave is quite obviously specimen independent, this is something that needs to be done only once for a given holographic setup.

Retrieving $g_{o,r}$ can be, however, more complicated. In the simplest case, that is, for a highly coherent light source such as HeNe lasers, $g_{o,r}$ is uniform and close to unity over the entire hologram. It can therefore be approximated to unity and I_o be deduced from ψ and I_r only. Unfortunately, this is no longer the case when working with femtosecond lasers. With such low temporal coherence light source, $g_{o,r}$ varies a lot over the hologram and has to be precisely determined. One calibration procedure used to measure $g_{o,r}$ consists in recording one hologram I and the intensities I_o and I_r of its associated object and reference waves. Then, by reconstructing the hologram to retrieve ψ, one can deduce $g_{o,r}$ from Equation 9.17. In practice, such calibration is made only once for a given setup.

9.5.3 Phase Images

The phase-contrast image is obtained by assigning the phase φ of the reconstructed wavefront to a colormap.

9.5.3.1 Bright-Field Digital Holography Phase Imaging

In bright-field digital holography, the phase provides a quantitative measure of the optical path length difference light underwent in the object arm. However, it is important to mention that the reconstructed phase distribution is defined modulo 2π, meaning that phase unwrapping techniques have to be applied for specimens producing optical path differences larger than one wavelength.

In reflected-light holography, and for a specimen immersed in a uniform medium of known refractive index $n(\lambda)$, the phase can be expressed as

$$\Delta\varphi = \left(\frac{4\pi}{\lambda}\right) n(\lambda)\Delta z, \tag{9.18}$$

and is directly related to the surface topography Δz of the specimen.

In transmitted-light holography, the detected phase is related to the optical length of the path traveled by the light and can be expressed as

$$\Delta\varphi = \frac{2\pi}{\lambda} \int_{z_I}^{z_F} n(z,\lambda)\,\mathrm{d}z, \tag{9.19}$$

where n is the refractive index and z_I, z_F are respectively the coordinates of the light source and the detector. For transmitted light, the phase cannot be easily related to either the refractive index distribution or the specimen thickness and advanced schemes to decouple the two are needed. Such schemes include changing the known value of refractive index of the surrounding medium (Rappaz et al., 2008), confining the specimen in a microchannel of known depth (Lue et al., 2006) or using air bubbles in the medium surrounding the specimen to determine its refractive index (Kemper et al., 2006).

We have seen that digital holography is especially suited for imaging moving specimens. It is therefore no surprise that phase imaging finds applications in topographic measurements of MEMS operating in the MHz frequency range and in live cell analysis. But phase imaging is also uniquely appropriated for characterization of microlenses and micro-optics in general.

9.5.3.2 Holographic SHG Phase Imaging

The interpretation of the phase signal in holographic SHG imaging is rather complicated, as the wavelength of the light changes due its interaction with the specimen. But while the wavelength changes, the phase remains continuous since SHG preserves the coherent nature of light. The detected phase thus reflects the optical path length, but is not as readily related to specimen height or thickness, as it is the case in bright-field digital holography.

An important aspect to consider is that the SHG phase is only determined where second harmonic is generated (Figures 9.9a and 9.9b). While this is perfectly logic, it has an undesired consequence: wherever no second harmonic is generated, the reconstruction algorithms introduce physically meaningless random phase fluctuations that might disturb the observer. Fortunately, regions of random phase can be eliminated by applying a binary mask, based on thresholding of the SHG amplitude, to the image (Figure 9.9c).

Let us consider the case of transmitted-light holographic SHG phase imaging. The detected SHG phase φ depends on the phase of the fundamental wave at the location z_{SHG} of SHG, as well as on the optical path length (at SHG wavelength) from that point to the detector, located at z_H. The SHG phase can thus be expressed as

$$\Delta\varphi = \int_{z_0}^{z_{SHG}} \frac{2\pi}{\lambda_0} n(z,\lambda_0)\,\mathrm{d}z + \int_{z_{SHG}}^{z_H} \frac{2\pi}{\lambda_0/2} n\left(z,\frac{\lambda_0}{2}\right) \mathrm{d}z. \tag{9.20}$$

Quantitatively relating the SHG phase to physical properties like refractive index, specimen thickness, or surface topography is not trivial. Some cases are nonetheless rendered quite accessible by making simple assumptions.

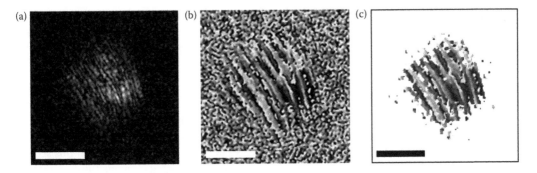

FIGURE 9.9 Holographic SHG imaging of a very small portion of *Caenorhabditis elegans* worm shell. Amplitude image is normalized, and phase images are wrapped, that is, they scale from $-\pi$ to π over the linear grayscale colormap. (a) SHG amplitude, (b) SHG phase, and (c) SHG phase (masked). Scale bars are 5 μm.

9.5.3.3 Uniform, Nondispersive Medium of Known Refractive Index

Let us suppose that the refractive index of the medium surrounding the specimen is uniform and wavelength-independent, that is, $n = n(\lambda_0) = n(\lambda_0/2)$. This is notably the case for vacuum and air environment, but could become valid in other materials, given a judicious choice of wavelengths λ_0 and $\lambda_0/2$.

Even if this assumption is not exactly verified, the following will provide a very good approximation, since refractive index changes due to dispersion at typical wavelengths of ultrafast lasers are generally very low and rarely exceed a few percent, unless the medium has a strong absorption band in that spectral region, in which case it is intrinsically not suited for imaging.

Under the assumption of nondisperive medium, Equation 9.20 simplifies a little and the variations in the observed SHG phase can be directly related to variations in the axial position of SHG:

$$\Delta\varphi(\Delta z) = \frac{2\pi}{\lambda_0} n \Delta z, \tag{9.21}$$

Interestingly, this relation also corresponds to the phase advance term of the fundamental field at wavelength λ_0 in a medium of refractive index n.

9.5.3.4 Uniform Medium of Known Refractive Index

Now, let us suppose that the medium surrounding the specimen is still uniform but dispersive. Its refractive index therefore is not the same for electromagnetic waves of wavelength λ_0 or $\lambda_0/2$, but can always be expressed as

$$n(\lambda_0) = a\, n\!\left(\frac{\lambda_0}{2}\right), \tag{9.22}$$

where a is a proportionality factor. Different $(\lambda_0, \lambda_0/2)$ couples will yield different coefficients a, but for a given λ_0, there will always be a coefficient a that satisfies the previous equation. Under this assumption, Equation 9.20 leads to

$$\Delta\varphi(\Delta z) = \frac{2\pi}{\lambda_0}(2a - 1)n\Delta z. \tag{9.23}$$

This more general equation suits any uniform medium. Water, for instance, has a refractive index $n(800\text{ nm}) = 1.339$ and $n(400\text{ nm}) = 1.329$, yielding a coefficient a of 1.0075. Here, the influence of dispersion is very small: it makes the phase-position relation differ by only 1.5% compared to Equation 9.21 that was developed for the case of nondispersive medium.

It is interesting to note that for $a > 1$, the sensitivity is increased, as the phase varies more rapidly for the same Δz displacement. In contrast, the medium for which $a < 1$ reduces the sensitivity. This suggests a way to tune the sensitivity by appropriate selection of the surrounding medium or laser wavelength.

9.6 Selected Applications

We have, up to now, discussed all the underlying bases of holographic SHG imaging: we have covered the principles of holography, the implementation of a holographic SHG microscope, the numerical reconstruction of digital holograms, and the various types of image contrast it yields. In this section, we describe a few selected applications in which holographic SHG imaging is among those reported in the actual literature.

9.6.1 Imaging of Biological Structures

A good imaging tool should provide contrast specific to the structure of interest with a satisfying signal-to-noise ratio (SNR), a high enough spatial resolution to resolve it, and a high enough temporal resolution to observe its dynamics.

In modern microscopy, contrast specificity of biological specimens is very often obtained via fluorescence imaging. Sometimes, the fluorescent substance (fluorophore) is the material of interest, but most often it consists of a marker. Markers may be of exogenous nature, for example, molecules biochemically functionalized to bind to the material of interest, or of endogenous nature, for example, genetically encoded to be expressed by the specimen. In any case, there are inconveniences related to the use of markers. One is that it generally increases the specimen preparation time. Another, more important, is that markers may alter the normal behavior of the specimen in unexpected ways. For these reasons, it is favorable, when possible, to exploit a contrast intrinsic to the structure of interest, and thus avoid the use of markers. One advantage often claimed by SHG imaging, and nonlinear coherent imaging in general, is that it does not necessarily require labeling. Indeed, many materials, especially biological structures like mitotic spindles, actomyosin complexes, microtubules, collagen, and muscles, have intrinsic nonlinear response to electromagnetic radiation (Campagnola and Loew, 2003; Mertz, 2004), which is rather fortunate since collagen is the most abundant protein in mammals. SHG microscopy has already proved especially appropriate for structural investigation of collagen and muscles (Freund et al., 1986; Kim et al., 1999; Lin et al., 2006; Plotnikov et al., 2006; Teng et al., 2006; LaComb et al., 2008; Matteini et al., 2009; Nucciotti et al., 2010; Xu et al., 2010a).

Additionally, SHG microscopy is based on an instantaneous nonsaturating interaction with the specimen, and may thus, in principle, achieve a much higher temporal resolution than its fluorescence counterpart. Unfortunately, technical limitations have until now prevented from fully exploiting this advantage. Originally, the limiting technology of light sources and detectors made impossible to record full-field SHG images with good SNR at high frame rates. For instance, in their first reports of SHG microscopy (Hellwarth and Christensen, 1974, 1975), Hellwarth and Christensen had to expose ASA 3000 Polaroid films for 20–60 min to produce SHG images. It must be emphasized that, at the time of SHG microscopy's birth, light sources were either continuous wave lasers or Q-switched lasers delivering pulses with durations in the hundred of nanosecond range (Hellwarth and Christensen, 1974; Sheppard et al., 1977). To overcome the insufficient SHG efficiency of such lasers for full-field imaging systems, scanning SHG microscopes were developed (Sheppard et al., 1977). Yet, because of their nature, scanning microscopes are intrinsically flawed for high-speed imaging. Recent developments of ultrafast lasers and digital sensors now make possible nonscanning SHG imaging.

Off-axis holographic SHG imaging is a nonscanning technique and, as such, it is especially suited for studying dynamics of live tissues or cells. Provided sufficient signal, it may truly exploit the instantaneous nature of SHG. Already, label-free holographic SHG imaging of static biological specimens has

$d_s = 48.9\ \mu m$

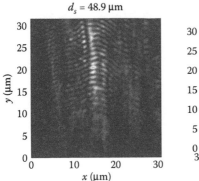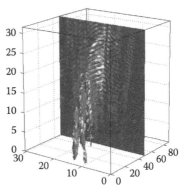

FIGURE 9.10   Hologram reconstruction showing several separate human muscle fibrils. The 3D representation shows the assembly of reconstruction planes into the three-dimensional depiction of the sample. (Reprinted from Masihzadeh, O., Schlup, P., and Bartels, R.A., 2010. Label-free second harmonic generation holographic microscopy of biological specimens, *Optics Express*, 18(10), 9840–9851. With permission of Optical Society of America.)

been reported (Masihzadeh et al., 2010; Shaffer et al., 2010b), and holographic SHG imaging of fast-moving specimens is expected in the near future.

The first of these label-free holographic SHG images of biological tissues was reported by Masihzadeh et al. (2010), and contained holographic SHG (amplitude contrast) images of several separate human muscle fibrils. In one experiment (Figure 9.10), they reconstruct the hologram at various depths to obtain a 3D stack of images, corresponding to cross-sections of the specimens. In another experiment, not presented here, they also investigate incident-polarization dependence of SHG, and observed that the intensity changes versus the polarization angle, in agreement with previous investigation, performed with scanning confocal SHG microscopy (Plotnikov et al., 2006; Tiaho et al., 2007; Nucciotti et al., 2010).

That same year, Shaffer et al. studied the SHG of connective tissue in mouse tail dermis, by investigating both amplitude and phase-contrast holographic SHG images (Figure 9.11). While the amplitude contrast holographic SHG images do not differ much from intensity images obtained by scanning SHG microscopes, the SHG phase-contrast images are a completely different matter. According to Section 9.5.3, the SHG phase can, in principle, be related to quantitative physical properties, like refractive index and axial coordinates of SHG. But even if without going into quantitative analyses that may require some a priori knowledge, the SHG phase is still qualitatively very interesting for investigation of biological structures. In the following example, we see how SHG phase imaging points to fulfilled phase-matching conditions in collagen.

9.6.1.1 SHG Phase Matching Conditions in Collagenous Tissues

For some time, it was believed that phase-matching played no role in SHG within collagenous tissues. Supporting this claim, Kim et al. (1999) predicted, based on values of refractive indices of mammalian tissues and birefringence properties of collagen available at the time, that phase-matching could not occur in the visible and near-infrared spectral ranges. That belief held for some time, but was later challenged (Theodossiou et al., 2006; LaComb et al., 2008; Xu et al., 2010a).

Among the challengers, Theodossiou et al. (2006) pointed out that the refractive indices used by Kim et al. are group refractive indices for whole tissues (taken from Bolin et al., 1989), not strictly for collagen, and that therefore, these values do not represent the extraordinary–ordinary interaction which could be involved in phase-matching considerations. A couple of years later, La Comb et al. explicitly postulated that the SHG measured in a tissue imaging experiment consists of both (quasi)-coherent and incoherent scattered components that need to be considered separately for full interpretation of the

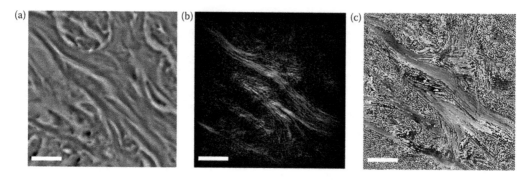

FIGURE 9.11 Mouse tail dermis and epidermis. (a) Bright-field image. (b) SHG amplitude (normalized) and (c) phase (wrapped) reconstructed from a single hologram. All images present the same region of the specimen and scale bars are 10 μm. (Reprinted from Shaffer, E. et al., 2010b. Label-free second harmonic phase imaging of biological specimen by digital holographic microscopy, *Optics Letters*, 35, 4102–4104. With permission of Optical Society of America.)

image data. This year, Xu et al. (2010a) suggested that single molecule SHG would be too weak to account for the detected SHG signal and that coherent scattering had to play an important role. Observation of the phase of the SHG signal generated by collagen fibers, originally published in Shaffer et al. (2010b) and reprinted here (Figure 9.11), tends to support this theory.

In this experiment, 5 μm-thick coronal sections of mouse tail were fixed with paraformaldehyde (PAF) 4%, embedded in paraffin, and observed with both bright-field (Figure 9.11a) and holographic SHG microscopy, leading to reconstruction of amplitude and phase-contrast images (respectively Figures 9.11b and 9.11c). All three images were recorded with the same microscope objective and imaging wavelength of 400 nm, using a light-emitting-device for the bright-field image and a 800 nm fundamental-wavelength femtosecond laser for the SHG holographic images. Under these conditions, the continuous and almost uniform phase in regions where second harmonic is generated is a possible indicator of coherent scattering in phase-matching conditions.

This short example simply points out one of the interest of the SHG phase, which can be obtained by holographic SHG imaging, without even going through quantitative assessments.

9.6.2 3D-Tracking of Nanoparticles

In Section 9.6.1, we insisted on label-free applications. However, there seems to be a growing trend to use labeling in SHG microscopy (Extermann et al., 2009; Pantazis et al., 2010). Compared to habitual fluorescent markers, SHG markers offer the many advantages attributed to the scattering nature of signal generation. For one, they do not blink nor bleach, and thus offer stable, nonsaturating signal of ultrafast response time, making possible observation of fast dynamics over long time periods. Another advantage, still related to the scattering nature of signal generation, is the flexibility in the choice of the excitation wavelength to which is linked that of the detected signal.

For the following discussion, we make a distinction between markers and (nano) probes. Markers are exogenous contrast agents whose purpose is to reveal specific structures upon imaging. They are generally very small, for example, molecule-sized, and are used in such large quantities that they become individually indistinguishable. Membrane potential-sensitive molecules capable of SHG (Peleg et al., 1999; Moreaux et al., 2000) are an example of SHG markers, and so would be metallic nanoparticles enhancing the signal of such markers (Campagnola et al., 2001). In opposition, nanoprobes are nanostructures, sometimes as large as a few hundreds of nanometers, which can be localized, hopefully with precision, and tracked in time to provide functional information about the specimen. They are generally used in much lower densities than markers, so as to be individually distinguished, localized, and tracked in both

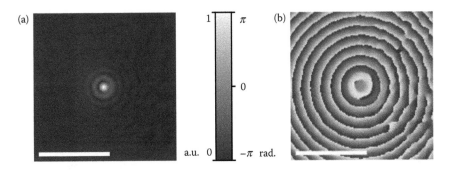

FIGURE 9.12   SHG amplitude (a) and phase (b) generated by a BaTiO$_3$ nanoparticle and retrieved by holographic SHG imaging. Scale bars are 5 μm. (Reprinted with permission from Shaffer, E. and Depeursinge, C., 2010. Digital holography for second harmonic microscopy, in: *Proceedings of the SPIE: Multiphoton Microscopy in the Biomedical Sciences X.* Copyright 2010, Society of Photo-Optical Instrumentation Engineers.)

space and time. To this date, nanocrystals of ZnO (Kachynski et al., 2008), polar Fe (IO$_3$)$_3$ (Bonacina et al., 2007), KNbO$_3$ (Nakayama et al., 2007), KTiOPO$_4$ (*KTP*) (Le Xuan et al., 2008), BaTiO$_3$ (Hsieh et al., 2009; Pantazis et al., 2010; Shaffer et al., 2010a), and of different types have been investigated for this purpose.

To be of interest, nanoprobes such as the one of Figure 9.12 must be used in conjugation with a technique capable of localizing them through space and time. Over the years, many algorithms were proposed for determining the lateral position of nanoparticles at presumably nanometer (or at least sub-pixel) precision—for a quantitative comparison, see Cheezum et al. (2001). But the real challenge has always been and remains the determination of the axial position, which is generally no better than the micrometer range, unless the full diffraction field can be accessed, in terms of amplitude and phase. Even in these cases, it does not reach the sub-micrometer without a priori knowledge of shape or size of the particle. Precise sub-micrometer tracking of nanoprobes is therefore a need to be addressed.

Holographic SHG imaging proposes two different methods for determination of the axial position of nanoprobes in the appropriate precision range. One relies on numerical field propagation and the other is based on SHG phase measurement.

Method 1: Determination of axial position by numerical field propagation: We have seen in Section 9.4.1 that numerical reconstruction of digital holograms, by allowing to bring in focus scatterers located at various depths in the specimen, gives holographic SHG imaging an extended depth of field that can be used to determine the axial position of these scatterers (Hsieh et al., 2009; Shaffer et al., 2010a). Here, we show how this is possible and discuss the performance of this method.

To make things as clear as possible, let us consider a distribution of SHG-emitting nanoparticles, spread in both lateral and axial positions, and imaged by a lens, for example, a microscope objective, at different positions along the optical axis, as illustrated in Figure 9.13. Let us suppose that a digital sensor records an hologram of the out-of-focus images (or Fresnel zone plates) in the indicated plane. Numerically reconstructing the hologram at distance d would bring in focus the image of one nanoparticle, while reconstructing the same hologram, but at a greater distance (here $d + \Delta d$) would bring in focus the image of the other nanoparticle. The Δd change in the reconstruction distance needed to bring the image of the second particle in focus is directly related to the Δz difference in the axial position of the objects by

$$\Delta d = M_L \Delta z = M_T^2 \Delta z, \tag{9.24}$$

where M_L and M_T are, respectively, the longitudinal and transverse magnification of the imaging system.

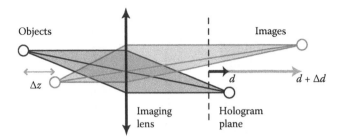

FIGURE 9.13 Determination of axial position by numerical field propagation. The reconstruction distance needed to bring the image in focus is directly related to the axial position of the object.

In bright-field digital holography, this method is not always easy to implement, mostly because it requires focus-detection algorithms to cope with high background scattering. In holographic SHG imaging, focus detection is made much simpler by the absence of background scattering, and as a simple rule, the focus position corresponds to the highest SHG intensity.

The efficiency of this method depends on the precision at which can be determined the reconstruction distance that brings the image in focus. While the reconstruction distance can be incremented by infinitely small values, the changes in focus might not be detectable. The sensitivity thus strongly depends on the longitudinal magnification of the imaging system, and is ultimately limited by the SHG wavelength, the pixel pitch and possibly the hologram bit depth that together determine the extent of the diffraction. Indeed, for very small reconstruction distance increments, numerical field propagation (or diffraction) spreads the intensity of reconstructed wavefront only to immediate neighbor pixels, thus limiting the sensitivity at which focus can be determined. For a 100× microscope objective, a precision of about a few hundreds of nanometers was evaluated for determination of axial position (Shaffer et al., 2010a). This method is however expected to perform better in the micrometer range.

The one major disadvantage of this method is its greedy use of resources that makes it quite time consuming. The problem is that one hologram has to be numerically reconstructed many times to detect the axial position of only one SHG scatterer. While this is not dramatic (it may take seconds or possibly minutes to localize a distribution of SHG scatterers), it prevents the method from working in real time at video frame rates.

Method 2: Determination of axial position from direct SHG phase value: There is another and more sensitive method to determine the axial position of an SHG scatterer that is also less time-consuming. It is based on the direct relation that exists between the SHG phase reconstructed from the hologram and the axial position of SHG.

We saw in Section 9.5.3 that the SHG phase in a given plane depends on the phase of the fundamental-wavelength illumination at the position of SHG, as well as on the optical path length (at SHG wavelength) from there to the plane of interest (Figure 9.14a). It follows that, in the plane of interest, the phase of second harmonic generated by different individual nanoparticles under plane-wave illumination directly depends on their respective axial position (Figure 9.14b).

For SHG scatterers dispersed in a nondispersive uniform medium, the SHG phase relates to the relative axial position by Equation 9.21

$$\Delta\varphi(\Delta z) = \frac{2\pi}{\lambda_0} n\Delta z, \tag{9.21}$$

which states that a change in axial position corresponding to the fundamental wavelength in the said medium λ/n sees the SHG phase vary by 2π. This makes the method very sensitive to small changes

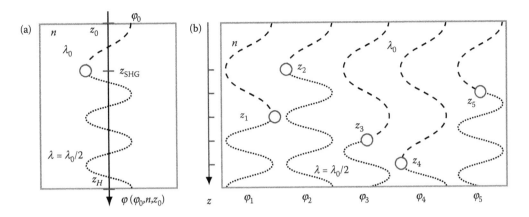

FIGURE 9.14 Determination of axial position from direct SHG phase value. (a) Dependence of the detected SHG phase on the phase of the fundamental-wavelength illumination at the position of SHG z_{SHG}, as well as on the optical path length (at SHG wavelength) from there to the plane of interest. (b) The SHG phase of different SHG scatterers relates to their respective axial position.

in the axial direction. In fact, the precision of this method was evaluated at roughly 10–50 nm for hologram-to-hologram comparison and 10 nm for comparison within the same hologram (Shaffer et al., 2010a). On the other hand, it also raises the problem of determining the position of two or more particles located more than one fundamental wavelength apart in the axial direction. Even worst, for discrete SHG scatterers, it is impossible to rely on unwrapping algorithms, because the wavefront is discontinuous. Without a priori knowledge, recourse to the first method presented here is the only way to lift the 2π phase ambiguity.

This method, unlike the first one, is not based on an imaging principle, but only on direct phase observation. Accordingly, there is, in principle, no need to numerically propagate the SHG field to form in focus images. The SHG phase might very well be compared right in the hologram plane, which reduces the required processing steps and considerably shortens the processing time. However, this method will work only as long as the respective phase patterns of the SHG scatterers can be spatially (laterally or axially) resolved. At high nanoparticle densities, this could require numerical field propagation. Even then, the hologram will most likely need to be reconstructed only once, which makes this method much faster than the other one, enough to work in real time at video frame rates.

9.6.3 Optical Phase Conjugation Imaging

Focusing or imaging through turbid media is actively sought for. However, the nature of turbid media prevents from doing so efficiently. By definition, a turbid medium is a medium that has very inhomogeneous, possibly random optical properties. Light propagating through such medium has its wavefront and polarization state seriously perturbed. Naive attempts to focus light or image a subject in a turbid medium result in very blurry, deformed focal spots or unrecognizable images.

One solution to overcome this problem is to use optical phase conjugation to precompensate for the effects of the turbid medium. Optical phase conjugation is a two-step technique that first requires to fully characterize the wavefront distortion induced by the medium, and then engineer an illumination that counterbalances it. In principle, this should allow focusing or imaging through turbid medium with results comparable to those achievable in a nondispersive, nonscattering, uniform environment.

Hsieh et al. have recently demonstrated the efficiency of an optical phase conjugation microscope relying on holographic SHG characterization of the turbid medium (Hsieh et al., 2010a,c). Their approach

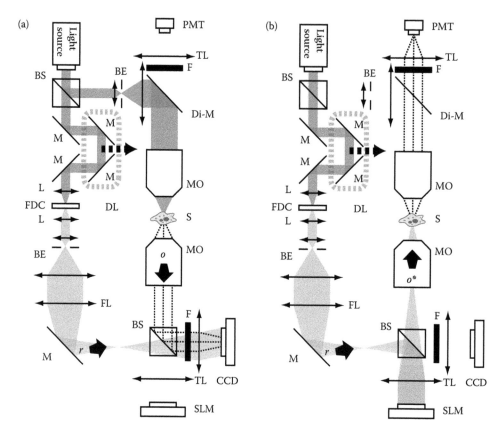

FIGURE 9.15 Schematics of holographic SHG microscope for recording the complex scattered SHG field (a) and of the phase-conjugate scanning microscope (b). Note that (a) and (b) show the same setup with different light illuminations for different steps of the experiment. The light not in use is blocked in the experiment and is not shown in the figures. BS, beam splitter; L, lens; M, mirror; Di-M, dichroic mirror; BE, beam expander; FDC, frequency doubler crystal; MO, ∞-corrected microscope objective; TL, tube lens; S, specimen; F, filter; CCD digital sensor; SLM phase-only reflective spatial light modulator and PMT photo-multiplier tube. Beam envelopes are represented by uniform gray shades (dark for fundamental and pale for the second harmonic), while image formation is represented by dashed rays.

is based on characterization of wavefront distortion by use of a coherent point source emitter located inside the turbid medium, at the position of desired focus or subject to image. In their work, a non-centrosymmetric second-harmonic generating $BaTiO_3$ nanoparticle (300 nm in diameter) served as the coherent point-like emitter.

Let us look into the details of holographic SHG imaging-based optical phase conjugation process. In a first step, a hologram of the second harmonic emitted by the nanoparticle and subsequently scattered throughout the turbid medium is recorded by its interference with an SHG reference wave, see Figure 9.15a. Full characterization of the wavefront distortion induced by the turbid medium is provided by the reconstructed SHG amplitude and phase. The reader may have noticed that first step is nothing but a straightforward application of imaging with a state-of-the-art holographic SHG microscope, as described in Section 9.3.

Then, in a second step, an illumination wave that is the phase-conjugate of the scattered SHG emission from the nanoparticle has to be generated and projected in the turbid medium. A simple way to generate the phase-conjugated illumination wave, at the SHG wavelength, is to shape the reference wave

  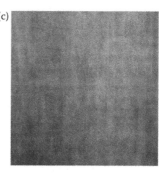

FIGURE 9.16 Phase-conjugate scanning images. (a) The wide-field transmission image of the target. The bright region in this figure indicates the transparent area while the dark region indicates the gold film. (b) The corresponding phase-conjugate scanning image of the target with the target clearly resolved. (c) The scanning image of the same target without phase conjugation. Since the focus is severely distorted by the turbid medium, the image is completely blurry. The size of all images is 115×115 μm^2. (Reprinted from Hsieh, C. et al. 2010a. Imaging through turbid layers by scanning the phase conjugated second harmonic radiation from a nanoparticle, *Optics Express*, 18(20), 20723–20731. With permission of Optical Society of America.)

used for hologram recording by using a phase-only reflective spatial light modulator (SLM), as illustrated in Figure 9.15b. When the phase-conjugate illumination is projected in the turbid medium, the latter converts the pre-distorted wavefront to a clean focal spot at the desired location, as demonstrated in Hsieh et al. (2010c).

However, the holographic SHG characterization of the wavefront distortion introduced by turbid medium is valid only for a point source emission originating from the position of the $BaTiO_3$ nanoparticle and scattered through a specific volume of the medium. If the specimen stage holding the physically interdependent nanoparticle and scattering medium is translated, the phase-conjugate illumination will propagate through a different volume of the scattering medium and will most likely not produce a clean focal spot anymore. To form an image, it is therefore essential that the phase-conjugate illumination be angularly scanned (in opposition to the specimen stage being translated) in order to propagate roughly through the same volume of the turbid medium, so that the characterization of the wavefront distortion remains valid. Angular scanning of the phase-conjugate illumination leads to a lateral scanning of the focus on the target plane where the nanoparticle is located, and a pixel-by-pixel image may thus be formed by using a point detector, for example, a photomultiplier tube (PMT), as in Hsieh et al. (2010a). Figure 9.16b presents such image of a glass slide target bearing a lithographied 130-nm thick gold pattern logo of École Polytechnique Fédérale de Lausanne (EPFL), obtained through a commercial ground glass diffuser serving as turbid medium. For comparison, Figure 9.16a shows a bright-field image of the target, and Figure 9.16c shows the image obtained through the turbid medium, without phase conjugation. The field of view of the scanning image is dependent on the thickness of the turbid medium and the distance between the turbid medium and the imaging target, and achievable frame rates are limited by either the speed of scanning mechanism or the refreshing rate of the SLM.

One of the main interests for such optical phase conjugation is to combine it with, for example, multiphoton fluorescence microscopy for tissue imaging.

9.7 Concluding Remarks

This chapter was intended to help the reader familiarize with holographic SHG imaging. We hope it provided the required knowledge and references to interpret holographic SHG images, and possibly even to set up a custom holographic SHG microscope.

9.7.1 Summary

The most distinctive advantage of holographic SHG imaging certainly is the simultaneous retrieval of both the SHG amplitude and the phase, instead of only its intensity. This feature makes digital holography stand apart from most, if not all, other imaging techniques and normally so, since detectors are generally sensitive only to light intensity. The unique nature of digital holography makes possible to retrieve multiple different contrast images from the same intensity-only hologram.

The numerical hologram reconstruction process—more specifically the numerical field propagation algorithms—makes possible numerical focusing of the images. Numerical focusing is especially useful to compensate drifts of focus over long-term experiments or in an unstable environment. Also, it makes possible to bring in focus different sections of a specimen separated by distances exceeding the depth of the field of the imaging system, which makes possible to assess the axial position of SHG scatterers.

Holographic SHG imaging, based on off-axis digital holography, is a nonscanning, single-shot image acquisition technique. As such, it is in principle limited in speed only by the camera frame rate and available SHG signal, which depends in a nonsaturating manner on the peak power density of the laser source. In other words, holographic SHG imaging is especially suited for real-time imaging, and has the potential to truly exploit the instantaneous response time of SHG.

9.7.2 Outlook and Perspectives

To this day, nonscanning holographic SHG cannot compete with scanning SHG microscopy in terms of sensitivity. Holographic SHG offers coherent amplification, low sensitivity to shot noise and very high-phase SNR, but, in the end, it all comes down to a matter of available light sources and detectors. It is actually not possible to deliver peak powers comparable to those offered by scanning microscopes over very large field of views. It is no wonder that the increasing enthusiasm around holographic SHG microscopy is fueled by the recent developments of both ultrafast lasers and electron-multiplying digital cameras.

Acknowledgments

The authors would like to thank Francesco S. Pavone and Paul J. Campagnola for their invitation to contribute to this book. In addition, the authors would also like to acknowledge all members of Professor Christian Depeursinge's group, especially Nicolas Pavillon, as well as Chia-Lung Hsieh, Pierre Marquet, and Randy Bartels, for their respective contributions.

References

Ansari, Z., Gu, Y., Tziraki, M., Jones, R., French, P., Nolte, D., and Melloch, M., 2001. Elimination of beam walk-off in low-coherence off-axis photorefractive holography, *Optics Letters*, 26(6), 334–336.

Bolin, F.P., Preuss, L.E., Taylor, R.C., and Ference, R.J., 1989. Refractive-index of some mammalian-tissues using a fiber optic cladding method, *Applied Optics*, 28(12), 2297–2303.

Bonacina, L., Mugnier, Y., Courvoisier, F., Le Dantec, R., Extermann, J., Lambert, Y., Boutou, V., Galez, C., and Wolf, J.P., 2007. Polar Fe(IO3)(3) nanocrystals as local probes for nonlinear microscopy, *Applied Physics B-Lasers and Optics*, 87(3), 399–403.

Campagnola, P.J., Clark, H.A., Mohler, W.A., Lewis, A., and Loew, L.M., 2001. Second-harmonic imaging microscopy of living cells, *Journal of Biomedical Optics*, 6(3), 277–286.

Campagnola, P.J. and Loew, L.M., 2003. Second-harmonic imaging microscopy for visualizing biomolecular arrays in cells, tissues and organisms, *Nature Biotechnology*, 21(11), 1356–1360.

Chang, R.K., Ducuing, J., and Bloembergen, N., 1965. Relative phase measurement between fundamental and second-harmonic light, *Physical Review Letters*, 15(1), 6–8.

Cheezum, M.K., Walker, W.F., and Guilford, W.H., 2001. Quantitative comparison of algorithms for tracking single fluorescent particles, *Biophysical Journal*, 81(4), 2378–2388.

Collier, R.J., Burckhardt, C.B., and Lin, L.H., 1971. *Optical Holography*, Academic Press, New York.

Colomb, T., Cuche, E., Charrire, F., Khn, J., Aspert, N., Montfort, F., Marquet, P., and Depeursinge, C., 2006a. Automatic procedure for aberration compensation in digital holographic microscopy and applications to specimen shape compensation, *Applied Optics*, 45(5), 851–863.

Colomb, T., Khn, J., Charrire, F., Depeursinge, C., Marquet, P., and Aspert, N., 2006b. Total aberrations compensation in digital holographic microscopy with a reference conjugated hologram, *Optics Express*, 14(10), 4300–4306.

Colomb, T., Pavillon, N., Khn, J., Cuche, E., Depeursinge, C., and Emery, Y., 2010. Extended depth-of-focus by digital holographic microscopy, *Optics Letters*, 35(11), 1840–1842.

Cuche, E., Marquet, P., and Depeursinge, C., 1999. Simultaneous amplitude-contrast and quantitative phase-contrast microscopy by numerical reconstruction of Fresnel off-axis holograms, *Applied Optics*, 38(34), 6994–7001.

Extermann, J., Bonacina, L., Cuna, E., Kasparian, C., Mugnier, Y., Feurer, T., and Wolf, J.P., 2009. Nanodoublers as deep imaging markers for multi-photon microscopy, *Optics Express*, 17(17), 15342–15349.

Ferraro, P., Grilli, S., Alfieri, D., Nicola, S., Finizio, A., Pierattini, G., Javidi, B., Coppola, G., and Striano, V., 2005. Extended focused image in microscopy by digital holography, *Optics Express*, 13(18), 6738–6749.

Freund, I., Deutsch, M., and Sprecher, A., 1986. Connective-tissue-polarity: Optical second harmonic microscopy, crossed-beam summation, and small-angle scattering in rat-tail tendon, *Biophysical Journal*, 50(4), 693–712.

Gabor, D., 1948. A new microscopic principle, *Nature*, 161(4098), 777–778.

Goodman, J.W., 1968. *Introduction to Fourier Optics*, McGraw-Hill, New York.

Goodman, J.W. and Lawrence, R.W., 1967. Digital image formation from electronically detected holograms, *Applied Physics Letters*, 11(3), 77–79.

Hariharan, P., 1996. *Optical Holography: Principles, Techniques and Applications*, Cambridge University Press, New York.

Hellwarth, R. and Christensen, P., 1974. Nonlinear optical microscopic examination of structure in polycrystalline ZnSe, *Optics Communications*, 12(3), 318–322.

Hellwarth, R. and Christensen, P., 1975. Nonlinear optical microscope using second-harmonic generation, *Applied Optics*, 14(2), 247–248.

Hsieh, C., Pu, Y., Grange, R., Laporte, G., and Psaltis, D., 2010a. Imaging through turbid layers by scanning the phase conjugated second harmonic radiation from a nanoparticle, *Optics Express*, 18(20), 20723–20731.

Hsieh, C.L., Grange, R., Pu, Y., and Psaltis, D., 2009. Three-dimensional harmonic holographic microcopy using nanoparticles as probes for cell imaging, *Opt. Express*, 17(4), 2880–2891.

Hsieh, C.L., Grange, R., Pu, Y., and Psaltis, D., 2010b. Bioconjugation of barium titanate nanocrystals with immunoglobulin G antibody for second harmonic radiation imaging probes, *Biomaterials*, 31(8), 2272–2277.

Hsieh, C.L., Pu, Y., Grange, R., and Psaltis, D., 2010c. Digital phase conjugation of second harmonic radiation emitted by nanoparticles in turbid media, *Optics Express*, 18(12), 12283–12290.

Kachynski, A.V., Kuzmin, A.N., Nyk, M., Roy, I., and Prasad, P.N., 2008. Zinc oxide nanocrystals for nonresonant nonlinear optical microscopy in biology and medicine, *Journal of Physical Chemistry C*, 112(29), 10721–10724.

Kemper, B., Carl, D., Schnekenburger, J., Bredebusch, I., Schäfer, M., Domschke, W., and von Bally, G., 2006. Investigation of living pancreas tumor cells by digital holographic microscopy, *Journal of Biomedical Optics*, 11(3), 034005.

Kim, B.M., Eichler, J., and Da Silva, L.B., 1999. Frequency doubling of ultrashort laser pulses in biological tissues, *Applied Optics*, 38(34), 7145–7150.

Kreis, T., 2005. *Handbook of Holographic Interferometry*, Wiley VCH, Weinheim.

Kronrod, M.A., Merzlyakov, N.S., and Yaroslavsky, L.P., 1972. Reconstruction of holograms with a computer, *Soviet Physics-Technical Physics*, 17, 333–334.

LaComb, R., Nadiarnykh, O., Townsend, S.S., and Campagnola, P.J., 2008. Phase matching considerations in second harmonic generation from tissues: Effects on emission directionality, conversion efficiency and observed morphology, *Optics Communications*, 281(7), 1823–1832.

Le Xuan, L., Zhou, C., Slablab, A., Chauvat, D., Tard, C., Perruchas, S., Gacoin, T., Villeval, P., and Roch, J.F., 2008. Photostable second-harmonic generation from a single KTiOPO4 nanocrystal for nonlinear microscopy, *Small*, 4(9), 1332–1336.

Leith, E. and Upatnieks, J., 1962. Reconstructed wavefronts and communication theory, *Journal of the Optical Society of America*, 52(10), 1123–1130.

Lin, S.J., Jee, S.H., Kuo, C.J., Wu, R.J., Lin, W.C., Chen, J.S., Liao, Y.H. et al. 2006. Discrimination of basal cell carcinoma from normal dermal stroma by quantitative multiphoton imaging, *Optics Letters*, 31(18), 2756–2758.

Lue, N., Popescu, G., Ikeda, T., Dasari, R., Badizadegan, K., and Feld, M., 2006. Live cell refractometry using microfluidic devices, *Optics Letters*, 31(18), 2759–2761.

Masihzadeh, O., Schlup, P., and Bartels, R.A., 2010. Label-free second harmonic generation holographic microscopy of biological specimens, *Optics Express*, 18(10), 9840–9851.

Matteini, P., Ratto, F., Rossi, F., Cicchi, R., Stringari, C., Kapsokalyvas, D., Pavone, F.S., and Pini, R., 2009. Photothermally-induced disordered patterns of corneal collagen revealed by SHG imaging, *Optics Express*, 17(6), 4868–4878.

Maznev, A., Crimmins, T., and Nelson, K., 1998. How to make femtosecond pulses overlap, *Optics Letters*, 23(17), 1378.

Mertz, J., 2004. Nonlinear microscopy: New techniques and applications, *Current Opinion in Neurobiology*, 14(5), 610–616.

Miccio, L., Alfieri, D., Grilli, S., Ferraro, P., Finizio, A., De Petrocellis, L., and Nicola, S., 2007. Direct full compensation of the aberrations in quantitative phase microscopy of thin objects by a single digital hologram, *Applied Physics Letters*, 90, 041104–3.

Mills, G.A. and Yamaguchi, I., 2005. Effects of quantization in phase-shifting digital holography, *Applied Optics*, 44(7), 1216–1225.

Moreaux, L., Sandre, O., and Mertz, J., 2000. Membrane imaging by second-harmonic generation microscopy, *Journal of the Optical Society of America B-Optical Physics*, 17(10), 1685–1694.

Nakayama, Y., Pauzauskie, P.J., Radenovic, A., Onorato, R.M., Saykally, R.J., Liphardt, J., and Yang, P.D., 2007. Tunable nanowire nonlinear optical probe, *Nature*, 447(7148), 1098–U8.

Nucciotti, V., Stringari, C., Sacconi, L., Vanzi, F., Fusi, L., Linari, M., Piazzesi, G., Lombardi, V., and Pavone, F.S., 2010. Probing myosin structural conformation *in vivo* by second-harmonic generation microscopy, *Proceedings of the National Academy of Sciences of the United States of America*, 107(17), 7763–7768.

Pache, C., Khn, J., Westphal, K., Toy, M.F., Parent, J., Bchi, O., Franco-Obregn, A., and Egli, M., 2010. Digital holographic microscopy real-time monitoring of cytoarchitectural alterations during simulated microgravity, *Journal of Biomedical Optics*, 15(2), 026021 (9 pages).

Pantazis, P., Maloney, J., Wu, D., and Fraser, S.E., 2010. Second harmonic generating (SHG) nanoprobes for *in vivo* imaging, *Proceedings of the National Academy of Sciences of the United States of America*, 107(33), 14535–14540.

Pavillon, N., Seelamantula, C., Khn, J., Unser, M., and Depeursinge, C., 2009. Suppression of the zero-order term in off-axis digital holography through nonlinear filtering, *Applied Optics*, 48(34), H186–H195.

Peleg, G., Lewis, A., Linial, M., and Loew, L.M., 1999. Nonlinear optical measurement of membrane potential around single molecules at selected cellular sites, *Proceedings of the National Academy of Sciences of the United States of America*, 96(12), 6700–6704.

Plotnikov, S.V., Millard, A.C., Campagnola, P.J., and Mohler, W.A., 2006. Characterization of the myosin-based source for second-harmonic generation from muscle sarcomeres, *Biophysical Journal*, 90(2), 693–703.

Pu, Y., Centurion, M., and Psaltis, D., 2008. Harmonic holography: A new holographic principle, *Applied Optics*, 47(4), A103–A110.

Pu, Y. and Psaltis, D., 2006. Digital holography of second harmonic signal, in: *Conference on Lasers and Electro-Optics/Quantum Electronics and Laser Science Conference and Photonic Applications Systems Technologies 2006 Technical Digest*, Optical Society of America, Washington, DC, 1–2.

Rappaz, B., Barbul, A., Emery, Y., Korenstein, R., Depeursinge, C., Magistretti, P.J., and Marquet, P., 2008. Comparative study of human erythrocytes by digital holographic microscopy, confocal microscopy, and impedance volume analyzer, *Cytometry Part A*, 73A(10), 895–903.

Schnars, U. and Juptner, W.P.O., 2002. Digital recording and numerical reconstruction of holograms, *Measurement Science and Technology*, 13(9), R85–R101.

Shaffer, E. and Depeursinge, C., 2010. Digital holography for second harmonic microscopy, in: *Proceedings of the SPIE: Multiphoton Microscopy in the Biomedical Sciences*, 7569, 75691K.

Shaffer, E., Marquet, P., and Depeursinge, C., 2010a. Real time, nanometric 3d-tracking of nanoparticles made possible by second harmonic generation digital holographic microscopy, *Optics Express*, 18(16), 17392–17403.

Shaffer, E., Moratal, C., Marquet, P., and Depeursinge, C., 2010b. Label-free second harmonic phase imaging of biological specimen by digital holographic microscopy, *Optics Letters*, 35, 4102–4104.

Shaffer, E., Pavillon, N., Kühn, J., and Depeursinge, C., 2009. Digital holographic microscopy investigation of second harmonic generated at a glass/air interface, *Optics Letters*, 34(16), 2450–2452.

Sheppard, C.J.R., Gannaway, J.N., Kompfner, R., and Walsh, D., 1977. Scanning harmonic optical microscope, *IEEE Journal of Quantum Electronics*, 13(9), D100–D100.

Teng, S.W., Tan, H.Y., Peng, J.L., Lin, H.H., Kim, K.H., Lo, W., Sun, Y. et al. 2006. Multiphoton autofluorescence and second-harmonic generation imaging of the ex vivo porcine eye, *Investigative Ophthalmology & Visual Science*, 47(3), 1216–1224.

Theodossiou, T.A., Thrasivoulou, C., Ekwobi, C., and Becker, D.L., 2006. Second harmonic generation confocal microscopy of collagen type I from rat tendon cryosections, *Biophysical Journal*, 91(12), 4665–4677.

Tiaho, F., Recher, G., and Rouede, D., 2007. Estimation of helical angles of myosin and collagen by second harmonic generation imaging microscopy, *Optics Express*, 15(19), 12286–12295.

Weng, J.W., Zhong, J.G., and Hu, C.Y., 2008. Digital reconstruction based on angular spectrum diffraction with the ridge of wavelet transform in holographic phase-contrast microscopy, *Optics Express*, 16(26), 21971–21981.

Xu, P., Kable, E., Sheppard, C.J.R., and Cox, G., 2010a. A quasi-crystal model of collagen microstructure based on SHG microscopy, *Chinese Optics Letters*, 8(2), 213–216.

Xu, Q., Shi, K.B., Li, H.F., Choi, K., Horisaki, R., Brady, D., Psaltis, D., and Liu, Z.W., 2010b. Inline holographic coherent anti-Stokes Raman microscopy, *Optics Express*, 18(8), 8213–8219.

Yamaguchi, I. and Zhang, T., 1997. Phase-shifting digital holography, *Optics Letters*, 22(16), 1268–1270.

Yazdanfar, S., Laiho, L.H., and So, P.T.C., 2004. Interferometric second harmonic generation microscopy, *Optics Express*, 12(12), 2739–2745.

10

Imaging Membrane Potential with SHG

Mutsuo Nuriya
Keio University

Rafael Yuste
Columbia University

10.1 Introduction: Membrane Potential Measurements with Imaging

The functioning of the central nervous system is arguably one of the major challenges in contemporary science. After more than a century of neuroscience research, it is clear that neurons are the fundamental units of the nervous system and that they integrate electrical signals, that is, changes in the membrane voltage, and in turn communicate with other neurons by generating electrical action potentials. Thus, methods that can measure these electrical input and output properties are crucial to the understanding of how the central nervous system functions.

The traditional method to investigate the electrical properties of neurons is to insert microelectrodes into neurons, or perform whole-cell recordings, or use extracellular probes, to measure the membrane potential and its change on receiving stimuli. Although electrical recordings remain the work horse of functional studies in neuroscience, they are not usable to probe small (<1 μm) neuronal structures or to record the activity of assemblies of hundreds or thousands of neurons. As an example of their limitation, it is not possible at this time to measure the membrane potential of dendritic spines, small structures (~1 fl in volume), which are of special significance because they are the major sites of excitatory inputs [1]. Indeed, even 120 years after their discovery by Cajal, it is still unclear whether or not they have an active electrical function [2]. A similar case could be made for the electrical measurement from axons or axonal terminals. Although their function is essential, aside from particularly large preparations, like the squid giant axon, it is still unclear exactly how they propagate and integrate electrical signals and their functional neurobiology is a practically unexplored territory in terms of experimental data [3].

The development of optical methods, which are less invasive than microelectrodes, can obviate the spatial restriction of electrical recordings and therefore have the potential to greatly advance neuroscience [4]. Although current optical methods, which are mostly fluorescence based, remain useful in the study of many biological systems, they are severely limited by their lack of selectivity. In fact, to optically measure membrane potential in neurons, it would be ideal to use an optical method that can selectively probe the plasma membrane, which is the only part of the neuron that sustains the biologically relevant membrane potential. But in fluorescence, one obtains fluorescence photons from the dye molecules wherever they are located, either cytoplasm, membrane, or neuropil, and this lack of selectivity generates a significant background signal that interferes with imaging the plasma membrane of the neuron, producing poor signal-to-noise, and imposing the need for extensive averaging. Indeed, although fluorescent voltage-sensitive chromophores have been used to image dendrites and axons [5,6], their lack of sensitivity, partly resulting from the lack of specificity when staining plasma versus intracellular membranes, and the large background fluorescence signals, which originate from other regions of the neuron, could severely limit their usefulness, for example, to measure dendritic spines or axonal voltage dynamics.

10.2 SHG: Theoretical Background

To circumvent this problem, one can apply the second-order nonlinear optical method of second-harmonic generation (SHG) to optically measure the membrane potentials. We will not review the theoretical basis of SHG, which is covered in the chapters in Part I of this volume, but will briefly provide some pointers to its essential features. In SHG, incident light at frequency ω generates light at 2ω on interacting with the neuron, which selectively images membranes, and not bulk regions because of symmetry requirements [7]. Therefore, by cancelation of the dipole moments of the chromophores, there is essentially no SHG signal from inside or outside the cell. Meanwhile, the plasma membrane contributes to position the SHG chromophores in the same orientation, either by lipophilic adsorption to the membrane leaflet, or by electrostatic forces. Moreover, if the SHG chromophore is delivered only to one side of the membrane (either extra or intracellularly), the plasma membrane then acts as a symmetry-breaking interface, generates therefore an array of oriented chromophores with a dipole asymmetry, which is an ideal circumstance for strong SHG. This means that when one measures SHG from neurons there is essentially no background SHG signal and all SHG photons are generated in the plasma membrane, exactly where the electrical field is located. This physicochemical "perfect storm" has not passed unnoticed by investigators: Indeed, in the last decade, following the pioneering work of Lewis and Loew [8], a number of groups, including ourselves, have successfully applied SHG and performed high-resolution optical measurements of membrane potential from living neurons [9–15].

10.3 SHG Voltage Chromophores

Successful measurement of membrane potential by SHG depends heavily on the choice of chromophores. So far, many chromophores have been examined for their use in SHG imaging of membrane potential and several of them are now being used routinely [9–13,16]. However, this by no means suggests that they are best chromophores for this purpose. In fact, there exist tremendous numbers of already available chromophores as well as chromophores being developed that have high potential to show good SHG response to membrane potential changes. Detailed review of chemical properties of these chromophores can be found elsewhere [17]. Here, we describe several key aspects of the SHG chromophores.

In general, chromophores used for SHG have electron donors and acceptors bridged by π-bonds provided by double bonds or aromatic rings. Chromophores successfully applied to SHG membrane potential measurements, including FM4-64 and ANEPP chromophores have these characteristics. These molecules are polarized and are highly susceptible to electromagnetic fields induced by laser illumination. When these chromophores are aligned in an ordered manner breaking the center of symmetry, they become good SHG materials to generate SHG photons. This hyperpolarizability of chromophores determines how many SHG photons can be recorded under basal condition and thereby partly determines the signal-to-noise ratio of the measurements. While SHG does not involve absorption of photons and therefore are not restricted by the matching of energy of photons (i.e., wavelength) to the energy gaps between the ground states and the excited states of the chromophores, it is enhanced around the wavelength that induces absorption. In addition to this, basal interactions of chromophores with photons, local electrical fields generated by the plasma membrane heavily influence the molecules that are inserted in the plasma membrane [4]. When neurons fire action potential, for example, the membrane potential changes from the resting state of ~-65 mV to the peak depolarized point with a magnitude of $\sim+100$ mV. This change in voltage is huge considering the thickness of the plasma membrane being <10 nm, which gives rise to $>10^7$ V/m at the plasma membrane. Under these conditions, several things are expected to occur for these chromophores inside the plasma membrane. First, hyperpolarizability of the chromophores changes under different electrical fields, which alters the hyperscattering of light by the chromophore array. This, so-called electrooptic mechanism is what is considered to be the main mechanism by which membrane potentials are measured by SHG imaging [15]. This is an instantaneous phenomenon and therefore there is no delay between actual membrane potential changes and SHG signal changes, which is crucial in measuring fast membrane potential fluctuations. In contrast, membrane potential changes may induce slow changes in chromophores by altering their distribution, especially between those on the plasma membrane and those outside the plasma membrane [4]. Obviously, changes in dye concentration at the plasma membrane affect the SHG photons generated by the same intensity of laser, but this redistribution of chromophores takes time and is not instantaneous and therefore, not suitable for measuring the fast membrane potential changes.

Another factor to consider is the balance between membrane affinity and solubility in the aqueous solution. Many chromophores, especially those designed to stain plasma membranes, are hydrophobic in nature. These chromophores can be dissolved in a nonpolar medium such as DMSO and be applied to the samples. However, biological samples are sensitive to these media and physiological properties of cells can be easily affected by such solvents. In this regard, chromophores should be soluble in aqueous solution under physiological ionic strength. To suffice these two contradictory requirements, amphiphilic chromophores have been chosen for SHG detection at the plasma membrane. These chromophores insert themselves into the plasma membrane with their nonpolar tail group while leaving the charged head group outside the membrane. Due to this charge group, they cannot freely pass through or flip inside the plasma membrane and thus when applied from one side of the plasma membrane, they accumulate on a single side of membrane leaflet in a highly ordered fashion, sufficing the requirement of SHG.

10.4 Experimental Setup

10.4.1 General Description of the SHG Setup

As the SHG is a two-photon phenomenon, it requires two-photon microscopy setup, as described in detail in Figure 10.1 and [18]. One thing that is special about SHG setup is that due to the forward propagating nature of SHG, detectors on the forward path are also required. The best wavelength of femtosecond laser for membrane potential measurement by SHG depends on the chromophores to be used in the experiment. Things to consider in determining the wavelength include basal SHG signal strength, membrane potential sensitivity, and phototoxicity. We have empirical evidence to suggest that wavelength around 1000 nm works well in these aspects for our experiments using FM4-64 as a dye. The laser power is then modulated by acoustooptical modulator, Pockels cell, or by other means before entering the microscope. The cells loaded with SHG chromophores will then be illuminated by the laser in the recording chamber. In contrast to the two-photon fluorescence (TPF) signals that are emitted isotropically in all directions, which allow the collection of signals both in forward (transmission) and backward (epi) directions, SHG photons can only be detected in the forward direction. While both TPF and SHG photons are collected through the condenser nonselectively, these signals can be easily separated by the combination of appropriate dichroic mirror and bandpass filter that allow only the

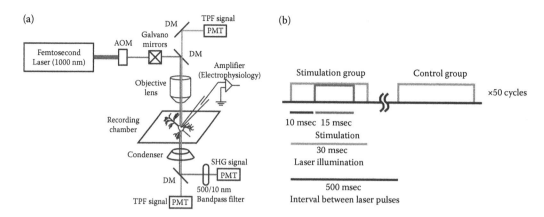

FIGURE 10.1 Schematic diagram of the experimental setup for imaging membrane potential with SHG. (a) Schematic diagram of the SHG setup. Neurons are placed in a recording chamber under the regular two-photon microscopy setup equipped with femtosecond laser and SHG detector in the transmission path. Electrophysiological and optical signals are recorded simultaneously. AOM, acousto-optic modulator; DM, dichroic mirror; PMT, photomultiplier tube. (b) Schematic diagram of the point-scan protocol for measuring fast electrical events. In the point-scan protocol, the laser position is fixed at a single point and the laser is illuminated on the sample for a short period of time with an interval between illuminations. In this diagram, laser is illuminated on the sample for 30 ms with 500 ms of interval in between illuminations. In the typical experiments, neurons are stimulated electrophysiologically during the laser illumination and the SHG signals during this period are compared to those without stimulation.

photons of exactly half the wavelength of illuminating laser to be detected in the SHG channel while longer wavelength TPF photons are detected in the other detector. The numerical aperture of the condenser should at least match that of the objective lens to collect all the SHG photons generated from the cone-shaped incoming photons [19]. Even higher NA does not dramatically help in theory as well as in practice under our experimental conditions, but one can expect that it may increase photon collection in detecting SHG signals from highly scattering samples.

10.4.2 Imaging Electrical Signals with SHG

SHG images can be obtained in any mode, but to image membrane potential dynamics, there are some factors that need to be considered. Although it varies on the preparation and temperature, action potentials are very fast phenomena, where the entire events can be finished in the order of milliseconds. As such, the detection and recordings of SHG signals need to be faster than a millisecond to capture these events. This is challenging for the regular two-photon imaging regime (frame-scan) as laser scanning is a relatively slow process. This can be overcome by acquiring whole areas of images and focusing the recording sites to a single line (line-scan) or even a single point (point-scan). With the advancement of devices, the line-scan can now be performed at a rate faster than 1 kHz (i.e., submillisecond). However, faster scanning means less pixel dwell time and therefore less signal photons from a single point at a single time, which, in the end, reduces the signal-to-noise ratio. Therefore, continuous recording from single point with laser illumination at a fixed point has a big advantage in this regard. In this case, the temporal resolution is limited only by the recording device itself. In fact, we normally use point-scan to detect fast SHG changes induced by fast electrical events such as action potential. However, longer pixel dwell time means more photon-induced damages at the site of laser focus. This is especially true for two-photon imaging as it utilizes high power lasers to achieve rare two-photon events. In theory, if laser illumination only produces SHG signals without absorption of photon energy by chromophores (i.e., fluorescence), it should not induce any photodamage. However, in reality, because of resonant enhancement nature of SHG as described earlier, SHG is performed with the dye at wavelengths that also generate two-photon fluorescence and, therefore, photodamage [8]. As such, one needs to limit the laser illumination time at a single point to avoid photodamage while collecting enough SHG photons. In practice, we found that neurons can sustain tens of milliseconds of laser illumination at a single point in a given time, which is enough to observe whole action potential event. In addition, to improve signal-to-noise ratio, we collect tens of events in single SHG recordings and average them. Furthermore, we provide laser pulses at an interval of 500 ms (2 Hz recordings) so that neurons have enough time to recover in between the laser pulses.

10.5 Applications

Here, we describe some examples of membrane potential measurements using SHG. As described in the introduction, the unique strength of membrane potential imaging by SHG becomes evident in measuring membrane potential dynamics at small targets in neurons, such as axons, distal dendrites, and dendritic spines that are hardly accessible to conventional electrophysiological techniques.

10.5.1 SHG Imaging of Somatic Voltages

The potential of SHG imaging for membrane potential measurement in cells was realized and investigated from the early 1990s [20,21]. With the development of techniques, it has been applied to mammalian neurons to measure the membrane potentials in these cells [10–12], and particularly to perform optical measurements of somatic voltage (Figure 10.2).

Shown in Figure 10.2 is the actual response of SHG to membrane potential changes taken with the regular frame scan with alternating membrane potential controlled by the voltage-clamp technique. As can be seen, mean SHG signals at the soma changes with membrane potential. Further analyses with different voltage steps reveal that SHG signals collected at the soma respond to membrane potential changes in a linear fashion within and beyond the physiological voltage fluctuation ranges. SHG chromophores (in this case, FM4-64) can be applied both from inside (intracellular loading) and outside (extracellular loading) to give rise to the same magnitude of membrane potential sensitivity (~10% per 100 mV), but with the opposite sign. This is because the chromophores fill the plasma membrane from exactly an opposite direction in these loading schemes and therefore sense the membrane potential changes in the opposite direction as well. Combined with the intrinsic high spatial resolution of SHG and high temporal resolution provided by point-scan protocol or other methods, this high magnitude of the linear response of SHG signals to membrane potential changes allows quantitative imaging of membrane potential changes in the areas where previous techniques had limited access.

10.5.2 SHG Imaging of Dendritic Spines

The imaging of voltage at dendritic spines with SHG constitutes one of the clearest cases where the advantages of SHG are exploited by neuroscientists (Figure 10.3). The membrane potential dynamics in dendritic spines had been a mystery as these structures are too small for conventional electrophysiological recordings. However, from previous experiments, especially from those using two-photon calcium imaging, it had been well recognized that these small structures play very important roles in neuronal physiology [1,22]. As direct measurements are difficult, researchers have utilized numerical simulations to explore the dynamics of electrical signaling in dendritic spines [23,24]. In addition, attempts have been made to use fluorescence signals of voltage-sensitive chromophores to measure the membrane potential changes in dendritic spines [5,25]. However, direct measurement of membrane potential changes in spines continues to be a big challenge due to background signals originating from intracellular dyes, as mentioned in the introduction. Taking advantage of SHG's ability to measure membrane

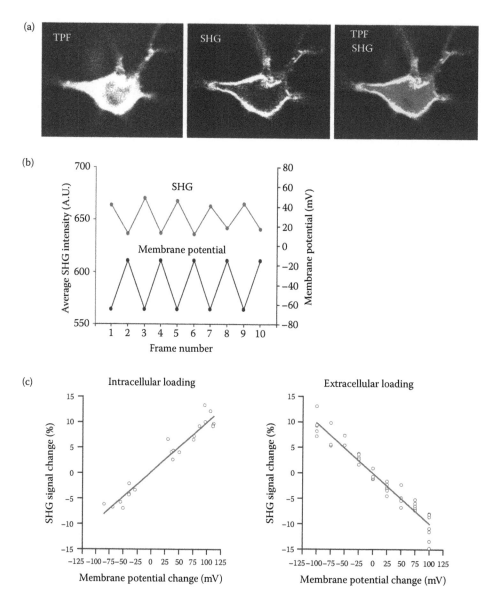

FIGURE 10.2 **(See color insert.)** Characterization of SHG responses to membrane potential. (a) Comparison of two-photon fluorescence (TPF) and SHG signals. TPF and SHG signals were collected simultaneously from a neuron intracellularly stained with FM4-64 through the patch pipette. While TPF signals were obtained from FM4-64 dyes distributed both at the plasma membrane and cytoplasm, SHG signals were only collected from the plasma membrane. (b) SHG signal responses to membrane potential changes. SHG signals were collected from soma of the neuron stained with FM4-64 held at different membrane potentials. Average SHG signal intensity (red) follows the membrane potential (black). (c) SHG signal has large and linear response to membrane potential changes. Neurons were stained with FM4-64 either intracellularly (left) or extracellularly (right) and the SHG signal changes upon membrane potential changes were plotted. In both cases, SHG signal shows linear response to membrane potential changes with the amplitude of ~10% per 100 mV, but with the opposite sign. The red lines show linear fit of the data.

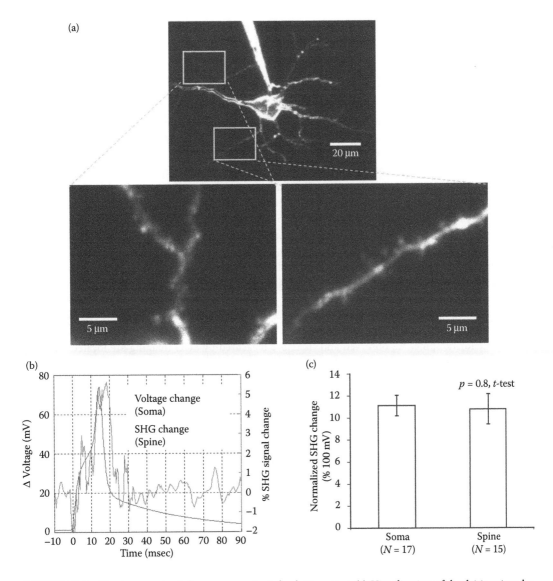

FIGURE 10.3 Membrane potential measurements at dendritic spines. (a) Visualization of dendritic spines by SHG. The neuron in an acute brain slice was intracellularly stained with FM4-64 through a patch pipette and SHG signals were visualized with laser illumination at 1064 nm. At higher magnifications, dendritic spines on the oblique dendrites can be clearly visualized by SHG. (b) Membrane potential measurements at dendritic spines. Point-scan protocol at the dendritic spine revealed the SHG signal changes at the target spine (light gray) upon induction of action potential at soma recorded electrophysiologically (dark gray). (c) Comparison of SHG responses at soma and dendritic spines. Peak amplitudes of SHG signal changes were compared between those obtained at soma and those from dendritic spines, which revealed no significant difference in the membrane potential changes in these locations. (Adapted, with permission, from Nuriya, M. et al. 2006. Imaging membrane potential in dendritic spines. *Proc Natl Acad Sci USA* 103:786–790. Copyright 2006, National Academy of Sciences, USA.)

potential changes in a quantitative manner with high temporal and spatial resolution, we recently succeeded to use SHG to measure action potential invasion into dendritic spines [13].

For those experiments, we loaded neurons in acute brain slices with the SHG dye FM4-64 through the patch-clamp pipette and illuminated neurons with 1064 nm femtosecond laser. FM4-64 diffused into the neurons and filled the inner leaflet of plasma membrane from the site of patch (i.e., soma). Over time, distal structures such as oblique dendrites and basal dendrites became visible by SHG imaging and later, dendritic spines were observed both by SHG imaging and two-photon fluorescence. After locating target spines, we illuminated the target spines continuously (point-scan) while inducing action potential backpropagation by injecting positive current at the soma. The generation of action potential at the soma was monitored by current clamp recording. When we compared SHG signals from the target spines in the nonstimulated status and with action potential backpropagation, we observed SHG signal changes that occurred simultaneously with the somatic action potential. In addition to this kind of qualitative argument, SHG imaging allows us to draw quantitative conclusions. For instance, a 10% change in relative SHG (ΔSHG/SHG) meant a 100 mV membrane potential in our hands regardless of the location or morphology of the target. Therefore, we compared peak SHG signal changes obtained at soma and dendritic spines to compare the amplitudes of action potentials at these locations. This analysis revealed that when the action potential invades into dendritic spines, it invades with full magnitude, without significant decay.

One might think that this observation of the lack of voltage attenuation at dendritic spines does not match with the prediction from cable theory [26]. However, one thing that we should keep in mind is that the dendritic spines we observed and therefore measured by SHG imaging are located relatively close to the soma, mostly within a distance of 50 μm. Although we have little knowledge in voltage dynamics in dendritic spines, previous studies have characterized backpropagation of action potential into dendritic shafts. These studies show that there is indeed an attenuation of peak voltage as it goes distant from the soma, but the decay is relatively mild with the length constant of 138 μm even at basal dendrite, where the attenuation is more severe than in the most studied apical dendrite [27]. If the attenuation is small enough, the difference in SHG signal change may come short of statistical differences. Another intriguing possibility is that the dendritic spines are endowed with active conductances mediated by various voltage-gated ion channels and boost action potential inside. While the presence of voltage-gated sodium channels in spines remain elusive, it is well known that spines have a rich variety of functional voltage-gated calcium channels [28]. Furthermore, active roles of these voltage-gated ion channels in dendritic spines have been suggested [29]. Therefore, it is of particular interest to study backpropagation of action potentials into dendritic spines and parent dendrites in distal locations by SHG imaging, in combination with pharmacological and genetic tools to dissect out molecular and cellular mechanisms of propagation of action potential into dendritic spines.

10.5.3 SHG Imaging of Axons

As another example of SHG measurements of voltage in neurons, we will discuss recent SHG imaging of axons (Figure 10.4). Axons are considered to be designed to deliver the sole output of neurons, action potentials to distal sites. In addition to the known roles of faithfully transmitting action potential to distal sites, however, more complex roles of axons have recently been proposed [30,31]. In fact, in some preparations, electrophysiological recordings were successfully obtained from axons of mammalian neurons, which provide clues to such additional analog information processing in axons [32,33]. However, it still remains a challenging task to perform electrophysiological experiments at axons in an intact preparation. Because of this, we utilized SHG to measure the action potential propagations into axonal arbors [34].

For some yet unknown reasons, axons cannot be visualized by SHG in brain slices by intracellular loading of SHG chromophores. Obviously, it is not difficult to imagine that passive diffusions of FM chromophores are extremely slow in axons that have much smaller diameters compared to dendrites. To

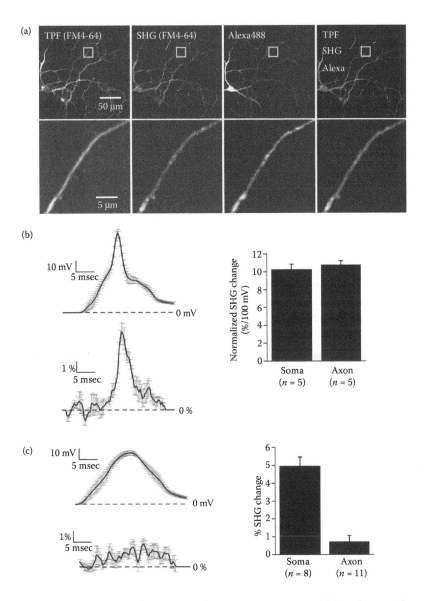

FIGURE 10.4 (See color insert.) Membrane potential measurements at axons. (a) Visualization of axons by SHG. Primary cultured dissociated neurons were extracellularly stained with FM4-64 and loaded with Alexa 488 intracellularly. The whole neuronal morphologies including fine processes of axons can be clearly visualized by SHG. (b) Membrane potential changes at axons with action potential. The upper left panel shows the voltage changes recorded at the soma and the lower left panel shows SHG changes simultaneously recorded at axons. Data are shown as the mean (black) ± SEM (red) from five neurons. The right panel shows a comparison of the peak amplitude of SHG signal changes normalized to the action potential with 100 mV amplitude. (c) Membrane potential changes at axons with nonregenerating somatic voltage change. Left: Neurons were held under current-clamp conditions in the presence of 1 μM TTX and injected with a depolarizing current pulse. The voltage change at the soma (upper left panel) and corresponding changes of SHG recorded at axons (lower left panel) are shown. Data shown is the mean (black) ± SEM (red) from 12 recordings from 7 neurons. Right: Neurons were held under voltage-clamp configuration and 50 mV voltage pulses were applied to the soma. SHG responses at the soma (n = 8 from 6 neurons) and axons (n = 11 from 6 neurons) were measured. Data in the panel presented is mean ± SEM. (Adapted from Nuriya, M. and M. Yasui. 2010. *J Biomed Opt* 15:020503. With permission.)

overcome this issue, we utilized primary cultured dissociated neurons obtained from mouse hippocampus and applied SHG chromophores from the outside of the cell, in the bath solution. The advantage of this system is that when the density of neurons is low enough, there is little overlap between membrane structures in the culture dish, which will alter the SHG signal itself and/or voltage sensitivity when the chromophore is applied in the bath. Indeed, such an uncertainty is one of the major challenges in applying SHG membrane potential imaging to membrane-rich densely packed samples such as brain slices. Using very low-density culture neurons, SHG signals could be observed from all over the neuronal structures including axons and dendrites, which were confirmed by immunocytochemical analysis with specific markers for axons and dendrites.

When the action potential was generated at the soma by positive current injection, reliable SHG signal changes were observed along axonal arbors. Again, as SHG imaging is a quantitative imaging technique, we could compare peak SHG signal changes at the axons and soma and conclude that there is no difference between voltage deflections at the soma and axons. Furthermore, we realized that the variability in voltage changes at the axons is very small, not like calcium signals reported previously [35,36]. These data suggest that the variability in calcium signals reported previously is likely due to the differences in calcium channels rather than voltage deflection per se.

In addition, we performed different sets of experiments to address the nature of this nonattenuating propagation of action potential into axonal arbors. Action potentials are generated by the sequential activations of voltage-gated sodium channel and potassium channels [37]. When axons are endowed with these channels in enough density, action potentials regenerate themselves on their propagation direction. In contrast to these regenerating active propagation of membrane potentials, passive propagation of voltage changes are attenuating in nature. To address to what extent this kind of passive propagation occurs in the axons, we induced nonregenerative voltage changes at the soma and measured SHG signal changes in the axons. In sharp contrast to what was observed with action potential, these nonregenerative membrane potential changes attenuated significantly along the axonal arbors. These data reveal two types of voltage propagations from soma to axonal arbors in these cultured hippocampal neurons: nonattenuating regenerative action potential and highly attenuating nonregenerative voltage fluctuations.

10.6 Conclusions and Perspectives

As mentioned earlier, membrane potential measurement by SHG imaging has a unique advantage in that it is selective for plasma membranes and it can also provide quantitative information that could not have been obtained from other methods. As this kind of information is crucial in understanding the physiology of neurons and other excitable cells, expectations are high for further applications of this technique. At the same time, the history of membrane potential measurement by SHG is still recent and therefore there is still plenty of room for improvement. Especially, at present, the signal-to-noise ratio is still quite low compared to more established fluorescence-based membrane potential imaging techniques. Therefore, as has been the case so far, efforts need to be made in parallel to further refine the technique itself and to apply the technique to physiological questions. For this goal, we believe that collaborative efforts between biologists and researchers from other fields such as synthetic chemists are crucial. We are hopeful that such collaborations will help further develop the technique and shed new light on the physiology of neurons and other excitable cells. Finally, we briefly describe other potential applications as well as two important aspects of researches in considering future developments of the field.

10.6.1 Applications

There are many targets and physiological events for which we have limited knowledge about the membrane potential dynamics. First of all, membrane potential dynamics in small structures in neurons such as dendritic spines and axons have just begun to be investigated. For example, there still is no direct experimental evidence about forward propagation of voltage signals upon synaptic input onto dendritic spines. Efforts have been made to elucidate the membrane potential changes upon synaptic stimulation using different techniques [25,38], but SHG recordings will be the ideal tools to answer this question. With regard to axons, changes in axonal electrical information processing have drawn much attention [39], but the direct observation of such processes needs to follow. Here again, membrane potential imaging by SHG will be a powerful tool that can be employed to meet these needs. With the improvement of this technique, we expect that SHG imaging will become a precious tool to complement other existing techniques in investigating the physiology of neurons and other cells.

10.6.2 Genetic Approaches

Fluorescent proteins have been modified to probe various aspects of cell physiology, including calcium and membrane potential [40]. While these are developed for fluorescence imaging, there are reports suggesting that GFP (green fluorescent protein) and their relative proteins generate SHG signals [41,42]. In addition, efforts have been made to generate membrane-tethered fluorescent proteins for the use of membrane potential measurements, which also should help increase SHG signals by aligning the molecules in an ordered manner [43]. This type of genetic approach has two potential benefits: (1) it would not require intracellular dye-loading processes and, therefore, may be able to circumvent the slow dye diffusion issues that FM4-64 and other organic chromophores have experienced; and (2) in combination with proper promoters, it would allow labeling of genetically defined subpopulation of cells. Furthermore, this genetic approach may pave the way to *in vivo* SHG imaging. Protein-based imaging of intracellular calcium concentration has been explored for a long time and improved indicators have been utilized to image neuronal activities *in vivo* [44,45]. Recently, a genetically targeted voltage-sensitive fluorescent protein was successfully applied to *in vivo* imaging [46]. Clearly, these are ideal methods to monitor neuronal behavior *in vivo* as they circumvent dye loading using invasive methods. Unfortunately, *in vivo* imaging of SHG is difficult as most of the SHG signals go to the transmission path and do not come back to the objective lenses. However, if the SHG signals are strong enough, back-scattered SHG signals can be detected and provide unique information about the membrane potential dynamics *in vivo*. Indeed, recent demonstrations of intrinsic SHG imaging of muscle and collagen *in vivo* clearly show that *in vivo* SHG is feasible, and further suggest that it can also be applied to membrane potential measurements [47,48].

For genetic approaches, however, time-resolution issues must be considered, as these techniques quite often have time lag between actual voltage change and signal response. It can be predicted that if the SHG has electrooptic response from these probes, the response can be near instantaneous and therefore does not have any time lag. However, if changes in SHG by membrane potential fluctuation involve realignment of the probe or conformational changes of the molecule, then these processes take certain time and therefore it cannot be instantaneous. Although it can practically be used if these changes are fast enough, careful characterization will be required for this type of genetic approach.

10.6.3 Hardware Devices

In addition to chromophores, modifications of the imaging devices are expected to improve the current SHG imaging of membrane potential and further expand its applications.

First, to overcome the inherent issue of time resolution in obtaining 2D images by laser scanning microscope, efforts have been made to utilize random access microscopy in obtaining SHG signals [49]. In this setting, the position of laser illumination is not regulated by a set of mirrors (galvano mirrors), as most laser scanning microscopes do. Instead, these utilize acoustooptic devices to direct laser in 2D space, or even in 3D. This technique has been successfully implemented into calcium imaging [50,51]. There still remains a problem of spatial resolution, but the temporal resolution and freedom of collection patterns are far superior to conventional mirror-based laser scan system [49]. This technique should be useful in collecting the membrane potential information from many spatially distributed points, either from a single neuron or a population of neurons.

As an alternative solution to faster scanning is the recently developed SLM microscopy [52], in which the laser beam is multiplexed in space by a spatial light modulator, so that every part of the sample of interest is illuminated simultaneously. Then, one can use a camera as a detector to collect photons that emerge from every area of interest. Thus, with SLM, one does not scan the sample any more, and this solves the problem of small pixel dwell time, introduced by faster scanning methods. "Scanless" SLM microscopy works well with femtosecond lasers and is therefore poised to be very helpful to advance fast SHG measurements.

Finally, similar to most imaging techniques, the signal-to-noise ratio needs to be improved for SHG imaging. This is especially crucial in voltage measurement since better signal-to-noise ratio is directly translated into better ability to detect smaller voltage fluctuations with higher confidence. Photon counter has been implemented to increase signal-to-noise ratio in the SHG recording of membrane potential in neurons, which is generally photon-limited [53]. Use of photon-counting mode has two advantages: (1) it reduces background signals arising from thermal noises, and (2) it allows accumulation of SHG photons in a given recording time. Increase in signal and reduction in noise would lead to better signal-to-noise ratio and therefore better recordings. Just as photon counting is shown to improve SHG imaging, we can expect that application of other techniques used in different imaging researches may improve the SHG imaging as well.

References

1. Yuste, R. 2010. *Dendritic Spines.* The MIT Press, Cambridge, MA.
2. Tsay, D. and R. Yuste. 2004. On the electrical function of dendritic spines. *Trends Neurosci* 27:77–83.
3. Stys, S. G. W., D. K. Jeffery, and K. Peter. 1995. *The Axon: Structure, Function and Pathophysiology.* Oxford University Press, Oxford, UK.
4. Peterka, D. S., H. Takahashi, and R. Yuste. 2011. Imaging voltage in neurons. *Neuron* 69:9–21.
5. Holthoff, K., D. Zecevic, and A. Konnerth. 2010. Rapid time course of action potentials in spines and remote dendrites of mouse visual cortex neurons. *J Physiol* 588:1085–1096.
6. Foust, A., M. Popovic, D. Zecevic, and D. A. McCormick. 2010. Action potentials initiate in the axon initial segment and propagate through axon collaterals reliably in cerebellar Purkinje neurons. *J Neurosci* 30:6891–6902.
7. Eisenthal, K. B. 1996. Liquid interfaces probed by second-harmonic and sum-frequency spectroscopy. *Chem Rev* 96:1343–1360.
8. Campagnola, P. J., M. D. Wei, A. Lewis, and L. M. Loew. 1999. High-resolution nonlinear optical imaging of live cells by second harmonic generation. *Biophys J* 77:3341–3349.
9. Millard, A. C., L. Jin, M. D. Wei, J. P. Wuskell, A. Lewis, and L. M. Loew. 2004. Sensitivity of second harmonic generation from styryl dyes to transmembrane potential. *Biophys J* 86:1169–1176.
10. Dombeck, D. A., M. Blanchard-Desce, and W. W. Webb. 2004. Optical recording of action potentials with second-harmonic generation microscopy. *J Neurosci* 24:999–1003.
11. Dombeck, D. A., L. Sacconi, M. Blanchard-Desce, and W. W. Webb. 2005. Optical recording of fast neuronal membrane potential transients in acute mammalian brain slices by second-harmonic generation microscopy. *J Neurophysiol* 94:3628–3636.
12. Nemet, B. A., V. Nikolenko, and R. Yuste. 2004. Second harmonic imaging of membrane potential of neurons with retinal. *J Biomed Opt* 9:873–881.
13. Nuriya, M., J. Jiang, B. Nemet, K. B. Eisenthal, and R. Yuste. 2006. Imaging membrane potential in dendritic spines. *Proc Natl Acad Sci USA* 103:786–790.
14. Sacconi, L., D. A. Dombeck, and W. W. Webb. 2006. Overcoming photodamage in second-harmonic generation microscopy: Real-time optical recording of neuronal action potentials. *Proc Natl Acad Sci USA* 103:3124–3129.

15. Jiang, J., K. B. Eisenthal, and R. Yuste. 2007. Second harmonic generation in neurons: Electro-optic mechanism of membrane potential sensitivity. *Biophys J* 93:L26–L28.

16. Theer, P., W. Denk, M. Sheves, A. Lewis, and P. B. Detwiler. 2011. Second-harmonic generation imaging of membrane potential with retinal analogues. *Biophys J* 100:232–242.

17. Reeve, J. E., H. L. Anderson, and K. Clays. 2010. Dyes for biological second harmonic generation imaging. *Phys Chem Chem Phys* 12:13484–13498.

18. Nikolenko, V., B. Nemet, and R. Yuste. 2003. A two-photon and second-harmonic microscope. *Methods* 30:3–15.

19. Moreaux, L., O. Sandre, S. Charpak, M. Blanchard-Desce, and J. Mertz. 2001. Coherent scattering in multi-harmonic light microscopy. *Biophys J* 80:1568–1574.

20. Ben-Oren, I., G. Peleg, A. Lewis, B. Minke, and L. Loew. 1996. Infrared nonlinear optical measurements of membrane potential in photoreceptor cells. *Biophys J* 71:1616–1620.

21. Bouevitch, O., A. Lewis, I. Pinevsky, J. P. Wuskell, and L. M. Loew. 1993. Probing membrane potential with nonlinear optics. *Biophys J* 65:672–679.

22. Yuste, R. and W. Denk. 1995. Dendritic spines as basic functional units of neuronal integration. *Nature* 375:682–684.

23. Segev, I. and W. Rall. 1998. Excitable dendrites and spines: Earlier theoretical insights elucidate recent direct observations. *Trends Neurosci* 21:453–460.

24. Tsay, D. and R. Yuste. 2002. Role of dendritic spines in action potential backpropagation: A numerical simulation study. *J Neurophysiol* 88:2834–2845.

25. Palmer, L. M. and G. J. Stuart. 2009. Membrane potential changes in dendritic spines during action potentials and synaptic input. *J Neurosci* 29:6897–6903.

26. Vetter, P., A. Roth, and M. Hausser. 2001. Propagation of action potentials in dendrites depends on dendritic morphology. *J Neurophysiol* 85:926–937.

27. Nevian, T., M. E. Larkum, A. Polsky, and J. Schiller. 2007. Properties of basal dendrites of layer 5 pyramidal neurons: A direct patch-clamp recording study. *Nat Neurosci* 10:206–214.

28. Yasuda, R., B. L. Sabatini, and K. Svoboda. 2003. Plasticity of calcium channels in dendritic spines. *Nat Neurosci* 6:948–955.

29. Araya, R., V. Nikolenko, K. B. Eisenthal, and R. Yuste. 2007. Sodium channels amplify spine potentials. *Proc Natl Acad Sci USA* 104:12347–12352.

30. Alle, H. and J. R. Geiger. 2008. Analog signalling in mammalian cortical axons. *Curr Opin Neurobiol* 18:314–320.

31. Segev, I. and E. Schneidman. 1999. Axons as computing devices: Basic insights gained from models. *J Physiol Paris* 93:263–270.

32. Shu, Y., A. Hasenstaub, A. Duque, Y. Yu, and D. A. McCormick. 2006. Modulation of intracortical synaptic potentials by presynaptic somatic membrane potential. *Nature* 441:761–765.

33. Alle, H. and J. R. Geiger. 2006. Combined analog and action potential coding in hippocampal mossy fibers. *Science* 311:1290–1293.

34. Nuriya, M. and M. Yasui. 2010. Membrane potential dynamics of axons in cultured hippocampal neurons probed by second-harmonic-generation imaging. *J Biomed Opt* 15:020503.

35. Koester, H. J. and B. Sakmann. 2000. Calcium dynamics associated with action potentials in single nerve terminals of pyramidal cells in layer 2/3 of the young rat neocortex. *J Physiol* 529 Pt 3:625–646.

36. Mackenzie, P. J., M. Umemiya, and T. H. Murphy. 1996. Ca^{2+} imaging of CNS axons in culture indicates reliable coupling between single action potentials and distal functional release sites. *Neuron* 16:783–795.

37. Bean, B. P. 2007. The action potential in mammalian central neurons. *Nat Rev Neurosci* 8:451–465.

38. Grunditz, A., N. Holbro, L. Tian, Y. Zuo, and T. G. Oertner. 2008. Spine neck plasticity controls postsynaptic calcium signals through electrical compartmentalization. *J Neurosci* 28:13457–13466.

39. Debanne, D. 2004. Information processing in the axon. *Nat Rev Neurosci* 5:304–316.

40. Miyawaki, A. 2005. Innovations in the imaging of brain functions using fluorescent proteins. *Neuron* 48:189–199.

41. Khatchatouriants, A., A. Lewis, Z. Rothman, L. Loew, and M. Treinin. 2000. GFP is a selective non-linear optical sensor of electrophysiological processes in *Caenorhabditis elegans*. *Biophys J* 79:2345–2352.

42. Asselberghs, I., C. Flors, L. Ferrighi, E. Botek, B. Champagne, H. Mizuno, R. Ando et al. 2008. Second-harmonic generation in GFP-like proteins. *J Am Chem Soc* 130:15713–15719.

43. Sjulson, L. and G. Miesenbock. 2008. Rational optimization and imaging *in vivo* of a genetically encoded optical voltage reporter. *J Neurosci* 28:5582–5593.

44. Mank, M., A. F. Santos, S. Direnberger, T. D. Mrsic-Flogel, S. B. Hofer, V. Stein, T. Hendel et al. 2008. A genetically encoded calcium indicator for chronic *in vivo* two-photon imaging. *Nat Methods* 5:805–811.

45. Tian, L., S. A. Hires, T. Mao, D. Huber, M. E. Chiappe, S. H. Chalasani, L. Petreanu et al. 2009. Imaging neural activity in worms, flies and mice with improved GCaMP calcium indicators. *Nat Methods* 6:875–881.

46. Akemann, W., H. Mutoh, A. Perron, J. Rossier, and T. Knopfel. 2010. Imaging brain electric signals with genetically targeted voltage-sensitive fluorescent proteins. *Nat Methods* 7:643–649.

47. Llewellyn, M. E., R. P. Barretto, S. L. Delp, and M. J. Schnitzer. 2008. Minimally invasive high-speed imaging of sarcomere contractile dynamics in mice and humans. *Nature* 454:784–788.

48. Bao, H., A. Boussioutas, R. Jeremy, S. Russell, and M. Gu. 2010. Second harmonic generation imaging via nonlinear endomicroscopy. *Opt Express* 18:1255–1260.

49. Sacconi, L., J. Mapelli, D. Gandolfi, J. Lotti, R. P. O'Connor, E. D'Angelo, and F. S. Pavone. 2008. Optical recording of electrical activity in intact neuronal networks with random access second-harmonic generation microscopy. *Opt Express* 16:14910–14921.

50. Otsu, Y., V. Bormuth, J. Wong, B. Mathieu, G. P. Dugue, A. Feltz, and S. Dieudonne. 2008. Optical monitoring of neuronal activity at high frame rate with a digital random-access multiphoton (RAMP) microscope. *J Neurosci Methods* 173:259–270.

51. Iyer, V., T. M. Hoogland, and P. Saggau. 2006. Fast functional imaging of single neurons using random-access multiphoton (RAMP) microscopy. *J Neurophysiol* 95:535–545.

52. Nikolenko, V., B. O. Watson, R. Araya, A. Woodruff, D. S. Peterka, and R. Yuste. 2008. SLM microscopy: Scanless two-photon imaging and photostimulation with spatial light modulators. *Front Neural Circuits* 2:5.

53. Jiang, J. and R. Yuste. 2008. Second-harmonic generation imaging of membrane potential with photon counting. *Microsc Microanal* 14:526–531.

III

Applications of SHG

11

Second-Harmonic Generation Imaging of Self-Assembled Collagen Gels

Christopher B. Raub
*University of California,
Irvine*

Bruce J. Tromberg
*University of California,
Irvine*

Steven C. George
*University of California,
Irvine*

11.1 Introduction

The biological molecules that exhibit second-harmonic generation (SHG) include fibrillar collagens, myosin, microtubules, silk, and cellulose [1–4]. Laser scanning microscopy (LSM) allows for noninvasive and nondestructive three-dimensional imaging of the SHG signal from biological samples possessing second-harmonic-generating molecules [5–9]. As an optical signal, SHG is uniquely sensitive to the spatial organization of generating dipoles [10–14], allowing for quantitative and selective structural characterization of second-harmonic-generating tissue. Numerous optical and structural parameters have been derived from the SHG signal, from which inferences can be made about tissue structural, compositional, optical, and mechanical properties [15–26].

This chapter focusses on SHG signal imaging studies of hydrogels composed of acid-solubilized type I collagen. Self-assembled silk and cellulose scaffolds that generate second-harmonic signal have also been characterized [4,17,18,27]. The prevalence of studies utilizing collagen hydrogels originates from the extremely important role and ubiquitous presence of collagen in connective tissues throughout the body. Collagen is the most common protein in the body, comprising 6% of body weight and ~25–33% of the total protein mass [28]. Self-assembled collagen hydrogels are not only useful *in vitro* models to study cell–matrix interactions [29–32], cellular modulation of wound healing [8,33–41], fibrosis [42–47], and microstructure–mechanics relations [21,22,48–52], but they also serve as starting scaffolds for many tissue-engineering experiments and applications [42,43,45,53–55].

Cell-seeded collagen hydrogels are particularly amenable to analysis by three-dimensional LSM, including SHG imaging, since the tissues may be imaged nondestructively at any time during *in vitro* culture [5,24,56–61]. Collagen self-assembly is controlled by polymerization conditions that influence polymer aggregation, creating gels of varied microstructures and network properties [21,22,62–68]. The cells seeded in or on such hydrogel scaffolds create tissue constructs that change dynamically due to force interactions between cells and the surrounding scaffold, and by proteolysis and new matrix deposition. Dynamic remodeling of cell-seeded collagen gels may be tracked in four dimensions by assembling z-stacks of SHG image frames from tissues at different locations and culture time points. Other optical signals, such as reflectance [69], optical coherence [70], one- and two-photon fluorescence (TPF) [56,71], and coherent anti-Stokes Raman scattering (CARS) [72] may be imaged simultaneously or nearly on a multimodal platform, increasing the structural, biochemical, and optical information derived from the imaged tissue regions and allowing the study of interactions between the signal-producing species within the tissue. SHG imaging of acellular and cell-seeded self-assembled gels is a powerful technique to address the fundamental questions regarding tissue mechanics and cell behavior within a three-dimensional matrix environment.

11.2 Background

11.2.1 Collagen Structural Properties

Over 20 unique types of collagen have been described, including the fibril-forming collagens (types I, II, III, V, and XI) [63,73]. Each of these fibril-forming collagens posses a similar noncentrosymmetric structure that is required for both SHG and for self-assembly into supramolecular aggregates and, finally, an entangled network. The three properties of SHG, self-assembly, and resistance to tension arise from the unique primary structure of fibril-forming collagens. This structure consists of repeats of the triple amino acid sequence glycine–X–Y, where X and Y are most frequently proline and hydroxyproline, respectively [73]. These repeats comprise ~10% of the collagen monomer sequence and enable the formation of left-handed helices, termed alpha chains tightly coiled with about three amino acids per turn. The three alpha chains interact with each other to form a right-handed, coiled-coil triple helix. This triple helical procollagen molecule is capped by nonhelical propeptide regions that promote solubility and whose presence and enzymatic removal are both necessary for fibrillogenesis to occur [65]. Procollagen molecules are synthesized near the endoplasmic reticulum and must be packaged and secreted by the Golgi apparatus before extracellular initiation of fibrillogenesis. Procollagen N- and C-peptidase are extracellular enzymes that cleave the 15 and 10 nm-long N and C propeptides, respectively, yielding a collagen monomer ~300 nm long and 1.5 nm wide [73]. Following procollagen cleavage, the collagen monomer contains only the helical region capped by nonhelical telopeptide regions that consist of 10–25 amino acids, and are important in cross-linking and in directing monomer packing and fibrillogenesis. The interchain spacing within the triple helix is ~0.286 nm, close enough for hydrogen bonding, hydrophobic interactions, and interchain cross-links to stabilize the monomer. After the conversion of procollagen into collagen monomers, fibrillogenesis may occur. The process is an entropy-driven self-assembly that may occur in a cell-free environment modulated by physical variables (pH, temperature, ions) or in living tissue where cell-secreted molecules and enzymes may modulate fibrillogenesis [73].

11.2.2 Collagen Gel Self-Assembly

Collagen for the *in vitro* construction of collagen gels is derived by soaking collagen-rich tissues (such as rat-tail tendons or calf skin) in acetic acid for several days, followed by dialysis to concentrate the solubilized, triple-helical collagen monomers [74]. The resulting soluble collagen contains few intramolecular cross-links, though aggregates of 5–17 monomers may exist in soluble form [63,73]. Raising the temperature, pH, and ionic strength to physiological levels typically initiates self-assembly, in which collagen fibrils form via lateral and linear fusion of monomer aggregates. Fibril length, diameter, and aggregation into fibers are affected by polymerization, pH, temperature, and ionic strength [21,22,62–68,75]. Self-assembly results in the formation of a collagen gel, that is, an entangled network of highly hydrated collagen fibrils surrounding fluid-filled pores. The gelation process proceeds with an initial lag phase, during which the monomer aggregates initiate fusion, followed by a rapid growth phase, and eventual plateau. The mechanical and optical properties of the gel change with gelation time in a similar, sigmoidal shape. For example, gel turbidity (a measure of light scattering) increases as

$$x = 1 - \exp(-Z_n t^n), \tag{11.1}$$

where x is the mass fraction of precipitated collagen (linearly related to turbidity), Z is a rate constant, t is the gelation time, and n is a constant related to collagen fiber nucleation [65]. Bulk shear modulus of the gel also increases rapidly after a lag and prior to a plateau value [73]. The collagen self-assembly process is hypothesized to occur through simultaneous nucleation and linear growth of fibrils [73]. In this model of collagen fibrillogenesis, collagen monomers exist in equilibrium with small, soluble aggregates of 5–17 monomers, termed microfibrils. The existence of these purported microfibrils is supported by x-ray diffraction and electron microscopy measurements of collagen fibrils, which suggest that fibrils exist with quantized diameters, of integer multiples of ~4 nm. Self-assembly proceeds by both lateral and axial accretion of these microfibrillar subunits [76]. Increasing ion concentration or decreasing temperature or pH tends to favor the lateral aggregation of microfibrils, leading to increased fibril diameters [73,77]. The physical parameters that increase fibril diameter tend to delay fibrillogenesis and prolong the lag phase, during which nucleation is the predominant process [63]. Other extracellular matrix constituents present during collagen self-assembly may also modulate the collagen fibril and network structure. Proteoglycan and glycosaminoglycan binding to collagen may either delay or accelerate fibrillogenesis, thus affecting the collagen fibril diameter. Hyaluronic acid and decorin tend to decrease the diameter of fibrils formed in their presence [23], whereas dermatan sulfate binding favors the formation of thicker fibers [78]. The physical and chemical parameters can alter *in vitro* fibrillogenesis that affects collagen fiber dimensions and, given a limited concentration of collagen monomers, can vary the fiber number density and therefore the pore size of the formed collagen network, thus impacting network mechanics. The final result of collagen self-assembly and fibrillogenesis is an ordered and hierarchical array of collagen monomers, forming an entangled biopolymer network of fibrils and fibril bundles.

The fibril level of organization consists of arrays of monomers that are ordered in an axially staggered pattern in which molecules are stacked and staggered by one-quarter of the molecular length, or about 68 nm. This staggered array allows for interchain cross-link formation, and results in native fibrils 20–500 nm wide and up to 1 cm long [79] displaying *D*-banding: 68 nm wide bands that stripe the collagen fibril, apparent in metal-stained fibrils and resulting from the quarter-staggered array of monomers [80]. Bundles of fibrils form collagen fibers, typically 1–20 μm wide; bundles of fibers form thicker fascicles that form bundles within tendons [81]. Between the monomer and the assembled collagen unit, physical and biological factors alter the assembly process. Importantly, collagen structure and cross-link content affect the mechanical properties on the level of single-collagen fibers, collagen networks, and bulk tissue. Knowledge of the mechanisms of collagen fibrillogenesis allows control over the formation of collagen-containing tissues *in vitro* and *in vivo*.

11.2.3 Optical Properties of Collagen Gels

SHG signal intensity depends upon the square of the incident laser intensity, the square of local collagen concentration, and, due to SHG coherence, on the spatial organization and scattering properties of the generating collagen dipole arrays [10–12,14,16]. SHG intensity, I_{SHG}, may be expressed as

$$I_{SHG} = 16\pi(\omega^2/n_\omega^2 n_{2\omega}^2 c^2)\kappa S_{2\omega}^2 \left|d_{eff}^2\right| I_\omega^2,$$

(11.2)

where ω is the fundamental (laser frequency), n_ω is the refractive index at the fundamental frequency, $n_{2\omega}$ is the refractive index at the second-harmonic frequency, c is the speed of light, κ is a function of particle size, $S_{2\omega}$ is the second-harmonic backscattering coefficient (SHG scattering cross-section), d_{eff} is the effective second-order nonlinear susceptibility, and I_ω is the laser intensity within the focal region [14,16,82,83].

The SHG signal from collagen gels arises from a coherent/quasi-coherent component due to direct detection of the generated signal and an incoherent component due to the detection of multiply-scattered signal. SHG signal depends, therefore, on collagen gel-scattering properties that influence the incoherent component as well as fibril size, aggregation, and orientation that influence the coherent signal component [84]. Collagen possesses a high refractive index ($n \sim 1.5$) [14], and thus scatters a significant amount of light in an aqueous environment ($n = 1.33$). Single-photon scattering by collagen occurs according to Mie theory, in which collagen fibrils smaller than the wavelength of incident light tend to scatter equally in the forward and backward direction. Collagen fibers and close-packed fibrils that approach and exceed the wavelength of scattered light possess increasing scattering cross-sections, and are predominantly forward scattering [14]. Similarly, SHG from a point source smaller than $\lambda_{2\omega}/10$ (or about 40 nm) radiates more homogenously, whereas SHG from a cluster of harmonophores larger than $\lambda_{2\omega} \sim 400$ nm, the second-harmonic wavelength, produces almost entirely forward-generated second harmonic [10,14,84]. Theoretical and experimental work suggest that collagen fibril aggregates can generate significant backward-generated second-harmonic signal due to relaxed phase-matching conditions imparted by interfibrillar spacing roughly equal to the coherence length in the backward direction [85]. Hence, SHG is primarily forward generated by large fibers and fibril bundles $>\lambda_{2\omega}$, but significant backward-generated SHG may result from small fibrils $\ll\lambda_{2\omega}$ with spacing on the order of the backward coherence length, which for collagen SHG is < 7 μm [85]. At a typical LSM optical resolution of ~450 nm, collagen fibers and fibril aggregates at least two pixels wide will be 900 nm $> \lambda_{2\omega}$, resulting in primarily forward SHG. Smaller fibrils, however, may produce up to 25% backward-generated SHG, augmenting signal detection in the epidirection [84,85].

Turbidity of a collagen gel solution has been shown to vary linearly with collagen concentration for low concentrations of collagen (<4 mg/mL) [56]. Not surprisingly, due to the reliance of SHG signal on backscattering events, a linear dependence of SHG on acellular collagen gel concentration has also been reported [14]. Thus, it is clear that the bulk optical properties of nonlinear scattering signals from a collagen-rich tissue depend on collagen concentration, microstructure, and fiber size [86].

11.2.4 Image Processing

Generally speaking, quantitative information from SHG images can result from analysis of signal levels or from analysis of image textural and spatial features. To the extent that second-harmonic image parameters are sensitive to collagen fiber and network structure (e.g., fiber, orientation, or network anisotropy), the parameters may be used as indices that track microstructure–mechanics relationships. The interpretation of some quantitative image parameters is unambiguous. For example, the diameter of collagen fibers or network pores measured manually or algorithmically from SHG images is a direct assessment of fiber and network structure. The mean SHG signal, on the other hand, is a function of fiber shape, orientation, image area fraction, bulk gel collagen content, and bulk gel scattering properties. Furthermore, some parameters such as mean SHG signal depend upon objective numerical aperture, detector sensitivity, laser parameters, and other external factors [1–3]. For such instrument-dependent parameters, absolute values are relatively meaningless without proper instrument calibration, and information is best gleaned by analysis of trends or by taking a ratio to effectively normalize the parameter, with the ratio being independent of instrument parameters. For example, careful calibration and optical setup allow for the measurement of forward-to-backward signal ratios [23,25,27].

11.3 SHG Imaging of Acellular Collagen Gels

11.3.1 Introduction

Acellular type I collagen gels are ideal constructs to study the relationship of collagen fiber and network structural characteristics to SHG signal. These gels are pure and polymerization conditions may be varied independently of collagen concentration to control the aspects of fiber and network microstructure. At low concentrations (<~9 mg/mL), the fiber network is sparse enough that individual fiber features can be resolved as well as network features. SHG image parameters correlate with fiber and network structural features, with implications for using SHG images to estimate bulk mechanical properties.

11.3.2 Quantification of Collagen Fiber Shape from Second-Harmonic Images

Altering the polymerization, pH, and temperature of collagen hydrogels significantly influences the gel microstructure and mechanical properties at a given collagen concentration. In the experiments described below, gel collagen concentration was kept constant at 4 mg/mL and polymerization was varied between 4°C and 37°C, or between pH 5.5 and 8.5. It was found that increasing pH or temperature tends to result in longer, thinner collagen fibers, a reduced pore area fraction and size, and an increased pore density. These characteristics are visible in both SHG and TPF images (Figures 11.1a through 11.1d, 11.1i through 11.1l) and scanning electron microscopy (SEM) images (Figures 11.1e through 11.1h, 11.1m through 11.1p) of acellular collagen gels. SHG signal to noise is larger than that of TPF signal, a difference that is likely reflective of the quadratic SHG versus linear TPF concentration dependence, and also on the generally weak autofluorescence signal from poorly cross-linked collagen. Therefore, incoherent and homogeneous emission allows the fiber cross-sections to be clearly seen by TPF signal, whereas the coherent nature of SHG disallows signal generation from dipoles within collagen oriented parallel to the laser propagation. SEM images have a higher intrinsic resolution and reveal that collagen fibers are actually closely packed bundles of fibrils, with especially large bundles containing many fibrils at the lower temperature and pH polymerization conditions. Fiber diameter varies across the polymerization conditions because of differences in the number of fibrils per fiber rather than large changes in fibril diameter.

The measurements of fiber diameter from SHG and SEM images reveal a linear correlation (Figure 11.2), although the diameters measured from SEM images tend to be smaller (due to dehydration of the fibrils during sample preparation). Small diameter fibers are visible in SHG (Figures 11.1a, 11.1b, 11.1i, and 11.1j) and SEM images (Figures 11.1e, 11.1f, 11.1m, and 11.1n) of gels polymerized at the lower temperature and pH values, sometimes independent of larger diameter fibers and sometimes emanating from the splayed ends of large diameter fibers. However, SHG and SEM images show that large diameter fibers dominate the space-filling characteristics of these gels. With increasing polymerization temperature and pH, the hydrogels display a finer and more homogeneous network of fibers.

11.3.3 Effects of Collagen Fiber Size on Second-Harmonic Signal

In acellular collagen hydrogels, in which collagen is the only significant scattering component, backward-detected SHG signal primarily results from scattering of forward-generated second-harmonic photons and from backward-generated SHG from small fibrils (diameters ~10% of $\lambda_{2\omega}$). The

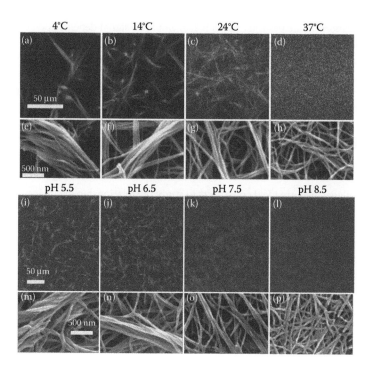

FIGURE 11.1 **(See color insert.)** Simultaneously collected SHG (blue signal) and TPF (green signal) images of 4 mg/mL acellular collagen hydrogels polymerized at various temperatures (a–d) and pH values (i–l), with corresponding SEM images (e–h, m–p). The scale bars are indicated in the figure. (Reprinted from *Biophys J.*, 92, Raub, C. B. et al., Noninvasive assessment of collagen gel microstructure and mechanics using multiphoton microscopy, 2212–2222, Copyright 2007; *Biophys J.*, 94, Raub, C. B. et al., Image correlation spectroscopy of multiphoton images correlates with collagen mechanical properties, 2361–2373, Copyright 2008, with permission from Elsevier.)

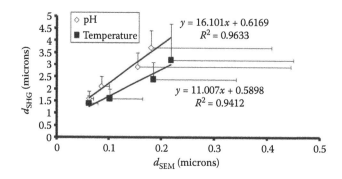

FIGURE 11.2 Correlation of manual fiber diameter measurement averages from SHG images (d_{SHG}) and SEM images (d_{SEM}) of acellular collagen gels polymerized at pH 5.5, 6.5, 7.5, and 8.5, or temperatures 4°C, 14°C, 24°C, and 37°C. The linear best-fit lines are shown, with slope and R^2 indicated in the figure. The error bars are standard deviation. (Reprinted from *Biophys J.*, 92, Raub, C. B. et al., Noninvasive assessment of collagen gel microstructure and mechanics using multiphoton microscopy, 2212–2222, Copyright 2007; *Biophys J.*, 94, Raub, C. B. et al., Image correlation spectroscopy of multiphoton images correlates with collagen mechanical properties, 2361–2373, Copyright 2008, with permission from Elsevier.)

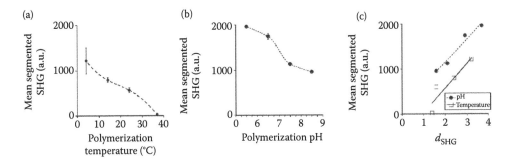

FIGURE 11.3 (a) Mean segmented SHG signal versus polymerization temperature of acellular collagen gels. (b) Mean segmented SHG signal versus polymerization pH of acellular collagen gels. (c) Mean segmented SHG signal plotted versus average fiber diameter measured from SHG images, d_{SHG}, for acellular gels polymerized at pH 5.5, 6.5, 7.5, and 8.5, or temperatures 4°C, 14°C, 24°C, and 37°C. The error bars are standard error of the mean. (Reprinted from *Biophys J.*, 92, Raub, C. B. et al., Noninvasive assessment of collagen gel microstructure and mechanics using multiphoton microscopy, 2212–2222, Copyright 2007; *Biophys J.*, 94, Raub, C. B. et al., Image correlation spectroscopy of multiphoton images correlates with collagen mechanical properties, 2361–2373, Copyright 2008, with permission from Elsevier.)

dependency of epi-detected SHG signal on collagen fiber size can be seen in Figures 11.3a and b, for 4 mg/mL acellular collagen gels polymerized from 4°C to 37°C and pH 5.5–11, respectively. The SHG signal has been segmented to exclude void regions, so that the measurements represent average signal values only from collagen. Plotting the segmented SHG signal versus fiber diameter shows a direct, linear correlation for both varying pH ($R^2 = 0.96$) and temperature ($R^2 = 0.85$) polymerization conditions, with an offset attributable to the differing detector gain (Figure 11.3c). The effect of increasing mean fiber diameter is to increase the scattering cross-section of the fiber as well as bulk scattering within the tissue, which is thus able to scatter more forward-directed SHG photons into the epi-configured detectors.

11.3.4 Effects of Acellular Collagen Gel Concentration on Second-Harmonic Signal

While the concentration dependence of second-harmonic signal scales with the square of generating dipole concentration, changes in the bulk collagen concentration of acellular gels may increase both the average density of collagen within signal-containing pixels and the relative volume fraction of collagen within the gel. The effect of increasing fiber number density can be seen in SHG images for fine- (Figures 11.4a–c) and coarse-structured gels (Figures 11.4d–f) with low collagen concentrations (1.5–9 mg/mL). For these collagen gels, SHG signal mean intensity (Figure 11.4g) and area fraction (Figure 11.4h) increase linearly with collagen content. The linearity of the signal increase is robust to changes in collagen fiber morphology observable from the SHG images and may be attributed simply to changes in collagen fiber number density and concomitant decreasing of void volume fraction, rather than increased fibril packing within pixels, which would introduce a nonlinear (second-order) dependence of the signal on collagen concentration. The scattering coefficient within collagen gels is expected to increase by ~2.9 cm⁻¹ per 1 mg/mL of collagen, from 4.3 cm⁻¹ at 1.5 mg/mL, and 26 cm⁻¹ at 9 mg/mL [87], and may contribute to linear increases in SHG signal intensity with collagen concentration. A similar linear trend in SHG area fraction versus collagen concentration (Figure 11.4h) suggests that for these acellular, low-density collagen gels, the microstructure determines SHG image parameters.

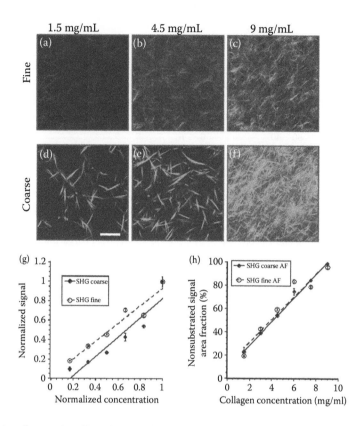

FIGURE 11.4 Simultaneously collected SHG and TPF (signals co-registered in grayscale) images of acellular collagen hydrogels polymerized at pH 8.5 (a–c) or 14°C (d–f) and collagen concentrations of (a,d) 1.5 mg/mL, (b,e) 4.5 mg/mL, and (c,f) 9 mg/mL. (g) Normalized SHG signal versus normalized collagen concentration for the fine- and coarse-structured gels, with linear best-fit lines shown. (h) SHG signal image area fraction versus collagen concentration for fine- and coarse-structured gels, with linear best-fit lines shown. The error bars represent standard deviation. The data markers are indicated in the figure legend. The scale bar represents 50 μm.

11.3.5 Effect of Collagen Fiber Orientation on SHG

Several studies have shown through theoretical and experimental methods that SHG depends upon the orientation of collagen monomers aggregated into fibrils and fibers with respect to both the laser polarization angle and the laser propagation direction [3,10–14,25,84,88]. Specifically, for a fibril perpendicular to the laser propagation direction, SHG is maximized when the dipoles within the fibril are aligned parallel to the incident electric field and is minimized when the dipoles are perpendicular. The coalignment of dipoles and the electric field allows for a maximum nonlinear polarization. This orientation dependence of SHG has been used to characterize the in-plane orientation of collagen fibrils since it has been shown that the dipoles within collagen align with the fibril long axis [13].

The orientation dependence of SHG signal from collagen fibrils, though a useful structural parameter, can interfere with accurate structural characterization of a collagen network from a single SHG image or image stack since fibrils would possess variable SHG intensity depending upon fibril orientation. For circularly polarized laser illumination, however, SHG signal does not depend upon fibril–dipole orientation within the image plane [14]. However, there still exists an axial (out-of-plane) dependence, in which SHG is maximized from fibrils perpendicular to the laser propagation direction (i.e., in the image plane), and is minimized from fibrils parallel to the laser propagation direction (i.e., perpendicular to the image plane). The following discussion addresses the axial dependence of SHG signal in collagen gels, showing that SHG from in-plane fiber cross-sections is more intense that from fibers orientated with long axes perpendicular to the image plane.

The theoretical orientation dependence of SHG signal was determined by utilizing the previously described expression for SHG intensity, $I_{2\omega}$ [86,89]

$$I_{2\omega} = \frac{p}{n_{2\omega}n_\omega^2}(I_\omega)^2 d_{eff}^2 \left(\int_{z_0}^{z_0+L} \frac{e^{i\Delta kz}}{1 + iz/z_R} dz \right)^2,$$ (11.3)

where in this case, p is a lumped term of fundamental constants and beam parameters, $n_{2\omega}$ and n_ω are the index of refraction at the SHG and fundamental wavelengths, I_ω is the laser intensity at the focal point, d_{eff} is collagen's orientation-dependent effective second-order nonlinear susceptibility, z_R is the Rayleigh distance, and Δk is the phase mismatch. Assuming that p, n_ω, $n_{2\omega}$, z_R, and Δk remain constant over any fiber orientation, the ratio of SHG intensities from fibers nearly perpendicular ($I_{2\omega}^\perp$) versus fibers nearly parallel ($I_{2\omega}^=$) to the laser propagation direction is then

$$I_{2\omega}^\perp / I_{2\omega}^= = (d_{eff}^\perp)^2 / (d_{eff}^=)^2,$$ (11.4)

where $d_{eff}^\perp$ and $d_{eff}^=$ are the effective second-order nonlinear susceptibilities of collagen fibers perpendicular and parallel to the laser propagation direction, respectively. For parallel polarized laser light, the nonlinear susceptibility d_{eff} is [90,91]

$$d_{eff} = 3d_{16}(\cos\beta\cos\delta - \cos^3\beta\cos^3\delta) + d_{22}\cos^3\beta\cos^3\delta,$$ (11.5)

where β is the angle of the (assumed randomly oriented) fiber axis with respect to the fundamental electric field and δ is the angle between the fiber axis and the imaging plane. The case of circularly polarized laser light corresponds to allowing β to vary between 0° and 360°, and averaging d_{eff}^2 over all values of β for a specific value of δ between 0° (fiber perpendicular to laser propagation) and 90° (fiber parallel to laser propagation). The ratio d_{22}/d_{16}, necessary to calculate d_{eff} was estimated to be ~2, based upon previous studies in collagen [89]. In this study, experimentally determined ratios of $I_{2\omega}^\perp / I_{2\omega}^=$ were compared to these calculations to determine their reasonableness. This ratio was calculated by carrying out a fiber segmentation, using TPF signal, into circular (out-of-plane) fiber cross-sections c, and elliptical (in-plane) fiber cross-sections e, and then determining the ratio of colocalized SHG signal in these segmented fiber regions.

Fibers parallel and perpendicular to the image plane are visible in TPF images (see Figures 11.1a, and 11.1i) and can be segmented after thresholding based upon particle circularity (Figure 11.5a, in-plane fibers based upon TPF signal from coarse-structured gel similar to that of Figure 11.1a; 11.5b, out-of-plane fibers based upon TPF signal). In contrast, only fibers with scattering interfaces more or less parallel to the image plane (and perpendicular to the laser propagation direction) produce strong backward-detected SHG signal (see Figures 11.1a, 11.1i). $TPF_e/TPF_c \pm SE$ was 1.1 ± 0.1 for the pH 5.5 condition and 0.90 ± 0.05 for the pH 6.5 condition; $SHG_e/SHG_c \pm SE$ was 3.1 ± 1.3 for the pH 5.5 condition and 2.9 ± 0.8 for the pH 6.5 condition (Figure 11.5c).

The squared effective nonlinear susceptibility of collagen was calculated using Equations 11.3 through 11.5 and was plotted as a function of δ, the angle of the fiber axis with respect to the image plane (Figure 11.5d). A ratio of 3 for SHG_e/SHG_c, for example, corresponds to d_{eff}^2 values of 2.64 and 0.88 for fibers oriented at $\delta = 10°$ and 63°, respectively (Figure 11.5d). These example values were chosen to show that the theoretical estimation of the ratio of d_{eff}^2 based upon fibers tilted at shallow versus steep angles tends to predict an SHG intensity ratio similar to experimentally determined values.

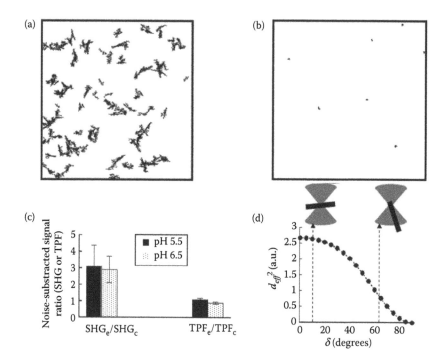

FIGURE 11.5 (a) Mask of in-plane-oriented collagen fiber cross-sections, taken from a TPF image of an acellular collagen hydrogel. (b) Mask of out-of-plane-oriented collagen fiber cross-sections, taken from a TPF image of an acellular collagen hydrogel. (c) The ratio of noise-subtracted SHG and TPF signal from the signal-containing regions masked in (a) (elliptical particles, *e*) and (b) (circular particles, *c*). (d) d_{eff}^2, in arbitrary units, was estimated for various angles of tilt of the collagen fiber axis with respect to the laser propagation direction. Fibers at 10° and 63° of tilt are diagrammed with respect to the laser focal region (gray). (Reprinted from *Biophys J.*, 94, Raub, C. B. et al., Image correlation spectroscopy of multiphoton images correlates with collagen mechanical properties, 2361–2373, Copyright 2008, with permission from Elsevier.)

The 1:1 ratio of TPF signal from fibers predominantly parallel versus transverse to the image plane confirms the isotropic angular distribution of TPF generation in collagen gels. In contrast, one would expect greater backward-detected SHG from fibers parallel to the image plane, both because of scattering of forward-generated SHG at the fiber bottom interface [14], and because of the collagen dipoles' optimal orientations with the laser propagation direction [73]. Indeed, a 3:1 ratio of SHG intensity from fibers roughly parallel versus roughly transverse to the image plane was measured, which is consistent with theoretical considerations using a lower numerical aperture (NA) [92]. A higher NA [1.3] objective was used to collect the SHG and TPF images, introducing a component of the electric field transverse to the image plane, which would change the SHG angular power distribution from fibers aligned with the *z*-axis. Specifically, these fibers should emit forward-generated SHG at a steeper angle (~40–47°) from the *z*-axis compared to the lower NA case, but no backward-generated SHG due to the Guoy phase shift and destructive interference{}. Fibers aligned parallel to the laser propagation should exhibit lower d_{eff} ($d_{eff}^= < d_{eff}^\perp$) as well as reduced single and multiple scattering of SHG photons from the highly forward-directed SHG emission compared to fibers oriented perpendicular to the laser propagation direction. Although a higher NA objective provides additional electric field contributions to SHG signal, the forward-directed generation and lack of scattering interfaces in transverse collagen fiber sections could explain the observed threefold difference in epi-detected SHG signal.

From Equation 11.5, a ratio of SHG signal intensities was calculated by estimating values of d_{eff}, which depend on the orientation of collagen dipoles with respect to the incident electric field. This calculation is valid if the phase mismatch Δk and the index of refraction remain constant regardless of fiber orientation. In fact, collagen's birefringence of $\Delta n \sim 0.003$ (where n along the collagen monomer's long axis $> n$ along the short axis) is assumed to be negligible. This seems a valid assumption since the calculation is compared to an experimental SHG signal averaged over an ensemble of collagen fibers only roughly aligned. Furthermore, Δk may be assumed constant since the SHG interaction volume (roughly the focal volume dimensions, ~ 200 nm lateral $\times$ 600 nm axial) is smaller than the mean fiber diameter and length in these cases, and can be assumed to be completely filled when centered on a collagen fiber regardless of fiber orientation. For an incident wavelength of 780 nm, $\Delta k \sim 0.48$ μm, with a corresponding coherence length of 6.5 μm, allowing efficient SHG from ~ 3 μm thick bundles of collagen fibrils.

In summary, second-harmonic-generating dipoles within a collagen fibril or fiber tend to align with the fibril long axis. For linearly polarized laser light, SHG depends on the angle between the laser's electric field polarization direction and the fibril long axis. For circularly polarized incident light, this dependence is removed. However, SHG still depends upon the angle of the fibril long axis with respect to the laser propagation direction, with maximal signal perpendicular and minimal signal parallel to the propagation direction. In essence, a circularly polarized LSM with SHG capability will capture the signal from collagen oriented within the plane of the image, but with less and less signal from collagen oriented increasingly perpendicular to the image plane. While an image of SHG signal from a collagen gel may thus underestimate the true collagen content, most collagen structures seem to emit detectable SHG signal. A superimposition of SHG and TPF signals from the same region of an acellular collagen gel typically reveals the full cross-sectional structure of the collagen network.

11.3.6 Quantification of Collagen Network Architecture from Second-Harmonic Images

The pores within collagen gels allow diffusion, facilitate cell migration, and affect gel mechanical properties. Collagen network pores larger than the optical resolution set for a given imaging experiment may be quantified from SHG images. The collagen network could become so dense that 100% of image pixels contain collagen SHG signal, in which case, an alternative method to determine the pore characteristics would have to be used, but for most acellular collagen gels, the collagen network is sparse enough that pores may be characterized from SHG images. The pore characteristics were quantified from SHG and TPF signal of collagen gels polymerized at 4 mg/mL and pH 5.5–8.5 (Figures 11.6a and 11.6b). The average pore size and pore area fraction tended to decrease, whereas pore number density increased with increasing polymerization, pH, as the available collagen tended to form more numerous, thinner fibers at higher pH values. Particle analysis of thresholded SHG images produced a mean pore size $\pm$SE of 81.7 ± 3.7 mm^2 for the pH 5.5 condition, decreasing $\sim 90\%$ to 7.8 ± 0.4 mm^2 for the pH 8.5 condition (Figure 11.6a). As expected, the number density of pores increased ~ 3.2-fold from 7.1 ± 0.2 to 23.0 ± 0.6 per 1000 mm^2 (Figure 11.6b). The trends in pore characteristics were confirmed by particle analysis of thresholded SEM images that demonstrated a similarly trending decrease in pore size and increase in pore number density with increasing polymerization pH. From this analysis, it appears that SHG images of acellular gels yield direct information about pore characteristics that correspond to measurements from SEM images. In case TPF images contain additional pore information due to the orientation independence of TPF signal, pores were quantified from thresholded SHG signal alone and also from thresholded SHG and TPF signals combined. The particle analysis method of measuring pore characteristics is fairly robust, as analysis of the combined SHG and TPF signal-masked images showed little change in the pore measurements.

11.3.7 Determining Mechanical Relationships in Acellular Gels from Second-Harmonic Images

In two separate studies described below, an attempt was made to correlate SHG and other image parameters to bulk gel mechanical properties from shear and indentation tests. In the first study, varying

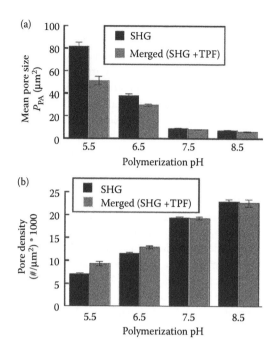

FIGURE 11.6 (a) Mean pore size, determined from particle analysis from SHG and combined SHG + TPF images, of acellular collagen gels polymerized at pH 5.5, 6.5, 7.5, and 8.5. (b) Mean pore number density, determined from particle analysis from SHG and combined SHG + TPF images, of acellular collagen gels polymerized at pH 5.5, 6.5, 7.5, and 8.5. The error bars represent standard deviation. (Reprinted from *Biophys J.*, 94, Raub, C. B. et al., Image correlation spectroscopy of multiphoton images correlates with collagen mechanical properties, 2361–2373, Copyright 2008, with permission from Elsevier.)

the polymerization temperature of 4 mg/mL collagen hydrogels systematically changed the fiber volume fraction and space-filling characteristics as well as bulk shear moduli, measured by rheology. The shear moduli G' and G'' were found to correlate positively with SHG image fraction for these gels (Figure 11.7a), although the correlation between the shear moduli and SHG image fraction is most linear for the 14–37°C polymerization conditions. In this case, the collagen fibers polymerized at 4°C formed an extremely weak, sparse gel, with most fibers not entangled and therefore not contributing to an elastic stress response. In contrast, the shear moduli correlate negatively with mean-segmented SHG signal for 4 mg/mL collagen gels with polymerization temperature-controlled microstructure (Figure 11.7b). Since all gels contained collagen at 4 mg/mL, gels with larger diameter fibers contained fewer (though larger with brighter SHG signal) fibers and larger pores, linking the shear moduli, SHG signal, and SHG image area fraction. This study was an initial attempt to correlate the mechanical properties with SHG image parameters, in an attempt to understand the origins of the shear moduli values. It was concluded that the network space-filling characteristics, specifically of the pores between collagen fibers, play an important role in determining the mechanical properties, and that SHG image area fraction is one way to describe collagen network space-filling characteristics.

The second study was designed to attempt to explain bulk gel mechanics with a microstructural model and SHG image-derived microstructural parameters. Acid-soluble collagen hydrogels consist of an uncross-linked, tightly entangled network of semiflexible (but rather rod-like) polymer chains (i.e., fibers), and as such, their viscoelastic properties may be modeled, employing only a handful of microstructural input parameters [93]. Specifically, the storage modulus G' scales as

$$G' \sim \rho^{7/5} L_p^{-1/5} \tag{11.6}$$

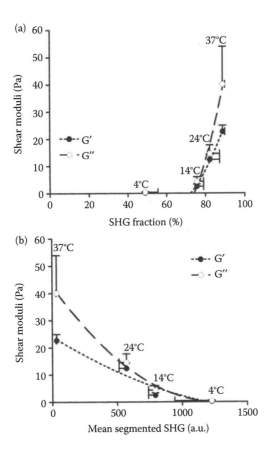

FIGURE 11.7 (a) Average shear moduli (storage modulus, G'; loss modulus, G'') versus average SHG image area fraction from 4 mg/mL acellular collagen gels polymerized at temperatures 4°C, 14°C, 24°C, and 37°C. (b) Average shear moduli (storage modulus, G'; loss modulus, G'') versus mean segmented SHG image intensity from 4 mg/mL acellular collagen gels polymerized at temperatures 4°C, 14°C, 24°C, and 37°C. (Reprinted from *Biophys J.*, 92, Raub, C. B. et al., Noninvasive assessment of collagen gel microstructure and mechanics using multiphoton microscopy, 2212–2222, Copyright 2007, with permission from Elsevier.)

where ρ is the density of polymer contour length per unit volume and L_p is the chain's persistence length. A network mesh size, L_m, relates to ρ as $\rho = L_m^{-2}$. A fiber's persistence length is proportional to its bending modulus, which for collagen scales with the fourth power of fiber diameter, d [94,95]. Recasting the scaling relationship of Equation 11.6 in terms of mesh size and fiber diameter,

$$G' \sim L_m^{-14/5}\, d^{-4/5} \tag{11.7}$$

To determine if SHG images can provide structural data that scales appropriately with G', the mesh size and fiber diameter were estimated from SHG images (Figure 11.8a) in two ways: using particle analysis and manual fiber diameter measurements (P_{PA} and d_{SHG} for fibers and pores, respectively), and using image correlation spectroscopy, which create characteristic pore and fiber diameters, P_{ICS} and d_{ICS}, based upon signal periodicity in the SHG image [22]. The data from SHG images were input into the scaling relationship and a plot of log G' versus log ($L_m^{-14/5}\, d^{-4/5}$) was generated. The correlation using the particle analysis and hand-measured parameters is good ($R^2 = 0.93$). The best-fit slope is of the order one ($m \sim 0.84$, Figure 11.8b). One conclusion to draw from this correlation is that mesh size and fiber diameter, parameterized from SHG images, may explain most of the variation in storage modulus of these acellular collagen gels, and that the gels behave as entangled networks of semiflexible fibers.

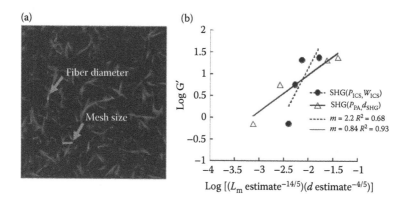

FIGURE 11.8 (**See color insert.**) (a) Combined SHG + TPF (SHG, blue signal; TPF, green signal) image of acellular collagen hydrogel showing examples of measurements of collagen fiber diameter and network mesh size. (b) Log–log plot of the scaling relationship of the storage modulus versus mesh size, L_m, and fiber diameter, d, estimated from SHG images of collagen gels polymerized at temperatures 4°C, 14°C, 24°C, and 37°C. The estimates were from image correlation spectroscopy (P_{ICS} for mesh size; W_{ICS} for fiber diameter), or particle analysis and manual image measurements (P_{PA} for mesh size; d_{SHG} for manual image measurements). The linear best-fit slopes and R^2 values are indicated in the figure. (Reprinted from *Biophys J.*, 94, Raub, C. B. et al., Image correlation spectroscopy of multiphoton images correlates with collagen mechanical properties, 2361–2373, Copyright 2008, with permission from Elsevier.)

11.3.8 Summary: SHG Imaging of Acellular Collagen Gels

In the studies described above, acellular collagen gels were polymerized at a range of concentrations (1.5–9 mg/mL), temperatures (4–37°C), and pH values (5.5–8.5). The segmented signal intensity, image area fraction, average fiber diameter, and average mesh size were measured from SHG images of the gels. In some cases, a scaling relationship from semiflexible network theory was applied, showing good correspondence to experimental data inputs (average fiber diameter, mesh size from SHG images, G' from rheology). While this correspondence shows the sensitivity of SHG imaging to mechanical properties, there are limits to the microstructural information captured by SHG signal. Most notably, the smallest structures that can be measured from SHG images are determined by the optical resolution of the system. Structural information on the order of single pixels can still be obtained through careful measurement of optical parameters—such as signal orientation dependence with respect to polarizers, or forward-to-backward signal ratios. Second, SHG signal decays exponentially with penetration depth into the tissue [1,2,96]; so, the interrogated microstructure must be within a resolvable distance (typically ~150 μm to several mm). Third, SHG is specific only for certain structural proteins, most notably collagen and myosin [3,97–100]; so, SHG imaging will fail to capture mechanically relevant matrix components that do not emit, such as elastin, proteoglycans, cells, and noncollagenous tissue structures. The matrix components that do not emit second harmonic may be characterized through other imaging modalities and signals, such as TPF [21,22,26,59,101,102]. Acellular collagen gels, however, possess a uniform microstructure and are free of noncollagenous matrix components, which allow for very thorough structural characterization of these gels from SHG images alone. Tissue-engineering experiments utilize collagen gels as matrix scaffolds for cells, which typically remodel the gel through stress generation, protease activity, and new matrix deposition. SHG imaging, especially in conjunction with other nonlinear optical signals, can provide detailed information about the dynamic changes during culture of cellularized collagen gels.

11.4 SHG Imaging of Cellularized Collagen Gels

11.4.1 Introduction

The studies described below use multiphoton microscopy (MPM) imaging and mechanical testing to study floating cellularized collagen gels, which are contracting to <5% of their original volume, attain physiological collagen and cell densities. During the contraction process, cellularized gels exhibit microstructure-dependent trends in mechanical and optical properties, and these dynamic changes may be captured through MPM imaging and analysis of the SHG signal. The analysis of microstructure–mechanics relationships in these simple engineered tissues is important to understand how engineered tissues develop and how cell-induced matrix remodeling occurs *in vitro*. Cellularized collagen gels were prepared in a method similar to that used to make acellular gels at pH 6.5 (coarse-structured gels) and 8.5 (fine-structured gels), except that 50,000 normal human lung fibroblasts (NHLFs), passages 3–7 were added per milliliter of a 4 mg/mL collagen solution. The gels were polymerized in 24-well plates at room temperature (24°C) for 1 h and were then pH equilibrated with excess culture media. After overnight tissue culture to allow fibroblasts to adhere and spread within the collagen gels, the constructs were released from the wells and were placed in floating culture in Petri dishes half-filled with culture media. These floating gels were cultured in standard conditions for up to 15 days, during which time the gels were periodically removed for imaging and mechanical testing. To monitor the cell location and interactions with collagen, the imaging wavelength was set to 780 nm, and SHG signal from collagen was collected at 390 nm, while TPF signal from endogenous fluorophores [56,96] within the fibroblasts was collected at 500–550 nm.

11.4.2 Effect of Gel Contraction on SHG Images

Fine and coarse gels retained distinct microstructures, revealed in MPM images by SHG signal, even after significant cell-induced contraction (Figures 11.9a through 11.9c, fine-structured gels; 11.9d through 11.9f, coarse-structured gels). Matrix defects and holes appear in the SHG images from denser cellularized gels, typically adjacent to areas with cellular TPF signal (Figure 11.9f). The final collagen concentration approached approximately 200 mg/mL for the most contracted gels, with a final volume of ~11 µL (Figure 11.9g). Collagen mass content and total cell content increased during the culture period, in a trend clearly visible from the coregistered SHG and TPF images.

Interestingly, the mean SHG signal intensity from both fine- and coarse-structured gels increased linearly with collagen concentration (Figure 11.9h, signal and concentration normalized), identically to acellular gels (Figure 11.4g). The linear trend of SHG signal with bulk collagen concentration may be interpreted as a function of increased collagen volume fraction and multiple backscattering of SHG signal within the gel.

11.4.3 SHG Image Texture Simulation

To determine whether SHG image parameters could accurately assess the collagen network microstructure from cellularized gels varying in concentration over two orders of magnitude (~4–200 mg/mL), simulated textural images with well-defined numbers of overlapping "collagen fibers" were constructed to recreate structural features from the SHG images of cellularized gels. Using the simulations of image texture, the relationship between image parameters and collagen fiber number density could be determined and trends could be compared between the simulation and SHG images.

The textural features of images of the SHG signal from coarse-structured collagen gels were simulated using a MATLAB® routine. The constructed images were meant to simulate the images of a randomly oriented collagen fiber network, to determine the precise relationship of robust, gain-independent image parameters of fiber number density. Collagen fiber segments within the MPM image plane were simulated as two-dimensional elliptical Gaussian functions. The length and width of the fiber segments

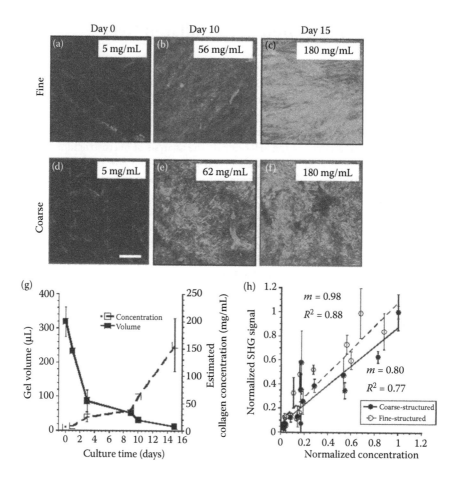

FIGURE 11.9 MPM images of (a–c) fine-structured and (d–f) coarse-structured cellularized collagen gels at three stages of contraction during floating culture. The estimated collagen concentration for each gel is indicated, and days of culture. SHG and TPF signals, co-registered in grayscale, reveal collagen and fibroblasts. The bar represents 50 μm. (g) Measured gel volume and the corresponding estimated collagen concentration estimate versus culture time. (h) Normalized SHG signal versus normalized collagen concentration for fine- and coarse-structured cellularized gels in various stages of contraction during free-floating culture. The linear best-fit slopes and R^2 values are indicated. (Reprinted from *Acta Biomater,* 6, Raub, C. B. et al., Predicting bulk mechanical properties of cellularized collagen gels using multiphoton microscopy, 4657–4665. Copyright 2010, with permission from Elsevier.)

were distributed normally, with mean and standard deviation determined from $n = 50$ line-segment measurements from SHG images of real collagen gels. Gaussian peak intensity was directly related to the length and width, so that larger fiber segments proportionally possessed more intense signal. Furthermore, the fiber edges were defined where the signal fell to $1/e^2$ times the maximum intensity of each Gaussian function. The simulated fiber areas and intensities were determined so that SHG images from cellularized gels at day 0 of culture would have similar mean intensity and signal area fraction to the simulated image of the corresponding fiber number density. Fiber orientations were distributed uniformly through 360° and were positioned at random locations within a 512 × 512 pixel matrix (the same size as the images from the cellularized gel imaging study). The intersecting fibers were allowed to superimpose, creating a linear relationship between the mean image intensity and fiber number density, as well as creating a reasonable approximation to the texture of SHG images from cellularized gels containing ~4–200 mg/mL collagen.

To relate the simulated images to SHG images from cellularized gels, simulated images were assigned collagen concentrations equal to the number of Gaussian "fibers" in the simulation times with a scaling factor, with units of milligram per milliliter fiber number. The scaling factor was determined by counting the number of fiber segments in SHG images of cellularized gels at day 0 of culture. These SHG images were thresholded at the noise cutoff, despeckled as before to remove the remaining noise, and a binary opening algorithm was performed in ImageJ to isolate the adjacent fibers. Then, particle analysis was carried out in ImageJ to count particles larger than 1 μm^2. It was determined that the day 0 gel images contained 178 ± 34 fibers ($\mu \pm$ s.d.), and the averaged gel concentration was 6.4 mg/mL. Therefore, the scaling factor used for simulated images containing 200–5000 fiber segments was 0.0356 mg/mL/fiber. A second scaling factor ensured similar brightness of Gaussian ellipses in simulated images to collagen fibers in SHG images, both on an 8-bit [0–255] scale. The Gaussian ellipse brightness scaling was chosen such that the intensity of the sparsest simulated image equaled SHG image intensity from day 0 cellularized gels.

11.4.4 Multiphoton Image Parameters Are Sensitive to Cellularized Gels Microstructure

To understand the microstructure–mechanics relationships of cellularized gels during cell-mediated matrix contraction, the image parameters were measured from SHG, TPF, and textural simulation images. Textural simulations of SHG images have a similar appearance to the SHG images of cellularized gels (Figures 11.10a through 11.10c, SHG images; 11.10d through 11.10f, texture simulation). Particle analysis of SHG images of cellularized gels on day 0 of culture revealed 178 ± 34 fiber segments (mean $\pm$ s.d.). On the basis of this measurement, simulation images containing 200–5000 fiber segments were assigned concentration values of 7–180 mg/mL. The visual comparison of SHG images and simulations shows a rough parity of texture and collagen fiber density for similar collagen concentrations.

Several signal and image parameters changed during the 16-day *in vitro* culture period. Trends from the textural simulations suggest that the functional form of SHG image parameters is largely due to

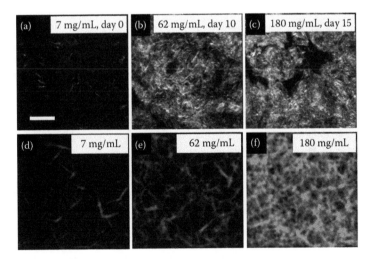

FIGURE 11.10 (a–c) Images of SHG signal from cellularized gels, and (d–f) simulated images from a randomly oriented fiber network of similar texture to the SHG images. The SHG images are from cellularized gels at three stages of contraction during floating culture, and the simulated images are of roughly corresponding collagen fiber density. The estimated collagen concentration for each gel/image is indicated, and the days of culture. The bar represents 50 μm. (Reprinted from *Acta Biomater*, 6, Raub, C. B. et al., Predicting bulk mechanical properties of cellularized collagen gels using multiphoton microscopy, 4657–4665. Copyright 2010, with permission from Elsevier.)

changes in collagen fiber concentrations (Figure 11.11, SHG versus simulation). For example, SHG signal intensity is a linear function of collagen concentration (Figure 11.11a, solid line, $m = 6.8$ a.u./mg/mL, $R^2 = 0.78$), which is corroborated by the linear relationship between image intensity and concentration in the simulated images (Figure 11.11a, dashed line, $m = 6.9$ a.u./mg/mL, $R^2 = 1.0$).

SHG signal area fraction increases quickly to a plateau near 100% by ~60 mg/mL (Figure 11.11b, solid markers), and depends upon collagen concentration in a logarithmic fashion ($R^2 = 0.66$ for the linear fit of ln(1-*area fraction*) versus *concentration*). This relationship is confirmed by the simulated signal area fraction (Figure 11.11b, open markers), which reaches a plateau near 100% by ~100 mg/mL ($R^2 = 0.99$ for the linear fit of ln(1-*area fraction*) versus *concentration*). SHG signal intensity is not instrument independent and therefore, without calibration, has little microstructural information to provide, other than the expected linearity of the signal intensity with collagen concentration. SHG signal area fraction is more robust to instrument parameters, and suggests that in this experiment, cellularized gels containing ~60 mg/mL collagen contain very few "pores" or void regions with cross-sections larger than a single pixel (in this study, ~0.2 μm²). Therefore, an image analysis algorithm to extract pore information or image area fraction of the signal will be unhelpful to characterize the range of collagen microstructures over all gel contraction levels, and other robust image parameters should be sought.

The image parameters such as skewness and speckle contrast (SC) are gain independent and are thus more robust parameters to potentially characterize the structural features that impact bulk mechanics. In contrast to the linear intensity and log area fraction dependences, we find that the skewness of the image pixel histograms relates to collagen concentration in SHG images (Figure 11.11c) with a power-law dependence (SHG, exponent $n = -0.6$, $R^2 = 0.90$; simulation, $n = -0.5$, $R^2 = 0.99$). The SC of SHG images (Figure 11.11d, solid markers) and texture simulation images (Figure 11.11d,

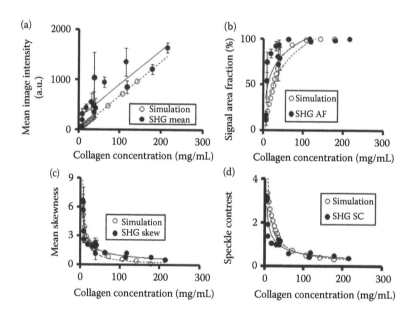

FIGURE 11.11 (a) Mean image intensity versus collagen concentration for SHG and texture simulation images. (b) SHG signal image area fraction versus collagen concentration for SHG and texture simulation images. (c) Mean image skewness versus collagen concentration for SHG and texture simulation images. (d) Mean speckle contrast versus collagen concentration for SHG and texture simulation images. SHG values are filled circles; simulation values are open circles. The data points from SHG data represent an average of five images per gel. R^2 coefficients for the linear best fits (a), logarithmic fits (b), and power-law fits (c,d) are given in the text. (Reprinted from *Acta Biomater*, 6, Raub, C. B. et al., Predicting bulk mechanical properties of cellularized collagen gels using multiphoton microscopy, 4657–4665. Copyright 2010, with permission from Elsevier.)

open markers) scale similarly to skewness (exponent $n = -0.6$, $R^2 = 0.90$ for SHG; $n = -1.0$, $R^2 = 0.91$ for the simulation). Finding instrument-independent SHG image parameters that are sensitive to a range of cellularized gel concentrations is challenging, but the results above provide some evidence that SHG skewness and SC are sensitive to collagen network microstructure over that wide concentration range.

11.4.5 Skewness and SC of SHG and TPF Signals Predict *E* of Cellularized Gels

The multiphoton image parameters that serve as robust predictors of cellularized gel mechanical properties such as Young's modulus (*E*) must be gain independent and sensitive to the changes in collagen concentration, network microstructure, crosslinking, and changes in cellularity. The skewness and SC of SHG and TPF signals may possess the desired gain independence and structural sensitivity. To determine the strength of these image parameters in predicting *E*, multiple regressions were performed of log-transformed *E* values on log-transformed SHG, cell-derived TPF, and matrix-derived TPF skewness (*skew*) and SC parameters. The data points were from cellularized gels that were imaged with LSM for SHG and TPF signal, and were mechanically tested, after both *in vitro* culture and after glutaraldehyde cross-linking, which introduces fluorescent cross-links into the collagen network (Figure 11.12) [21]. The TPF signal was divided into cell and matrix components with a particle-based masking procedure. The cell-derived TPF parameters were found to covary with SHG parameters, being unable to

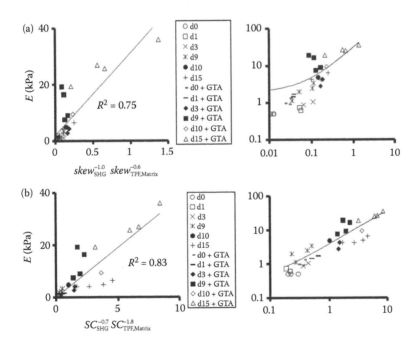

FIGURE 11.12 (a) Nonlinear best-fit model for *E* on SHG signal skewness and matrix-derived TPF signal skewness. The markers indicate days in culture of the collagen gels and the presence of glutaraldehyde (GTA) cross-linking. The inset on the right is a log–log plot of the data. (b) Nonlinear best-fit model for *E* on SHG signal speckle contrast and matrix-derived TPF signal speckle contrast. The markers indicate days (notation "d1" for day 1) in culture of the collagen gels and the presence of GTA cross-linking. The inset on the right is a log–log plot of the data. Power-law best-fit exponents and R^2 values are indicated in the figure. *E* was averaged from five measurements per gel for $n = 16$ gels. (Reprinted from *Acta Biomater,* 6, Raub, C. B. et al., Predicting bulk mechanical properties of cellularized collagen gels using multiphoton microscopy, 4657–4665. Copyright 2010, with permission from Elsevier.)

explain the additional variation in E. We found that the relationship $E \sim skew_{SHG}^{-1.0} skew_{TPF,matrix}^{-0.6}$ provided a best fit that explained the most variation in E (Figure 11.12a, $R^2 = 0.80$). Similarly, the relationship $E \sim SC_{SHG}^{-0.7} SC_{TPF,matrix}^{-1.8}$ provided a best fit that explained the most variation in E (Figure 11.12b, $R^2 = 0.83$). The observation of the linear and log–log plots of the multiple regressions shows that the nonlinear model using the skewness parameters tends to overestimate E of gels with sparse matrix (days 0–3, Figure 11.12a, inset), whereas the SC nonlinear model tends to overestimate E of uncross-linked gels cultured for 16 days (day 15, Figure 11.12b, inset).

11.4.6 Summary

Microstructural parameters change systematically during cell-mediated gel contraction: pores become smaller, fiber bundles become larger, and cells occupy holes in a dense three-dimensional collagen network. For cellularized collagen gels, SHG image parameters such as skewness and SC change with collagen fiber density in a predictable manner. Cellularized gel E can largely be predicted by the variation in SHG and matrix-derived TPF image parameters (skewness and SC), which depend upon collagen fiber and cross-link spatial patterns.

11.5 Conclusions

MPM is a promising imaging tool that is currently being adapted for use with fiber-optic handheld probes and scanning modules. It has become the premier technique for imaging of living cells and cultures of excised and engineered tissues. There is a great deal of interest in the link between extracellular matrix microstructure and bulk tissue mechanical properties. Researchers are beginning to apply the knowledge of tissue microstructure derived from MPM for understanding the development of mechanical properties in tissues such as cellularized and acellular silk, the visceral pericardium [103], and the fibrous cap of atherosclerosis.

In this burgeoning research environment, there is a need to develop simple and robust methods for data mining of MPM images to extract all mechanically relevant information. Such mechanically relevant information includes the concentration of mechanically relevant species (e.g., collagen), the volume fraction of the species, the pore size of polymer networks, the fluorescent cross-link content of tissues, the diameter and length of fibers, and in general, the size and abundance of mechanically relevant structures, and the spatial distribution of species in three dimensions. This chapter focused on the measurement of mechanically relevant image parameters from SHG signals of acellular and cellularized gels.

Collagen gels of similar concentrations may nonetheless possess varied microstructure and resulting bulk mechanics. Signal area fractions, pore size measurements, and fiber diameter measurements from SHG images robustly and effectively estimate the aspects of collagen network microstructure that influence the bulk shear moduli.

MPM imaging of cellularized gels contracting through a range of collagen concentrations from 4 to 200 mg/mL revealed problems with the robustness of signal image fraction and pore size measurements. The optical resolution of LSM limits microstructural information extractable from SHG images, especially from concentrated gels (>60 mg/mL). The pores for these gels were smaller than the pixel resolution limit (~0.3 μm^3) and could not be quantified. Nevertheless, SHG signal and image parameters are sensitive to a wide range of collagen network concentrations. MPM imaging of SHG signal provides an unparalleled noninvasive approach for studying microstructure–mechanics and cell–matrix interactions in collagen gel-based engineered tissues.

Acknowledgments

This work was supported, in part, by the National Heart, Lung, and Blood Institute (R01 HL067954, SCG), the Air Force Office of Scientific Research (FA9550-04-1-0101). CBR was supported by a Kirschstein predoctoral fellowship from the National Institute of Biomedical Imaging and Bioengineering (F31 EB006677, CBR) and by the ARCS Foundation, Orange County Chapter (ARCS Fellowship). This work was made possible, in part, through access to the Laser Microbeam and Medical Program (LAMMP) at the University of California, Irvine. The LAMMP facility is supported by the National Institutes of Health under a grant from the National Center for Research Resources (NIH no. P41RR01192, BJT). The support from the Arnold and Mabel Beckman Foundation is gratefully acknowledged. The authors wish to thank Dr. Bernard Choi, Dr. Tatiana Krasieva, and Dr. Andrew J. Putnam for excellent technical and research advice.

References

1. Campagnola, P. J. and L. M. Loew. 2003. Second-harmonic imaging microscopy for visualizing biomolecular arrays in cells, tissues and organisms. *Nat Biotechnol* 21:1356–1360.

2. Helmchen, F. and W. Denk. 2005. Deep tissue two-photon microscopy. *Nat Methods* 2:932–940.

3. Campagnola, P. J., A. C. Millard, M. Terasaki, P. E. Hoppe, C. J. Malone, and W. A. Mohler. 2002. Three-dimensional high-resolution second-harmonic generation imaging of endogenous structural proteins in biological tissues. *Biophys J* 82:493–508.

4. Brown, R. M. Jr., A. C. Millard, and P. J. Campagnola. 2003. Macromolecular structure of cellulose studied by second-harmonic generation imaging microscopy. *Opt Lett* 28:2207–2209.

5. Agarwal, A., M. L. Coleno, V. P. Wallace, W. Y. Wu, C. H. Sun, B. J. Tromberg, and S. C. George. 2001. Two-photon laser scanning microscopy of epithelial cell-modulated collagen density in engineered human lung tissue. *Tissue Eng* 7:191–202.

6. Chakir, J., N. Page, Q. Hamid, M. Laviolette, L. P. Boulet, and M. Rouabhia. 2001. Bronchial mucosa produced by tissue engineering: A new tool to study cellular interactions in asthma. *J Allergy Clin Immunol* 107:36–40.

7. Chetty, A., P. Davis, and M. Infeld. 1995. Effect of elastase on the directional migration of lung fibroblasts within a three-dimensional collagen matrix. *Exp Lung Res* 21:889–899.

8. Choe, M. M., P. H. Sporn, and M. A. Swartz. 2003. An *in vitro* airway wall model of remodeling. *Am J Physiol Lung Cell Mol Physiol* 285:L427–433.

9. Thompson, H. G., J. D. Mih, T. B. Krasieva, B. J. Tromberg, and S. C. George. 2006. Epithelial-derived TGF-(β)2 modulates basal and wound healing subepithelial matrix homeostasis. *Am J Physiol Lung Cell Mol Physiol* 291:L1277–1285.

10. Stoller, P., P. M. Celliers, K. M. Reiser, and A. M. Rubenchik. 2003. Quantitative second-harmonic generation microscopy in collagen. *Appl Opt* 42:5209–5219.

11. Stoller, P., B. M. Kim, A. M. Rubenchik, K. M. Reiser, and L. B. Da Silva. 2002. Polarization-dependent optical second-harmonic imaging of a rat-tail tendon. *J Biomed Opt* 7:205–214.

12. Stoller, P., K. M. Reiser, P. M. Celliers, and A. M. Rubenchik. 2002. Polarization-modulated second harmonic generation in collagen. *Biophys J* 82:3330–3342.

13. Odin, C., T. Guilbert, A. Alkilani, O. P. Boryskina, V. Fleury, and Y. Le Grand. 2008. Collagen and myosin characterization by orientation field second harmonic microscopy. *Opt Express* 16:16151–16165.

14. Erikson, A., J. Ortegren, T. Hompland, C. de Lange Davies, and M. Lindgren. 2007. Quantification of the second-order nonlinear susceptibility of collagen I using a laser scanning microscope. *J Biomed Opt* 12:044002.

15. Balu, M., T. Baldacchini, J. Carter, T. B. Krasieva, R. Zadoyan, and B. J. Tromberg. 2009. Effect of excitation wavelength on penetration depth in nonlinear optical microscopy of turbid media. *J Biomed Opt* 14:010508.

16. Theodossiou, T. A., C. Thrasivoulou, C. Ekwobi, and D. L. Becker. 2006. Second harmonic generation confocal microscopy of collagen type I from rat tendon cryosections. *Biophys J* 91:4665–4677.

17. Georgakoudi, I., W. L. Rice, M. Hronik-Tupaj, and D. L. Kaplan. 2008. Optical spectroscopy and imaging for the noninvasive evaluation of engineered tissues. *Tissue Eng Part B Rev* 14:321–340.

18. Rice, W. L., S. Firdous, S. Gupta, M. Hunter, C. W. Foo, Y. Wang, H. J. Kim, D. L. Kaplan, and I. Georgakoudi. 2008. Non-invasive characterization of structure and morphology of silk fibroin bio-materials using non-linear microscopy. *Biomaterials* 29:2015–2024.

19. Rubbens, M. P., A. Driessen-Mol, R. A. Boerboom, M. M. Koppert, H. C. van Assen, B. M. TerHaar Romeny, F. P. Baaijens, and C. V. Bouten. 2009. Quantification of the temporal evolution of collagen ori-entation in mechanically conditioned engineered cardiovascular tissues. *Ann Biomed Eng* 37:1263–1272.

20. Schenke-Layland, K. 2008. Non-invasive multiphoton imaging of extracellular matrix structures. *J Biophotonics* 1:451–462.

21. Raub, C. B., V. Suresh, T. Krasieva, J. Lyubovitsky, J. D. Mih, A. J. Putnam, B. J. Tromberg, and S. C. George. 2007. Noninvasive assessment of collagen gel microstructure and mechanics using multi-photon microscopy. *Biophys J* 92:2212–2222.

22. Raub, C. B., J. Unruh, V. Suresh, T. Krasieva, T. Lindmo, E. Gratton, B. J. Tromberg, and S. C. George. 2008. Image correlation spectroscopy of multiphoton images correlates with collagen mechanical properties. *Biophys J* 94:2361–2373.

23. Legare, F., C. Pfeffer, and B. R. Olsen. 2007. The role of backscattering in SHG tissue imaging. *Biophys J* 93:1312–1320.

24. Hu, J. J., J. D. Humphrey, and A. T. Yeh. 2009. Characterization of engineered tissue development under biaxial stretch using nonlinear optical microscopy. *Tissue Eng Part A* 15:1553–1564.

25. Williams, R. M., W. R. Zipfel, and W. W. Webb. 2005. Interpreting second-harmonic generation images of collagen I fibrils. *Biophys J* 88:1377–1386.

26. Kirkpatrick, N. D., J. B. Hoying, S. K. Botting, J. A. Weiss, and U. Utzinger. 2006. *In vitro* model for endogenous optical signatures of collagen. *J Biomed Opt* 11:054021.

27. Nadiarnykh, O., R. B. Lacomb, P. J. Campagnola, and W. A. Mohler. 2007. Coherent and incoherent SHG in fibrillar cellulose matrices. *Opt Express* 15:3348–3360.

28. Seeley, R., T. Stephens, and P. Tate. 2003. *Anatomy and Physiology*. New York: McGraw-Hill Publishers.

29. Schor, S. L. 1980. Cell proliferation and migration on collagen substrata *in vitro. J Cell Sci* 41:159–175.

30. Larsen, M., V. V. Artym, J. A. Green, and K. M. Yamada. 2006. The matrix reorganized: Extracellular matrix remodeling and integrin signaling. *Curr Opin Cell Biol* 18:463–471.

31. Wolf, K., S. Alexander, V. Schacht, L. M. Coussens, U. H. von Andrian, J. van Rheenen, E. Deryugina, and P. Friedl. 2009. Collagen-based cell migration models *in vitro* and *in vivo. Semin Cell Dev Biol* 20:931–941.

32. Ngo, P., P. Ramalingam, J. A. Phillips, and G. T. Furuta. 2006. Collagen gel contraction assay. *Meth Mol Biol* 341:103–109.

33. Bell, E., B. Ivarsson, and C. Merrill. 1979. Production of a tissue-like structure by contraction of collagen lattices by human fibroblasts of different proliferative potential *in vitro. Proc Natl Acad Sci USA* 76:1274–1278.

34. Ehrlich, H. P., D. J. Buttle, and D. H. Bernanke. 1989. Physiological variables affecting collagen lat-tice contraction by human dermal fibroblasts. *Exp Mol Pathol* 50:220–229.

35. Buttle, D. J. and H. P. Ehrlich. 1983. Comparative studies of collagen lattice contraction utilizing a normal and a transformed cell line. *J Cell Physiol* 116:159–166.

36. Bell, E., H. P. Ehrlich, D. J. Buttle, and T. Nakatsuji. 1981. Living tissue formed *in vitro* and accepted as skin-equivalent tissue of full thickness. *Science* 211:1052–1054.

37. Bell, E., H. P. Ehrlich, S. Sher, C. Merrill, R. Sarber, B. Hull, T. Nakatsuji, D. Church, and D. J. Buttle. 1981. Development and use of a living skin equivalent. *Plast Reconstr Surg* 67:386–392.

38. Redden, R. A. and E. J. Doolin. 2003. Collagen crosslinking and cell density have distinct effects on fibroblast-mediated contraction of collagen gels. *Skin Res Technol* 9:290–293.

39. Tomasek, J. J. and S. K. Akiyama. 1992. Fibroblast-mediated collagen gel contraction does not require fibronectin-α 5-β 1 integrin interaction. *Anat Rec* 234:153–160.

40. Zhu, Y. K., T. Umino, X. D. Liu, H. J. Wang, D. J. Romberger, J. R. Spurzem, and S. I. Rennard. 2001. Contraction of fibroblast-containing collagen gels: Initial collagen concentration regulates the degree of contraction and cell survival. *In Vitro Cell Dev Biol Anim* 37:10–16.

41. Turley, E. A., C. A. Erickson, and R. P. Tucker. 1985. The retention and ultrastructural appearances of various extracellular matrix molecules incorporated into three-dimensional hydrated collagen lattices. *Dev Biol* 109:347–369.

42. Auger, F. A., M. Rouabhia, F. Goulet, F. Berthod, V. Moulin, and L. Germain. 1998. Tissue-engineered human skin substitutes developed from collagen-populated hydrated gels: Clinical and fundamental applications. *Med Biol Eng Comput* 36:801–812.

43. Lee, C. H., A. Singla, and Y. Lee. 2001. Biomedical applications of collagen. *Int J Pharm* 221:1–22.

44. Wallace, D. G., J. M. McPherson, L. Ellingsworth, L. Cooperman, R. Armstrong, and K. A. Piez. 1988. Injectable collagen for tissue augmentation. In *Collagen* M. E. Nimni, ed. CRC Press, Inc., Boca Raton, FL, pp. 117–144.

45. Wallace, D. G. and J. Rosenblatt. 2003. Collagen gel systems for sustained delivery and tissue engineering. *Adv Drug Deliv Rev* 55:1631–1649.

46. Rosenblatt, J., B. Devereux, and D. G. Wallace. 1992. Effect of electrostatic forces on the dynamic rheological properties of injectable collagen biomaterials. *Biomaterials* 13:878–886.

47. Silver, F. H. and G. Pins. 1992. Cell growth on collagen: A review of tissue engineering using scaffolds containing extracellular matrix. *J Long Term Effects Med Implants* 2:67–80.

48. Arora, P. D. and C. A. McCulloch. 1994. Dependence of collagen remodelling on α-smooth muscle actin expression by fibroblasts. *J Cell Physiol* 159:161–175.

49. Arora, P. D., N. Narani, and C. A. McCulloch. 1999. The compliance of collagen gels regulates transforming growth factor-β induction of α-smooth muscle actin in fibroblasts. *Am J Pathol* 154:871–882.

50. Yang, Y. L. and L. J. Kaufman. 2009. Rheology and confocal reflectance microscopy as probes of mechanical properties and structure during collagen and collagen/hyaluronan self-assembly. *Biophys J* 96:1566–1585.

51. Yang, Y. L., L. M. Leone, and L. J. Kaufman. 2009. Elastic moduli of collagen gels can be predicted from two-dimensional confocal microscopy. *Biophys J* 97:2051–2060.

52. Grinnell, F. 2000. Fibroblast–collagen–matrix contraction: Growth-factor signalling and mechanical loading. *Trends Cell Biol* 10:362–365.

53. Cukierman, E., R. Pankov, and K. M. Yamada. 2002. Cell interactions with three-dimensional matrices. *Curr Opin Cell Biol* 14:633–639.

54. Young, R. G., D. L. Butler, W. Weber, A. I. Caplan, S. L. Gordon, and D. J. Fink. 1998. Use of mesenchymal stem cells in a collagen matrix for Achilles tendon repair. *J Orthop Res* 16:406–413.

55. Ramachandran, G. and A. Reddi, eds. 1976. *Biochemistry of Collagen.* Plenum Press, New York.

56. Zoumi, A., A. Yeh, and B. J. Tromberg. 2002. Imaging cells and extracellular matrix *in vivo* by using second-harmonic generation and two-photon excited fluorescence. *Proc Natl Acad Sci USA* 99:11014–11019.

57. Lee, P. F., A. T. Yeh, and K. J. Bayless. 2009. Nonlinear optical microscopy reveals invading endothelial cells anisotropically alter three-dimensional collagen matrices. *Exp Cell Res* 315:396–410.

58. Thompson, H. G., J. D. Mih, T. B. Krasieva, B. J. Tromberg, and S. C. George. 2006. Epithelial-derived TGF-β-2 modulates basal and wound-healing subepithelial matrix homeostasis. *Am J Physiol Lung Cell Mol Physiol* 291:L1277–1285.

59. Raub, C. B., A. J. Putnam, B. J. Tromberg, and S. C. George. 2010. Predicting bulk mechanical properties of cellularized collagen gels using multiphoton microscopy. *Acta Biomater* 6:4657–4665.

60. Abraham, T., J. Carthy, and B. McManus. Collagen matrix remodeling in 3-dimensional cellular space resolved using second harmonic generation and multiphoton excitation fluorescence. *J Struct Biol* 169:36–44.

61. Bowles, R. D., R. M. Williams, W. R. Zipfel, and L. J. Bonassar. Self-assembly of aligned tissue-engineered annulus fibrosis and intervertebral disc composite via collagen gel contraction. *Tissue Eng Part A* 16:1339–1348.

62. Christiansen, D. L., E. K. Huang, and F. H. Silver. 2000. Assembly of type I collagen: Fusion of fibril subunits and the influence of fibril diameter on mechanical properties. *Matrix Biol* 19:409–420.

63. Silver, F. H., J. W. Freeman, and G. P. Seehra. 2003. Collagen self-assembly and the development of tendon mechanical properties. *J Biomech* 36:1529–1553.

64. Wood, G. C. 1960. The formation of fibrils from collagen solutions. 2. A mechanism of collagen–fibril formation. *Biochem J* 75:598–605.

65. Wood, G. C. and M. K. Keech. 1960. The formation of fibrils from collagen solutions. 1. The effect of experimental conditions: Kinetic and electron-microscope studies. *Biochem J* 75:588–598.

66. Provenzano, P. P., K. W. Eliceiri, D. R. Inman, and P. J. Keely. Engineering three-dimensional collagen matrices to provide contact guidance during 3D cell migration. *Curr Protoc Cell Biol* 10:10–17.

67. Sung, K. E., G. Su, C. Pehlke, S. M. Trier, K. W. Eliceiri, P. J. Keely, A. Friedl, and D. J. Beebe. 2009. Control of 3-dimensional collagen matrix polymerization for reproducible human mammary fibroblast cell culture in microfluidic devices. *Biomaterials* 30:4833–4841.

68. Wozniak, M. A. and P. J. Keely. 2005. Use of three-dimensional collagen gels to study mechanotransduction in T47D breast epithelial cells. *Biol Proced Online* 7:144–161.

69. Chernyavskiy, O., L. Vannucci, P. Bianchini, F. Difato, M. Saieh, and L. Kubinova. 2009. Imaging of mouse experimental melanoma *in vivo* and *ex vivo* by combination of confocal and nonlinear microscopy. *Microsc Res Tech* 72:411–423.

70. Tang, S., T. B. Krasieva, Z. Chen, and B. J. Tromberg. 2006. Combined multiphoton microscopy and optical coherence tomography using a 12-fs broadband source. *J Biomed Opt* 11:020502.

71. Sun, Y., H. Y. Tan, S. J. Lin, H. S. Lee, T. Y. Lin, S. H. Jee, T. H. Young, W. Lo, W. L. Chen, and C. Y. Dong. 2008. Imaging tissue engineering scaffolds using multiphoton microscopy. *Microsc Res Tech* 71:140–145.

72. Lim, R. S., A. Kratzer, N. P. Barry, S. Miyazaki-Anzai, M. Miyazaki, W. W. Mantulin, M. Levi, E. O. Potma, and B. J. Tromberg. Multimodal CARS microscopy determination of the impact of diet on macrophage infiltration and lipid accumulation on plaque formation in ApoE-deficient mice. *J Lipid Res* 51:1729–1737.

73. Nimni, M. E., ed. 1988. *Collagen*. CRC Press, Boca Raton, FL.

74. Forgacs, G., S. A. Newman, B. Hinner, C. W. Maier, and E. Sackmann. 2003. Assembly of collagen matrices as a phase transition revealed by structural and rheologic studies. *Biophys J* 84:1272–1280.

75. Pins, G. D., D. L. Christiansen, R. Patel, and F. H. Silver. 1997. Self-assembly of collagen fibers. Influence of fibrillar alignment and decorin on mechanical properties. *Biophys J* 73:2164–2172.

76. Shoulders, M. D. and R. T. Raines. 2009. Collagen structure and stability. *Annu Rev Biochem* 78:929–958.

77. Hulmes, D. J. 2002. Building collagen molecules, fibrils, and suprafibrillar structures. *J Struct Biol* 137:2–10.

78. Maurice, D. M. 1970. The transparency of the corneal stroma. *Vision Res* 10:107–108.

79. Jen, S. H., G. Gonella, and H. L. Dai. 2009. The effect of particle size in second harmonic generation from the surface of spherical colloidal particles. I: Experimental observations. *J Phys Chem A* 113:4758–4762.

80. Badylak, S., D. Freytes, and T. Gilbert. 2009. Extracellular matrix as a biological scaffold material: Structure and function. *Acta Biomater* 5:1–13.

81. Brown, E., T. McKee, E. diTomaso, A. Pluen, B. Seed, Y. Boucher, and R. K. Jain. 2003. Dynamic imaging of collagen and its modulation in tumors *in vivo* using second-harmonic generation. *Nat Med* 9:796–800.

82. Guo, Y., P. P. Ho, H. Savage, D. Harris, P. Sacks, S. Schantz, F. Liu, N. Zhadin, and R. R. Alfano. 1997. Second-harmonic tomography of tissues. *Opt Lett* 22:1323–1325.

83. Guo, Y., P. P. Ho, H. Savage, D. Harris, P. Sacks, S. Schantz, F. Liu, N. Zhadin, and R. R. Alfano. 1998. Second-harmonic tomography of tissues: Errata. *Opt Lett* 23:733.

84. Mertz, J. and L. Moreaux. 2001. Second-harmonic generation by focused excitation of inhomogeneously distributed scatterers. *Opt Commun* 196:325–330.

85. Lacomb, R., O. Nadiarnykh, S. S. Townsend, and P. J. Campagnola. 2008. Phase matching considerations in second harmonic generation from tissues: Effects on emission directionality, conversion efficiency and observed morphology. *Opt Commun* 281:1823–1832.

86. Yew, E. Y. S. and C. J. R. Sheppard. 2006. Effects of axial field components on second harmonic generation microscopy. *Opt Express* 14:1167–1174.

87. Levitz, D., M. T. Hinds, N. Choudhury, N. T. Tran, S. R. Hanson, and S. L. Jacques. Quantitative characterization of developing collagen gels using optical coherence tomography. *J Biomed Opt* 15:026019.

88. Chu, S. W., S. Y. Chen, G. W. Chern, T. H. Tsai, Y. C. Chen, B. L. Lin, and C. K. Sun. 2004. Studies of chi(2)/chi(3) tensors in submicron-scaled biotissues by polarization harmonics optical microscopy. *Biophys J* 86:3914–3922.

89. Cox, G., E. Kable, A. Jones, I. Fraser, F. Manconi, and M. D. Gorrell. 2003. 3-Dimensional imaging of collagen using second harmonic generation. *J Struct Biol* 141:53–62.

90. Maitland, D. J. and J. T. Walsh, Jr. 1997. Quantitative measurements of linear birefringence during heating of native collagen. *Lasers Surg Med* 20:310–318.

91. Stanworth, A. and E. Naylor. 1953. Polarised light studies in cornea. I. The isolated cornea. *J Exp Biol* 30:160–163.

92. Lilledahl, M. B., O. A. Haugen, C. de Lange Davies, and L. O. Svaasand. 2007. Characterization of vulnerable plaques by multiphoton microscopy. *J Biomed Opt* 12:044005.

93. MacKintosh, F. C., J. Kas, and P. A. Janmey. 1995. Elasticity of semiflexible biopolymer networks. *Phys Rev Lett* 75:4425–4428.

94. Yang, L., K. O. van der Werf, B. F. Koopman, V. Subramaniam, M. L. Bennink, P. J. Dijkstra, and J. Feijen. 2007. Micromechanical bending of single collagen fibrils using atomic force microscopy. *J Biomed Mater Res A* 82:160–168.

95. Poirier, M. G., S. Eroglu, and J. F. Marko. 2002. The bending rigidity of mitotic chromosomes. *Mol Biol Cell* 13:2170–2179.

96. Zipfel, W. R., R. M. Williams, R. Christie, A. Y. Nikitin, B. T. Hyman, and W. W. Webb. 2003. Live tissue intrinsic emission microscopy using multiphoton-excited native fluorescence and second harmonic generation. *Proc Natl Acad Sci USA* 100:7075–7080.

97. Plotnikov, S. V., A. M. Kenny, S. J. Walsh, B. Zubrowski, C. Joseph, V. L. Scranton, G. A. Kuchel et al. 2008. Measurement of muscle disease by quantitative second-harmonic generation imaging. *J Biomed Opt* 13:044018.

98. LaComb, R., O. Nadiarnykh, S. Carey, and P. J. Campagnola. 2008. Quantitative second harmonic generation imaging and modeling of the optical clearing mechanism in striated muscle and tendon. *J Biomed Opt* 13:021109.

99. Plotnikov, S. V., A. C. Millard, P. J. Campagnola, and W. A. Mohler. 2006. Characterization of the myosin-based source for second-harmonic generation from muscle sarcomeres. *Biophys J* 90:693–703.

100. Plotnikov, S., V. Juneja, A. B. Isaacson, W. A. Mohler, and P. J. Campagnola. 2006. Optical clearing for improved contrast in second harmonic generation imaging of skeletal muscle. *Biophys J* 90:328–339.

101. Raub, C. B., S. Mahon, N. Narula, B. J. Tromber, M. Brenner, and S. C. George. 2010. Linking optics and mechanics in an *in vivo* model of airway fibrosis and epithelial injury. *J Biomed Opt* 15:15004-1–15004-9.

102. Kirkpatrick, N. D., S. Andreou, J. B. Hoying, and U. Utzinger. 2007. Live imaging of collagen remodeling during angiogenesis. *Am J Physiol Heart Circ Physiol* 292:H3198–H3206.
103. Jobsis, P. D., H. Ashikaga, H. Wen, E. C. Rothstein, K. A. Horvath, E. R. McVeigh, and R. S. Balaban. 2007. The visceral pericardium: Macromolecular structure and contribution to passive mechanical properties of the left ventricle. *Am J Physiol Heart Circ Physiol* 293:H3379–H3387.

12

Chiu-Mei Hsueh
National Taiwan University

Po-Sheng Hu
*National Cheng-Kung
University*

Wen Lo
National Taiwan University

Shean-Jen Chen
*National Cheng-Kung
University*

Hsin-Yuan Tan
National Taiwan University

Chang Gung University

Chen-Yuan Dong
National Taiwan University

SHG and Multiphoton Fluorescence Imaging of the Eye

12.1 Introduction

As the organ responsible for vision, the eye plays a vital role in our communication with the external world. Among the intricate components that constitute the eye, collagen-rich cornea is the key optical element responsible for most of the eye's refractive properties. Although the cornea only represents one-sixth of the outer coating of the eye, its dome-shaped and optically transparent structure contributes to more than two-thirds of the eye's focusing power. In addition to its collagen content, corneal epithelium is maintained by stem cells located at limbus, which constitutes the peripheral boundary of the cornea. Beyond the limbus, the sclera forms the remaining ocular surface and is responsible for structural integrity and protection of intraocular contents. Like the cornea, the sclera is covered by the epithelial conjunctiva and is mainly composed of collagen fibers. Nonetheless, it is the unique alignment of the collagen fibers that is responsible for corneal transparency [1,2].

The unique transparency and important visual function of the cornea is of intensive interest to researchers. Structurally, the cornea can be divided into the following five layers: epithelium, Bowman's membrane, stroma, Descemet's membrane, and the endothelium [3]. The epithelium is composed of five to six thin layers of squamous, nonkeratinized epithelial cells. Below the epithelium exists a randomly and densely arranged collagen fibrous layer (type I, III, V, and VI) known as Bowman's membrane, which separates the epithelium from the stroma. The stroma, which makes up 90% of the cornea, consists of orthogonally stacked lamella layers. Each lamella layer is composed of long, parallel-aligned collagen fibrils, whose main composition is type I. Interspersed within the collagen lamella are the keratocytes whose main functions are maintaining corneal transparency, synthesizing cellular components, and promoting wound healing. The Descemet's membrane connected to the posterior stroma is an acellular and homogeneous layer rich in basement membrane glycoproteins, laminin, and type IV collagen. The deepest layer of the cornea is the endothelium composed of a single squamous cell layer. In ophthalmology, physiological studies at the organ level have depended heavily on histological techniques such as biopsy examination and electron micrographs. However, potential artifacts associated with the process of fixation and labeling are always concerns. In addition, examination of fixed specimens cannot provide dynamic information of the ophthalmological systems under study.

Since the cornea is transparent and exteriorly located, optical microscopy is an appropriate technique for direct imaging and visualization of its structures. In the last few decades, a number of optical imaging modalities have been developed for these purposes. Techniques such as slit-lamp examination, optical coherence tomography (OCT), reflective confocal microscopy, and harmonic generation microscopy have been successfully developed for improving the understanding of corneal physiology and diagnosis of pathological conditions for potential clinical applications [4–14]. In addition to the widely used slit-lamp technique, the interferometry-based OCT methodology has been used in a variety of medical applications, including those in ophthalmology [7,15–19]. It was shown that OCT is capable of providing cross-sectional images at micron-level resolution [6,20]. In this manner, the layered structure from the anterior to the posterior segments of the eye, including that of the cornea, can be visualized [21,22]. The application of OCT upon the human cornea was first reported in 1994, which mapped the contours of the epithelium and endothelium, and quantitatively analyzed corneal thickness [22]. This technique has also been used in characterizing corneal pathologies, such as bullous keratopathy [17], corneal edema [23], and macular diseases [24]. In one study, the potential of OCT, as a medical diagnosis tool, was investigated by imaging cornea morphology before and after phototherapeutic keratectomy [25].

Confocal microscopy is another optical imaging technique that has been applied to cornea imaging. Since its initial introduction by Marvin Minsky demonstrating the improvement in image quality by the use of confocal aperture [26], subsequent developments have led to tandem scanning reflective confocal microscopy with multiple scanning points [27] elucidating the full thickness of human cornea *ex vivo* and of rabbit cornea *in situ* [28]. In the reflection mode, this noninvasive imaging modality has been applied clinically in ophthalmology. In previous reports, *in vivo* and *ex vitro* confocal images of corneas have proven to be effective in the visualization of epithelia, endothelial cells, and the stromal keratocytes [9,29–31]. Wound healing processes following refractive surgeries were also visualized, which helped to determine the potential complications arising from surgeries [32–34]. Nowadays, reflective confocal microscopy is being routinely applied in clinical ophthalmological observation and the detection of corneal diseases [9,10,35,36], such as infectious diseases, keratitis [37–39], and keratoconus [40,41].

While OCT and confocal imaging are effective in the visualization of corneal microstructures, imaging of the stromal collagen, the main component of the cornea, has been challenging with these two techniques. In confocal imaging, reflection is not an effective image contrast mechanism for visualizing the collagen fibers in the transparent cornea. On the other hand, although OCT has been demonstrated to be capable of imaging corneal stroma [23,42], it cannot provide structural information at submicron-level resolution. Finally, since both OCT and reflected confocal microscopy use reflection as the contrast mechanism, intrinsic molecular changes, which may be important for basic studies and disease diagnosis, cannot be investigated. In recent years, the development of nonlinear optical microscopic imaging modalities such as multiphoton fluorescence or second harmonic generation (SHG) microscopy has contributed to the repertoire of tools that researchers can use in addressing biomedical questions. Like confocal imaging, the intrinsic optical sectioning capabilities of nonlinear optical interaction can result in images with excellent axial depth discrimination. Furthermore, by limiting specimen excitation to the focal volume, specimen longevity is greatly prolonged. Finally, the near-infrared wavelengths used in multiphoton and SHG microscopy are less absorbed and scattered by tissues, thus allowing greater imaging depth to be achieved [43,44]. The development of multiphoton fluorescence and harmonic generation microscopy has led to tissue imaging applications in a number of areas, including dermatology [45,46], hepatology [47,48], neurology [49–51], and ophthalmology [52,53]. In ophthalmological imaging, multiphoton microscopy holds particular promise for the diagnosis of corneal diseases. Unlike other tissue types, the cornea cannot be routinely removed for histological examination and the development of a high-resolution, *in vivo* monitoring technique will be of significant value for clinical diagnosis of corneal pathologies. In this chapter, we will discuss the principles and instrumentation of multiphoton microscopy that is relevant to corneal imaging, and present results that demonstrate promises of multiphoton imaging in corneal diagnostic imaging.

12.2 Principles of Second Harmonic Generation Microscopy

The recent development of second harmonic generation microscopy (SHGM) has led to the exciting possibilities for pathological diagnostics. Like multiphoton fluorescence excitation, SHG is a nonlinear process offering similar advantages such as optical sectioning capability, improved penetration depth, label-free imaging, and significantly reduced photo-damage [45,54,55]. In SHGM, a near-infrared, ultrafast laser is chosen as the excitation source since it can provide light beam with high, instantaneous intensity below the tissue destruction threshold. Furthermore, the femtosecond source is effective in generating the nonlinear signal (multiphoton-excited fluorescence and SHG) within the focal volume [52,56]. In addition, the noncentrosymmetric structure of stromal and scleral collagen is a strong generator of SHG signal. Therefore, SHGM is an ideal technique for imaging the intrinsic triple-helical, noncentrosymmetric structures of cornea and sclera collagen fibril without extrinsic labeling [53,57,58].

12.2.1 Second Harmonic Generation Microscopy

Higher harmonic generation is a nonlinear polarization process related to the interaction of intense light with matters. In general, the polarization, P_i, of a material can be expressed as

$$P_i = \chi_{ij}E_j + \chi_{ijk}E_jE_k + \chi_{ijkl}E_jE_kE_l + \cdots \tag{12.1}$$

where χ_{ij}, χ_{ijk}, and χ_{ijkl} are, respectively, the first-, second-, and third-order susceptibility tensors, and E is the applied electric field amplitude. The second term of this expression $\chi_{ijk}E_jE_k$ represents the induced SHG polarization, which contributes to the generation of radiation field at one half wavelength of the excitation source. Thus, in the SHG process, two near-infrared incident photons are converted into one visible photon at twice the energy. As a result, there is no energy deposited in the illuminated specimen during the SHG process. This further allows microscopic imaging to be achieved with minimal invasion [12]. In addition, when the electric field is reversed, it corresponds to the reversal of polarization direction but does not include the second- and even higher-order terms because the SHG signal depends on the square of the electric field. As a result, the second harmonic signal can only be produced in noncentral symmetric material, such as structures composed of collagen molecules.

12.2.2 Multiphoton Microscopy Instrumentation

A typical multiphoton imaging system that can be utilized for corneal imaging is shown in Figure 12.1. For excitation, a titanium–sapphire (ti-sa) laser (Tsunami, Spectra Physics, Mountain View, CA) pumped by a diode-pumped, solid-state (DPSS) laser system (Millennia X, Spectra Physics) is used. The 780 nm output wavelength of the ti-sa laser is used for sample excitation and can be guided into a commercial upright microscope (E800, Nikon, Japan) by a pair of galvanometer-driven, x–y mirrors (Model 6220, Cambridge Technology, Cambridge, MA). After entering the microscope, the laser light is beam-expanded and reflected by the short-pass, primary dichroic mirror (700 dcspruv-3p, Chroma Technology, Rockingham, VT) onto the back aperture of the focusing objective (such as S Flour, 40 × / NA 0.8, water-immersion WI, Nikon). Multiphoton images can then be acquired by scanning the focal spot across the specimen. To obtain three-dimensional images at both high resolution and large scale, a motorized stage (H101, Prior Scientific, UK) is adopted to the microscope for specimen translation after each optical scan. After signal collection by the focusing objective in the epi-illuminated geometry,

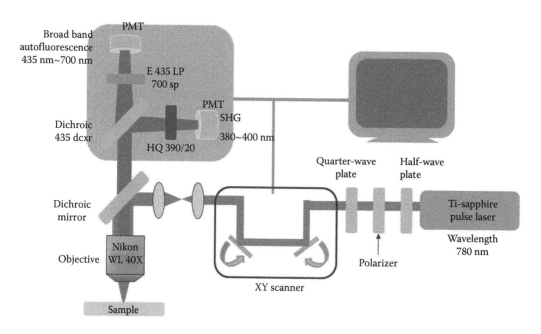

FIGURE 12.1 The experimental setup of multiphoton microscope.

the signals pass through the primary dichroic mirror. Next, the broadband multiphoton autofluorescence and SHG signals are separated by a secondary dichroic mirror (435dcxr, Chroma Technology) and respectively filtered by two band-pass filters (MAF: E435lp-700sp, SHG: HQ390/20, Chroma Technology) before reaching the detectors. For signal detection, photon-counting photomultiplier tubes (R7400P, Hamamatsu, Japan) can be employed. The typical detection bandwidths of the broadband fluorescence and SHG used in our laboratory are 435–700 and 380–400 nm, respectively.

12.3 Visualization of Normal Corneal and Surrounding Tissues

In the pioneering work published by Tromberg's group, it was shown that SHG signal propagating in the backward direction can be registered for visualization of corneal structures under *ex vivo* conditions [57]. Our experimental results and that of others demonstrated the efficacy of multiphoton microscopy for corneal imaging [13,59,60]. As shown in Figure 12.2, the large-area, multiphoton autofluorescence and backward SHG images of the *ex vivo* porcine cornea were acquired at different depths. At the corneal surface (0 μm), corneal epithelium can be identified by its intrinsic autofluorescence originated from the cytoplasmic portion of cells and the lack of the signal from the nuclei. Owing to the intrinsic curvature of the eye, backward SHG (BWSHG) signal from stromal collagen can be observed at the same depth as the epithelium. At the greater depth of approximately 862 μm, autofluorescence can be used to characterize the Descemet's membrane and endothelial region while BWSHG can be detected at the deepest layer of stromal collagen. The BWSHG pattern reveals the presence of the collagen fibers and indicates the difference of orientation between anterior and posterior regions of the cornea. Other ocular components adjacent to the cornea, such as the limbus and sclera, can also be effectively imaged by autofluorescence and SHG microscopy (Figure 12.3). Near the surface of corneal–scleral junction (0 μm), conjunctival and limbal epithelium can be identified by its intrinsic autofluorescence, and BWSHG signal emanated from corneal and scleral collagen can be used for imaging purposes. At the greater depth of approximately 50 μm, autofluorescence can be

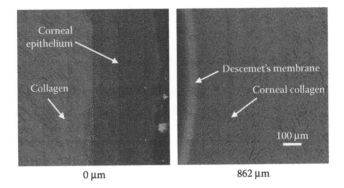

0 µm   862 µm

FIGURE 12.2 Large-area, multiphoton autofluorescence and backward SHG images of the *ex vivo* porcine cornea at different depths. At the corneal surface (0 mm), corneal epithelium can be identified by its intrinsic autofluorescence while backward SHG signal from stromal collagen can be used for imaging purposes. At the greater depth of approximate 862 mm, autofluorescence can be used to characterize the Descemet's membrane and endothelial region while backward SHG can be detected at the deepest layer of stromal collagen. (Adapted from Teng, S. W. et al. 2006. *Invest Ophthalmol Vis Sci* 47:1216–1224.)

used to characterize the limbal cells. The ability of acquiring limbus structure information and its biochemical state is of considerable significance because of the proliferation capability of the stem cell within the limbus and its potential application in tissue engineering. Unlike corneal collagen, BWSHG shows that scleral collagen is more randomly organized. In addition, owing to the shorter coherent length of BWSHG radiation, the forward SHG (FWSHG) signal is usually stronger than BWSHG and exhibits the fibrous morphology in corneal imaging better [61]. As a result, FWSHG has been frequently used to investigate the corneal collagen fibril orientation [62,63]. Nonetheless, the FWSHG signal in the clinical or *in vivo* observation is practically unavailable. Therefore, understanding the nature of BWSHG is of considerable significance in both normal and pathological imaging of the cornea.

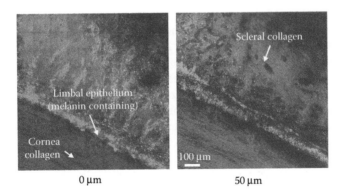

0 µm   50 µm

FIGURE 12.3 Large-area, multiphoton autofluorescence and backward SHG images of the *ex vivo* porcine cornea at different depths. Near the surface of corneal–scleral junction (0 mm), conjunctival and limbal epithelium can be identified by its intrinsic autofluorescence while backward SHG signal from corneal and scleral collagen can be used for imaging purposes. At the greater depth of approximate 50 mm, autofluorescence can be used to characterize the limbal cells between corneal and scleral collagen. Backward SHG shows that unlike corneal collagen, scleral collagen is more randomly organized. (Adapted from Teng, S. W. et al. 2006. *Invest Ophthalmol Vis Sci* 47:1216–1224.)

12.4 Imaging and Characterization of Pathological Corneas

An example of the application of multiphoton microscopy in the diagnosing of corneal diseases is the detection of infectious keratitis. Shown in Figure 12.4a is the large-area, multiphoton autofluorescence (green) and BWSHG (blue) images of the *ex vivo* human cornea that have been infected with *Acanthamoeba castellinii* and *Pseudomonas aeruginosa*. Intrinsic autofluorescence shows the presence of *Acanthamoeba* cysts (yellow arrowheads) and *Pseudomonas* bacteria (red arrowhead), and the deviation of the BWSHG image pattern from its normal counterpart indicates the degradation of corneal collagen. An increase in autofluorescence was also observed in the infected cornea. Magnified images from the selected regions of interests (ROIs) in Figure 12.4a are shown in Figures 12.4b through 12.4d. The presence of pathogens can be clearly visualized from autofluorescence. For the purpose of comparison, multiphoton autofluorescence image of isolated *Pseudomonas aeruginosa* in Figure 12.4e shows that the spots in Figure 12.4b are individual bacteria (red arrowheads), which does not resemble the appearance of the fluorescent *Acanthamoeba* cysts (yellow arrowheads) in Figures 12.4b and 12.4c. Besides, it was found that the less affected region with preserved collagen structure as demonstrated by the SHG signal (Figure 12.4d) is less likely to be infiltrated by inflammatory cells (yellow arrowheads). This observation implies the presence of quiescent *Acanthamoeba* cysts within the clinically clear area. These results show that cornea infected with pathogens may be imaged, and delineation of infectious pathogens and corneal collagen can be achieved by the use of label-free multiphoton microscopy [64].

The other example we presented here is the *ex vivo* multiphoton imaging of fungal keratitis in human cornea [64]. The large-scale multiphoton image (Figure 12.5a) taken at the surface of the ulcerated area was found to be composed of fluorescent signals and irregularly distributed SHG signals remnants. The parallel fluorescence pattern and residual SHG generating collagen can be observed in the magnified images (Figures 12.5b and 12.5c). Meanwhile, the laboratory examination represented that the responsible pathogen of this infection was the *Alternaria* sp. Compared with the multiphoton image of purified *Alternaria* (Figure 12.5e), the tube-like structure found in Figures 12.5a and 12.5d suggests that these structures are most likely the hyphae of the infecting fungus. This result again demonstrates the advantage of multiphoton microscopy in identifying some infecting pathogens without additional histological processing.

Another important corneal pathology that has been examined with multiphoton imaging is keratoconus. For reasons that are not completely understood, corneas of patients with keratoconus undergo structural transformation in the corneal lamella and result in significant vision degradation. In recent years, researchers have realized the potentials of multiphoton imaging in diagnosing this condition [63,65], and a demonstration of the use of multiphoton microscopy in visualizing structural changes associated with this pathology is shown in Figure 12.6. The large-area, multiphoton autofluorescence and BWSHG images of the *ex vivo* human keratoconical cornea markedly identify the global architectural change in such a specimen. Intrinsic autofluorescence shows the presence of the keratoconical apex while BWSHG signal identified the global organization of collagen fibers around the apex. Corneal topography in Figure 12.6b shows that the location of the keratoconical apex is consistent with that found from multiphoton imaging. To visualize detailed and localized changes of the pathological tissue, magnified regions of selected ROIs are shown in Figure 12.6c. Specifically, I and III are enlarged images from the selected ROIs indicated in Figure 12.6a. Further enlargement of these regions shows that the fluorescent mass near that apex is composed of epithelial cells with elongated and spindle-like shape (I-1). Furthermore, BWSHG patterns near the apex are aligned in parallel, suggesting that the altered corneal collagen structure can affect the morphological appearance of the nearby epithelial cells. Moreover, from SHG images shown in III and III-1, the patterns of centripetal and thickened stromal collagen directed toward the apical domain were found. The reorganization of collagenous stroma may be due to pathological structural modification in response to intraocular pressure. This observation demonstrates the ability of multiphoton microscopy not only in detecting the detailed collagen structural alteration but also in examining changes in corneal global morphology.

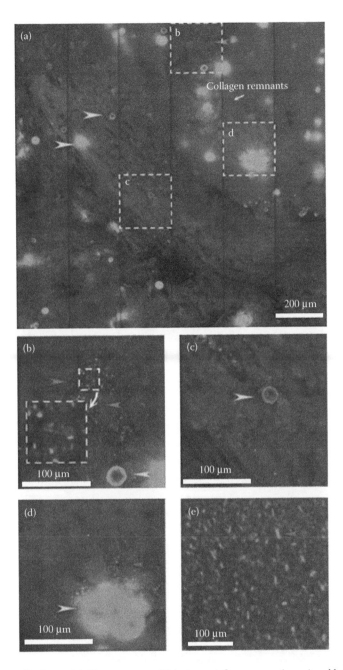

FIGURE 12.4 (See color insert.) (a) Large-area, multiphoton autofluorescence (green) and backward SHG (blue) images of the *ex vivo* human cornea that have been infected with *Acanthamoeba castellinii* and *Pseudomonas aeruginosa*. Intrinsic autofluorescence shows the presence of *Acanthamoeba* cysts (yellow arrowhead) and *Pseudomonas* bacteria (red arrowhead) while backward SHG signal shows the degradation of corneal collagen. Increase in autofluorescence was also observed in infected cornea. (b), (c), (d) Magnified images from the selected regions of interests in (a). The presence of pathogens can be clearly visualized from autofluorescence. (e) Multiphoton autofluorescence image of isolated *Pseudomonas aeruginosa*. (Adapted from Tan, H. Y. et al. 2007. *J Biomed Opt* 12:024013.)

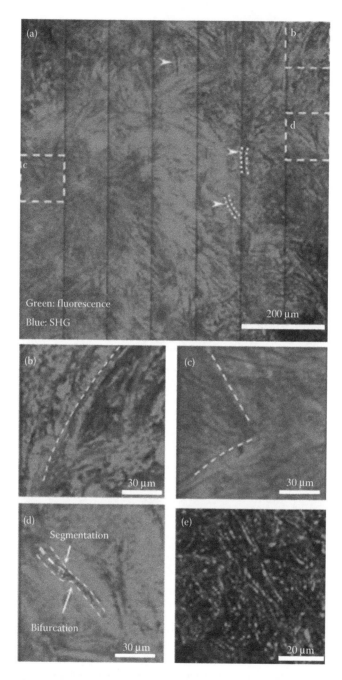

FIGURE 12.5 **(See color insert.)** (a) Large-area, multiphoton autofluorescence (green) and backward SHG (blue) images of the *ex vivo* human cornea that have been infected with *Alternaria* sp. (b), (c), and (d) represent the magnified areas b, c, and d in (a). Parallel distributed fluorescence in the background and residual SHG-generating collagen remnants were found in (b) and (c). Possible fungal hyphae with characteristic morphology of bifurcation and segmentation (white arrows) can be visualized both in the large-scale (a) and detailed image (d). (e) Multiphoton autofluorescence image of isolated *Alternaria* sp. (Adapted from Tan, H. Y. et al. 2007. *J Biomed Opt* 12:024013.)

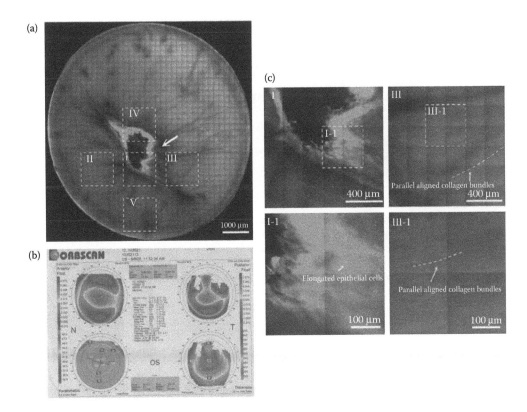

FIGURE 12.6 (a) Large-area, multiphoton autofluorescence and backward SHG images of the *ex vivo* human keratoconical cornea. Intrinsic autofluorescence shows the presence of the keratoconical apex while backward SHG signal identified the global organization of collagen fibers around the apex. (b) Corneal topography shows that the location of the keratoconical apex is consistent with that found from multiphoton imaging. (c) I and III are magnified images from the selected regions of interests in (a). Further enlargement of the images show that epithelial cells near that apex are elongated (I-1) and that backward SHG patterns near the apex are parallel aligned. (Adapted from Tan, H. Y. et al. 2006. *Invest Ophthalmol Vis Sci* 47:5251–5259.)

Physical trauma to the cornea represents another condition that multiphoton imaging can be useful for diagnostic purposes. Shown in Figure 12.7 are large-area, multiphoton autofluorescence and BWSHG images of an excised human cornea that is known to contain a scar. Near the corneal surface (0 μm), corneal epithelium can be identified by its intrinsic autofluorescence, and backward SHG signal illustrates corneal collagen protruding across the Bowman layer. At the depth of approximately 1200 μm, wound regions lacking SHG signal were observed. Along the wound edge, intense autofluorescence and BWSHG pattern aligned in parallel to the wound edge were both observed. The ability of multiphoton microscopy identifying the structural alteration of the cornea is again demonstrated.

In addition to diagnosing corneal pathologies of human cornea specimens, multiphoton microscopy may also be used to study the processes associated with corneal pathologies under controlled conditions. One experiment that we performed was to simulate infection caused by *Pseudomonas aeruginosa* in bovine corneas *in vitro* [14]. Following artificial injection of the pathogen into the cornea, the specimens were kept at 37°C and multiphoton images were acquired at fixed intervals following pathogen injection. In this manner, we were able to follow the temporal effect of pathogen infection in corneas. Shown in Figure 12.8 are our results, large-area, multiphoton autofluorescence and BWSHG images of excised

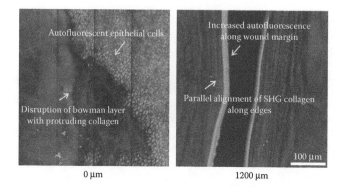

FIGURE 12.7 Large-area, multiphoton autofluorescence and backward SHG images of the scarred human cornea (*ex vivo*). Near the corneal surface (0 mm), corneal epithelium can be identified by its intrinsic autofluorescence while backward SHG signal shows corneal collagen protruding across the Bowman layer. At the depth of approximate 1200 mm, autofluorescent wound edges and parallel aligned backward SHG patterns are found along the wound edge. (Adapted from Teng, S. W. et al. 2007. *Arch Ophthalmol* 125(7):977–978.)

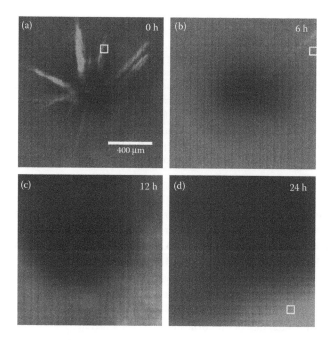

FIGURE 12.8 Large-area, multiphoton autofluorescence and backward SHG images of excised bovine corneas that have been injected with *Pseudomonas aeruginosa*. Images acquired at different times (0, 6, 12, and 24 h) following injection show the effected of simulated infection on corneal stromal collagen with time. At 0 h, injected *Pseudomonas aeruginosa* in radiated pattern are clearly visible by autofluorescence. At the same time, the stromal collagen is largely intact with no signs of degradation. With progression of the simulated infection process, a decrease in backward SHG signal and increase in autofluorescence are observed. By 24 h, the corneal stroma is almost filled entirely with strong autofluorescence. (Adapted from Chang, Y. L. et al. 2010. *Appl Phys Lett* 97:183703.)

bovine corneas that have been injected with *Pseudomonas aeruginosa*. Images acquired at different time-points (0, 6, 12, and 24 h) following injection show the effects of simulated infection on corneal stromal collagen as a function of time. At 0 h, injected *Pseudomonas aeruginosa* in radiated pattern can be observed by autofluorescence imaging. At this time, the stromal collagen is largely intact with no signs of degradation. With progression of the simulated infection process, a decrease in BWSHG signal and increase in autofluorescence are observed. In 24 h, the corneal stroma is almost filled entirely with intense autofluorescence, whereas little SHG signal was observed. Therefore, qualitatively, increased damage of cornea is accompanied with increased autofluorescence and decrease in the SHG signal.

To visualize our observation in greater details, selected regions of interest at 0, 6, and 24 h following pathogen injection were selected from Figure 12.8. The respective magnified images are shown in Figures 12.9a through 12.9c, where injected *Pseudomonas aeruginosa* are clearly visible (arrows). With increase in infection time, there is a decrease in BWSHG signal and corresponding increase in autofluorescence. Therefore, similar to the images acquired from excised human corneas infected with fungous amoeba and bacteria, severely infected corneas containing weak SHG signals and significantly increased autofluorescence were observed [64].

Multiphoton microscopy has also been employed in experiments imaging and characterizing the extent of bovine cornea glycation. The glycation process is thought to contribute to many of the complications as seen in aging and diabetes mellitus [66]. Hence, early diagnostics of the influence of glycation-induced ophthalmological pathologies is of primary importance for their therapeutics.

Cornea glycation was induced by incubating cornea in 0.5 M ribose solution to mimic physiological hyperglycemia [67]. The control group was treated with the same chemical composition as the glycation solution with ribose being excluded. All corneas were incubated at 37°C with condition of 5% CO_2 for different periods of 2, 4, 6, 8, and 10 days. Time-lapsed multiphoton imaging on glycated and control samples of bovine cornea is illustrated in Figure 12.10, in which autofluorescence and BWSHG intensities extracted from glycated bovine cornea increase and decrease, respectively, as glycation prolonged in number of days. In contrast, for control group, no significant variance of the BWSHG intensity was observed. The analytic result of MPM images, shown in Figure 12.11, delineates the trends of autofluorescence and SHG intensities as a function of duration of glycation. The rise of autofluorescence accompanied with the formation of the autofluorescent advanced glycation end products (AGEs), and the decline of BWSHG indicates glycation-induced disorder in the collagen structure [68]. The consequences of MPM images and its analysis are twofold: indication of the rate of glycation in bovine cornea and elucidation of the drastic responsiveness of collagen over elastic

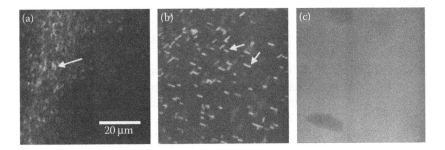

FIGURE 12.9 (a), (b), and (c) are the respective magnified images of selected regions of interest at 0, 6, and 24 h following simulated infection images in Figure 12.7. (a,b) Injected *Pseudomonas aeruginosa* are clearly visible (arrows). With increase in infection time, there is a decrease in backward SHG signal and corresponding increase in autofluorescence. (c) The corneal stroma is almost filled entirely with strong autofluoroescence. (Adapted from Chang, Y. L. et al. 2010. *Appl Phys Lett* 97:183703.)

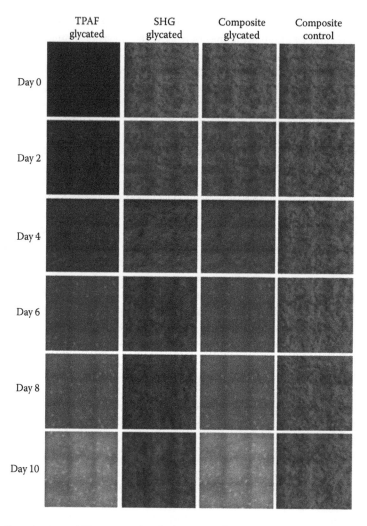

FIGURE 12.10 (**See color insert.**) Time-coursed multiphoton autofluorescence (green) and SHG (red) images of glycated and control (nonglycated) samples of bovine cornea. Frame size is 620×620 mm². (Adapted from Tseng, J. Y. et al. 2011. *Biomed Opt Express* 2(2):218–230.)

fiber in the formation of autofluorescent AGEs, which may be useful for early detection of diabetes mellitus-induced ophthalmological pathologies.

As a final example of the ability of multiphoton imaging in visualizing pathological corneas, bovine corneas were immersed in de-ionized water for 2 h to simulate cornea edema [69]. Following induced edematous conditions, the corneas were mounted for FWSHG and BWSHG imaging. Shown in Figures 12.12a 12.12b, 12.12e, and 12.12f are FWSHG and BWSHG image stacks of untreated bovine corneas. On the other hand, Figures 12.12c, 12.12d, 12.12g, and 12.12h represent image stacks from edematous corneas. In both cases, SHG images were acquired from both the anterior and posterior stroma. Note that in edematous corneas, lamella oriented obliquely to the corneal surface (arrowheads in Figure 12.12d) and increased lamellae spacing (arrowheads in Figure 12.12h) can be visualized.

Enlarged *en face* forward SHG (FWSHG) and backward SHG (BWSHG) images of untreated and edematous bovine corneas at depths of 100 and 1000 μm were illustrated in Figure 12.13. As noted in the

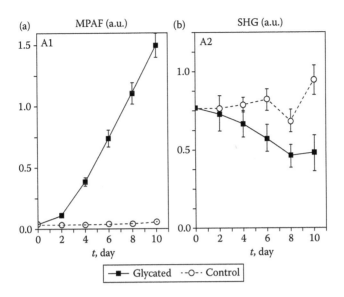

FIGURE 12.11 The dependence of autofluorescence (a) and BWSHG (b) signal intensity of bovine cornea. (From Tseng, J. Y. et al. 2011. *Biomed Opt Express* 2(2):218–230. With permission.)

anterior part (100 μm), SHG images reveal similar structures as the tightly interwoven collagen fibers in both normal and edematous corneas. However, in the posterior region (1000 μm), the BWSHG image of the edematous cornea becomes similar to the FWSHG image, which indicates the increased back scattering of the FWSHG signal due to a decrease of corneal transparency.

Comparison of the anterior and posterior of the edematous cornea showed that the reduced structural alteration in the anterior cornea as compared to that in the posterior region may be due to the more tightly interwoven collagenous structure in anterior cornea. This study demonstrated the ability of multiphoton microscopy in recognizing corneal morphology, and the potential of BWSHG signal characteristic in revealing detailed collagen structure in corneal edema.

12.5 Future in Clinical Diagnosis

Cornea is a key organ responsible for vision. Unlike many other organs, the cornea cannot be removed for histological examination in pathological diagnostics. Therefore, the development of a label-free, high-resolution imaging modality that is capable of visualizing different corneal structural motifs will be of significant value in the clinical diagnostics of corneal diseases. In this chapter, we have described the principles and instrumentation of multiphoton microscopy and provided specific examples of how nonlinear optical phenomena, such as multiphoton autofluorescence and second harmonic generation, can be used to visualize the elements of cornea, which is transparent to linear optical sources. Moreover, we have shown that numerous corneal pathologies, such as corneal edema, keratoconus, physical trauma, infection keratitis, intraocular pressure-induced structural disorder, and diabetes mellitus-induced ophthalmological complications, can be imaged and discriminated from normal cornea without extrinsic labeling. With additional development in instrument miniaturization and increased image acquisition speed, multiphoton microscopy has the potential to be developed into a clinically viable tool for the diagnostics of corneal pathologies.

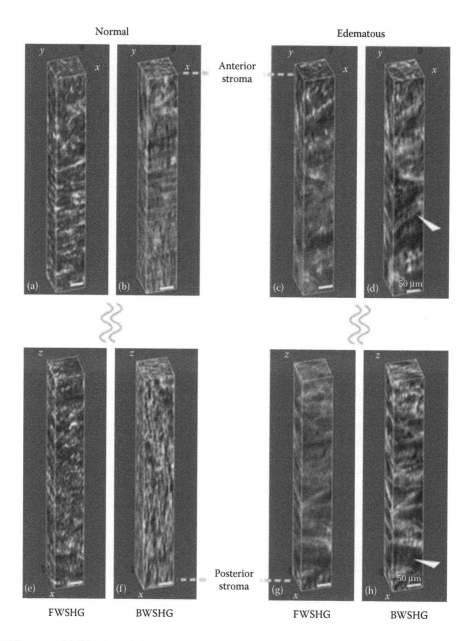

Normal   Edematous

FWSHG   BWSHG   FWSHG   BWSHG

FIGURE 12.12   (a), (b), (e), and (f) are forward SHG (FWSHG) and backward SHG (BWSHG) image stacks of untreated bovine corneas. (c), (d), (g), and (h) have been treated with de-ionized water immersion for 2 h to simulate corneal edema. In both cases, SHG images were acquired from both the anterior and posterior stroma. Note that in edematous corneas, lamella oriented obliquely to the corneal surface (arrowheads in (d)) and that increased lamellae spacing (arrowheads in (h)) can be visualized. (Adapted from Hsueh, C. M. et al. 2009. *Biophys J* 97(4):1198–1205.)

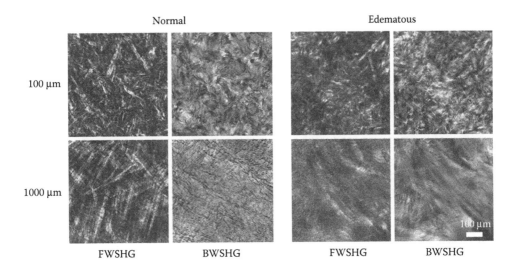

FIGURE 12.13 En face forward SHG (FWSHG) and backward SHG (BWSHG) images of untreated and edematous bovine corneas at depths of 100 and 1000 mm. Note that in anterior part (100 mm), SHG images reveal similar structures in both normal and edematous corneas. However, in the posterior region (1000 mm), a deviation from the normal lamella orientation can be observed. (Adapted from Hsueh, C. M. et al. 2009. *Biophys J* 97(4):1198–1205.)

References

1. Maurice, D. M. 1957. The structure and transparency of the cornea. *J Physiol* 136:263–286.
2. Hart, R. W. and R. A. Farrell. 1969. Light scattering in the cornea. *J Opt Soc Am* 59:766–774.
3. Forrester, J. V. 2002. *The Eye: Basic Sciences in Practice*. W.B. Saunders, Edinburgh.
4. Bengtsson, B. 1991. Glaucoma case detection. *Acta Ophthalmol* 69:288–292.
5. Fernandez, A. G. A., H. Demirci, D. A. Darnley-Fisch, and D. W. Steen. 2010. Interstitial keratitis secondary to severe hidradenitis suppurativa: A case report and literature review. *Cornea* 29:1189–1191.
6. Ko, T. H., J. G. Fujimoto, J. S. Schuman, L. A. Paunescu, A. M. Kowalevicz, I. Hartl, W. Drexler, G. Wollstein, H. Ishikawa, and J. S. Duker. 2005. Comparison of ultrahigh- and standard-resolution optical coherence tomography for imaging macular pathology. *Ophthalmology* 112:1922–1935.
7. Li, G., M. Zhou, H. J. Wu, and L. Lin. 2010. The research status and development of noninvasive glucose optical measurements. *Spectrosc Spect Anal* 30:2744–2747.
8. MollerPedersen, T., M. Vogel, H. F. Li, W. M. Petroll, H. D. Cavanagh, and J. V. Jester. 1997. Quantification of stromal thinning, epithelial thickness, and corneal haze after photorefractive keratectomy using *in vivo* confocal microscopy. *Ophthalmology* 104:360–368.
9. Jalbert, I., F. Stapleton, E. Papas, D. F. Sweeney, and M. Coroneo. 2003. *In vivo* confocal microscopy of the human cornea. *Br J Ophthalmol* 87:225–236.
10. Guthoff, R. F., A. Zhivov, and O. Stachs. 2009. *In vivo* confocal microscopy, an inner vision of the cornea—A major review. *Clin Exp Ophthalmol* 37:100–117.
11. Aptel, F., N. Olivier, A. Deniset-Besseau, J. M. Legeais, K. Plamann, M. C. Schanne-Klein, and E. Beaurepaire. 2010. Multimodal nonlinear imaging of the human cornea. *Invest Ophthalmol Vis Sci* 51:2459–2465.
12. Chen, S. Y., H. C. Yu, I. J. Wang, and C. K. Sun. 2009. Infrared-based third and second harmonic generation imaging of cornea. *J Biomed Opt* 14:04412.

13. Hao, M., K. Flynn, C. Nien-Shy, B. E. Jester, M. Winkler, D. J. Brown, O. La Schiazza, J. Bille, and J. V. Jester. 2010. *In vivo* non-linear optical (NLO) imaging in live rabbit eyes using the Heidelberg Two-Photon Laser Ophthalmoscope. *Exp Eye Res* 91:308–314.

14. Chang, Y. L., W. L. Chen, W. Lo, S. J. Chen, H. Y. Tan, and C. Y. Dong. 2010. Characterization of corneal damage from *Pseudomonas aeruginosa* infection by the use of multiphoton microscopy. *Appl Phys Lett* 97:183703.

15. Huang, D., E. A. Swanson, C. P. Lin, J. S. Schuman, W. G. Stinson, W. Chang, M. R. Hee, T. Flotte, K. Gregory, and C. A. Puliafito. 1991. Optical coherence tomography. *Science* 254:1178–1181.

16. Fukuchi, T., K. Takahashi, K. Shou, and M. Matsumura. 2000. Optical coherence tomographic (OCT) findings on normal retina and laser induced choroidal neovascularisation (CNV) in rat. *Invest Ophthalmol Vis Sci* 41:S174–S174.

17. Hirano, K., Y. Ito, T. Suzuki, T. Kojima, S. Kachi, and Y. Miyake. 2001. Optical coherence tomography for the noninvasive evaluation of the cornea. *Cornea* 20:281.

18. Fujimoto, J. G. 2003. Optical coherence tomography for ultrahigh resolution *in vivo* imaging. *Nat Biotechnol* 21:1361–1367.

19. Welzel, J. 2001. Optical coherence tomography in dermatology: A review. *Skin Res Technol* 7:1–9.

20. Brezinski, M. E. and J. G. Fujimoto. 1999. Optical coherence tomography: High-resolution imaging in nontransparent tissue. *IEEE J Sel Top Quantum Electron* 5:1185–1192.

21. Swanson, E. A., J. A. Izatt, M. R. Hee, D. Huang, C. P. Lin, J. S. Schuman, C. A. Puliafito, and J. G. Fujimoto. 1993. *In vivo* retinal imaging by optical coherence tomography. *Opt Lett* 18:1993–1911.

22. Izatt, J. A., M. R. Hee, E. A. Swanson, C. P. Lin, D. Huang, J. S. Schuman, C. A. Puliafito, and J. G. Fujimoto. 1994. Micrometer-scale resolution imaging of the anterior eye *in vivo* with optical coherence tomography. *Arch Ophthalmol-Chic* 112:1584–1589.

23. Wang, J., T. L. Simpson, and D. Fonn. 2004. Objective measurements of corneal light-backscatter during corneal swelling, by optical coherence tomography. *Invest Ophthalmol Vis Sci* 45:3493.

24. Puliafito, C. A., M. R. Hee, C. P. Lin, E. Reichel, J. S. Schuman, J. S. Duker, J. A. Izatt, E. A. Swanson, and J. G. Fujimoto. 1995. Imaging of macular diseases with optical coherence tomography. *Ophthalmology* 102:217–229.

25. Hoerauf, H., C. Wirbelauer, C. Scholz, R. Engelhardt, P. Koch, H. Laqua, and R. Birngruber. 2000. Slit-lamp-adapted optical coherence tomography of the anterior segment. *Graefes Arch Clin Exp Ophthalmol* 238:8–18.

26. Minsky, M. 1988. Memoir on inventing the confocal scanning microscope. *Scanning* 10:128–138.

27. Petran, M., M. Hadravsky, M. D. Egger, and R. Galambos. 1968. Tandem-scanning reflected-light microscope. *J Opt Soc Am* 58:90–93.

28. Lemp, M. A., P. N. Dilly, and A. Boyde. 1985. Tandem-scanning (confocal) microscopy of the full-thickness cornea. *Cornea* 4:205.

29. Patel, S. V., J. W. McLaren, D. O. Hodge, and W. M. Bourne. 2001. Normal human keratocyte density and corneal thickness measurement by using confocal microscopy *in vivo*. *Invest Ophthalmol Vis Sci* 42:333–339.

30. Masters, B. R. and M. Bohnke. 2002. Three-dimensional confocal microscopy of the living human eye. *Annu Rev Biomed Eng* 4:69–91.

31. Chen, W. L., Y. Sun, W. Lo, H. Y. Tan, and C. Y. Dong. 2008. Combination of multiphoton and reflective confocal imaging of cornea. *Microsc Res Tech* 71:83–85.

32. Moller-Pedersen, T., H. F. Li, W. M. Petroll, H. D. Cavanagh, and J. V. Jester. 1998. Confocal microscopic characterization of wound repair after photorefractive keratectomy. *Invest Ophthalmol Vis Sci* 39:487–501.

33. McLaren, J. W., W. M. Bourne, and S. V. Patel. 2010. Standardization of corneal haze measurement in confocal microscopy. *Invest Ophthalmol Vis Sci* 51:5610–5616.

34. Jester, J. V., W. M. Petroll, and H. D. Cavanagh. 1999. Corneal stromal wound healing in refractive surgery: The role of myofibroblasts. *Prog Retin Eye Res* 18:311–356.

35. Erie, J. C., J. W. McLaren, and S. V. Patel. 2009. Confocal microscopy in ophthalmology. *Am J Ophthalmol* 148:639–646.
36. Hollingsworth, J. G., R. E. Bonshek, and N. Efron. 2005. Correlation of the appearance of the kera-toconic cornea *in vivo* by confocal microscopy and *in vitro* by light microscopy. *Cornea* 24:397.
37. Beuerman, R. W., S. J. Chew, L. Pedroza, M. Assouline, B. Barron, J. Hill, and H. E. Kaufman. 1992. Early diagnosis of infectious keratitis with *in vivo* real-time confocal microscopy. *Invest Ophthalmol Vis Sci* 33:1234–1234.
38. Nakano, E., M. Oliveira, W. Portellinha, D. de Freitas, and K. Nakano. 2004. Confocal microscopy in early diagnosis of Acanthamoeba keratitis. *J Refract Surg* 20:S737–S740.
39. Takezawa, Y., A. Shiraishi, E. Noda, Y. Hara, M. Yamaguchi, T. Uno, and Y. Ohashi. 2010. Effectiveness of *in vivo* confocal microscopy in detecting filamentous fungi during clinical course of fungal kerati-tis. *Cornea* 29:1346–1352.
40. Balestrazzi, A., G. Martone, C. Traversi, G. Haka, P. Toti, and A. Caporossi. 2006. Keratoconus associated with corneal macular dystrophy: *In vivo* confocal microscopic evaluation. *Eur J Ophthalmol* 16:745–750.
41. Mazzotta, C., S. Baiocchi, O. Caporossi, D. Buccoliero, F. Casprini, A. Caporossi, and A. Balestrazzi. 2008. Confocal microscopy identification of keratoconus associated with posterior polymorphous corneal dystrophy. *J Cataract Refract Surg* 34:318–321.
42. Fercher, A. F., W. Drexler, C. K. Hitzenberger, and T. Lasser. 2003. Optical coherence tomography-principles and applications. *Rep Progr Phys* 66:239–303.
43. Helmchen, F. 2005. Deep tissue two-photon microscopy. *Nat Methods* 2:932.
44. Denk, W., J. H. Strickler, and W. W. Webb. 1990. 2-Photon laser scanning fluorescence microscopy. *Science* 248:73–76.
45. Tsai, T. H., S. H. Jee, C. Y. Dong, and S. J. Lin. 2009. Multiphoton microscopy in dermatological imaging. *J Dermatol Sci* 56:1–8.
46. Lin, S. J., S. H. Jee, and C. Y. Dong. 2007. Multiphoton microscopy: A new paradigm in dermatologi-cal imaging. *Eur J Dermatol* 17:361–366.
47. Liang, Y., T. Shilagard, S. Y. Xiao, N. Snyder, D. Lau, L. Cicalese, H. Weiss, G. Vargas, and S. M. Lemon. 2009. Visualizing hepatitis C virus infections in human liver by two-photon microscopy. *Gastroenterology* 137:1448–1458.
48. Sun, T. L., Y. Liu, M. C. Sung, H. C. Chen, C. H. Yang, V. Hovhannisyan, W. C. Lin, W. L. Chen, L. L. Chiou, and G. T. Huang. 2009. Label-free diagnosis of human hepatocellular carcinoma by mul-tiphoton autofluorescence microscopy. *Appl Phys Lett* 95:193703.
49. Dombeck, D. A., A. N. Khabbaz, F. Collman, T. L. Adelman, and D. W. Tank. 2007. Imaging large-scale neural activity with cellular resolution in awake, mobile mice. *Neuron* 56:43–57.
50. Losavio, B. E., Y. Liang, A. Santamaria-Pang, I. A. Kakadiaris, C. M. Colbert, and P. Saggau. 2008. Live neuron morphology automatically reconstructed from multiphoton and confocal imaging data. *J Neurophysiol* 100:2422.
51. Kim, K. H., C. Buehler, K. Bahlmann, T. Ragan, W. C. A. Lee, E. Nedivi, E. L. Heffer, S. Fantini, and P. T. C. So. 2007. Multifocal multiphoton microscopy based on multianode photomultiplier tubes. *Opt Express* 15:11658–11678.
52. Hsueh, C., W. E. N. Lo, S. Lin, T. Wang, F. R. Hu, H. Y. Tan, and C. Y. Dong. 2009. Multiphoton microscopy: A new approach in physiological studies and pathological diagnosis for ophthalmol-ogy. *J Innov Opt Health Sci* 2:45–60.
53. Jester, J. V., M. Winkler, B. E. Jester, C. Nien, D. Chai, and D. J. Brown. 2010. Evaluating corneal colla-gen organization using high-resolution nonlinear optical macroscopy. *Eye Contact Lens* 36:260–264.
54. Carriles, R., D. N. Schafer, K. E. Sheetz, J. J. Field, R. Cisek, V. Barzda, A. W. Sylvester, and J. A. Squier. 2009. Invited Review Article: Imaging techniques for harmonic and multiphoton absorption fluorescence microscopy. *Rev Sci Instrum* 80:081101-081101-081123.
55. Campagnola, P. J. and L. M. Loew. 2003. Second-harmonic imaging microscopy for visualizing bio-molecular arrays in cells, tissues and organisms. *Nat Biotechnol* 21:1356–1360.

56. Zipfel, W. R., R. M. Williams, and W. W. Webb. 2003. Nonlinear magic: Multiphoton microscopy in the biosciences. *Nat Biotechnol* 21:1369–1377.

57. Yeh, A. T., N. Nassif, A. Zoumi, and B. J. Tromberg. 2002. Selective corneal imaging using combined second-harmonic generation and two-photon excited fluorescence. *Opt Lett* 27:2082–2084.

58. Lyubovitsky, J. G., J. A. Spencer, T. B. Krasieva, B. Andersen, and B. J. Tromberg. 2006. Imaging corneal pathology in a transgenic mouse model using nonlinear microscopy. *J Biomed Opt* 11:014013.

59. Morishige, N., W. M. Petroll, T. Nishida, M. C. Kenney, and J. V. Jester. 2006. Noninvasive corneal stromal collagen imaging using two-photon-generated second-harmonic signals. *J Cataract Refract Surg* 32:1784–1791.

60. Teng, S. W., H. Y. Tan, J. L. Peng, H. H. Lin, K. H. Kim, W. Lo, Y. Sun, W. C. Lin, S. J. Lin, S. H. Jee, P. T. C. So, and C. Y. Dong. 2006. Multiphoton autofluorescence and second-harmonic generation imaging of the *ex vivo* porcine eye. *Invest Ophthalmol Vis Sci* 47:1216–1224.

61. Han, M., G. Giese, and J. F. Bille. 2005. Second harmonic generation imaging of collagen fibrils in cornea and sclera. *Opt Express* 13:5791–5797.

62. Morishige, N., T. Nishida, and J. V. Jester. 2009. Second harmonic generation for visualizing 3-dimensional structure of corneal collagen lamellae. *Cornea* 28:S46–S53.

63. Morishige, N., A. J. Wahlert, M. C. Kenney, D. J. Brown, K. Kawamoto, T. Chikama, T. Nishida, and J. V. Jester. 2007. Second-harmonic imaging microscopy of normal human and keratoconus cornea. *Invest Ophthalmol Vis Sci* 48:1087–1094.

64. Tan, H. Y., Y. Sun, W. Lo, S. W. Teng, R. J. Wu, S. H. Jee, W. C. Lin et al. 2007. Multiphoton fluorescence and second harmonic generation microscopy for imaging infectious keratitis. *J Biomed Opt* 12:024013.

65. Tan, H. Y., Y. Sun, W. Lo, S. J. Lin, C. H. Hsiao, Y. F. Chen, S. C. M. Huang, W. C. Lin, S. H. Jee, and H. S. Yu. 2006. Multiphoton fluorescence and second harmonic generation imaging of the structural alterations in keratoconus *ex vivo. Invest Ophthalmol Vis Sci* 47:5251–5259.

66. Ahmed, N. 2005. Advanced glycation endproducts—Role in pathology of diabetic complications. *Diabetes Res Clin Pract* 67:3–21.

67. Hadley, J., N. Malik, and K. Meek. 2001. Collagen as a model system to investigate the use of aspirin as an inhibitor of protein glycation and crosslinking. *Micron* 32:307–315.

68. Kim, B. M., J. Eichler, K. M. Reiser, A. M. Rubenchik, and L. B. Da Silva. 2000. Collagen structure and nonlinear susceptibility: Effects of heat, glycation, and enzymatic cleavage on second harmonic signal intensity. *Laser Surg Med* 27:329–335.

69. Hsueh, C. M., W. Lo, W. L. Chen, V. A. Hovhannisyan, G. Y. Liu, S. S. Wang, H. Y. Tan, and C. Y. Dong. 2009. Structural characterization of edematous corneas by forward and backward second harmonic generation imaging. *Biophys J* 97(4):1198–1205.

70. Teng, S. W. et al. Multiphoton fluorescence and second-harmonic-generation microscopy for imaging structural alterations in corneal scar tissue in penetrating full-thickness wound. 2007. *Arch Ophthalmol* 125(7):977–978.

71. Tseng, J. Y. et al. 2011. Multiphoton spectral microscopy for imaging and quantification of tissue glycation. *Biomed Opt Express* 2(2):218–230.

13

Multiphoton Excitation Imaging of the Arterial Vascular Bed

Edward B. Neufeld
National Institutes of Health

Bertrand M. Lucotte
National Institutes of Health

Robert S. Balaban
National Institutes of Health

13.1 Introduction

Multiphoton excitation microscopy provides a powerful means to study the macromolecular microstructure of the arterial vascular bed. The multiphoton effect restricts excited light to the focal spot, allowing optical sectioning without the need for a confocal pinhole. This provides greatly improved photon collection efficiency (compared with single-photon confocal imaging), while maintaining three-dimensional submicrometric spatial resolution. Moreover, owing to the reduced tissue scattering of near-infrared excitation beams, deeper tissue penetration with minimal tissue damage can be attained. Though second- and higher-order harmonic generation (HG) is the major focus of this book, other nonlinear optical contrast mechanisms are simultaneously generated such as multiphoton excitation fluorescence (MEF), and with a second excitation laser, coherent anti-Stokes Raman scattering (CARS). These nonlinear optical microscopy techniques can be used to record three-dimensional, fully registered images of the major macromolecular elements of the arterial wall without stains or dyes. These three readout mechanisms, together with appropriate exogenous probes, can provide a wealth of information concerning the three major elements of the diseased vascular wall, collagen (HG), elastin (MEF), and fat (CARS). These imaging technologies provide *en face* imaging of wall structures relative to the blood, allowing for improved evaluation of interactions of wall structures with the vascular space, and, in specialized cases, can provide information *in vivo*.

In this chapter, we review (i) the fundamental properties of the different nonlinear optical technologies used to study the arterial wall; (ii) their application to identify the different macromolecular microstructures that comprise the arterial vascular bed; and (iii) the insights gained from these studies regarding the role that these microstructures play in arterial health and disease. In addition to second harmonic generation (SHG), we will review these other multiphoton excitation schemes since the information they provide is critical for the interpretation of vascular wall images.

13.2 Methodology and Preparations

13.2.1 Nonlinear Optical Applications in the Vessel Wall

As mentioned in this chapter's Introduction, optical sectioning in multiphoton microscopy [1,2] allows improved collection efficiency over single-photon confocal microscopy for several reasons. First, the primary excitation light at a lower energy (generally IR) is less susceptible to scattering and absorbance, which allows deeper tissue penetration. Second, the multiphoton effect intrinsically restricts emission to the focal spot and maintains optical sectioning without confocal detection. Thus, every emitted photon that escapes from the tissue can be attributed to the focal spot, and thereby participates in image formation [4,5]. From an image reconstruction perspective, the *en face* imaging capability of multiphoton excitation techniques allows a unique 3D visualization of the vessel wall relative to the blood space that is far more informative than the multiple perpendicular slices obtained in conventional histological studies. As discussed in this chapter's Introduction, the microstructural components in the normal and diseased arterial wall are particularly well suited to multiphoton excitation imaging schemes since they are highly scattering, thereby compromising high-energy, single-photon excitation studies. In addition, these primary macromolecules of clinical interest can be detected without exogenous probes, fixation, or tissue sectioning by MEF, HG, or CARS. Such a multimodal imaging scheme is illustrated in Figure 13.1 [3], where the various components of the multiphoton emission spectrum of porcine skin are shown. MEF and HG are routinely used in vascular studies, with CARS quickly gaining interest as a means of detecting lipids. Thus, we believe a review of these three approaches is warranted in the discussion of vascular wall imaging.

13.2.1.1 Multiphoton Excitation Fluorescence

MEF is an incoherent process that involves the absorption of two or more photons and the re-emission of a single photon with a spectral density similar to that in single-photon excitation (see Figure 13.2).

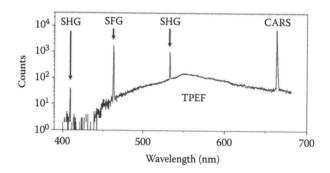

FIGURE 13.1 A representative multiphoton emission spectrum of porcine skin generated using two spatially and temporally overlapped pulsed light sources with $\lambda_1 = 816.7$ nm and $\lambda_2 = 1064$ nm. The four sharp spectral lines were resulted from SHG (408.4 and 532 nm), SFG (462 nm), and CARS (663 nm), respectively, whereas the broad spectral feature on which the four sharp lines are superimposed was from the TPEF. (Reproduced from Jhan, J. W. et al. 2008. Integrated multiple multi-photon imaging and Raman spectroscopy for characterizing structure-constituent correlation of tissues. *Opt Express* 16:16431–16441. With permission of Optical Society of America.)

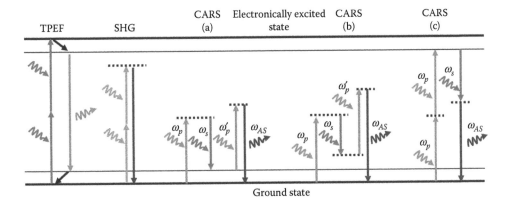

FIGURE 13.2 Energy diagram of TPEF, SHG, and CARS. CARS (a) resonant contribution, (b) nonresonant contribution, and (c) two-photon-enhanced nonresonant contribution.

The major intrinsic probe observed in the arterial wall is elastin, which apparently has multiple fluorophores. This will be discussed in more detail in Section 13.4.1. Conventional organic fluorophores (e.g., Cy-, Alexa) have been used to study the vascular endothelium, but suffer from photobleaching and broad emission spectra, which cause cross-talk between channels. We have labeled blood vessels *i.v.* with di-8-ANEPPS [4], a fluorescent dye that is restricted to the blood compartment, to outline vessel diameters. In a similar manner, many have used fluorescent dextrans [5]. Semiconductor quantum dots (QD) on the other hand provide many advantages compared to conventional organic dyes. They are much brighter due to their large action cross section, which can be up to three orders of magnitude higher than with conventional probes. In addition, increased contrast and sample viability is achieved because of the increased multiphoton excitation probability of QD compared to the autofluorescence background associated with intrinsic probes in tissues [6,7]. Detection sensitivity is further increased due to the large Stokes shift of QD, up to 300–400 nm, which helps to spectrally resolve QD signal from the autofluorescence background [8]. Real-time visualization of single-molecule movement in single living cells was demonstrated by Dahan et al. [9], an extremely difficult task to achieve with organic dyes. QD are more resistant to photobleaching, typically a 100-fold, and can be attached to antibodies for specific labeling [10]. QD have a broad absorption spectrum that allows single-wavelength excitation of multicolored QD and have a narrow emission spectrum, which can be tuned with varying particle size and chemical composition [10]. This facilitates spectral unmixing and improves the sensitivity of the fluorescence quantitation [11]. Polyethylene glycol (PEG)-coated QD have been shown to remain in blood vessels for extended periods, their half-life is ~3 h as opposed to a few minutes for conventional probes, and are therefore highly suitable for imaging the arterial wall [6]. A limiting feature of QD for biological labeling and other applications is their irregular temporal fluorescence fluctuations. This blinking process reduces their quantum yield (the ratio of emitted to absorbed photons) and is the object of widespread studies that generally aim to suppress it [10,12–14]. Potentially toxic effects of semiconductor QDs need to be further studied. Although some results have shown QD with stable polymer coatings to be nontoxic to cells and animals [15], all engineered QDs cannot be considered alike and their toxicity will have to be evaluated individually [16].

13.2.1.2 Harmonic Generation

HG is a coherent process that involves the destruction of two photons that are scattered into a single photon at the second harmonic frequency (see Figure 13.2). The reader is referred to other chapters in this book for a detailed description of HG (see also [17]). Within the vessel wall, the major source of HG is collagen, which is a major structural component. CARS microscopy (described in the next section) uses two laser beams, and thus concurrently generates two SHG signals. Each SHG signal provides redundant data, which may increase the signal-to-noise ratio. The CARS setup also allows sum frequency generation (SFG) between the two laser lines. SFG refers to the process of scattering of two photons at different frequencies ω_1 and ω_2 into a photon at the frequency $\omega_1 + \omega_2$. Though similar to SHG, SFG can provide higher signals from noncentrosymmetric molecules like collagen [18].

13.2.1.3 Coherent Anti-Stokes Raman Scattering

CARS is a nonlinear optical imaging method that exhibits chemical specificity by probing the vibrational mode of a molecule [17,19]. Excellent reviews of the various techniques and processes involved in CARS microscopy have been published; see, for example, [20–22]. We summarize here the fundamental aspects and advantages of this imaging modality, which was described in more detail in Chapter 4. In addition to the benefits associated with nonlinear optical microscopy mentioned above, CARS presents several other advantages for biological studies [19,22]: (1) Vibrational contrast allows imaging with chemical specificity and suppresses the need for staining. Hence, the risk of perturbing the biological function of the molecule under study with a fluorescent probe is removed. Note that SFG also has vibrational contrast but only with surface sensitivity rather than volume and surface sensitivity for forward-CARS (F-CARS) and epi-CARS (E-CARS), respectively [20,21]; (2) Similar to harmonic processes, there is no photobleaching or photodamage at reasonable laser powers [22] since there are no transitions to an electronically excited state; (3) There is reduced saturation compared with MEF imaging, which can saturate at moderate power levels. This is due to the quasi-instantaneous nature of the nonresonant scattering process, which has a decay time of <1 ps, as opposed to the relatively long fluorescence lifetimes of the order of several nanoseconds; (4) Coherent summation of the CARS fields from the sample volume results in a quadratic signal increase with the number of oscillators, in contrast to the linear increase obtained with incoherent processes such as MEF or spontaneous Raman scattering. This can be problematic at low concentrations since the sensitivity equally reduces quadratically with the number of oscillators; (5) Resonant CARS is highly directional, a feature that improves collection efficiency and allows real-time imaging. Real-time measurements of water diffusion rate in living cells using CARS imaging have been reported [23]; and (6) Because the CARS signal is blue shifted, it can be easily separated from the fluorescence.

CARS is a third-order process in which the interaction of a pump field at the frequency ω_p and a red-shifted Stokes field at the frequency ω_s results in the emission of an anti-Stokes field at the frequency $\omega_{as} = 2\omega_p - \omega_s$. When the beat frequency $\omega_p - \omega_s$ is tuned to a Raman-active vibrational band Ω of the probed molecule, the CARS signal is enhanced by at least five orders of magnitude compared to spontaneous Raman scattering [21]. Thus, CARS microscopy provides vibrational contrast and allows imaging with chemical specificity. Unfortunately, CARS images are not background free and are degraded by the presence of a nonresonant signal from the solvent that effectively limits the sensitivity (see the energy diagram in Figure 13.2). The CARS signal intensity I_{CARS} is proportional to the third-order susceptibility $\chi^{(3)}$ of the molecule, $I_{CARS} \propto |\chi^{(3)}(\delta)|^2 I_p^2 I_s$, where $\delta = \Omega - (\omega_p - \omega_s)$ is the detuning from the vibrational band Ω, and I_p, and I_s refer to the intensity of the pump and Stokes beams, respectively. $\chi^{(3)}(\omega)$ is a complex quantity, which is the sum of a resonant component $\chi_R^{(3)}(\delta)$ and a constant nonresonant component $\chi_{NR}^{(3)}$ in the absence of two-photon electronic resonance (see the energy diagram in Figure 13.2). The CARS intensity is therefore proportional to [24–26]

$$I_{CARS} \propto \left|\chi_R^{(3)}(\delta)\right|^2 + \left(\chi_{NR}^{(3)}\right)^2 + 2\mathrm{Re}\left[\chi_R^{(3)}(\delta)\right]\chi_{NR}^{(3)} \tag{13.1}$$

The first and second terms represent the Lorentzian line signal and an offset, respectively. The third term is due to the interference between the resonant and nonresonant components and causes the CARS spectrum to disperse [21,24]. These components are shown in Figure 13.3 and explain the well-known peak and dip of the CARS spectrum.

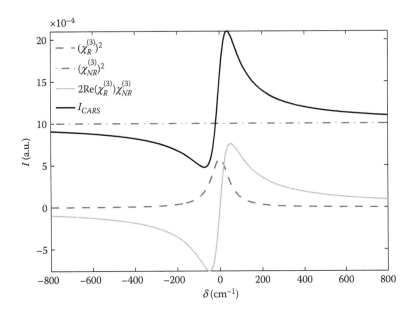

FIGURE 13.3 Spectrum of the CARS signal and of its different components for a single Raman-active line.

Efforts to remove the nonresonant background have led to the development of various methods, polarization-sensitive detection, time-resolved CARS [27], epidetection, but at the expense of an attenuation of the resonant signal [20–22]. This disadvantage was suppressed in a method proposed in Ref. [24], where an intermediate image Δ *is* formed by subtracting the image I_- at the dip frequency from the image I_+ at the peak frequency. The image I_{bg} of the surrounding medium is recorded and the corrected image is obtained by calculating the image $\Delta I/\sqrt{I_{bg}}$. However, this method increases the overall acquisition time since two images must be recorded. Ganikhanov et al. proposed an efficient method to rapidly acquire the image ΔI using frequency-modulated CARS (FM-CARS) [25]. In FM-CARS, the vibrational band probed is frequency-modulated using a Pockels cell to rapidly switch between two pump beams at the peak and dip frequencies of the CARS spectrum. Highly sensitive detection of the resulting amplitude-modulated CARS signal is then performed after amplitude demodulation with a lock-in amplifier. Sensitivity improvements by three orders of magnitude over conventional CARS have been reported. Simultaneous sensing of two Raman bands was also proposed as a means to suppress the nonresonant background [26], where this is achieved by calibrating the nonresonant background ratio between the two CARS signals with a sample that exhibits no vibrational resonance within these two bands. Contrast improvements by a factor of three were reported. Similar to dual CARS, differential CARS uses the probing of two Raman bands to obtain background-free images. This is efficiently achieved using a single detector and a single pump laser however [28]. A replica of the pump-Stokes pulse train is delayed by half the repetition period and its beat frequency is adjusted by mean of glass dispersion. The differential-CARS (D-CARS) image between these two Raman bands is then extracted after lock-in amplification.

While HG and MEF can readily be recorded with the same femtosecond pulsed laser, CARS necessitates two tightly synchronized picosecond pulsed lasers at different frequencies. Picosecond pulses are preferable to femtosecond pulses for generating CARS because the spectral width of a picosecond pulse matches that of a Raman line. The energy is thus concentrated on the Raman line width, which helps improve the resonant to nonresonant background ratio. Precise temporal synchronization between the pump and Stokes pulses must be attained to create the multiphoton effect and can be achieved with adjustable delay lines.

Strong CARS signal is generated at the C–H and CH_2 stretch vibrational resonance, $\omega_p - \omega_s = 2840$ cm^{-1}, in lipid-rich cells and extracellular lipid droplets present in atherosclerotic plaques because of their rich content of these bonds [29]. CARS microscopy is easily amenable to multimodal imaging via the SFG generated between its pump and Stokes beams as demonstrated in spinal tissues [18]. Addition of a femtosecond pulsed laser enables to visualize collagen with HG and elastin with MEF and therefore cover the three major components of the arterial wall [29]. It is important to note that elastin and collagen are also rich in CH_2 bonds and that these components will contribute to the CARS signal when imaging lipid-rich cells [30]. While this may prevent accurate quantification of these three components, multimodal microscopy including CARS, MEF, and SHG is potentially an invaluable tool for image-based diagnosis [31].

CARS could also prove useful in characterizing water diffusion across the arterial wall. CARS imaging of water diffusion can be achieved using D_2O as a contrast agent due to the 10-fold increase in CARS intensity at 3220 cm^{-1} of the O–H stretch vibration over that of the O–D stretch vibration. The O–D stretch vibration on the other hand exhibits two broad peaks at 2385 and 2515 cm^{-1}. Single-cell water dynamics were observed in real time with a CARS microscope by recording intracellular H_2O/D_2O exchanges [23]. The diffusion coefficient across the cell and the membrane permeability were estimated by fitting a water diffusion model to the spatiotemporal evolution of the CARS signal from a cell during D_2O perfusion.

13.3 Arterial Sample Preparations

13.3.1 *In Vitro*

Excised arteries have been mounted in perfusion chambers that allow the application of intraluminal pressure, thereby providing a means to systematically study effects of pressure on arterial wall structure and function *in vitro* [32]. In such studies, the arterial wall is imaged from the adventitia (outer wall); thus, arterial wall thickness limits the imaging depth, restricting the use of this method to vessels with small thickness (i.e., mouse arteries). Many studies have imaged excised arterial samples *en face*, either as (i) aortic ring preparations, allowing cross-sectional imaging from the luminal face to the adventitia, or (ii) longitudinally cut vessels, allowing imaging from the luminal surface down to a depth of ~200 μm. Arterial preparations may be imaged *in vitro* with or without chemical fixation. Most studies use fresh samples maintained in saline or various physiological buffers. In our experience, formalin or formaldehyde fixation does not alter the collagen HG and elastin MEF signal amplitude. However, cross-links in tissue components produced by chemical fixation can generate autofluorescent signals not present in fresh tissue that can interfere with the collagen SHG and elastin autofluorescence signals [33]. Thus, it is prudent to compare imaging of fresh and fixed tissues to assure that no artifactual fluorescent signals are generated by chemical fixation.

13.3.2 *In Vivo*

Observation of the dynamic subcellular processes in their normal physiological environment is highly desirable and can be achieved with *in vivo* microscopy [34–36]. A major complication of high-resolution *in vivo* imaging is physiological tissue motion, which clearly limits the resolution and, to some extent, signal-to-noise ratio by effectively limiting signal averaging. Physiological motions are linked to cardiac and respiratory activity and may be enhanced by fluid redistribution during physiological perturbations [4,35]. These complications have limited the application of multiphoton microscopy to *in vivo* vascular studies.

One solution to address this issue is to mechanically immobilize thin tissues. In a recent study of atherosclerotic plaques *in vivo*, Yu et al. physically constrained surgically exposed carotid arteries in living mice between a stainless-steel vessel holder and a cover glass [36]. In this model of atherosclerosis,

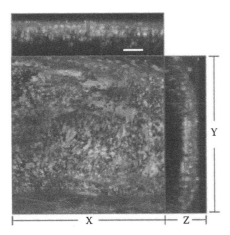

FIGURE 13.4 (**See color insert.**) Atherosclerotic lesion in a ligated mouse carotid artery. Elastin autofluorescence is green, lipid droplets labeled with Nile Red is yellowish green, and collagen SHG is red. Scale bar: 40 μm. (Reproduced from Yu, W. M. et al. 2007. *J Biomed Opt* 12. With the permission of Lippincott Williams & Wilkins.)

lesions are induced by ligation of the carotid artery in apolipoprotein E-deficient mice 2 weeks prior to imaging. The induced lesions had increased collagen content and lipid deposition as shown in Figure 13.4 (collagen (SHG, red), elastin (two-photon excitation fluorescence, green), and lipid droplets labeled with Nile Red (TPEF, yellow)). As in the *in vitro* perfused preparations [32], imaging the clinically interesting intimal surface requires imaging through the adventitia, limiting the approach to small arteries (see Figure 13.5 and Section 13.4 for a description of the vessel wall layer). Moreover, blood flowing through the vessel completely eliminates any possibility of observing the opposing wall. Although these constrained preparations have provided new insights, it is desirable to observe unperturbed tissue *in vivo*.

Dynamic microscopy is an emerging technology that allows imaging of tissue *in vivo* in its natural undisturbed state. This method can reduce blur or distortion-induced motion by (i) maximizing the acquisition rate, (ii) synchronizing the image trigger with the respiratory and cardiac cycles [35], and (iii) fast tracking of image features, using closed-loop schemes for stage control [4]. Such schemes rely on a metric of displacement, such as the normalized cross correlation, between a reference image/Z-stack and the updated image/Z-stack. Real-time tracking can be achieved with the use of a dedicated graphics processing unit. Motion-corrected *in vivo* images of mouse tibialis anterior muscle and associated capillaries were recently obtained using a motion tracking stage in free-breathing anesthetized mice. The lower leg was immobilized by clamping above the foot without obstructing proximal blood flow and the skin and surface fascia over the muscle were removed. Two channel images were recorded and consisted of intrinsic muscle fiber NAD(P)H fluorescence and the dye di-8-ANEPPS to visualize vasculature endothelium, centered at 460 and 590 nm, respectively. The spatial information in the vascular channel was used for tracking 3D motion on the order of 20 μm/min and increased image sharpness was clearly demonstrated in both channels [4]. With cyclic triggering, data are only collected during short acquisition times periodically. An improvement over cyclic triggering is adaptive motion filtering wherein the image is collected with continuous, adaptive precompensation for the periodic motion. The pseudoperiodic motion is first analyzed via conventional tracking. Once learned, compensation is applied and regularly updated. The coupling of a dynamic microscope with adaptive filtering to compensate for the repetitive motion associated with the pulse pressure and respiratory motion may overcome this major limitation in the future.

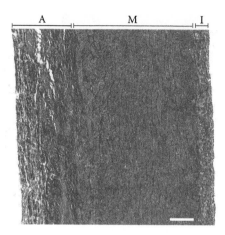

FIGURE 13.5 **(See color insert.)** Structure of the porcine arterial wall. I, intima; M, media; A, adventitia. 25× magnification, scale bar = 400 µm.

13.4 Investigation of Macromolecular Structure

The arterial wall is composed of three layers: the intima, media, and adventitia (see Figure 13.5). The intima consists of (i) a layer of thin endothelial cells lining the lumen of the vessel and its underlying basement membrane, below which is (ii) a layer of connective tissue containing fibroblasts, macrophages, and lymphocytes embedded in extracellular matrix. The major components of the extracellular matrix (ECM) include collagen, elastin, and proteoglycans. The medial layer of the arterial wall is rich in elastic fibers and smooth muscle cells. The adventitia is composed of collagen and elastic fibers. The intima and media are separated by a thick layer of elastic fibers called the internal elastic lamina. Similarly, the media and adventitia are separated by the external elastic lamina, which is also composed of elastic fibers. The macromolecular structure of collagen, elastin, and proteoglycans determines arterial wall mechanical properties and susceptibility to pathological remodeling such as atherosclerosis. In the sections that follow, we will review the advances that SHG and other nonlinear optical imaging technologies have made in our understanding of the role that macromolecular components of the arterial vascular bed play in these processes.

13.4.1 Elastin

Elastin fibers assemble to form layers that provide long-range deformability and passive recoil to arteries and are the major determinant of vascular resilience. Elastin fibers are composed of an amorphous core of the hydrophobic protein elastin surrounded by a mantle of fibrillin-rich microfibrils. Elastin amino acids are nonpolar (60%), and lysine-residue-linked pyridinoline groups form covalent cross-links between the chains. Pyridinoline groups exhibit ~400 nm emission maximum when excited in the UV [37] and may be responsible for elastin two-photon excitation fluorescence [38]. Elastin powder exhibits an emission maximum at ~480 nm that is also seen in spectra obtained from elastin fibers in human skin and may correspond to pyridinoline aggregates [38]. Kwon et al. reported that fluorescence emission of elastin shifted with excitation frequency, suggesting that multiple chromophores are responsible for elastin fluorescence emission [39].

Optical signal intensities from collagen and elastin can be optimized and spectrally separated (see Figure 13.6) using an appropriate excitation wavelength and bandpass filters for detection [29,39–42]. An excitation wavelength of 840–860 nm appears to be ideal insofar as it (i) is above the lower end of SHG detection, (ii) is at the upper limit of elastin detection, (iii) optimizes depth of penetration in the tissue, and (iv) reduces photodamage due to adsorption.

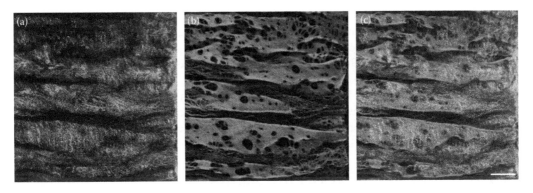

FIGURE 13.6 (**See color insert**.) Porcine renal artery wall. *En face* view of 3D reconstructed artery. (a) Collagen SHG. (b) Elastin autofluorescence. (c) Merged image. Scale bar = 50 μm.

Several investigators have used two-photon excited elastin autofluorescence to quantify various physical characteristics of arterial elastic laminae. Megens et al. found the size (1.2 versus 2.1 μm) and density (0.045 versus 0.57 μm^{-2}) of internal elastic lamina fenestrae differed between murine elastic (carotid) and muscular (uterine and mesenteric) arteries [32]. Kwon et al. found holes in surface elastin in the porcine carotid artery with diameters that ranged in size from 3 μm (circular) to 25 × 9 μm (oval) with a density of 0.002–0.004 μm^{-2} [39].

13.4.2 Collagen

Collagen fibers in the arterial wall provide tensile strength and thereby are a major determinant of vascular integrity. In the many types of collagen, including the fibril-forming collagens Types I, II, III, V, and XI, three protein chains align to form triple helical molecules. Secreted triple helical molecules (~1.5 nm diameter and 300 nm long) spontaneously assemble into five-stranded microfibrils (~4.5 nm diameter). The microfibers aggregate end to end and laterally in a radial pattern to form fibrils (10–100 nm diameter), which then cluster to form much larger fibers (1–100 μm), that are arranged in large bundles readily seen by light microscopy [43].

As mentioned in Section 13.2.1.2, based on its intrinsic properties, collagen can be imaged without the need for fixation or staining. Collagen generates both two-photon excited fluorescence and second harmonic photon emissions. Two-photon excitation of collagen at shorter excitation wavelengths (<800 nm) can generate both SHG and two-photon excited fluorescence (with a weak and broad spectrum), whereas at higher excitation wavelengths, the collagen signal is due to SHG alone [44].

13.4.3 Macromolecular Exogenous Probes

Exogenous fluorescent probes provide an alternate means to detect arterial wall macromolecular microstructures by two-photon excitation microscopy. The collagen-binding adhesion protein 35-Oregon Green 488 (CNA35-OG488) is a fluorescently labeled molecular imaging agent developed for optical imaging of collagen [43]. CNA binds to fibrillar collagen types I, III, and IV [43], all of which are abundant in the arterial wall, and are associated with atherosclerosis and plaque progression. Studies using CNA have provided insights into the sites where diffusion barriers are present in the arterial wall. In healthy muscular arteries, all layers are labeled *ex vivo*; however, in elastic arteries, medial and intimal labeling appears to be prevented by endothelium and elastic laminae [45]. After mechanical damage to the carotid artery, and in blood vessels with discontinuous endothelial vascular coverage (liver, spleen), or fenestrated endothelium (kidney), subendothelial collagen was strongly labeled but not in vessels in organs with continuous endothelium (heart, lungs). This finding is consistent with endothelium and elastic laminae providing a diffusion barrier. Interestingly, atherosclerotic lesions contained large amounts of labeled collagen. Based on these studies, van Zandvoort and colleagues suggest that CNA can be used to detect atherosclerotic lesions in the arterial wall. CNA has also been used to reveal platelet binding to exposed subendothelial collagen in a mouse model of atherothrombosis [46].

Sulforhodamine B has recently been reported to specifically stain elastic fibers intravitally in blood vessels, and to improve elastin fiber detection as a result of the high cross section and quantum yield of the dye with two-photon excitation [47]. The spectral properties of sulforhodamine B allows for simultaneous detection of collagen SHG or other vital dyes.

13.5 Investigations of Vessel Mechanics

13.5.1 Residual Strain

Several studies have investigated the residual strain present in the microstructural components in the arterial wall. Under conditions where no pressure is applied to the arterial vessel wall, the fibers are thicker, resulting in a greater overall thickness of the vessel wall. Residual strain in dissected vessels presents as longitudinal folds, which reflect the residual circumferential stress, or residual strain within the intimal-medial layers (see Figure 13.7). With no intraluminal pressure, the surface of dissected murine aorta is compressed in the radial direction into longitudinal wavy folds, indicating that both radial and longitudinal strains occur in the elastin lamellae [39]. Coiled collagen fibers were arranged radially within the compressed inner folds of the elastin lamellae and between each lamellar collagen unit. Individual collagen fibers were coiled with a mean periodicity of 2–5 μm [39].

Arteries at different anatomical sites and species reveal different spacings in the longitudinal folds [39]. In the mouse, the spacing of the lamellar folds was ~50–100 μm in the upper thoracic aorta and 30–50 μm in the lower abdominal aorta. The porcine carotid artery was compressed in the radial direction into longitudinal wavy folds spaced ~40–80 μm apart. Similar to the murine aorta, collagen fibers in the porcine carotid artery were radially arranged within the inner folds of the elastin layer and coiled with a mean periodicity of 5–10 μm. Residual strain in the porcine coronary artery resulted in longitudinal folds with a spacing of 30–50 μm, implying a dominant radial stress.

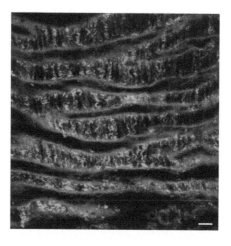

FIGURE 13.7 (**See color insert.**) Collagen and elastin microstructure of the porcine carotid artery. Two-photon image 24 mm below the surface of the porcine carotid intima. Elastin autofluorescence is red; collagen SHG is green. A wavy sheet of elastin with circular holes throughout envelops the luminal surface of the artery. Individual, bunched-up collagen fibrils are radially arranged within the inner folds of the elastin lamellae. Scale bar = 20 μm.

13.5.2 Deformation

The different microstructural components in the vessel wall have different mechanical properties and take up loads at different stress levels. Changes in the structure and organization of the vessel constituents under different mechanical loading conditions have been observed in a nondestructive manner by two-photon excitation microscopy. Zoumi et al. monitored the effect of transmural pressure (0, 30, and 180 mmHg) on excised porcine coronary artery microstructure [41]. With increasing pressure, the fibers became thinner and more elongated. These investigators concluded that the vessel wall is compressed during pressurization, while the elastic lamina is circumferentially stretched. Collagen fiber width, measured by SHG, was ~3, 3.6, 1.3, and 0.9 μm, in the zero-stress (radially cut no-load vessel), no-load, 30 mmHg distension, and 180 mmHg distension, respectively. Interestingly, two-photon excitation microscopy revealed that, with 30 mmHg distension, the fibers were thinner toward the lumen and became thicker toward the outer wall of the vessel, suggesting that the intimal-medial portion of the vessel takes up more of the load than the adventitia. The applied pressure had a more uniform effect throughout the vessel wall at 180 mmHg distension, however, as thin fibers were observed to span the entire wall thickness.

Megens et al. investigated pressure-induced changes in transmural collagen and elastin organization in large elastic and small muscular arteries mounted in a perfusion chamber [32]. After increasing the transmural pressure from 0 to 80 mmHg, the carotid artery thickness decreased from ~57 to ~33 μm, and the elastic laminae appeared to unfold. Collagen imaged by SHG was tortuous in mounted, pressurized elastic arteries (carotid), but appeared more stretched out in similarly mounted muscular arteries (uterine).

13.6 Investigations of LDL Interactions with the Vascular Wall

Low-density lipoprotein (LDL) retention in the arterial wall plays a pivotal role in the initiation and progression of cardiovascular heart disease. The initial step in atherosclerotic lesion formation is currently thought to involve LDL binding to ECM components (proteoglycans) that are associated with collagen and elastin [48]. During atherosclerotic lesion development, increasing amounts of LDL and LDL-derived cholesterol accumulate in the extracellular space, and in macrophages that ingest matrical LDL. The lipid-rich necrotic core characteristic of advanced lesions renders the vessel prone to rupture and thrombosis formation, with consequent risk of myocardial infarction, stroke, and death. Thus, the interaction of LDL with vascular wall components is a subject of great interest in cardiovascular research. Nonlinear optical microscopy technologies provide powerful tools to study these interactions both *in situ* and *in vivo*. Endogenous LDL in arterial cells and extracellular matrix can be imaged by CARS microscopy, while exogenously added fluorescent-tagged LDL can be imaged by two-photon excitation microscopy.

Our laboratory has used multimodal nonlinear microscopy to identify the atherosclerosis-prone anatomical sites and arterial components that bind exogenous fluorescent LDL. Based on SHG, Kwon et al. found that collagen fibrils are circumferentially organized into a knotted ring surrounding atherosclerosis-prone intervertebral, aortic arch, and coronary artery branch points (see Figure 13.8) [39]. Quantification of superficial collagen content by SHG revealed a 24% increase in collagen density in aortic branch regions relative to the aortic-free wall. Examination of elastin autofluorescence revealed that, unlike the free wall, the branch points lack an elastin layer, thus exposing collagen/proteoglycan complexes. As assessed by two-photon microscopy, (i) the luminal elastin layer limited penetration of fluorescent-tagged probes, whereas its absence at branch points resulted in extensive LDL binding; and (ii) fluorescent LDL colocalized with immunostained proteoglycans. These studies revealed that at atherosclerosis-prone branch points in nondiseased tissue, the absence of a luminal elastin barrier and the presence of a dense collagen/proteoglycan matrix contribute to increased retention of LDL.

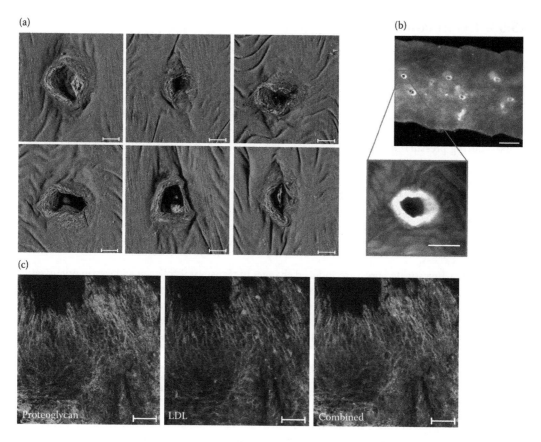

FIGURE 13.8 (**See color insert.**) LDL binds to collagen- and proteoglycan-rich, elastin-poor aortic branch points. (a) Three-dimensional surface reconstructions of atherosclerosis-susceptible intervertebral branch points along the mouse thoracic aorta revealing with a ring of exposed collagen immediately surrounding the ostia. Elastin autofluorescence is red; collagen SHG is green. Scale bar = 50 μm. (b) LDL binding along the mouse aorta detected by epifluorescence. Alexa 647-tagged LDL bound extensively to a circular region around intervertebral branch points. Scale bar = 500 μm. Zoom scale bar = 100 μm. (c) Colocalization analysis of LDL binding at porcine coronary arterial branch points. Two-photon image of LDL (red) colocalization with immunolabeled proteoglycans (green). (Adapted from Kwon, G. P. et al. 2008. *Circulation* 117:2919–2927. With permission from Wolters Kluwer.)

Further, two-photon microscopic studies of nondiseased aortic wall microstructure led to the discovery of a heretofore unidentified anatomical structure, the renal artery ostial diverter (see Figure 13.9) [42]. In these studies, maximum projection images of adjacent z-stacks (~180 μm deep) were tiled together to form a high-resolution, low-magnification, composite *en face* image of the entire renal ostium (~1 mm² area). The renal flow diverter was initially identified, by SHG imaging, as a crescent-shaped region of increased collagen density on the caudal (downstream) side of the aortic entrance to porcine renal artery (renal ostium). Subsequent histological analysis revealed the renal diverter to be a knife-like, intimal collagenous cap on top of a thickened tunica media, which together form a process that protrudes into the aorta. High-resolution color Doppler ultrasound studies *in vivo* confirmed that the structure serves to divert arterial blood flow into the renal artery. Interestingly, the anatomical site where the diverter resides has been shown to be the site of initiation of renal arterial atherosclerosis. Two-photon microscopy revealed that, relative to the cranial (upstream side) of the renal ostium, the diverter has a thin luminal elastin layer, and a 25-fold increase in fluorescent LDL binding, consistent with our observations at other atherosclerosis-prone arterial branch points

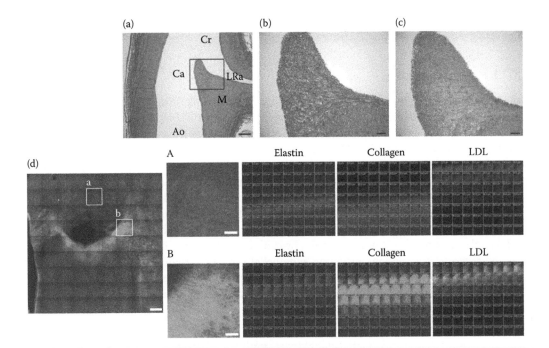

FIGURE 13.9 **(See color insert.)** LDL binds to the atherosclerosis-susceptible renal artery ostial diverter at the aortic entrance to the renal artery. (a, b, c) Histological analysis. Sagital sections of renal diverter stained with Masson's trichrome stain (a, b) for collagen (blue) and smooth muscle (red), and Movat stain (green) for proteoglycans (c). The caudal aspect (Ca) of the renal ostium (a) is thickened, angulated, and more rigid than the cranial aspect (Cr). Both the intimal (rectangle) and medial (*M*) layers of the caudal renal ostium are thickened, forming a flow diverter at the entrance to the renal artery. The thickened intimal layer on the caudal side of the renal ostium forms an elongated cap (rectangle in a). A zone of proteoglycan enrichment (c, green) just below the surface of the caudal ostium surrounds a densely collagenous core (b, blue) in the intimal "fibrous cap"-like structure. (a) Scale bar = 500 μm; (b and c) Scale bars = 10 μm. (d) LDL binding to renal diverter microstructural components assessed by two-photon multimodal microscopy. (d) Tiled (9 × 9) two-photon maximum projection of z-series images through 180 μm of the luminal surface of the aortic entrance to the left renal artery. Merged image. Collagen SHG is green, elastin autofluorescence is red, and Alexa 647-LDL is blue. LDL binds eccentrically to the collagen-enriched, elastin-poor, atherosclerosis-susceptible, caudal side of the renal ostium. The tiles (A) and (B) outlined in the cranial and caudal regions of the renal ostium in (d) are shown to the right in (A) and (B), respectively. Scale bar = 1 mm. (Right) Z-series images at 0–124 μm from the luminal surface of the aortic wall, for tiles (A) and (B), respectively. Elastin autofluorescence is red, collagen SHG is green, and Alexa 647-LDL fluorescence is white. Collagen is enriched in the caudal compared to the cranial ostium. LDL binding is also greatly enhanced in the caudal compared to the cranial ostium. LDL binding to the surface occurs prior to the appearance of collagen at both the cranial and caudal sites, but overlaps to a considerable extent with collagen in the caudal ostium. Scale bars = 250 μm. (Reprinted from *Atherosclerosis* 211, Neufeld, E. B. et al. The renal artery ostium flow diverter: Structure and potential role in atherosclerosis, 153–158, Copyright (2010), with permission from Elsevier Ireland LTD.)

[39], as noted above. These studies have shown that arterial branch points are predisposed to atherosclerotic lesion formation due to the intrinsic macromolecular composition required for structural integrity at these sites.

13.7 Investigations of the Diseased Vascular Wall

Nonlinear optical microscopy has been used to expand our current understanding of disease processes that occur in the arterial wall, including atherosclerotic disease progression, cardiomyopathy, and arterial aneurysm. We will focus only on studies of arterial wall disease that involve HG imaging in conjunction with other imaging modalities.

13.7.1 Atherosclerosis

Atherosclerotic lesion progression is a complex process that invokes overlapping mechanisms that lead to characteristic sequential remodeling of extracellular matrix components in the arterial wall. A number of studies have investigated atherosclerotic disease, including several using multimodal nonlinear optical imaging with CARS microscopy to identify lipids. Initial studies of atherosclerotic lesions established that nonlinear optical imaging methods could detect cells, lipid deposits, and matrical microstructures found in lesions by conventional histological methods [29,36,49–51]. Lilledahl et al. imaged vulnerable plaques in human aortic autopsy samples and found that the fibrous cap emits primarily SHG due to collagen, in contrast to the necrotic core and healthy artery, which emits primarily two-photon excited fluorescence from elastin [49]. Collagen in the cap appeared in thick bundles, in a nondirectional structure. Le et al. imaged unstained components of the arterial wall and atherosclerotic lesions including endothelial cells, extracellular lipid droplets, lipid-rich cells, LDL aggregates, collagen, and elastin using multimodal nonlinear optical microscopy [29]. Collagen fibers in porcine iliac atheromas appeared disordered from a luminal view, and perpendicular to those in the arterial wall from a cross-sectional view. Based on integrated image intensities, collagen density in the atheroma was increased as much as fourfold compared to the arterial wall. Yu et al. imaged collagen (SHG), elastin (autofluorescence), leukocytes (EGFP), cell nuclei (Hoechst 33342), and neutral lipids (Nile Red) in immobilized carotid artery atherosclerotic plaques *in vivo* [36]. Ko et al. reported a loss of the internal elastic lamina and the appearance of scattered collagen and lipid-rich structures (using CARS) in early WHHLMI rabbit atherosclerotic lesions [51]. In advanced plaques, thicker, directional collagen fibers were seen along with increased lipid accumulation. Megens et al. observed adhesion of inflammatory cells to the endothelium and increased intimal collagen labeling with CNA in 15-week-old carotid arterial lesions in apo E−/− mice, consistent with endothelial activation [50]. Parasassi et al. monitored endothelial adhesion and internalization of lipid hydroxyperoxide-containing LDL particles labeled with the lipophilic fluorescent probe 2-dimethylamino-6-lauroyl-naphthalene in rat aorta preparations [52]. In these studies, oxidized LDL was shown to induce fragmentation of autofluorescent matrix fibers in the arterial wall that could be prevented by antioxidants. Although these studies have confirmed histological descriptions of some stages of atherosclerotic lesion development, they have done little to further our understanding of the underlying mechanisms.

Recently, investigators have provided new insights into atherosclerosis lesion progression using multimodal nonlinear optical imaging to systematically assess the cellular and matrical changes that occur during the different stages of the disease [31,53,54]. Atherosclerotic lesions are classified by the American Heart Association as Type (I) initial lesion, (II) fatty streak, (III) intermediate lesion, (IV) atheroma, (V) fibrous atheroma, and (VI) calcific atheroma [55,56]. Lim et al. characterized cellular and structural changes in early stage II/III atherosclerotic plaques in apo E−/− mice using multimodal optical imaging with CARS microscopy to identify lipids [53]. A high-fat, high-cholesterol Western diet increased the intimal plaque area twofold, as defined by CARS signals of lipid-rich macrophages. Quantitative analysis of the collagen SHG signal revealed a nearly fourfold decrease in collagen distribution in the lipid-rich plaque regions (~13% versus ~4%, in standard versus Western diet). Collagen content in the surrounding matrix decreased in a similar manner (~7% versus 3%, in standard versus Western diet). Wang et al. systematically analyzed diet-induced atherosclerotic lesion development (Type I–VII) in porcine iliac arteries by CARS-based multimodal nonlinear microscopy [31]. Foam cells, lipid droplets, collagen, elastin, and fibrous caps were visualized with 3D submicron resolution. All stages of lesion development could be visualized by nonlinear microscopy and correlated with standard histological analysis. Adaptive intimal thickening and foam cell lipid accumulation was seen in early

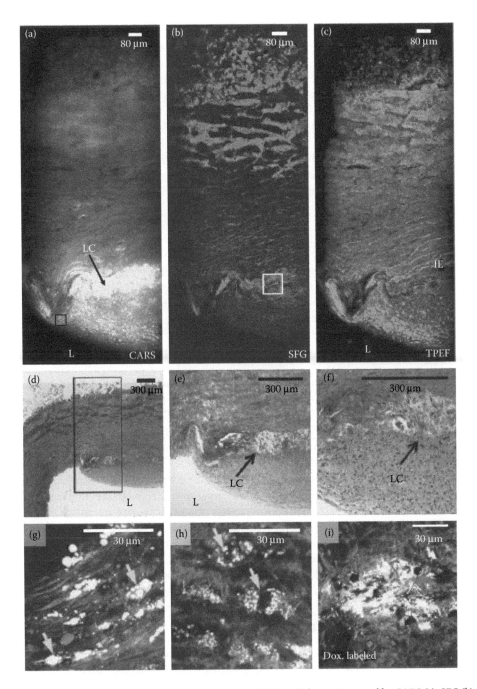

FIGURE 13.10 **(See color insert.)** Representative images of a Type IV lesion inspected by CARS (a), SFG (b), and TPEF (c). The lipid core (LC) was identified by CARS. (d and e) Masson's trichrome staining. Black rectangle: the corresponding area in (a) through (c). (f) H&E staining. (g and h) Colocalized NLO images corresponding to the red and yellow squares in (a), respectively. (i) Doxorubicin-labeled (red) image, around lipid core, colocalized with CARS and SFG signals. (Reproduced from Wang, H. W. et al. 2009. *Arterioscl Throm Vas* 29:1342–1348. With the permission of SPIE.)

lesions (types I and II). Intermediate lesions (type III) contained scattered interstitial lipid pools and further intimal thickening. Disordered collagen fibrils around the lipid pools were detected. Type IV lesions had a well-defined lipid core identified by a strong CARS signal starting at the shoulder of the lesion. Abundant lipid-laden foam cells, identified by CARS and cellular doxorubicin signals, were seen in the shoulder and in luminal regions of the atherosclerotic plaque. Extracellular lipid accumulation was also observed (see Figure 13.10). Type V lesions were identified based on characteristic dense core of lipid (CARS) and surrounding fibrous cap (SHG). Lipid and collagen content of the different lesion types were quantified based on CARS and SHG signals, respectively. Lipid accumulation in thickened intima culminated in Type IV whereas the highest collagen deposition was found in Type V lesions. Recently, Kim et al. have been able to classify four different morphologies of atherosclerotic lipids using multiplex CARS imaging; intracellular and extracellular lipid droplets as well as needle-shaped and plate-shaped lipid crystals [54].

13.7.2 Other Diseases

Human heart valve allografts from patients with cardiomyopathy are often harvested for human valve replacement. Two-photon excitation microscopy and SHG have been used to assess the structural integrity of heart valves harvested from human ischemic (ICM) and dilated cardiomyopthic (DCM) hearts, and from model acute and chronic ischemic porcine hearts [57]. Degradation of collagen bundle structures was seen within the collagen-rich outflow side of human aortic and pulmonary ICM leaflets and was even more pronounced on the inflow side of ICM leaflets. Depletion and disintegration of elastin-containing structures was observed mainly within the outflow side of aortic and pulmonary leaflets of DCM heart valves. Compared with normal porcine valve tissues, acute ICM specimens showed no significant changes in extracellular matrix components. Similar to the changes seen in valves from human ICM and DCM hearts, multiphoton imaging of chronic porcine ICM tissues demonstrated weak collagen SHG and elastin autofluorescence, indicating marked ECM remodeling. Quantification of collagen SHG signals revealed that normal leaflets were approximately six- to sevenfold higher in normal leaflets compared to similar tissues from chronic ICM pigs. Since the total amount of collagen measured biochemically was found to be unaltered in diseased tissues, these investigators suggest that the changes in SHG signal may be due to abnormalities in extracellular matrix architecture. In a similar manner, cryopreservation of heart valves also appears to diminish collagen SHG, suggesting this process causes structural alterations in collagen [58]. Nonlinear imaging has also been used to assess changes in the aortic arch in a rat model of dissecting aortic aneurysm [59]. Decreased SHG signal was seen from medial and adventitial collagen of affected fetal and newborn pups, in the absence of any change in elastin autofluorescence, suggesting that altered collagen structure plays a key role in this model of aortic dissection.

13.8 Summary and Future Directions

Nonlinear optical imaging has provided important new insights into the structure and functional interrelationships of macromolecular microstructures in the arterial vascular bed in both healthy and diseased states. Initial studies led to the development of improved spectral separation of collagen SHG and elastin autofluorescence, and to the application of multimodal nonlinear microscopy to allow imaging of other arterial wall components, including lipid deposits. Currently, essentially all of the components of the arterial wall can now be imaged, including cells, ECM components, and pathological lipid deposits, using a combination of intrinsic signals (collagen SHG, elastin autofluorescence) and exogenous fluorescent probes. Investigations of vessel mechanics have revealed the changes in distribution of extracellular matrix components that occur with changes in transluminal pressure as well as the inherent artifacts in collagen and elastin distribution seen *in vitro*. Studies of LDL interactions with the vascular wall have provided important new insights into the role that macromolecular microstructures play in the initiation of atherosclerosis. Examination of the diseased vascular wall have expanded the repertoire of imaging modalities that can be used to characterize atherosclerotic lesion initiation and progression, as well as other diseases of the wall, including cardiomyopathy and aneurysm formation. Application of the recent advances in *in vivo* nonlinear optical imaging [4] holds promise to allow live, three-dimensional, real-time imaging of normal and pathological processes in the arterial vascular bed in the absence of motion artifacts.

References

1. Denk, W., J. H. Strickler, and W. W. Webb. 1990. Two-photon laser scanning fluorescence microscopy. *Science* 248:73–76.
2. Gan, X. S. and M. Gu. 2000. Spatial distribution of single-photon and two-photon fluorescence light in scattering media: Monte Carlo simulation. *Appl Optics* 39:1575–1579.
3. Jhan, J. W., W. T. Chang, H. C. Chen, Y. T. Lee, M. F. Wu, C. H. Chen, and I. Liau. 2008. Integrated multiple multi-photon imaging and Raman spectroscopy for characterizing structure-constituent correlation of tissues. *Opt Express* 16:16431–16441.
4. Schroeder, J. L., M. Luger-Hamer, R. Pursley, T. Pohida, C. Chefd'Hotel, P. Kellman, and R. S. Balaban. 2010. Subcellular motion compensation for minimally invasive microscopy, *in vivo* evidence for oxygen gradients in resting muscle. *Circ Res* 106:1129–U1271.
5. Helmchen, F. and D. Kleinfeld. 2008. *In vivo* measurements of blood flow and glial cell function with two-photon laser-scanning microscopy. *Method Enzymol* 444:231–254.
6. Larson, D. R., W. R. Zipfel, R. M. Williams, S. W. Clark, M. P. Bruchez, F. W. Wise, and W. W. Webb. 2003. Water-soluble quantum dots for multiphoton fluorescence imaging *in vivo*. *Science* 300:1434–1436.
7. Pantazis, P., J. Maloney, D. Wu, and S. E. Fraser. 2010. Second harmonic generating (SHG) nanoprobes for *in vivo* imaging. *Proc Natl Acad Sci USA* 107:14535–14540.
8. Gao, X. H., L. L. Yang, J. A. Petros, F. F. Marshal, J. W. Simons, and S. M. Nie. 2005. *In vivo* molecular and cellular imaging with quantum dots. *Curr Opin Biotech* 16:63–72.
9. Dahan, M., S. Levi, C. Luccardini, P. Rostaing, B. Riveau, and A. Triller. 2003. Diffusion dynamics of glycine receptors revealed by single-quantum dot tracking. *Science* 302:442–445.
10. Michalet, X., F. F. Pinaud, L. A. Bentolila, J. M. Tsay, S. Doose, J. J. Li, G. Sundaresan, A. M. Wu, S. S. Gambhir, and S. Weiss. 2005. Quantum dots for live cells, *in vivo* imaging, and diagnostics. *Science* 307:538–544.
11. Ferrara, D. E., D. Weiss, P. H. Carnell, R. P. Vito, D. Vega, X. H. Gao, S. M. Nie, and W. R. Taylor. 2006. Quantitative 3D fluorescence technique for the analysis of en face preparations of arterial walls using quantum dot nanocrystals and two-photon excitation laser scanning microscopy. *Am J Physiol-Reg I* 290:R114–R123.
12. Ma, X. D., H. Tan, T. Kipp, and A. Mews. 2010. Fluorescence enhancement, blinking suppression, and gray states of individual semiconductor nanocrystals close to gold nanoparticles. *Nano Lett* 10:4166–4174.
13. Vela, J., H. Htoon, Y. F. Chen, Y. S. Park, Y. Ghosh, P. M. Goodwin, J. H. Werner, N. P. Wells, J. L. Casson, and J. A. Hollingsworth. 2010. Effect of shell thickness and composition on blinking suppression and the blinking mechanism in "giant" CdSe/CdS nanocrystal quantum dots. *J Biophotonics* 3:706–717.
14. Pelton, M., G. Smith, N. F. Scherer, and R. A. Marcus. 2007. Evidence for a diffusion-controlled mechanism for fluorescence blinking of colloidal quantum dots. *Proc Natl Acad Sci USA* 104:14249–14254.
15. Ballou, B., B. C. Lagerholm, L. A. Ernst, M. P. Bruchez, and A. S. Waggoner. 2004. Noninvasive imaging of quantum dots in mice. *Bioconjugate Chem* 15:79–86.
16. Hardman, R. 2006. A toxicologic review of quantum dots: Toxicity depends on physicochemical and environmental factors. *Environ Health Perspect* 114:165–172.
17. Boyd, R. W. 2008. *Nonlinear Optics*. Elsevier.

18. Fu, Y., H. F. Wang, R. Y. Shi, and J. X. Cheng. 2007. Second harmonic and sum frequency generation imaging of fibrous astroglial filaments in *ex vivo* spinal tissues. *Biophys J* 92:3251–3259.

19. Zumbusch, A., G. R. Holtom, and X. S. Xie. 1999. Three-dimensional vibrational imaging by coherent anti-Stokes Raman scattering. *Phys Rev Lett* 82:4142–4145.

20. Cheng, J. X. and X. S. Xie. 2004. Coherent anti-Stokes Raman scattering microscopy: Instrumentation, theory, and applications. *J Phys Chem B* 108:827–840.

21. Volkmer, A. 2005. Vibrational imaging and microspectroscopies based on coherent anti-Stokes Raman scattering microscopy. *J Phys D Appl Phys* 38:R59–R81.

22. Cheng, J.-X. 2007. Coherent anti-Stokes Raman scattering microscopy. *Appl Spectrosc* 61:197A–208A.

23. Potma, E. O., W. P. de Boeij, P. J. M. van Haastert, and D. A. Wiersma. 2001. Real-time visualization of intracellular hydrodynamics in single living cells. *Proc Natl Acad Sci USA* 98:1577–1582.

24. Li, L., H. F. Wang, and J. X. Cheng. 2005. Quantitative coherent anti-Stokes Raman scattering imaging of lipid distribution in coexisting domains. *Biophys J* 89:3480–3490.

25. Ganikhanov, F., C. L. Evans, B. G. Saar, and X. S. Xie. 2006. High-sensitivity vibrational imaging with frequency modulation coherent anti-Stokes Raman scattering (FM CARS) microscopy. *Opt Lett* 31:1872–1874.

26. Burkacky, O., A. Zumbusch, C. Brackmann, and A. Enejder. 2006. Dual-pump coherent anti-Stokes-Raman scattering microscopy. *Opt Lett* 31:3656–3658.

27. Ly, S., G. McNerney, S. Fore, J. Chan, and T. Huser. 2007. Time-gated single photon counting enables separation of coherent anti-Stokes Raman scattering (CARS) microscopy data from multiphoton-excited tissue autofluorescence. *Abstr Pap Am Chem S* 234 :16839–16851.

28. Rocha-Mendoza, I., W. Langbein, P. Watson, and P. Borri. 2009. Differential coherent anti-Stokes Raman scattering microscopy with linearly chirped femtosecond laser pulses. *Opt Lett* 34:2258–2260.

29. Le, T. T., I. M. Langohr, M. J. Locker, M. Sturek, and J. X. Cheng. 2007. Label-free molecular imaging of atherosclerotic lesions using multimodal nonlinear optical microscopy. *J Biomed Opt* 12 (5):1–20.

30. Wang, H. W., T. T. Le, and J. X. Cheng. 2008. Label-free imaging of arterial cells and extracellular matrix using a multimodal CARS microscope. *Opt Commun* 281:1813–1822.

31. Wang, H. W., I. M. Langohr, M. Sturek, and J. X. Cheng. 2009. Imaging and quantitative analysis of atherosclerotic lesions by CARS-based multimodal nonlinear optical microscopy. *Arterioscl Throm Vas* 29:1342–1348.

32. Megens, R. T. A., S. Reitsma, P. H. M. Schiffers, R. H. P. Hilgers, J. G. R. De Mey, D. W. Slaaf, M. G. A. O. Egbrink, and M. A. M. J. van Zandvoort. 2007. Two-photon microscopy of vital murine elastic and muscular arteries—Combined structural and functional imaging with subcellular resolution. *J Vasc Res* 44:87–98.

33. Schenke-Layland, K. 2008. Non-invasive multiphoton imaging of extracellular matrix structures. *J Biophotonics* 1:451–462.

34. Bouchard, M., S. Ruvinskya, D. Boas, A., and E. Hillman, M. 2006. *Video-Rate Two-Photon Microscopy of Cortical Hemodynamics In Vivo*. Optical Society of America. MI1.

35. Megens, R. T. A., S. Reitsma, L. Prinzen, M. G. A. O. Egbrink, W. Engels, P. J. A. Leenders, E. J. L. Brunenberg et al. 2010. *In vivo* high-resolution structural imaging of large arteries in small rodents using two-photon laser scanning microscopy. *J Biomed Opt* 15 (1):1–10.

36. Yu, W. M., J. C. Braz, A. M. Dutton, P. Prusakov, and M. Rekhter. 2007. *In vivo* imaging of atherosclerotic plaques in apolipoprotein E deficient mice using nonlinear microscopy. *J Biomed Opt* 12 (5):1–10.

37. Bridges, J. W., D. S. Davies, and R. T. Williams. 1966. Fluorescence studies on some hydroxypyridines including compounds of vitamin B6 group. *Biochem J* 98:451–468.

38. Zipfel, W. R., R. M. Williams, R. Christie, A. Y. Nikitin, B. T. Hyman, and W. W. Webb. 2003. Live tissue intrinsic emission microscopy using multiphoton-excited native fluorescence and second harmonic generation. *Proc Natl Acad Sci USA* 100:7075–7080.

39. Kwon, G. P., J. L. Schroeder, M. J. Amar, A. T. Remaley, and R. S. Balaban. 2008. Contribution of macromolecular structure to the retention of low-density lipoprotein at arterial branch points. *Circulation* 117:2919–2927.

40. Boulesteix, T., A. M. Pena, N. Pagès, G. Godeau, M. P. Sauviat, E. Beaurepaire, and M. C. Schanne-Klein. 2006. Micrometer scale *ex vivo* multiphoton imaging of unstained arterial wall structure. *Cytometry Part A* 69A:20–26.

41. Zoumi, A., X. A. Lu, G. S. Kassab, and B. J. Tromberg. 2004. Imaging coronary artery microstructure using second-harmonic and two-photon fluorescence microscopy. *Biophys J* 87:2778–2786.

42. Neufeld, E. B., Z. X. Yu, D. Springer, Q. Yu, and R. S. Balaban. 2010. The renal artery ostium flow diverter: Structure and potential role in atherosclerosis. *Atherosclerosis* 211:153–158.

43. Krahn, K. N., C. V. C. Bouten, S. van Tuijl, M. A. M. J. van Zandvoort, and M. Merkx. 2006. Fluorescently labeled collagen binding proteins allow specific visualization of collagen in tissues and live cell culture. *Anal Biochem* 350:177–185.

44. Zoumi, A., A. Yeh, and B. J. Tromberg. 2002. Imaging cells and extracellular matrix *in vivo* by using second-harmonic generation and two-photon excited fluorescence. *Proc Natl Acad Sci USA* 99:11014–11019.

45. Megens, R. T. A., M. G. A. O. Egbrink, J. P. M. Cleutjens, M. J. E. Kuijpers, P. H. M. Schiffers, M. Merkx, D. W. Slaaf, and M. A. M. J. van Zandvoort. 2007. Imaging collagen in intact viable healthy and atherosclerotic arteries using fluorescently labeled CNA35 and two-photon laser scanning microscopy. *Mol Imaging* 6:247–260.

46. Kuijpers, M. J. E., K. Gilio, S. Reitsma, R. Nergiz-Unal, L. Prinzen, S. Heeneman, E. Lutgens et al. 2009. Complementary roles of platelets and coagulation in thrombus formation on plaques acutely ruptured by targeted ultrasound treatment: A novel intravital model. *J Thromb Haemost* 7: 152–161.

47. Ricard, C., J. C. Vial, J. Douady, and B. van der Sanden. 2007. *In vivo* imaging of elastic fibers using sulforhodamine B. *J Biomed Opt* 12(6):1–8.

48. Nakashima, Y., T. N. Wight, and K. Sueishi. 2008. Early atherosclerosis in humans: Role of diffuse intimal thickening and extracellular matrix proteoglycans. *Cardiovasc Res* 79:14–23.

49. Lilledahl, M. B., O. A. Haugen, C. D. Davies, and L. O. Svaasand. 2007. Characterization of vulnerable plaques by multiphoton microscopy. *J Biomed Opt* 12(4):1–12.

50. Megens, R. T. A., M. G. A. O. Egbrink, M. Merkx, D. W. Slaaf, and M. A. M. J. van Zandvoort. 2008. Two-photon microscopy on vital carotid arteries: Imaging the relationship between collagen and inflammatory cells in atherosclerotic plaques. *J Biomed Opt* 13(4):1–10.

51. Ko, A. C. T., A. Ridsdale, M. S. D. Smith, L. B. Mostaco-Guidolin, M. D. Hewko, A. F. Pegoraro, E. K. Kohlenberg et al. 2010. Multimodal nonlinear optical imaging of atherosclerotic plaque development in myocardial infarction-prone rabbits. *J Biomed Opt* 15(2):1–3.

52. Parasassi, T., W. M. Yu, D. Durbin, L. Kuriashkina, E. Gratton, N. Maeda, and F. Ursini. 2000. Two-photon microscopy of aorta fibers shows proteolysis induced by LDL hydroperoxides. *Free Radical Bio Med* 28:1589–1597.

53. Lim, R. S., A. Kratzer, N. P. Barry, S. Miyazaki-Anzai, M. Miyazaki, W. W. Mantulin, M. Levi, E. O. Potma, and B. J. Tromberg. 2010. Multimodal CARS microscopy determination of the impact of diet on macrophage infiltration and lipid accumulation on plaque formation in ApoE-deficient mice. *J Lipid Res* 51:1729–1737.

54. Kim, S. H., E. S. Lee, J. Y. Lee, E. S. Lee, B. S. Lee, J. E. Park, and D. W. Moon. 2010. Multiplex coherent anti-Stokes Raman spectroscopy images intact atheromatous lesions and concomitantly identifies distinct chemical profiles of atherosclerotic lipids. *Circ Res* 106:1332–U1358.

55. Stary, H. C., A. B. Chandler, S. Glagov, J. R. Guyton, W. Insull, M. E. Rosenfeld, S. A. Schaffer, C. J. Schwartz, W. D. Wagner, and R. W. Wissler. 1994. A definition of initial, fatty streak, and intermediate lesions of atherosclerosis—A report from the committee on vascular-lesions of the council on arteriosclerosis, American-Heart-Association. *Arterioscler Thromb* 14:840–856.

56. Stary, H. C., A. B. Chandler, R. E. Dinsmore, V. Fuster, S. Glagov, W. Insull, M. E. Rosenfeld, C. J. Schwartz, W. D. Wagner, and R. W. Wissler. 1995. A definition of advanced types of atherosclerotic lesions and a histological classification of atherosclerosis—A report from the committee-on-vascular-lesions of the council-on-arteriosclerosis, American-Heart-Association. *Arterioscl Throm Vas* 15:1512–1531.

57. Schenke-Layland, K., U. A. Stock, A. Nsair, J. S. Xie, E. Angelis, C. G. Fonseca, R. Larbig, A. Mahajan, K. Shivkumar, M. C. Fishbein, and W. R. MacLellan. 2009. Cardiomyopathy is associated with structural remodelling of heart valve extracellular matrix. *Eur Heart J* 30:2254–2265.

58. Schenke-Layland, K., J. S. Xie, S. Heydarkhan-Hagvall, S. F. Hamm-Alvarez, U. A. Stock, K. G. M. Brockbank, and W. R. MacLellan. 2007. Optimized preservation of extracellular matrix in cardiac tissues: Implications for long-term graft durability. *Ann Thorac Surg* 83:1641–1650.

59. Gong, B., J. Sun, G. Vargas, Q. Chang, Y. Xu, D. Srivastava, and P. J. Boor. 2008. Nonlinear Imaging study of extracellular matrix in chemical-induced, developmental dissecting aortic aneurysm: Evidence for defective collagen type III. *Birth Defects Res A* 82:16–24.

14

Combined SHG/THG Imaging

Szu-Yu Chen
National Taiwan University

Chi-Kuang Sun
National Taiwan University

Currently, two-photon fluorescence (2PF) microscopy is the most common technique used in combination with second harmonic generation (SHG) microscopy. Making use of the various endogenous fluorophores found in bio-tissues, 2PF microscopy can provide more cellular, morphological, and molecular information regarding these tissues in addition to the SHG-revealed information; moreover, it can also help localize the SHG signals and identify the contrast sources of these signals. However, since 2PF microscopy is a fluorescence-based technique, issues of photodamage and photobleaching due to multiphoton absorption are always of concern in the imaging process. To avoid such fluorescence-induced concerns, third harmonic generation (THG) microscopy, a third-order nonlinear optical microscopy, could be used instead of 2PF microscopy. THG microscopy is well known to have interface sensitivity and can be used as a general-purpose type of microscopy to provide structural information of the tissues. Based on the characteristics of virtual-level transition and energy conservation, which are the same as in SHG microscopy, with a combination of SHG with THG, the problems of fluorescence-induced photodamage and photobleaching could be avoided so as to reduce the invasiveness. Meanwhile, owing to its higher-order nonlinearity, a higher spatial resolution can be achieved with THG microscopy. In this chapter, combined SHG/THG microscopy will be introduced, including the principles, system setup, and the biomedical applications accomplished.

14.1 Excitation Laser Sources for SHG/THG Microscopy

14.1.1 Selection of Wavelength

For studies in bio-tissues based on two-photon microscopic, combining SHG, and 2PF microscopies, femtosecond mode-locked Ti:sapphire (Ti:S) lasers are the most commonly used standard laser sources. Ti:S lasers can be broadly tuned from 700 to 1000 nm and have the ability to excite a wide range of fluorophores. By using a Ti:S laser, a more efficient excitation of SHG can be achieved in the bio-tissues

(Guo et al. 1996, 1999; Campagnola et al. 1999; Moreaux et al. 2000b; Brown et al. 2003). At the same time, the broad tuning range of 700–1000 nm covers the range of efficient two-photon excitation of most endogenous fluorophores in bio-tissues. Thus, Ti:S lasers have been adopted for various 2PF microscopic applications (Moreaux et al. 2000a; Campagnola et al. 2002; Zoumi et al. 2002). However, since THG radiation has a higher frequency (3ω) and a shorter wavelength ($\lambda/3$) than SHG radiation (2ω; $\lambda/2$) due to its higher nonlinearity, pulse lasers with longer wavelengths than that used in SHG microscopy are more appropriate and often used for excitation. For instance, Yelin et al. used a Ti:S laser with a wavelength of 800 nm for THG microscopy (Yelin et al. 2002). The generated THG at 267 nm fell within the deep UV region and suffered strong absorption and scattering in the bio-tissues. In 1996, Alfano (Guo et al. 1996) applied a Nd:YAG laser with an emission wavelength of 1064 nm for THG excitation in chicken tissues. THG signals at 354 nm were found in the chicken skin. To shift the wavelength of THG to the visible region so as to avoid serious absorption, lasers with longer wavelengths, including an optical parametric amplifier (OPA) (wavelength: 1200 nm; repetition rate: 250 KHz) pumped by a Ti:S amplifier (Muller et al. 1998; Squier et al. 1998; Debarre et al. 2006), a Cr:F laser at 1230 nm with a repetition rate of 110 MHz (Chu et al. 2001, 2003; Sun et al. 2003; Yu et al. 2007; Lee et al. 2009), an optical parametric oscillator (OPO) (wavelength: 1500 nm; repetition rate: 80 MHz) synchronously pumped by a Ti:S laser (Yelin and Silberberg 1999; Canioni et al. 2001), and a fiber laser at 1560 nm with a repetition rate of 50 MHz (Millard et al. 1999). Under the excitation of these lasers, the generated THG signals all fall within the visible range of 400–520 nm and showed reduced absorption and scattering in the bio-tissues. However, in addition to the attenuation of the THG signals, the attenuation of the excitation laser radiation in bio-tissues should also be taken into consideration since it can lead to the degradation of the excitation intensity, decreasing the penetrability and absorption-induced photodamage.

In an earlier study, Anderson and Parish measured the curves of both the scattering and absorption constant of human skin (Anderson and Parish 1981). The attenuation (combination of scattering and absorption) was found to reach a minimum value around 1200–1300 mm, which is the so-called penetration window, while the attenuation demonstrated a serious increase with increasing wavelength (>1300 nm) due to strong water absorption. This indicates that lasers with wavelengths above 1300 nm may suffer higher attenuation and cannot effectively help to improve the penetrability in human skin. Therefore, a femtosecond Cr:F laser with a wavelength of 1230 nm, well within the penetration window, could be an optimal laser source for reducing the attenuation of both THG signals and laser radiation to increase the imaging penetrability. Using the Cr:F laser for the excitation of combined SHG/THG microscopy, both the wavelengths of SHG (615 nm) and THG (410 nm) fall within the visible range (Sun et al. 2004), which also makes signal detection much easier.

14.1.2 Reduction of Photodamage

Using a high-intensity femtosecond laser source for excitation, the issue of photodamage induced by multiphoton absorption has to be seriously taken into consideration. It is important to reduce the photodamage in bio-tissues, especially for clinical trials. However, if the maximum applicable light intensity is limited to reduce photodamage, the signal intensity and penetrability can also be compromised. For the *in vivo* Ti:S-based (730–960 nm) 2PF techniques (Konig 2008), an average power of 30 mW was needed for imaging up to a depth of 100 μm. The irradiation of living cells with 730–800 nm beams of >1 mW average power (total exposure per cell = 0.2 J) was found to inhibit cloning efficiency (Konig et al. 1997). Excitation energy levels much higher than 1 mW can be assumed to be invasive. Besides, even when using a light intensity as high as 30 mW, the imaging depth of the Ti:S-based 2PF techniques is limited to 200 μm, meaning that Ti:S lasers do not meet the requirements for clinical trials. In a previous study of mammalian embryos, Squirrell et al. moved the excitation wavelength for 2PF microscopy to 1047 nm (Squirrell et al. 1999). The imaged embryos were found to maintain their viability through time-lapsed observations (five optical scanning sections collected every 15 min) with a 13 mW average power and 2 J total exposure. In previous studies of mouse embryos (Hsieh et al. 2008; Chen et al. 2010), as the Cr:F laser (1230 nm) was used for excitation, similar survival rates were found in nonimaged embryos versus embryos under 10 min continuous observation with >120 mW average power and >21.6 J total exposure per embryo. The total number of tested embryos was 146, which is greater than the minimum of 30, and is enough to provide statistical significance for this study. Compared with the Ti:S and 1047 nm excitation, the maximum tolerance of 140 mW excitation power is much higher than 1 mW for Ti:S and 13 mW for 1047 nm, and this indicates much reduced photodamage under the Cr:F excitation due to much reduced multiphoton absorption with a lower excitation photon energy (1230 nm; 1.01 eV) (Chen et al. 2002).

In addition to the average power of the laser radiation, the pulse energy and peak intensity of a laser pulse are also determined by the repetition rate and the pulse duration (Equations 14.1 and 14.2)

$$P_{peak} = \frac{\text{energy per pulse}}{\text{pulse duration}} \tag{14.1}$$

$$P_{ave} = \text{energy per pulse} \times \text{number of pulses per second}$$
(repetition rate of laser pulses = 1/number of pulses per second). $\tag{14.2}$

To maintain cell vitality, the repetition rate and the pulse duration should be carefully controlled to reduce the pulse energy and peak intensity to a safe level. Increasing the repetition rate can decrease the pulse energy, while simultaneously decreasing pulse width can help retrieve enough peak intensity for sufficient excitation of nonlinear signals. In addition, if the peak intensity is consistent with the safety requirement, it may limit the signal intensity, penetrability, and the frame rate of the imaging. This problem can also be solved by increasing the repetition rate, since under the same scanning speed, a single point on a sample can be excited more times and more signals can be collected and integrated to increase the signal intensity. This is different from the fluorescence-based technique, where a relaxation time is required for the excited upper-state elections to return to the ground state. There is no relaxation time restricting the selection of the repetition rate for virtual-transition harmonic generation processes. We chose a Cr:F laser with a repetition rate of 110 MHz and pulse duration of 140 fs for maintaining both the excitation efficiency and cell vitality of combined SHG/THG microscopy. A compact fiber pumped femtosecond Cr:F laser with more than 1 GHz repetition rate and 500 mW average laser power has been demonstrated (Liu et al. 2005) to a suitable laser source for nonlinear optical endoscopy (Chan et al. 2005).

The Cr:F laser is applied with the 110 MHz repetition rate and 140 fs pulse duration for the excitation of combined SHG/THG microscopy. The noninvasiveness of this imaging system is also tested in *in vivo* studies of various animal models in addition to the viability test of mammalian embryos. Previously, in Cr:F-based SHG/THG imaging of zebrafish embryos (Sun et al. 2004; Chen et al. 2006), the zebrafish embryonic brain development has been continuously observed in the same embryo. After 20 h of nonstop continuous imaging (100–140 mW, >7000 J exposure per embryo), all observed embryos were shown to have developed normally to their larval stages. In a previous *in vivo* SHG/THG virtual biopsy study of Syrian hamster oral mucosa (Tai et al. 2006), after 3 h of continuous observation of the same area (150 mW, 1620 J), the observed hamster buccal tissues were then excised immediately for pathological examination. There was no evidence of coagulation necrosis in the buccal squamous epithelium and no subepithelial stroma appeared under examination of all the studied animals. All the results of previous *in vivo* studies strongly indicate that the harmonic generation microscopy (HGM) system is noninvasiveness in nature and can satisfy the safety requirements for clinical trials.

14.1.3 Improvement of Penetrability

The reduction of the attenuation of both excitation irradiation and generated harmonic generation signals should improve the penetrability of the Cr:F-based SHG/THG microscopy in bio-tissues. In a work of comparing the penetration depth between Ti:S-based and Cr:F-based SHG microscopy (Yasui et al.

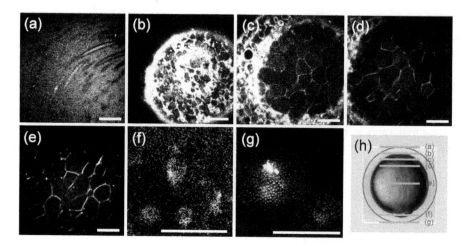

FIGURE 14.1 *In vivo* THG images of a zebrafish embryo obtained at different depths beneath the top chorion surface. The structures of the (a) top chorion, (b) top cellular layer, (c)–(e) yolk cells, (f) bottom cellular layer, and (g) bottom chorion can be observed within a depth of ~1.5 mm. The corresponding imaging planes are indicated in (h). Scale bar: 50 μm.

2009), setting the average excitation power of both lasers to 40 mW for imaging of porcine skin, the Ti:S SHG light almost disappeared beyond a depth of 200 μm, while the Cr:F SHG signals were still detectable even at a depth of 350 μm. This result strongly indicates improved penetrability under the Cr:F excitation. In previous studies of various animal models, a high penetrability (several hundred microns) can be found in most cases, but this varies widely among different tissues. Especially for transparent tissues, like zebrafish embryos (Chu et al. 2003; Chen et al. 2006) and mouse cornea (Chen et al. 2008), an extremely high penetrability can be obtained, enabling imaging deep inside the tissues, while still preserving a high spatial resolution. For example, Cr:F-based SHG/THG microscopy (forward-collection geometry) can be used for *in vivo* imaging of zebrafish embryo with a diameter of 1.5 mm (Chu et al. 2003). The upper chorion structures, top cellular layer, yolk cells, bottom cellular layer, and bottom chorion structures can all be observed throughout the whole embryo (Figures 14.1a through 14.1g). The corresponding imaging planes are indicated in Figure 14.1h. Even at an imaging depth of ~1.5 mm beneath the top chorion surface, the <1 μm diameter granular canals of the bottom chorion surface can be clearly distinguished, demonstrating the preservation of submicron resolution (500 nm resolution at a depth of 1.5 mm). Cr:F-based SHG/THG microscopy (backward-collection geometry) has been used to image an excised mouse eye. It can be seen that based on the transparency of the mouse cornea, a penetration depth of ~700 μm can be achieved, and the structures of the outermost corneal epithelium, corneal stroma, corneal endothelium, and even the lens fibers can be observed within the imaging depth inside the mouse eye (Chen et al. 2009b). In addition to the high penetrability demonstrated in the zebrafish embryos and mouse cornea, Cr:F-based SHG/THG microscopy has demonstrated a ~300-μm high penetrability of fixed human skin (Tai et al. 2005).

14.2 Cr:F-Based SHG/THG Microscopy

14.2.1 Principle of SHG and THG

In contrast to the linear process, SHG and THG are both higher-order nonlinear processes. As implied by the name "nonlinear optics," the polarization intensity of a molecule produced in the nonlinear optical process has a nonlinear dependence on the electric field of the excitation light. The polarization intensity of a molecule has to be described by using a power series expansion in the electrical field (Haus 1984).

$$\tilde{P}(t) = \varepsilon_0\chi^{(1)}\tilde{E}(t) + \varepsilon_0\chi^{(2)}\tilde{E}(t)^2 + \varepsilon_0\chi^{(3)}\tilde{E}(t)^3 + \cdots$$
$$\equiv \tilde{P}^{(1)}(t) + \tilde{P}^{(2)}(t) + \tilde{P}^{(3)}(t) + \cdots, \tag{14.3}$$

where $\chi^{(2)}$ and $\chi^{(3)}$ are the second-order and third-order nonlinear susceptibilities, respectively. Since both the polarization $\tilde{P}(t)$ and electric field $\tilde{E}(t)$ are vectors, the high-order nonlinear susceptibilities are tensors with different dimensions. For example, $\chi^{(2)}$ is a $3 \times 3 \times 3$ tensor with 27 elements and $\chi^{(3)}$ is a $3 \times 3 \times 3 \times 3$ tensor with 81 elements.

In Equation 14.3, $\tilde{P}(t)^{(2)} = \varepsilon_0\chi^{(2)}\tilde{E}(t)^2$ is referred to as the second-order nonlinear polarization and $\tilde{P}(t)^{(3)} = \varepsilon_0\chi^{(3)}\tilde{E}(t)^3$ is referred to as the third-order nonlinear polarization. Typically, second-order nonlinear processes, SHG, for example, can only occur in noncentrosymmetric materials that do not display inversion symmetry. For those media with inversion symmetry, each element of tensor $\chi^{(2)}$ vanishes and no second-order nonlinear process can be induced. On the other hand, third-order nonlinear process, such as THG, can occur in both centrosymmetrical and noncentrosymmetrical media. Since the radiation of the induced nonlinear polarizations and excitation light may constructively or destructively interfere due to the position dependence of the electric field, a phase-matching condition has to be satisfied to induce efficient nonlinear optical process. Moreover, according to Equation 14.3, the intensity of second-order nonlinear signals has a quadratic dependence on the electric field of excitation light, while the intensity of third-order nonlinear signals has a cubic dependence. This dependence indicates that (1) multiphoton excitation is needed for nonlinear optical process and the excitation is confined right near the focal spot; (2) compared with the second-order process, the third-order process is less efficient and a higher excitation intensity is required; and (3) the third-order process can provide higher spatial resolution than the second-order process.

14.2.1.1 Second Harmonic Generation

SHG is a second-order nonlinear optical process described by terms $\chi^{(2)}$ involving two electric fields with the same frequency, that is, $\chi^{(2)}$ $(2\omega: \omega, \omega)$. A single laser beam can be used for excitation and the generated signals have a second-harmonic frequency 2ω $(\lambda/2)$. The process of SHG can be considered as an up-conversion process, in which incident radiation with lower frequency ω is up-converted to radiation with higher-frequency 2ω. As illustrated in Figure 14.2a, two photons with a frequency ω are destroyed and a photon with a frequency 2ω is created. We find that the incident photon energy $2h\omega$ is equal to the generated photon energy $h(2\omega)$, which indicates that this up-conversion process obeys the energy-conservation rule and no energy is deposited in the interacting material. Moreover, it is noted that the solid line in the figure represents the "real" atomic ground state, while the dashed lines represent the so-called "virtual" levels. During the process of SHG, only the virtual-level transition with no real energy transition is involved in the process (Figure 14.2b).

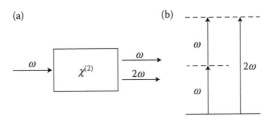

FIGURE 14.2 (a) Illustration of second-harmonic generation. (b) Energy-level diagram of second-harmonic generation.

SHG is well known to occur only in a material that is noncentrosymmetric. For a material that is centrosymmetric (i.e., possesses inversion symmetry), each element of the second-order nonlinear susceptibility $\chi^{(2)}$ must vanish and no SHG can be induced. This can be explained by changing the sign of the electric field $\tilde{E}(t)$ applied for the second-order nonlinear polarization, which is given by

$$\tilde{P}(t)^{(2)} = \varepsilon_0 \chi^{(2)} \tilde{E}(t)^2. \tag{14.4}$$

If the sign of the electric field is changed, the sign of the induced polarization also has to be changed because the medium is assumed to possess inversion symmetry, and Equation 14.4 must be replaced by (Boyd 1992)

$$-\tilde{P}(t)^{(2)} = \varepsilon_0 \chi^{(2)} [-\tilde{E}(t)]^2 = \varepsilon_0 \chi^{(2)} \tilde{E}(t)^2, \tag{14.5}$$

which shows that

$$\tilde{P}(t)^{(2)} = -\varepsilon_0 \chi^{(2)} \tilde{E}(t)^2. \tag{14.6}$$

Comparing Equations 14.4 and 14.6, the induced polarization described in these two equations must be equivalent, which occurs only if $\chi^{(2)}$ vanishes, that is

$$\chi^{(2)} = 0. \tag{14.7}$$

From this result, it is known that the contrast of the SHG is primarily provided by the noncentrosymmetry created due to the nonlinear meta material effect (Chu et al. 2002). In bio-tissues, SHG can arise from structural proteins, like collagen fibers (Campagnola et al. 2002; Cox et al. 2003; Mohler et al. 2003), nerve fibers (Dombeck et al. 2003; Chen et al. 2006), spindle fibers (Chu et al. 2003), and muscles (Chu et al. 2004; Plotnikov et al. 2006; Nucciotti et al. 2010), but not from the tooth enamel, which has a hexagonal symmetry and a vanishing $\chi^{(2)}$ (Chen et al. 2008).

14.2.1.2 Third Harmonic Generation

Similar to the SHG process, THG is a third-order nonlinear optical process described by terms $\chi^{(3)}$ involving three electrical fields with the same frequency, that is, $\chi^{(3)}$ (3ω: ω, ω, ω). The process of THG is illustrated in Figure 14.3. It can be considered an up-conversion process, in which incident radiation with a lower-frequency ω is up-converted to radiation with a higher frequency 3ω. During

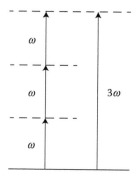

FIGURE 14.3   Energy-level diagram of third-harmonic generation.

the process, three photons with frequency ω are destroyed and a photon with frequency 3ω is created. We find that the incident photon energy $3h\omega$ is equal to the generated photon energy $h(3\omega)$. As with the process of SHG, there is only virtual-level transition involved in the process of THG. No energy is deposited in the interacted material since the energy-conservation rule is fulfilled in the up-conversion process.

Unlike SHG, which only occurs in noncentrosymmetric materials, THG is principally allowed in all materials, since the third-order susceptibility $\chi^{(3)}$ is nonvanishing regardless of the symmetry of materials. It is supposed that THG can be induced because the phase-matching condition ($\Delta k = 3k_\omega - k_{3\omega} = 0$) is satisfied. However, owing to the Gouy phase shift under strong focusing conditions, positive phase mismatching is needed to compensate the phase shift for efficient generation of the THG radiation, while THG is found to vanish in isotropic materials with a negative phase mismatch (normal dispersion), that is, $\Delta k = 3k_\omega - k_{3\omega} \leq 0$. The Gouy phase shift, also called a phase anomaly, is a well-known particularity of a focused light beam, in which the phase of the light beam shifts by a half-cycle when propagating through the focal center, and an effective dilation of light wavelength near the focal center is produced (Born and Wolf, 1999). Based on the coherent nature of THG, the phase anomaly can lead to destructive interference between the THG radiation induced before and after the focus center and cause the vanishing of the THG. Therefore, an efficient THG can only be produced at interfaces, where constructive interference possibly occurs due to changes of dispersion or to the nonlinear susceptibility of materials. Since a negative phase mismatch is common for most of the natural materials, including bio-tissues, THG is rarely observed in an isotropic bulk tissue. Based on the interface-sensitive nature of THG, THG microscopy is useful for imaging transparent objects that are difficult to be observed with a conventional microscope. Additionally, more and more important endogenous THG contrasts are being discovered in bio-tissues. THG microscopy is progressively emerging as a useful tool for morphological investigation, molecular imaging, and diagnosis in biology and medicine.

14.2.2 3D Spatial Resolution

In SHG microscopy, image formation can be considered as a convolution result of the excitation point spread function (PSF) and the features of the specimens. Since the SHG intensity has a quadratic dependence on the excitation light intensity, as mentioned above, the excitation of SHG is restricted to just near the focal spot and the size of the effective PSF can be reduced by a factor of $\sqrt{2}$ (Squier and Müller, 2001). The lateral resolution can be given by

$$r = \frac{1}{\sqrt{2}} \cdot \frac{0.51\lambda}{NA} \approx \frac{0.36\lambda}{NA}, \tag{14.8}$$

and the axial resolution can be given by

$$d_z = \frac{1}{\sqrt{2}} \cdot \frac{n\lambda}{NA^2} \approx \frac{0.7n\lambda}{NA^2}. \tag{14.9}$$

In contrast to the confocal pinhole in a confocal microscope, the restriction of SHG excitation can be considered to be a virtual pinhole, by which not only can the out-of-focus information be eliminated to increase the spatial resolution, but also the out-of-focus excitation can be avoided to reduce the photodamage.

Image formation in THG microscopy can be considered in a similar way to that of SHG microscopy. Since the THG intensity has a cubic dependence on the excitation light intensity (Squier and Muller 2001), the excitation is restricted to a narrower region near the focal spot than with SHG. The size of the effective PSF can be reduced by a factor of $\sqrt{3}$ instead of $\sqrt{2}$ used for SHG (Squier and Muller 2001). The lateral resolution can be given by

$$r = \frac{1}{\sqrt{3}} \cdot \frac{0.51\lambda}{NA} \approx \frac{0.3\lambda}{NA},$$ (14.10)

and the axial resolution can be given by

$$d_z = \frac{1}{\sqrt{3}} \cdot \frac{n\lambda}{NA^2} \approx \frac{0.57n\lambda}{NA^2}.$$ (14.11)

Again, the restriction of THG excitation can both help to increase the spatial resolution and reduce the out-of-focus photodamage. The resolution of THG microscopy is higher than with either confocal or SHG microscopy due to its higher-order nonlinearity. The virtual optical sectioning of bulky tissues can be easily achieved by making use of this high spatial resolution. A comparison of the 3D spatial resolution and excitation properties among conventional, confocal, SHG, and THG microscopy is shown in Table 14.1.

High 3D resolution is required to provide cellular or subcellular information for performing virtual optical sectioning of bio-tissues. Using a Cr:F laser for excitation and a high-NA objective for tight focusing, submicron lateral and axial resolution can be theoretically achieved with both SHG and THG microscopy (Table 14.1). In practice, the spatial resolution depends not only on the wavelength and NA, but also the spatial mode of the laser beam and the system alignment. With the optimized system alignment, a submicron 3D resolution can be achieved by our HGM system, for performing depth-resolved optical sectioning, and 3D images can be reconstructed from the obtained stack of HGM images versus

TABLE 14.1 Comparisons of Some Important Properties of Conventional Microscopy, Confocal Microscopy, SHG Microscopy, and THG Microscopy

	Conventional Microscopy	Confocal Microscopy	SHG Microscopy	THG Microscopy
Order of optical process	Linear	Linear	Second-order nonlinearity	Third-order nonlinearity
Illumination	Wide-field	Point (scanning)	Point (scanning)	Point (scanning)
Out-of-focus excitation	Yes	Yes	No	No
Spatial filter	No	Confocal pinhole	Localized excitation	Localized excitation
Lateral resolution	$\dfrac{0.61\lambda}{NA}$	$\dfrac{0.37\lambda}{NA}$	$\dfrac{0.36\lambda}{NA}$	$\dfrac{0.3\lambda}{NA}$
Axial resolution	$\dfrac{n\lambda}{NA^2}$	$\dfrac{0.64\lambda}{n-\sqrt{n^2-NA^2}}$	$\dfrac{0.7\lambda}{NA^2}$	$\dfrac{0.57n\lambda}{NA^2}$
Optical sectioning of bulky tissues	No	Yes	Yes	Yes
Out-of-focus photodamages	Yes	Yes	No	No
Energy deposition	Exists for single-photon fluorescence	Exists for single-photon fluorescence	No	No
Level transition	Real-level transition for single-photon fluorescence	Real-level transition for single-photon fluorescence	Virtual-level transition	Virtual-level transition

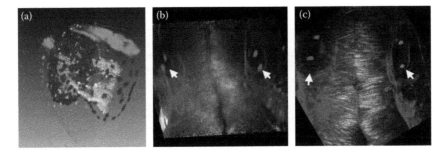

FIGURE 14.4 **(See color insert.)** *In vivo* 3D images of (a) a zebrafish heart and (b)–(c) a zebrafish brain. The 3D images are reconstructed from stacks of HGM images versus depth. (a) The 3D structures of the cardiac muscles (SHG; green), cardiac cells (2PF; red), and red blood cells (THG; yellow); and (b) the 3D structures of the neural tube (THG; purple), otic vesicles (arrows; THG; purple), and the nerve fibers (SHG; green) can be observed with a submicron resolution. Image size: 240×240 μm^2.

depth. For example, Figure 14.4a shows a 3D image of the zebrafish heart (Kung et al. 2007), including the 3D structures of the cardiac muscle, cardiac cells, and red blood cells (RBC), while Figures 14.4b and 14.4c show the 3D structures of the nerve fibers, neural tubes, and otic vesicles (arrows) of the zebrafish brain (Chen et al. 2006) with a submicron spatial resolution. Additionally, in another study (Tsai et al. 2006), the submicron spatial resolution in fixed human skin and its degradation versus depth beneath the skin surface have been analyzed. Based on these submicron resolution results, HGM is shown to have the capability of performing depth-resolved optical sectioning, providing cellular and subcellular information, and performing 3D imaging. Therefore, HGM has the potential for assisting or even replacing physical biopsies and pathohistological analysis.

14.2.3 System Setup

Depending on the applications, the Cr:F-based SHG/THG microscopic system can be designed to use either a forward-collection (Figure 14.5a) or a backward-collection (Figure 14.5b) geometry. Whatever the geometry, an SHG/THG microscope is composed of four main parts—excitation laser, scanning units, microscope, and detectors. For pixel-by-pixel imaging, the laser beam should be guided into a scanning unit to achieve real-time 2D beam scanning. Before being guided into a scanning system, the excitation beam has to be shaped and collimated by a pair of telescopes to avoid power loss at the scanning mirrors and to fill the back aperture of the focusing objective. Depending on the applications, an upright microscope or an inverted microscope is connected to the scanning system with an aperture fitting tube lens. After passing through the tube lens and the optics in the microscope, the scanned laser beam is focused onto the specimen by a high-NA objective. With the forward-collection geometry, the generated SHG and THG signals are collected by a condenser and guided into photomultipliers (PMTs) for detection. To obtain separate SHG (615 nm) and THG (410 nm) images, a dichroic beamsplitter with a suitable cut-off wavelength of around 500 nm should be used to separate the different signals. Two individual PMTs are used for detection. To increase the signal-to-noise ratio (SNR), bandpass filters with different center wavelengths and bandwidths are inserted in front of the PMTs to filter out the background noise. In addition, a color filter is used to filter out the laser radiation to avoid noise or damage resulting from the relatively strong laser radiation. On the other hand, for the backward-collection geometry, the excited epi-SHG and epi-THG signals are epi-collected by the same objective. The collected signals are then reflected by a dichroic beamsplitter, which can let 1230 nm laser beams pass through, reflect collected signals, and then direct them into PMTs for detection. The signal detection geometry is the same as that used in the forward-collection imaging system. To obtain simultaneous SHG and THG images, two PMTs are synchronized with the scanning system for intensity mapping.

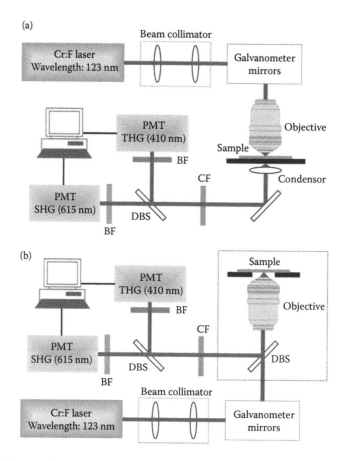

FIGURE 14.5 Schematic diagrams of the combined SHG/THG microscope with (a) forward-collection and (b) backward-collection geometry. DBS: dichroic beamsplitter; CF: color filter; BF: bandpass filter; PMT: photomultiplier.

As the laser beam is 2D scanned point by point over the specimen, the PMTs simultaneously record the excited harmonic generation signals point by point to form 2D SHG and THG images.

All the combined SHG/THG results discussed in this chapter are obtained using the home-made 1230 nm laser for excitation. The imaging system is adapted from a commercial confocal scanning system (Olympus, FV300) combined with an upright microscope (Olympus BX51) or an inverted microscope (Olympus IX71). All optics in the system are modified for the excitation of infrared light (~1230 nm). All the objectives used in our studies are listed in Table 14.1 along with their performances, including the ideal spatial resolution and viewing area. For the backward-collection system, a dichroic beamsplitter (FF665-Di02 produced by Semrock or 865dcxru from Chroma Technology) is used to separate the excitation laser beams and generate higher harmonic signals. A dichroic beamsplitter (490DRXR from Chroma Technology, with a cut-off wavelength at 490 nm) is used for separating the SHG and THG signals, while R4220P and R943-02 (or R928P) (from Hamamatsu) PMTs are used for the detection of SHG and THG, respectively. A color filter (CG-KG-5, CVI) is used before the PMTs to filter out the laser radiation, while the bandpass filters (HQ615/30X and D410/30 from Chroma) are used to filter the SHG and THG signals, respectively. The commercial software, Fluoview 4.3a (from Olympus) is used for programming, and a frame rate ranging from 1 to 4 Hz is applicable for a 512 × 512 image.

14.3 Biomedical Applications

14.3.1 Imaging Contrasts of SHG and THG

In our previous *ex vivo* and *in vivo* studies, various imaging contrasts of SHG and THG microscopy have been studied on different animal models and excised human tissues. In previous *in vivo* studies of zebrafish embryos (Chu et al. 2003; Sun et al. 2004; Chen et al. 2006) (Figures 14.6a through 14.6d), SHG was found to arise from noncentrosymmetric structures like spindle fibers (arrow in Figure 14.6a), skeleton muscles (arrow in Figure 14.6b), cardiac muscles (arrow in Figure 14.6c), and nerve fibers (arrow in Figure 14.6d) in the embryos, while THG was found to arise from cell membranes (arrowhead in Figure 14.6a) and tissue inhomogeneity, and the structural changes during the embryo development can be revealed by THG microscopy. In previous studies of mouse embryos (Hsieh et al. 2008) (Figure 14.6e), SHG can reveal zona pellucida (arrow in Figure 14.6e) of the mouse embryos, while THG contrasts were proved to be contributed by cytoplasmic organelles (arrowheads in Figure 14.6e). The previous studies of mouse skin (Lee et al. 2009) (Figures 14.6f through 14.6h) show that THG contrasts are provided by cytoplasm of keratinocytes in epidermis and the cellular morphology of epidermis (ED) can be revealed by THG microscopy. On the other hand, in dermis (D), SHG contrasts can be found in collagen fibers

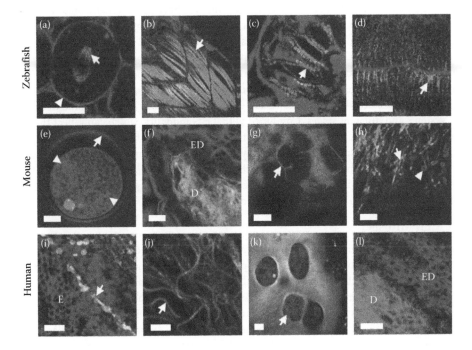

FIGURE 14.6 **(See color insert.)** (a)–(d) *In vivo* HGM images of zebrafish embryos. THG contrasts are provided by cell membranes (arrowhead in (a)) and tissue inhomogeneity; SHG contrasts are provided by spindle fibers (arrow in (a)), skeleton muscles (arrow in (b)), cardiac muscles (arrow in (c)), and nerve fibers (arrow in (d)). (e) *In vivo* HGM image of a mouse embryo. THG contrasts are contributed by cytoplasmic organelles (arrowheads in (e)); SHG contrasts are contributed by zona pellucida (arrow in (e)). (f)–(g) *Ex vivo* and (h) *in vivo* HGM image of mouse skin. THG contrasts are provided by cytoplasm of keratinocytes in epidermis (ED), and adipocytes (arrow in (g)) and red blood cells (arrowhead in (h)) in dermis; SHG contrasts are provided by collagen fibers (arrow in (h)) in dermis (D). (i)–(l) *Ex vivo* HGM images of excised human (i) teeth, (j) lung, (k) cartilage, and (l) skin. THG microscopy reveals rod structures of the tooth enamel (i), elastic fibers in the lung tissues (arrow in (j)), chondrocytes in the cartilage (arrow in (k)), and cellular morphology of epidermis (ED) of the skin (l); strain status of the abnormal enamel (arrow in (i)), type II collagen in the cartilage (k) and collagenous structures of dermis (D) of the skin (l) can be revealed by SHG microscopy. THG and SHG are represented by purple and green colors, respectively. Scale bar: 20 μm.

(arrow in Figure 14.6h), while THG contrasts can be found in adipocytes (arrow in Figure 14.6g) and RBC (arrowhead in Figure 14.6h). In addition to the animal models, our combined SHG/THG microscopy has been preliminarily applied on excised human tissues like teeth (Chen et al. 2008), lung (Yu et al. 2007), cartilage (Tsai et al. 2009), and skin (Tai et al. 2005; Chen et al. 2009a). The strain status of the abnormal enamel (arrow in Figure 14.6i), type II collagen in the cartilage (Figure 14.6k), and collagenous structures of dermis (D) of the skin (Figure 14.6l) can be revealed by SHG microscopy, while THG microscopy was shown to be able to reveal the rod structures of the tooth enamel (Figure 14.6i), elastic fibers in the lung tissues (arrow in Figure 14.6j), chondrocytes in the cartilage (arrow in Figure 14.6k), and cellular morphology of epidermis (ED) of the skin (Figure 14.6l). These preliminary studies not only show the rich imaging contrasts of the combined SHG/THG system, but also indicate the appreciated imaging capability of the combined SHG/THG system for clinical trails.

14.3.2 Animal Models: Zebrafish

Complex developmental processes of vertebrate embryos are difficult to observe noninvasively with high penetrability and high spatial resolution at the same time. In our previous *in vivo* studies (Chu et al. 2003; Sun et al. 2004; Chen et al. 2006), zebrafish embryos were used for investigating the complex embryonic development process, since the zebrafish embryos have much genetic material the same as humans' and have similar but simpler embryonic developmental programs. In addition, rapid developing rate, precisely defined developing stages, its transparency, small size, and external development also facilitate the neurology study and microscopic observation. Within the zebrafish embryos, SHG modality can provide various and valuable information, including mitosis spindle fibers (Chu et al. 2003; Sun et al. 2004; Chen et al. 2006), nerve fibers (Chen et al. 2006), muscle fibers (Chu et al. 2003; Sun et al. 2004), and stacked membranes (Chen et al. 2006), to study the embryonic development of different systems and different stages. Combined with THG modality, which can reflect the structural information of the embryos, can help to localize the SHG signals and identify and contrast sources of the SHG signals. Owing to embryos' transparency and ~1.5 mm thickness, the SHG/THG microscope with forward-collection geometry was used for zebrafish embryos' *in vivo* investigations. In the following subsections, previous SHG/THG studies of cell mitosis, nervous system development, and somite development of the zebrafish embryos will be introduced.

14.3.2.1 Cell Mitosis

In the *in vivo* investigation of cell mitosis in zebrafish embryos at 1-k-cell stage (2.5-hpf), strong SHG can be observed from centrosomes and mitotic spindles, which are known to be made up of spatially organized dynamic microtubules and of which the optical centro-symmetry is broken (Campagnola et al. 2002). On the other hand, based on the sensitivity to optical inhomogeneity, THG contrast can reflect various interfaces inside a single cell (Yelin and Silberberg 1999; Chu et al. 2001; Schins et al. 2002) such as nuclear membranes, cell membranes, and also the cytoplasmic organelles (Hsieh et al. 2008). Therefore, the cell structures in an embryo and the distribution of the cytoplasmic organelles can be revealed by THG modality. By using the combined SHG/THG microscope, the dynamic changes of spindle and membrane between two daughter cells can be imaged *in vivo* without any exogenous markers (Figure 14.7). At the initial prophase stage, two centrosomes can be revealed through SHG signals at the opposite sites of the nucleus, while a circular cell nuclear membrane is clearly visualized through THG signals (Figure 14.7a). At the prometaphase, the microtubules elongating from the centrioles to form the spindle can be observed through strong SHG signals, and the alignment of chromosomes in the center of the cell during the metaphase is disclosed by the broadening and flattening of the spindle fibers (Figure 14.7b). During the anaphase, the separation of spindle fibers and the alternation of the cell contour are picked up by the SHG and THG modalities, respectively (Figure 14.7c), while at the end of the mitosis, the telophase, SHG signals vanish as the spindle microtubules disperse into the cells and exhibit no more crystalline characteristic (Figure 14.7d).

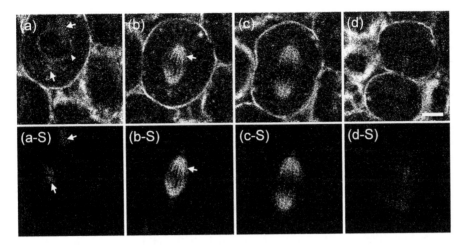

FIGURE 14.7 SHG/THG time series of the mitosis process in the 1-k-cell-stage zebrafish embryo. During (a) the prophase stage, (b) the metaphase, (c) the anaphase, and (d) the telophase, the cell nuclear membrane (arrowhead), centrosomes (arrows in (a)), and mitotic spindles (arrow in (b)) can be visualized through THG and SHG, respectively. Scale bar: 10 μm. SHG images corresponding to (a)–(d) are shown in (a-S)–(d-S), respectively.

14.3.2.2 Nervous System Development

Without any fluorescence markers, nonstop long-term observation of the brain development in the same live zebrafish embryo for more than 20 h has been accomplished by using SHG/THG microscopy (Chen et al. 2006). Through SHG modality, the dynamics of the cell mitosis, neuron differentiation, and the nerve fibers in the brain can be recorded. On the other hand, structure-sensitive THG modality can help to pick up the morphogenetic changes of the brain from neural plate to neural keel, neural rod, and then neural tube (Figure 14.8). At the right beginning of the brain development, bud stage, there is only a cluster of cells, which are still under differentiation and division, forming a plate-like structure named neural plate but no neural fibers in the neural plate. Through strong SHG signals from spindles, the cells under mitosis for differentiation or division can be clearly recognized, while THG images can help to localize the position of those cells (Figure 14.8a). After about 1 h, the embryo was

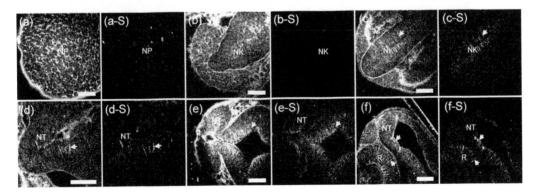

FIGURE 14.8 Continuous 20 h *in vivo* HGM imaging of the brain development in the same zebrafish embryo from the bud stage to the prim-15 stage. The structural changes of brain from (a) neural plate (NP) at the bud stage; (b) neural keel (NK) at the three-somite stage; (c) NK at the five-somite stage; (d) neural tube (NT) at the 14-somite stage; (e) midbrain at the 22-somite stage; to (f) NT and eye at the prim-15 stage can be revealed by THG microscopy. The nerve fibers (arrows) in both the NT and the retina (R) can be revealed by SHG. SHG images corresponding to (a)–(f) are shown in (a-S)–(f-S), respectively. Scale bar: 50 μm.

at the three-somite stage and the thickened neural plate with a middle line, named neural keel, can be revealed through THG modality (Figure 14.8b). However, because the rate of cell mitosis at this stage is retarded and there are no nerve fibers existing, almost no SHG signals were observed inside the neural keel. As the embryo developed into the five-somite stage, the first neuron inside the neural keel began to differentiate at the five-somite stage and elongated the nerve fibers (Kimmel et al. 1995), while meanwhile, SHG signals arising from the nerve fibers appeared again in the middle of the neural keel and they gradually extended to be linear-like (Figure 14.8c). As the development went on from the five-somite stage to prim-15 stage (Figures 14.8d through 14.8f), SHG modality dynamically recorded the growth of the nerve fibers and SHG signals became much stronger and more linear-like. On the other hand, the morphological changes from neural keel to a rod-like structure called neural rod, and the formation of the neural tube, which is a hollow structure with several lumens, can all be revealed by THG modality. During the whole process of the brain development, interface-sensitive THG provides the 3D sketch of different structures from a neural plate to a hollow neural tube; nerve-fiber- and mitotic-spindle-sensitive SHG can tell us much more stories about how cells behave during the development process. In addition to the brain, the axons elongated from the ganglia in the retina were also observed through the SHG and THG also showed the structure of the eyes, including the retina and the lens.

By enlarging the SHG/THG images of the neural tube (Figure 14.9a), different SHG contrast sources can be clearly recognized. With the structures of the neural tube revealed by the THG modality, in the enlarged image (Figure 14.9b), SHG signals can be identified to arise from the mitotic spindles (arrowhead) and the nerve fibers (arrow) according to their specific distribution and dynamic changes of the distribution. In the neural tube, SHG signals from the mitotic spindles were found to be highly concentrated in the region near the middle line, called ventricular zone or gray matter (Figure 14.9c). This completely matches the fact that stem cells or precursor cells only exist in the ventricular zone and mitosis can only occur in this zone (Kimmel et al. 1995). On the other hand, the linear-like SHG signals

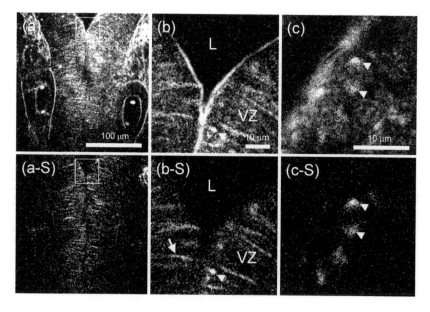

FIGURE 14.9 SHG signals in the neural tube. (a) THG shows the outline of the neural tube and SHG reveals the nerve fiber distribution in the neural tube. (b) Inset from (a) shows the SHG in the ventricular zone (VZ) near the lumens (L). In this region, SHG arises from both mitotic spindles (arrowhead) and nerve fibers (arrow). (c) Enlarged image corresponding to the area of inset in (b) shows clearly the mitotic spindles in the ventricular zone of the neural tube. (d) Cell mitosis (arrow) at the bottom of the retina (R). SHG images corresponding to (a)–(c) are shown in (a-S)–(c-S), respectively.

from the nerve fibers were found to elongate from the ventricular zone to the outer region, called marginal zone or white matter, and this also matches the fact that the differentiated ganglia cells elongate their nerve fibers from the ventricular zone to the marginal zone in the direction perpendicular to the neural tube (Kimmel et al. 1995). By applying the combined SHG/THG microscope, the mitosis process, the polarized neuron formation, and the neural stem cell behaviors can thus be dynamically tracked without the help of fluorescence markers.

In addition to the brain, the eye is also one important part of the peripheral nervous system worth being studied. By making use of SHG contrasts in nerve fibers and stack membranes, embryonic development of the zebrafish eyes can be carried out to see the elongating of the nerve fibers and the maturation of the retina. Meanwhile, the THG modality can help to show the morphological changes during the development process. At the 14-somite stage, a flat tissue called optic vesicle and a slit called the optic lumen (Li et al. 2000) in the middle of the optic vesicle can already be observed through the THG modality (Figure 14.10a), while the SHG signals are originated from mitotic spindles (Figure 14.10b). At the prim-5 stage, this flat tissue is found to transform into the shape of a cup, called eye cup, and the lens can be distinguished then. From the prim-12 to the prim-14 stage, the first neuron in retina is then born, SHG signals are found at the bottom of the eye cup to show the newborn nerve fibers (Figure 14.10c). After the prim-14 stage, the SHG signals gradually became denser and stronger in the retina, which showed the elongation and maturation of the nerve fibers. Meanwhile, the outline of the lens and the retina can all be revealed distinctly by the THG modality. By *in vivo* observing a 60-hpf embryo with its left side upward and optically sectioning the eyes downward from the lens to the retina and then the neural tube, the optic nerves, which consist of myelin-sheathed nerve fibers (arrow), is revealed by the SHG modality extending from the lens to the neural tube (Figure 14.10d). In addition to the optical nerves, strong SHG signals are also found at the bottom of the retina. In the SHG/THG image of a paraffin section of a 4-day-embryo eye (Figure 14.10e), the SHG signals are shown to have good contrast in every layer full of nerve fibers, including inner and outer plexform layers (arrows) and optical nerve (dotted arrow). Besides, the strong SHG signals at the bottom of the retina are found to arise from the outer segment of the photoreceptors (arrow). As the same origin for the SHG signals in grana (Huang et al. 1989), the strong SHG signals observed in the outer segment are the result of the stacked membranes.

Owing to the least-invasive nature of SHG/THG microscopy, the important morphogenetic changes of the zebrafish brain from the bud stage to the prim-15 stage were *in vivo* monitored continuously in the same live embryo without any fluorescence markers. Through the SHG modality, we are able to dynamically record the cell mitosis, cell differentiation, and the nerve fiber developments in the brain and the retina. Combined with THG modality, more information about the morphological changes during the development can be obtained and the valuable SHG signals from cell mitosis and nerve fiber development can be easily localized and recognized. Therefore, the combined SHG/THG microscopy is shown to have strong capability for various studies of the nervous system.

14.3.2.3 Somite Development

Based on the strong SHG contrasts provided by muscle fibers that are composed of collaterally organized myosin and actin filaments and have crystalline nanostructure (Chu et al. 2002), the combined SHG/THG microscopy can also be applied to studies of embryonic somite development. Since numerous mysteries about the somite development, including pinching off of somites, mytomes development, and differentiation and migration of muscle fibers, have not been revealed yet, the SHG/THG system may provide a useful tool for solving those problems. Through SHG modality, the detailed distribution of muscle fibers in a somite can be observed, where individual sarcomere composed of A- and I-bands with a periodicity <2 μm inside muscle fibers can be readily resolved; the general morphological structures, including the boundary of somite and notochord, can be revealed through THG modality. Starting from about 10.3 hpf, when it is the beginning of the segmentation period, the somites will appear sequentially in the trunk and tail and provide a staging index of the embryonic development. At

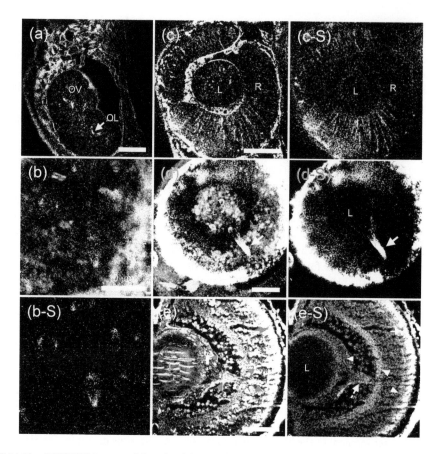

FIGURE 14.10 SHG/THG images of the zebrafish eye. (a) The left eye of a live 14-somite-stage embryo with the left side to the top. The flat tissue named optic vesicle (OV) and the silt, optic lumen (OL, arrow), are shown by THG; at this stage (b) only few SHG from mitotic spindles can be found near the optic lumen. (c) The left eye of a live prim-15-stage embryo. The lens (L) and retina (R) are clearly observed through THG and the nerve fibers elongated across the retina are picked up by SHG, respectively. (d) The eye of a live 60-hpf embryo. By superimposing each section at different depths, the optic nerve from the lens (L) to the neural tube is revealed by SHG (arrow). (e) SHG/THG image obtained from a paraffin section of a 4-day eye. Strong SHG can be found in layers full of nerve fibers (arrowheads), optic nerve (dotted arrow), and the outer segment of the photoreceptor (arrow). Scale bar: 100 μm for (a), (c), (d), and (e); 10 μm for (b). SHG images corresponding to (b)–(e) are shown in (b-S)–(e-S), respectively.

the end of the segmentation period, 24 hpf, 30 pairs of somites will be completely formed. By observing an 11-hpf (three-somite stage) zebrafish embryo from its dorsal side (Figure 14.11a), the anterior-most completely formed three somites (somite 1–3) and also the pinching off of the fourth somite (arrow) are clearly revealed by THG modality. At this stage, the fourth somite is just beginning to form and appears as a circumscribed region called "somitomere." Figure 14.11b shows the left-side view of the structures of the readily formed five somites at the five-somite stage. Since no muscle fibers have developed yet at the three-somite and five-somite stages, no SHG signals can be observed. In an 18-somite-stage (18 hpf) embryo with left side to the top (Figure 14.11c), THG signals show the small vacuoles (arrow) within the notochord cells and the structures of the somites, while in a 20-somite-stage (19 hpf) embryo with dorsal side to the top (Figure 14.11d), SHG signals appear to show muscle fibers in the somites. At the prim-15 stage (30 hpf), the SHG-revealed muscle fibers are found to become much straighter and denser, while the THG-revealed vacuoles are observed to become larger (Figure 14.11e). At 48 hpf, the straight

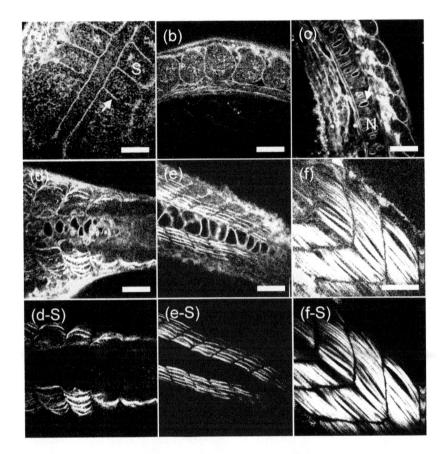

FIGURE 14.11 The structures of somites (S), notochords (N), and muscle fibers obtained from embryos at different development stages. (a) THG image of somites of a three-somite-stage embryo with dorsal side to the top and the pitching-off of the fourth somite (arrow) can be observed through THG; (b) THG image of a five-somite-stage embryo with left side to the top; (c) THG image of a 18-somite-stage embryo with left side to the top and the small vacuoles (arrow) within the notochord cells can be observed; (d) SHG/THG image of a 20-somite-stage embryos with dorsal side to the top, and the newborn muscle fibers in the somites are revealed by SHG; (e) In a 48-hpf embryo, matured muscle fibers are by SHG with the sarcomere clearly resolved. Scale bar: 100 μm. SHG images corresponding to (d)–(f) are shown in (d-S)–(f-S), respectively.

structures of the matured muscle fibers can be revealed by SHG and the sarcomere are also clearly resolved (Figure 14.11f). By applying SHG/THG microscopy, both the morphological changes of the somites and notochord and the development of the muscle fibers can be recorded at different stages and this can greatly help to study the somite development of the zebrafish embryo.

14.3.3 Animal Models: Mouse

Mouse is a popularly used standard animal model of mammals for many biomedical studies. Applying SHG/THG microscopy to mouse tissues can help to identify various image contrasts of SHG and THG modalities and find the imaging capability of SHG/THG microscopy in mammal tissues. According to previous studies of mouse tissues like skin, oral mucosa, and eye, SHG contrasts are found to be mainly provided by collagen fibers, especially type I collagen, which is the most widespread structural protein in mammals and is the main component of connective tissues. The strong SHG signals arising from the collagen fibers can be utilized to investigate the fibril structures in the connective tissues like the dermis of skin (Tai et al. 2005; Chen et al. 2009a), the stroma layer of cornea (Chen et al. 2009b), and the submucosa of oral cavity (Tai et al. 2006). On the other hand, the THG modality can generally provide morphological information to reveal the cellular structures of the skin, cornea, lens, oral mucosa, cartilage, and so on. Moreover, it is found that THG can be enhanced by specific kinds of molecule like hemoglobin (Clay et al. 2006; Tai et al. 2007; Chang et al. 2010), lipid (Debarre et al., 2006), and elastin (Yu et al. 2007) through the mechanisms of real-level absorption resonance enhancement. Based on these imaging contrasts, THG modality has the capability to reveal the RBC, adipocytes, and elastic fibers and perform molecular imaging. In the following studies of the mouse tissues, SHG/THG microscope with backward-collection geometry was used for investigation.

14.3.3.1 Mouse Skin

By optically sectioning excised mouse back skin at different depths, the cellular structures of the epidermis and the collagenous structures of the dermis can be revealed by THG and SHG modality, respectively. Figure 14.12 shows an exampled series of epi-HGM images at different depths beneath the skin surface of the mouse back skin (Figures 14.12a through 14.12h). Based on the strong SHG contrasts in the dermal collagen fibers, the collagenous structures at the dermo–epidermis junction (Figure 14.12c), in the papillary dermis (Figures 14.12d and 14.12e), and in the reticular dermis can be easily revealed (Figures 14.12f through 14.12h). From the papillary dermis to the reticular dermis, the collagen fiber bundles are found to become thicker and the collagen fibers become denser. At a depth of 80 μm (Figure 14.12g), the submicron spatial resolution is still preserved to distinctly reveal the collagen fibers. With the SHG-revealed collagenous structures, the sketch of the hair follicle can be shown at a depth of 120 μm (Figure 14.12h). In the hair follicle, the hair root (arrow in Figure 14.12h) and the sebaceous gland around it (arrowheads in Figure 14.12h) can be shown by THG modality. Since the sebaceous gland is made of lipid-filled cells, the strong THG signals are contributed by the lipid in the lipid-filled

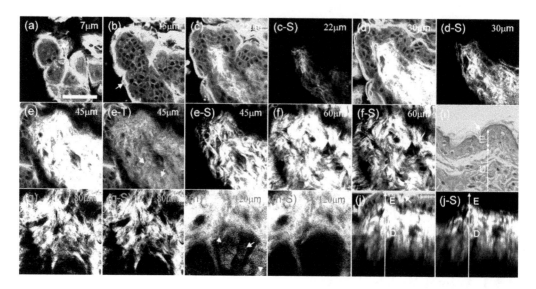

FIGURE 14.12   (a)–(h) An exampled series of SHG/THG images of the mouse back skin obtained at (a) 7 μm, (b) 15 μm, (c) 22 μm, (d) 30 μm, (e) 45 μm, (f) 60 μm, (g) 80 μm, and (h) 120 μm beneath the skin surface. The stratum corneum (arrows) with ultrastrong THG contrast is shown in (a) and (b). The hair follicle with the hair root (arrow) in it and the sebaceous glands (arrowheads) around it is shown in (h). (e-T) and (e-S) show the THG and SHG images corresponding to (e), respectively, and the elastic fibers (arrows) in the dermis can be found in (e-T). (i) Axial histology section with H&E staining and (j) axial epi-HGM image of the mouse back skin. The layer structures of the skin, including the stratum corneum (arrows), epidermis (E), and dermis (D) are shown in both images. SHG images corresponding to (c)–(h) and (j) are shown in (c-S)–(h-S) and (j-S), respectively. Scale bar: 50 μm.

cells. In addition to the SHG-revealed collagenous information, more cellular, morphological, and even molecular information can be provided by THG modality. Since THG is well known to be sensitive to local optical inhomogeneity (Squier et al. 1998; Peleg et al. 1999; Yelin and Silberberg 1999) and lipid (Debarre et al., 2006), the outermost stratum corneum, composed of multilayers of lipid and corneocytes, can be revealed by THG modality with ultrastrong THG contrast (arrows in Figures 14.12a and 14.12b). Beneath the stratum corneum, the honeycomb architectures of the stratum granulosum (arrowheads in Figure 14.12a), the stratum spinosum (Figure 14.12b), and the stratum basale (Figure 14.12c) can be revealed by THG modality based on THG contrasts arising from the cytoplasmic organelles (Hsieh et al. 2008). As shown in the THG images, the cytoplasm of the keratinocytes appears THG-bright, while the nuclei of the keratinocytes appear THG-dark. In contrast to the strong THG intensity in the stratum corneum, the THG intensity in the keratinocytes is shown to be much weaker, but can still provide clear cellular morphological information of the epidermis. At the dermo–epidermis junction (Figure 14.12c), the collagen structures of the dermal papillae are shown by SHG modality. Surrounding the SHG-revealed dermal papillae, THG reflects the basal cells with a circular locus, and the size of the locus progressively increases as the imaging plan moves deeper (Figures 14.12d and 14.12e). Splitting the different channels of Figure 14.12e into separate THG (Figure 14.12e-T) and SHG (Figure 14.12e-S) images, abundant THG signals can be found to arise from the elastic fibers (arrows in Figure 14.12e-T) and the fibroblasts collagen fiber boundaries, which are shown in the regions overlapping the simultaneous SHG signals. Figure 14.12i shows the axial histology section of the mouse back skin, while the axial SHG/THG image is shown in Figure 14.12j. In the axial SHG/THG image, the layer structures of the skin, including the THG-brighter stratum corneum (arrow in Figures 14.12i and 14.12j), epidermis (E), and dermis (D), are shown with a high axial resolution and have strong correspondences to the histology image. Moreover, the average thickness of the epidermis measured from the SHG/THG image is found to coincide with that measured from the histology section; this consistency indicates the capability of THG modality for investigating the pathological changes of the epidermis thickness.

Although the basic morphological structures and chemical composition of mouse skin are similar in different parts of mouse body, slight differences can be found among the mouse ear, back, and abdomen skin. Generally, the ear skin is a little thinner than the back skin and the fat tissues are more abundant in the abdomen skin. Owing to the regional variations, the imaging capability of SHG/THG microscopy may vary in different parts of mouse skin, and the application scope may thus be broadened. Figures 14.13a through 14.13f show a series of SHG/THG images of the mouse ear skin obtained at different depths beneath the skin surface. In the ear skin and back skin, only a slight difference exists in the thickness of epidermis. However, in the ear skin, the sebaceous glands around the hair follicles appear at a more superficial layer of the dermis and a higher density of the sebaceous glands can be found in contrast to the back skin. Since the lipid-filled cells in the sebaceous glands progressively accumulate lipids until the plasma membrane breaks down, the strong THG contrast in the sebaceous glands can be mostly contributed by the lipid (Debarre et al. 2006). As shown in Figures 14.13e and 14.13f, the sebaceous glands can extend to the deeper dermis with a >60 µm total length in axial. In skin, it is believed that there are discrete populations of epidermal stem cells in the stratum basale, hair follicles, sebaceous glands, apocrine glands, and eccrine glands. To study the activation of these stem cells, an imaging tool with the ability for *in vivo* observation is desired. Our results of the mouse skin not only show that THG microscopy has strong imaging contrast in the hair follicles and the sebaceous glands, but also indicate the capability of THG microscopy for studies of the epidermal stem cells. On the other hand, in the mouse abdomen skin, where more abundant fat cells can be found, the polygonal fat cells are revealed by THG modality at about 70 µm (Figure 14.13g), while the loose collagen fibers surrounding the fat cells are revealed by SHG modality. Even at 120 µm (Figure 14.13h), the fat cells can still be shown with a submicron spatial resolution.

14.3.3.2 Mouse Eye

According to the histological results of the cornea (Maurice 1984; Ramaesh et al. 2004), the cornea can be roughly divided into three main components—the corneal epithelium (EP), corneal stroma (CS),

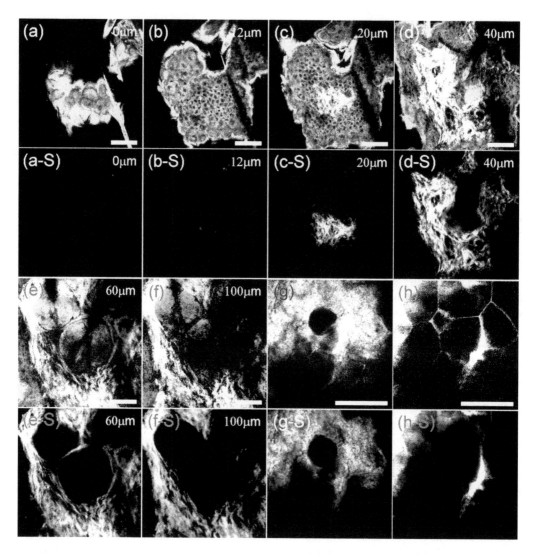

FIGURE 14.13 (a)–(f) An exampled series of SHG/THG images of the mouse ear skin obtained at (a) 0 μm, (b) 12 μm, (c) 20 μm, (d) 40 μm, (e) 60 μm, and (f) 100 μm beneath the skin surface. The sebaceous glands surrounding the hair follicles in the ear skin are observed at a more superficial layer of the dermis and are shown with stronger THG intensity and higher density. (g)–(h) The fat cells in the hypodermis are observed at (g) 70 μm and (h) 120 μm in the abdomen skin of a fat mouse. Even at 120 μm, the submicron of THG microscopy was still preserved. SHG images corresponding to (a)–(h) are shown in (a-S)–(h-S), respectively. Scale bar: 50 μm.

and corneal endothelium (ED) (Figure 14.14a). The EP and ED are cellular layers and the CS between them is composed of connective tissues. The corneal stroma consists of hundreds of layers of regularly organized collagen fibers, including mainly type I collagen, but also types III, V, and VI (McCally and Farrell 1990). The collagen fibers run parallel to each other but at large angles to the fibers in the next layer (McCally and Farrell 1990) accounting for the transparency of the cornea. Since SHG modality has high sensitivity to the collagen fibers, especially type I collagen (Campagnola and Loew 2003; Sun 2005; Wang et al. 2007), the arrangement of the collagen fibers in the stroma can be revealed. In addition, some more structural and cellular information in the CS can be given by THG modality, as shown in the SHG and THG images of the CS (Figures 14.14a and 14.14b). In the combined SHG/THG image (Figure

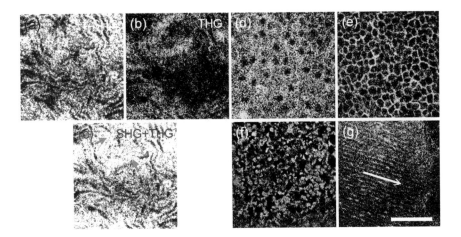

FIGURE 14.14 SHG/THG images of an excised mouse eye obtained at different depths beneath the corneal surface. (a) and (b) show the SHG and THG images of the corneal stroma and (c) shows the corresponding combined SHG/THG image. The collagenous structures can be revealed by SHG, while the keratocytes lying within the collagen fibers can be recognized from the non-SHG-overlapped THG signals (arrows). Through THG modality, the cellular structures of the (d) upper corneal epithelium, (e) deeper corneal epithelium, and (f) corneal endothelium can be revealed, while (g) the lens fibers in the lens are shown clearly along the direction of the arrow. Scale bar: 50 μm.

14.14c), THG and SHG are found to overlap in some regions (shown white in Figure 14.14c) but do not overlap in other regions. Since THG contrast can also be provided by the optical inhomogeneity of the collagen fibers, the SHG-overlapped THG signals can be recognized to arise from the collagen fibers (shown white in Figure 14.14c). On the other hand, the non-SHG-overlapped THG signals reflect the keratocytes lying within the collagen fiber meshes (arrows in Figure 14.14c). By combining SHG with THG modality, not only can the collagenous structures be observed, but also the cellular information in the CS can be distinguished and obtained more easily.

In addition to the CS, the outer and inner cellular layers, EP and ED, also play important roles in ophthalmology. ED covers the anterior surface of the cornea and has five to six cell layers thick, while the ED is a monolayer of flattened and polygonal ED cells. The EP cells of the superficial layer (close to the anterior surface of the cornea) are squamous and the EP cells of deeper layers are columnar. The EP and ED are responsible for protection and governing the fluid transportation of the eye, respectively. For diagnosis of many eye diseases, to highly resolve the cell morphology of these two layers is quite significant and an imaging tool with high sectioning power is required. Based on the THG contrast arising from the cytoplasmic organelles (Hsieh et al. 2008), the cytoplasm of the EP cells appeared bright in contrast to the dark nuclei. By optically sectioning the cornea at different depths, the morphology of the EP cells can be successfully revealed (Figures 14.14d through 14.14e) to be consistent with the histology results (Ramaesh et al. 2004). From the superficial layer (Figure 14.14d) to the deeper layer (Figure 14.14e), the shape of the EP cells is found to change from squamous shape to columnar shape. Figure 14.14f shows the THG image of the ED at 120 μm, and the uniformly sized polygonal cells in this monolayer could be revealed. The nuclei of the endothelial cells appear dark (arrowheads in Figure 14.14f), while the cytoplasm appears bright. Beneath the cornea is the aqueous humor (AH), which is mainly composed of water and has no THG contrast due to its optical homogeneity. Passing through the AH is the lens (L), consisting of three main parts: the membrane-like lens capsule (LC), cellular lens epithelium (LE), and lens fibers (LF). With a high penetrability of greater than 700 μm (Chen et al. 2009b), at a depth of 430 μm beneath the anterior corneal surface, the lens fibers (Figure 14.14g) with a width of ~5 μm (along the arrow in Figure 14.14g) can be highly resolved through THG signals but no significant SHG signals can be observed from the lens fibers. Even at a depth of >700 μm, the structure of the lens fibers can still be revealed through THG. Since the total thickness of the human cornea is about 535 μm (Jalbert et al. 2003), a penetration depth greater than 700 μm indicates the ability to investigate even the deepest part of the human cornea. With different imaging capabilities of SHG and THG modalities, not only the collagenous structures of the CS but also the cellular morphology of the EP and ED, and the structures of the lens fibers can be resolved with high resolution, and the combined SHG/THG microscopy is a suitable tool for cornea diagnoses.

14.3.4 Human Tissues

From the SHG/THG results of animal models, the imaging capabilities of SHG and THG modalities are demonstrated and SHG/THG microscopy is shown to be a potential tool for diagnoses of various diseases. With the strong SHG contrasts in the collagen fibers, SHG modality is able to reveal the changes of the collagenous structures of various human connective tissues like the dermis of skin (Chen et al. 2009a, 2010; Cicchi et al. 2010), the submucosa of oral cavity (Tsai et al. 2010a), cardiac muscles (Tsai et al. 2010b), the cartilage (Tsai et al. 2009), and the bone (Tsai et al. 2009) in human bodies. Based on these SHG-revealed changes, SHG modality can be applied to diagnoses of significant diseases with pathological changes of collagenous structures involved, for example, skin cancers. On the other hand, the SHG contrasts are also found to arise from the strains in the tooth enamel (Chen et al. 2008) and these imaging contrasts can help to reflect the changes of the enamel rods due to white spot lesions, thermal damages, and physical cracks, and has potentials for enamel diagnoses. By combining SHG with THG, more morphological, cellular, and molecular information can be provided by THG modality. The valuable THG-revealed information is useful for recognizing and localizing the observed SHG signals and helps to increase the diagnosis accuracy. In the following subsection, *ex vivo* SHG/THG imaging of human skin, including normal human skin and various diseased human skin, is demonstrated to show the capability of SHG/THG microscopy for skin disease diagnosis (Chen et al. 2010). For further clinical applications, the SHG/THG microscope with backward-collection geometry was used for these preliminary studies.

14.3.4.1 Normal Human Skin

As observed in the mouse skin, SHG microscopy can also show the collagenous structures of the dermis of human skin, while THG microscopy can provide more cellular and morphological information of both the dermis and epidermis of human skin. Figure 14.15 shows a series of SHG/THG images obtained at the epidermis (Figures 14.15a through 14.15c), the dermo–epidermal junction (Figure 14.15d), and the dermis (Figures 14.15e through 14.15h) of an excised normal human skin. Similarly, the dermis of the human skin can also be divided into two layers—the papillary dermis and reticular dermis. In the papillary dermis (Figures 14.15d and 14.15e), the collagen fibers are shown to be loose and areolar, while the collagen fibers in the thick reticular dermis are found to be dense and irregular (Figures 14.15f through 14.15h). Although the different collagenous structures in the papillary and reticular dermis can be clearly distinguished through the SHG images, the important cellular information in the dermis still lacks. Combining SHG with THG modality, the intradermal cellular information, including the inactivate melanocytes (arrow in Figure 14.15d) and the fibroblasts (arrowheads), can be revealed based on the THG contrasts arising from the optical inhomogeneity and cytoplasmic organelles. In addition, the RBC in the dermal capillaries are able to be shown by THG based on resonantly enhanced THG contrast with oxyhemoglobin (Clay et al. 2006; Tai et al. 2007; Chang et al. 2010). However, owing to the loss of the blood during excision, no RBC can be observed in the excised human skin specimen. By applying THG modality for human skin imaging, more than the dermis, the cellular morphology of the epidermis can also be revealed to reflect lots of diagnostic information about cell disorders in the epidermis. The epidermis of the human skin can be divided into four layers—the stratum corneum, stratum granulosum, stratum spinosum, and the stratum basale. At the skin surface of the normal human skin, the dead and cornified stratum corneum (arrow in Figure 14.15a) produces strong THG contrast caused

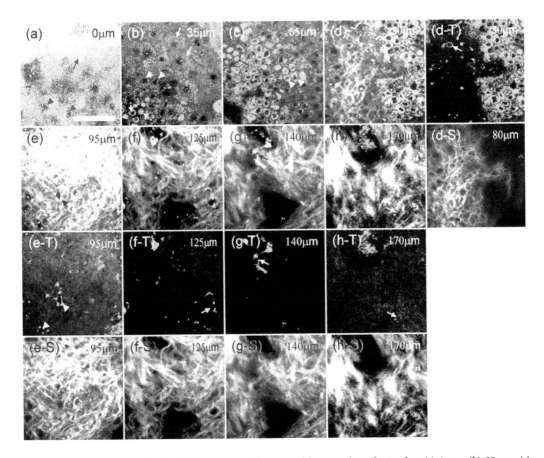

FIGURE 14.15 A series of SHG/THG images of the normal human skin obtained at (a) 0 µm, (b) 35 µm, (c) 65 µm, (d) 80 µm, (e) 95 µm, (f) 120 µm, (g) 140 µm, and (h) 170 µm beneath the skin surface. In (a), the stratum corneum (arrow) with ultrastrong THG contrast and THG-dark nuclei of the granular cells (arrowheads) are shown. (b) shows the granular cells (arrows), spinous cells (arrowheads), and basal cells (dashed arrows). (c) The THG-brighter basal cells (arrowheads) covering the peak of the dermal papilla. (d) At the dermo–epidermal junction, the collagen fibers in the peak of the dermal papilla surrounded by the basal cells are shown by SHG microscopy and the inactivated melanocyte (arrow) can be found in the dermis. (e) Fibroblasts (arrowheads) can be observed within the collagen fibers in the papillary dermis. (f)–(h) The collagenous structures of the reticular dermis and the fibroblasts (arrows) within the fibers. THG and SHG images corresponding to (d)–(h) are shown in (d-T)–(h-T) and (d-S)–(h-S), respectively. Scale bar: 50 µm.

by the interfaces of the multilayered structure (Tsang 1995; Barad et al. 1997) and by lipids (Debarre et al. 2006) within the corneocytes. Based on the THG contrast from cytoplasmic organelles (Hsieh et al. 2008), the nuclei of the epidermal cells appear dark (arrowheads in Figure 14.15a) in contrast to the bright cytoplasm in the THG images. Through THG modality, all the granular cells (arrows in Figure 14.15b), spinous cells (arrowheads in Figure 14.15b), and basal cells (dashed arrows in Figure 14.15b) can be found with honeycomb architectures throughout all levels of the epidermis. The progressive changes of nuclear diameter, cell diameter, and cell density in different layers can be clearly revealed to indicate the keratinization process. In the stratum basale (Figure 14.15c), clusters of basal cells (arrowheads) show stronger THG contrast relative to the surrounding cells and cover the peak of the dermal papilla at the "bumpy" dermo–epidermal junction. These strong THG contrasts in the cytoplasm of the basal cells are found to strongly correlate with the high concentration of the melanin in the basal cells and are identified to be contributed by the absorption resonance enhancement of the melanin (Chen et al. 2010).

At the dermo–epidermis junction (Figure 14.15d), the collagen fibers at the peak of the dermal papilla are revealed by SHG microscopy. The THG-bright basal cells surround the collagen fibers with a circular locus and the size of the locus increases as we move deeper into the dermis (Figure 14.15e). Depending on the conditions (age, sex, position, color, etc.) of different normal skin specimens, the penetrability can vary within 270–300 μm. The penetrability is simply defined as the maximum depth at which the fine collagenous structures or the subcellular structures can still be distinguished.

In addition to the normal human skin, the *ex vivo* SHG/THG imaging was also demonstrated on the excised diseased human skin, including three kinds of pigmented skin lesions—compound nevus, pigmented basal cell carcinoma (BCC), and superficial spreading melanoma (SSM). Although the three cases are all pigmented skin lesions, the pathological changes are greatly diverse and the major diagnostic characteristics are quite different. However, in clinical diagnosis, melanoma and other benign or malignant pigmented skin tumors can significantly overlap in their clinical and dermoscopical presentations and thus, pigmented skin lesions may be misdiagnosed (Makino et al. 2007). For example, melanoma at early stages may seem identical to benign nevi, while pigmented BCC and melanoma are sometimes misdiagnosed due to their similar appearance. By using SHG/THG microscopy for investigation, the typical characteristics of each skin lesion can easily be revealed for diagnosis of different skin diseases and the ability to accurately distinguish between benign nevi from malignant melanoma and between different types of malignant skin tumors can be provided.

14.3.4.2 Diseased Human Skin: Compound Nevus

According to the previous pathological evidences (Okun 1997; Gonzalez et al. 2003), compound nevus is a common melanocytic lesion that demonstrates nevomelanocytes (i.e., nevus cells) at both the dermo–epidermal junction and the superficial dermis. Figures 14.16a through 14.16h show a series of SHG/THG images obtained at different depths in a compound nevus specimen, while Figures 14.16i through 14.16l show a series of SHG/THG images obtained in the normal skin surrounding the compound nevus. In the SHG images obtained in the dermis, the collagen fiber bundles can be revealed to be thicker in the nevus (Figures 14.16d and 14.16e) than those in the normal skin (Figures 14.16k and 14.16l). Since the thickening of fiber bundles and the increasing of fibroblasts are both characteristics of the hosts response with fibrosis, the differences found in the dermis of the nevus and normal skin can indicate the higher degree of fibrosis in the nevus. In addition to the collagenous structures, through THG modality, the increasing number of the fibroblasts can be revealed as the other index for scoring the degree of fibrosis. Even at 180 μm, the fibroblast (arrowhead in Figure 14.16e) can be easily observed in the nevus but no obvious fibroblast can be found in the normal skin (Figure 14.16l). In the deeper dermis (Figures 14.16f through 14.16h), intradermal nevomelanocytes, the most significant characteristic for diagnosis of the compound nevus, are shown by THG modality to group in round clusters (arrows) and be surrounded by the SHG-revealed collagen fibers. These results are in complete agreement with previous histological conclusions. Even deep inside the reticular dermis at 300 μm, the grouping nevomelanocytes can still be observed by THG modality, while the much stronger THG contrast is provided by the melanin in the nevomelanocytes. Through the THG images, the individually distributed nevomelanocytes can also be found at the dermo–epidermis junction (Figure 14.16d) of the nevus (arrowhead) but not in the normal skin (Figure 14.16k), and this is another important characteristic of the compound nevus. In the epidermis, the THG-revealed epidermal architectures are shown to remain unchanged in the nevus (Figures 14.16a through 14.16c).

14.3.4.3 Diseased Human Skin: Superficial Spreading Melanoma

In contrast, SSM, the most common form of melanoma (Langley et al. 1998; Forman et al. 2008), is characterized by large epithelioid melanocytes, called melanoma cells, distributed singularly or in nests within all levels of epidermis. The normal architectures of epidermis are usually disrupted and even replaced by the melanoma cells, while the pattern of the rete ridges is often effaced. SSM often tends to be flat and asymmetric with varying colors. By performing SHG/THG imaging in two differently

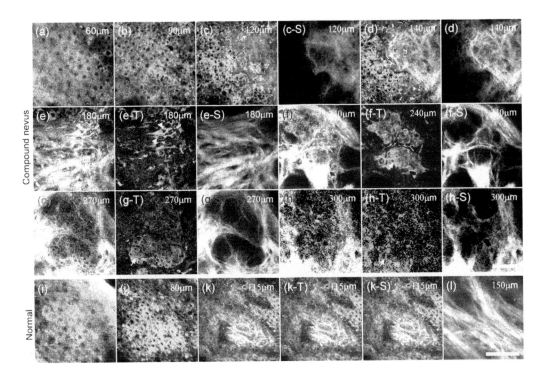

FIGURE 14.16 (a)–(h) *Ex vivo* SHG/THG images of a freshly excised compound nevus specimens obtained at (a) 60 μm; (b) 90 μm; (c) 120 μm; (d) 140 μm; (e) 180 μm; (f) 240 μm; (g) 270 μm; and (h) 300 μm beneath the skin surface. (i)–(l) *Ex vivo* SHG/THG images of the normal skin specimen obtained at (i) 45 μm; (j) 80 μm; (k) 115 μm; and (l) 150 μm. In contrast to the normal skin, individually distributed nevomelanocytes (arrowhead in (d)) at the dermo–epidermal junction, clustered nevomelanocytes in the dermis (arrows), and dermal fibrosis with thickened collagen fiber bundles ((d) and (e)) and the increased number of the fibroblasts (arrowheads in (e)) can be observed in the compound nevus specimen. SHG images corresponding to (c)–(h) and (k) are shown in (c-S)–(h-S) and (k-S) and THG images corresponding to (e)–(h) and (k) are shown in (e-T)–(h-T) and (k-T), respectively. Scale bar: 50 μm.

colored regions, not only the typical characteristics of SSM can be clearly revealed, but also the distinctive differences can be found between two regions. Figures 14.17a through 14.17d and Figures 14.17e through 14.17h show the SHG/THG images of the freshly excised SSM specimen obtained in the darker brownish region and the lighter brownish region, respectively, while Figures 14.17i through 14.17l show the SHG/THG images of the normal skin surrounding the SSM. In the normal skin, the collagen fibers in the papillary dermis begin to appear at only 55 μm (Figure 14.17j) and both the collagenous structures of the papillary dermis (Figure 14.17k) and reticular dermis (Figure 14.17l) can be clearly observed. In contrast to the normal human skin, the imaging depth in the darker region of this SSM specimen is limited to around 130 μm because of stronger attenuation of THG signals resulting from the strong absorption of the melanin in this darker region. Since the epidermis of the SSM greatly thickens due to the accumulation of the melanoma cells, in this darker region, the dermis is too deep to be observed and no SHG signals can be observed within the penetration depth. However, due to slighter invasion of the melanoma cells, the epidermis of the SSM in the lighter region is found to be thinner than that in the darker region and the collagen fibers in the papillary dermis can be observed by SHG modality (Figure 14.17h).

Since SHG signals from the dermal collagen fibers are usually beyond the visible range in the SSM, the THG-revealed information becomes much more significant and desirable for diagnosis. Based on the melanin-induced enhancement of THG, the THG intensity of the melanoma cells appears much higher than that of the normal keratinocytes. In the stratum granulosum within the darker region, the

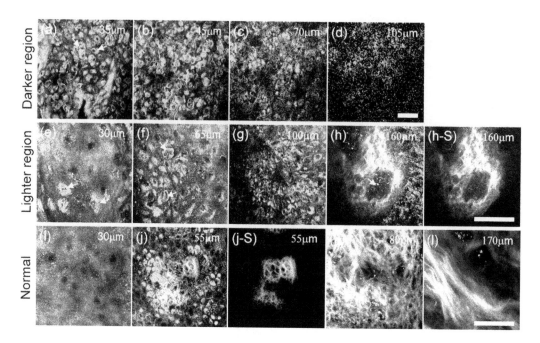

FIGURE 14.17 *Ex vivo* SHG/THG images of the freshly excised SSM specimen obtained in the (a)–(d) darker brownish region and (e)–(h) lighter brownish region at different depths beneath the skin surface. (a)–(d) Throughout the first 100-μm layer in the darker region, THG-bright melanoma cells (arrows in (a)) with varied shapes and larger nuclei were irregularly distributed to replace the normal keratinocytes and (d) no collagenous structure can be observed due to increased thickness of epidermis and limited penetrability. (e)–(h) In the lighter region, (e) no melanoma cells but some sparsely distributed grainy particles (arrows) with a stronger THG contrast were found in the stratum granulosum. (f) A few dendritic and THG-bright melanoma cells (arrows) were found within the keratinocytes. (g) In the stratum basale, THG-bright cells were found without uniform brightness and distinct cell borders, while the honeycomb morphology was disrupted. (h) Owing to milder invasion in this region, the collagenous structures of the papillary dermis and the capillary (arrow) were observed. (i)–(l) *Ex vivo* SHG/THG images of the freshly excised normal skin removed from the same patient, obtained at (i) stratum granulosum; (j) dermo–epidermal junction; (k) papillary dermis; and (l) reticular dermis. SHG images corresponding to (h) and (j) are shown in (h-S) and (j-S), respectively. Scale bar: 50 μm.

melanoma cells occupying this layer are shown by THG modality with strong THG intensity (arrows in Figure 14.17a). Through the THG signals, the larger nuclear diameter and varied cell shapes of the superficially spreading melanoma cells can be clearly revealed as diagnostic characteristics. Owing to a more severe invasion of the melanoma cells in the darker region, throughout the first 100 μm layer (Figures 14.17a through 14.17d), the regular keratinocytes are replaced by single or clustered melanoma cells and the honeycomb architectures of the normal epidermis (Figures 14.17i and 14.17j) are almost thoroughly destroyed. On the other hand, in the lighter region, only sparsely distributed grainy particles with strong THG contrast can be found to accumulate around the nuclei of the granular cells (arrows in Figure 14.17e), and most of the regular patterns of the granular cells are preserved. Owing to a less severe invasion of melanoma cells within this lighter region, in the shallower layers of the epidermis, only a few melanoma cells (arrows in Figure 14.17f) can be found and parts of the honeycomb architectures of the epidermis can still be observed. When moving the imaging plane into the stratum basale (Figure 14.17g), the THG-revealed cells in this layer show loss of distinct cell borders and appear more spindle-like, different from the basal cells with clearly defined cell borders and uniform THG brightness observed in the normal skin (Figure 14.17j). Notice that in the case of SSM, the regions with SHG contrasts are often beyond the penetrability. Since only the SHG modality is not sufficient for diagnosis, instead, the cellular and morphological information provided by THG modality plays a significant and indispensible role in the diagnosis of the SSM.

14.3.4.4 Diseased Human Skin: Pigmented Basal Cell Carcinoma

BCC is the most common type of skin cancer and pigmented BCC is a variant of nodular BCC. Owing to the abundant pigment in this lesion, it is sometimes misdiagnosed as malignant melanoma and an *in vivo* virtual biopsy tool with the ability to pathologically distinguish pigmented BCC and pigmented nevi from melanoma is required. In histology studies of pigmented BCC (Agero et al. 2006), tightly packed tumor cells bud from epidermis to papillary dermis with a nodular pattern enclosed by collagen fibers. The nuclei of the tumor cells appear as a parallel arrangement (palisading) while melanin pigment is nonuniformly distributed throughout the tumor nodules. Sometimes, a polarized cell pattern can be seen in epidermis with loss of the normal progressive changes of keratinocytes, loss of the honeycomb pattern, and loss of the dermal papillae structure. In previous researches (Cicchi et al. 2008), two-photon imaging modality, combined both SHG and 2PF microscopy, has been applied to investigate human *ex vivo* BCC samples, and the SHG/autofluorescence ratio of the dermis was found to be able to discriminate between normal and BCC tissues. Figure 14.18 shows the *ex vivo* SHG/THG images of the freshly excised pigmented BCC (Figures 14.18a through 14.18d), normal skin (Figures 14.18e through 14.18h), and suspicious tissue around the tumor (Figures 14.18i through 14.18l). At the dermo–epidermal junction, no dermal papilla architectures can be observed by SHG modality and the collagen fibers at the junction are shown with a much wavier pattern (arrows in Figure 14.18c); this is in contrast to the normal areolar pattern (Figure 14.18g). In contrast to the collagenous structures of the reticular dermis observed in the normal skin (Figure 14.18h), the tumor nodules enclosed by the collagen fibers, which

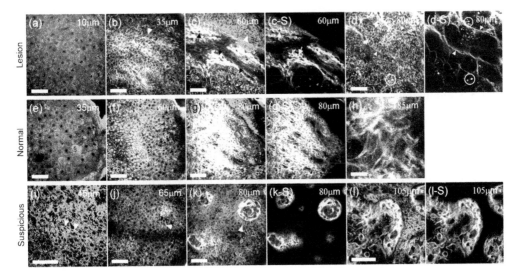

FIGURE 14.18 *Ex vivo* SHG/THG images of freshly excised (a)–(d) pigmented BCC specimen; (e)–(h) normal skin specimen; and (i)–(l) suspicious cancerous or precancerous specimen obtained at different depths beneath the skin surface. In lesional skin, the spindle-like keratinocytes (arrowheads in (b) and (c)) were usually found in epidermis and collagen fibers in a much wavier pattern (arrows in (c) and (d)) can be found at the dermo–epidermal junction. Without the normal pattern of rete ridges, tumor nodules (dashed arrow in (c)) enclosed by collagen fibers (arrowheads in (d)) were found to occupy the normal dermis. The parallel arrangement of the tumor cells (palisading; arrows in (d)) can be identified at the edges of the nodules and bright HG spots (circled in (d)) can suggest the high melanin contents. In the suspicious specimen, the spindle-like keratinocytes were observed in the epidermis (arrowheads in (i)). Elongated basal cells can be found around the dermal papilla and were parallel to one another (arrowheads in (j) and (k)), while (l) ring-shaped instead of (g) areolar collagen fibers were found in the papillary dermis. SHG images corresponding to (c)–(d), (g), and (k)–(l) are shown in (c-S)–(d-S), (g-S), and (k-S)–(l-S), respectively. Scale bar: 50 μm.

are the most significant diagnostic characteristic of BCC, are found to replace the normal papillary and reticular dermis in the pigmented BCC specimen (Figures 14.18c and 14.18d). Through SHG modality, the loose collagen fibers can be revealed to enclose the tumor cells, while the specific nodular pattern and parallel arrangement (palisading) of the tumor cells are clearly resolved by THG modality. Through the THG signals, the tumor nodules can be resolved to be composed of tightly packed oval cells and have parallel arrangements of the tumor cells (palisading) at the edge of the tumor nodules (arrows in Figure 14.18d). Moreover, within the tumor nodules, bright THG spots (circled in Figure 14.18d) are frequently found to be randomly distributed and these spots with isolated THG signal enhancements indicate the high melanin contents and also the melanophages in the nodules.

On the other hand, in the superficial epidermis, the stratum granulosum of the lesional specimen (Figure 14.18a), the honeycomb pattern of the granular cells is shown by THG modality to remain unchanged and the same as what is observed in the normal skin (Figure 14.18e); in the deeper epidermis, the disruption of the normal epidermal architecture in the lesional specimen (Figure 14.18b) can be revealed by THG in contrast to normal skin (Figure 14.18f). Through the THG signals, the cells with elongated and polarized shapes (arrowhead in Figures 14.18b and 14.18c) instead of the columnar basal cells (Figure 14.18f) can be observed in the lesional specimen, and this cellular morphology is one of the diagnostic characteristic of BCC. In addition, the skin around the lesion is also investigated. In contrast to the normal skin, the spindle-like and disarrayed keratinocytes in the stratum spinous and stratum basale of this surrounding skin specimen can be revealed by THG (arrowheads in Figure 14.18i). At the dermo–epidermal junction, the basal cells around the dermal papilla are found to be unusually elongated and parallel with one another (arrowheads in Figures 14.18j and 14.18k), while anomalous collagenous structures, ring-shaped (Figure 14.18l) but not areolar (Figure 14.18g) collagen fibers, are revealed in the papillary dermis by SHG modality. Since the cellular morphology and collagenous structures disagree with the findings in the normal skin, the skin specimen is suggested to be suspicious cancerous or precancerous tissue. In the *ex vivo* study of the BCC, SHG/THG microscopy is shown to possess the ability to reveal the most important diagnostic characteristic of BCC and also have the potential for early diagnosis of the precancerous tissues.

14.3.4.5 Conclusions

Comparing among the SHG/THG images of three different types of lesional skin, SHG modality can help to reveal the pathological changes of the collagenous structures of the dermis. However, in the cases with thicker epidermis and higher melanin contents in the epidermis like melanoma, even no SHG signals can be observed and the morphological information provided by THG modality plays an important role in the diagnoses. With abundant THG contrasts in both epidermis and dermis, THG modality has the ability to reveal the typical diagnostic characteristics of each kind of skin lesion with high spatial resolution. Combining SHG with THG modality, it is able to histopathologically distinguish among different skin diseases and between benign and malignant lesions. Based on the THG contrast from melanin, THG microscopy demonstrates the unique capability for molecular imaging of melanin distributions and for diagnosing and screening early melanocytic lesions. As shown in Table 14.2, the significant diagnostic characteristics of three different types of skin diseases observed in the previous histology studies and revealed by our epi-HGM system have been listed.

14.3.5 Clinical Applications

Based on the noninvasiveness, high spatial resolution, and high penetrability of Cr:F-based SHG/THG microscopy and abundant imaging contrasts of both SHG and THG modalities, the Cr:F-based SHG/THG microscopy has been applied to clinical (*in vivo*) imaging of human skin. Through clinical damage evaluation, the Cr:F SHG/THG imaging system has been further confirmed to possess essential safety requirement. More than the static morphological and collagenous information, valuable dynamic information like red blood flow in dermal capillaries can also be obtained from the *in vivo* observation.

TABLE 14.2 THG-Revealed and SHG-Revealed Diagnostic Characteristics of Three Types of Skin Diseases, Including the Compound Nevus, Superficial Spreading Melanoma, and Pigmented Basal Cell Carcinoma

Histopathology Evidences	Harmonic Generation Microscopy	
	THG	SHG
	Diagnostic Characteristics	
Compound nevus	Honeycomb patterns of keratinocytes Normal keratinization process Increased number of melanocytes in epidermis Individually distributed NMs at DEJ Clustered NMs in dermis Increased number of fibroblasts Accumulating melanin in clustered NMs and fibroblasts	Thickening of the collagen fiber bundles
Superficial spreading melanoma (SSM)	Clearly-revealed superficially spreading MCs with stronger THG contrast (without staining) Disrupted honeycomb patterns in epidermis Scoring different levels of invasions Distinguishing normal keratinocytes from MCs Sparsely distributed THG-bright spots in epidermis (higher contents of melanin)	Collagenous structures in papillary dermis can be observed in mildly invaded regions (lighter brownish region)
Pigmented basal cell carcinoma (BCC)	Elongating and spindle-like keratinocytes Higher cell density in epidermis (anomalous keratinization process) Tumor nodules in dermis Parallel arrangement of tumor cells at the edge of the nodules (palisading) Sparsely THG-bright spots in both epidermis and tumor nodules (accumulated melanin or melanophages)	Much wavier collagen fibers at DEJ Normal collagenous structures are replaced by tumor nodules Collagen fibers enclose the tumor nodules

Throughout a ~300 μm imaging depth, the *in vivo* THG imaging is shown to retain submicron lateral resolution and have less signal degradation than in fixed human skin. Without any staining, the combined SHG/THG microscopy is an excellent tool for *in vivo* optical skin biopsy and it can provide lots of significant information for skin disease diagnosis. In the following subsections, the *in vivo* SHG/THG imaging of Asian human skin will be introduced, including studies of the *in vivo* SHG/THG results, damage evaluation, and spatial resolution.

14.3.5.1 *In Vivo* SHG/THG Imaging of Asian Human Skin

In the *in vivo* studies of Asian human skin, investigations were focused on the forearm skin of healthy Asian volunteers and SHG/THG microscope with backward-collection geometry was applied (Chen et al. 2009a, 2010). Compared with the *ex vivo* SHG/THG results of the normal human skin shown above, similar morphological and collagenous information can be observed from the *in vivo* investigation. As the *in vivo* SHG/THG biopsy images shown in Figure 14.19, in epidermis, stratum corneum (Figure 14.19a) is shown not to contain nuclei and have a much stronger THG contrast than other layers of the epidermis. This strong THG contrast is mainly contributed by the multilayer structure of the stratum corneum (Tsang 1995; Muller et al. 1998) and the lipid within the corneocytes (Debarre et al. 2006). Based on the THG contrasts in the cytoplasmic organelles, the cytoplasm of cells in the stratum granulosum (Figure 14.19b), stratum spinosum (Figure 14.19c), and stratum basale (Figure 14.19d) appear THG-bright, while the nuclei of the cells appear THG-dark. Making use of the contrast between the cytoplasm and nuclei, the cellular morphology of the epidermis can be easily revealed and the progressive changes of the nuclear size, cell size, and cell density can all be clearly revealed to show the normal keratinization process in the epidermis. In the stratum basale, which is rich in melanin contents, the cell borders can be defined much easier due to much stronger THG signals in the cytoplasm of the basal cells (arrowheads in Figure 14.19d) and the THG-darker intercellular spaces between neighboring basal cells, where the

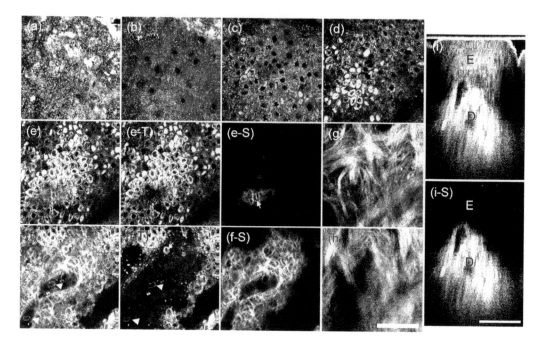

FIGURE 14.19 *In vivo* SHG/THG images of an Asian volunteer's forearm skin, obtained at (a) stratum corneum; (b) stratum granulosum; (c) stratum spinosum; (d) stratum basale; (e) dermo–epidermal junction; (f) papillary dermis; and (g), (h) reticular dermis. In epidermis, (b)–(d) the cell nuclei appeared dark under THG in contrast to the bright cytoplasm, while much stronger THG intensities were found in the basal cells (arrowheads in (d)). (e) At the dermo–epidermal junction, the collagen fibers at the peak of the dermal papilla (arrow) were revealed by SHG microscopy, surrounded by THG-bright basal cells (arrowhead). (f) In papillary dermis, the red blood cells (arrowheads) in the capillary were clearly recorded by THG modality. (i) *In vivo* axial SHG/THG image of human skin. SHG images corresponding to (e)–(f) and (i) are shown in (e-S)–(f-S) and (i-S) and THG images corresponding to (e)–(f) are shown in (e-T)–(f-T), respectively. Scale bar: 50 μm.

strong THG contrasts are contributed by the melanin (Matts et al. 2007). At the dermo–epidermal junction, the collagen fibers at the peak of the dermal papilla (arrow in Figure 14.19e) begin to be revealed by SHG microscopy and are surrounded by THG-bright basal cells (arrowhead in Figure 14.19e). In the papillary dermis, capillaries can be easily identified by the surrounding SHG-sensitive collagen fibers (Figure 14.19f) and the THG-revealed RBC in the capillary (arrowheads in Figure 14.19f). In contrast to the *ex vivo* results with lack of dynamic information, the dynamic movement and deformation of the RBC in the capillary can be continuously recorded by *in vivo* THG imaging. Even in the deep reticular dermis, the RBC can also be observed based on the THG contrast enhanced by hemoglobin (Tai et al. 2007). In the reticular dermis (Figures 14.19g and 14.19h), the reticular collagen fibers and much thicker fiber bundles can be revealed by SHG microscopy. As seen in the *in vivo* axial SHG/THG image shown in Figure 14.19i, the axial structures of epidermis and dermis can be resolved based on the high axial resolution and a penetrability of 300 μm beneath the skin surface of live human skin can be achieved by the Cr:F-based SHG/THG microscopy. In the *in vivo* SHG/THG biopsy, the skin is investigated under its most natural condition and more complete information can be obtained without artifacts resulting from excision or tissue loss. Moreover, in contrast to the physical biopsy, in which only a small region of the tissue is sampled, large-area searching can be achieved and the probability of misdiagnosis can be reduced.

In the dermis of human skin, in addition to the SHG-revealed collagenous structures, there are also lots of intradermal THG contrasts that can help to reveal the inactive melanocytes, fibroblasts, and the accumulated melanin in the dermis. In the Asian volunteers' skin, inactive melanocytes are often found in the papillary dermis, close to the dermo–epidermal junction. As shown in the five examples obtained at the dermo–epidermal junction of five different Asian volunteers' forearm skin (Figures 14.20a-1 through 14.20a-5), the inactive melanocytes (arrows) can easily be identified according to their oval shape and large nuclei, of which the mean diameter is about 7 μm. On the other hand, in the dermis of the Asian volunteers' skin, cells with strong THG contrasts and varied shapes are found within the collagen fibers (arrows in Figures 14.20b-1 through 14.20b-5). According to their locations and varied shapes, these cells can be recognized as fibroblasts, melanophages, or lymphocytes depending on the skin conditions (inflammation or pigmentation, etc.), but it still has difficulty to clearly distinguish these intradermal cells only by SHG/THG microscopy. As observed in the *ex vivo* results, in the *in vivo* SHG/THG imaging of the human skin, dermal capillaries can be regularly observed in both papillary and reticular dermis with SHG-sensitive collagen fibers surrounded (Figures 14.20c-1 through 14.20c-5). However, different from the *ex vivo* imaging, the RBC (arrows) in the capillaries can be revealed by THG

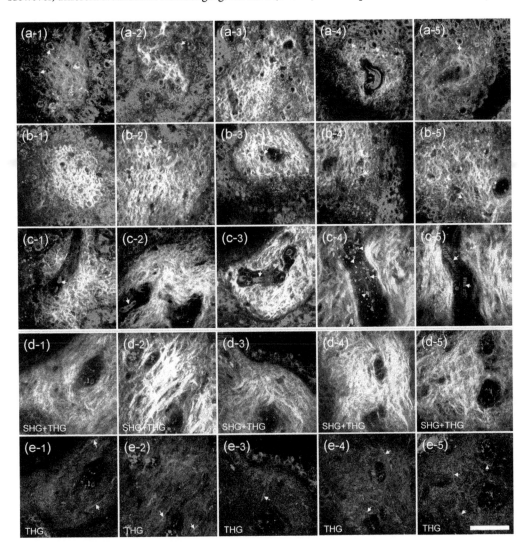

FIGURE 14.20 *In vivo* SHG/THG images of the dermis obtained in different volunteers' skin. In the dermis, THG microscopy can reveal the (a-1)–(a-5) inactive melanocytes (arrows) with oval shape and large nuclei; (b-1)–(b-5) THG-bright cells with varied shapes (arrows), probably the fibroblasts; (c-1)–(c-5) moving red blood cells (arrows) and the THG-bright spots (arrowheads in (c-4)), which are possibly the accumulating lipids in capillary walls; and (d-1)–(d-5) elastic fibers (arrows in separated THG images (e-1)–(e-5)). Scale bar: 50 μm.

modality with the THG contrast arising from hemoglobin (Tai et al. 2007), and both the movement and deformation of the RBC can be dynamically recorded. In the previous THG study (Yu et al. 2007), elastic fibers were revealed by THG modality in the human lung tissues and the THG contrast of the elastic fibers has been proven through H&E staining. In the *in vivo* SHG/THG images of the dermis, the THG-revealed elastic fibers and SHG-revealed collagen fibers are found to interlace (Figures 14.20d-1 through 14.20d-5), while in the separated THG images, the elastic fibers can be identified more clearly (arrows in Figures 14.20e-1 through 14.20e-5). Compared with the reflection confocal microscopic imaging of dermis, the cellular information and elastic fibers are all more distinguishable from the collagen fibers in the SHG/THG biopsy. This is due to a much improved spatial resolution and through the assistance of the simultaneous SHG contrast of collagen fibers.

14.3.5.2 Dynamic Information Recorded by *In Vivo* SHG/THG Imaging

In contrast to *ex vivo* SHG/THG imaging, similar static morphological information and imaging contrasts can be obtained from *in vivo* SHG/THG imaging, while valuable dynamic skin information like blood flow can only be recorded through *in vivo* SHG/THG imaging. Based on the THG contrast provided by hemoglobin, the RBC in the capillary can be clearly observed, and the movement, aggregation, and deformation of the RBC in the same capillary with a diameter of 12.6 μm are recorded at different seconds, as shown in Figures 14.21a through 14.21h. According to previous reports (Noguchi and Gompper 2005), human RBC have a biconcave-disk shape with a diameter of 8 μm and are easily deformed from a nonaxisymmetric discocyte to an axisymmetric parachute shape (coaxial with the flow axis) (Boryczko et al. 2003; Noguchi and Gompper 2005) to reduce the flow resistance of blood in capillaries. A reduction of RBC deformability and an enhanced flow resistance of blood can be found in some diseases, such as diabetes mellitus and sickle cell anemia (Tsukada et al. 2001; Havell et al. 2006). In Figures 14.21a through 14.21c, the RBC moving at a velocity lower than 0.04 mm/s are shown with a diameter of 7.5 μm, as expected from histological results, and the aggregation of the RBC, which is often seen in the capillaries, can be observed in Figure 14.21c. In Figure 14.21a, the RBC are shown in a biconcave-disk shape, while the axisymmetric parachute shape of the RBC can be observed in

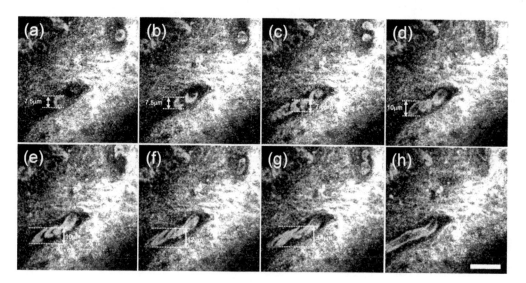

FIGURE 14.21 *In vivo* HGM images of blood flow in a capillary, dynamically recorded at different seconds. The red blood cells were found to appear in a (a) biconcave-disk shape or in a (b) axisymmetric parachute shape. (a)–(c) Red blood cells moved at a velocity lower than 0.04 mm/s and aggregation of the red blood cells can be observed in (c). (d)–(h) When moving faster, the imaged red blood cells began to be distorted due to limited frame rate. Scale bar: 20 μm.

Figures 14.21b and 14.21c. As the velocity of the RBC exceeds 0.04 mm/s, the image distortion begins to occur due to limited frame rate (Figures 14.21d through 14.21h). The velocity of the RBC can be roughly estimated from the lengths of the extended RBC to range from 0.05 mm/s (Figure 14.21d) to 0.1 mm/s (Figure 14.21g), but the velocity of the RBC shown in Figure 14.21h cannot be estimated due to a much higher velocity and serious image distortion. By using THG microscopy, the motion, shape transition, and velocity of the RBC under flow in the dermal capillaries can be easily *in vivo* monitored. THG microscopy can provide the potential for monitoring the velocity of RBC, determine the deformability of RBC, diagnose the abnormal RBC (sickle shape, for example), find the abnormal aggregation of the RBC, and so on in capillaries. Combined with SHG, which is sensitive to collagen fibers, pathological changes of capillaries in diseases like diabetes mellitus or angiogenesis in the cancerous tissues may possibly be easily investigated *in vivo*.

14.3.5.3 Damage Evaluation

In the *in vivo* SHG/THG imaging of human skin, volunteers including 10 Caucasian skins (type I and II), 37 Asian skins (type III and IV), and 1 African-American skin (type VI) have been investigated. According to different appearances and tanning ability of skin, six types of skin can be distinguished following the Fitzpatrick skin classification (Fitzpatrick and Breathnach 1963; Fitzpatrick 1988). During the clinical trials, the following protocols were applied: (1) the total exposure time was limited to 30 min for each volunteer in the same area; and (2) the average excitation power was limited by 90 mW; and (3) the frame rate was limited by 0.37 Hz. The corresponding accumulated photon energy was around 180–200 J. Before, during, and after the SHG/THG biopsy, the tested site—ulnar, ventral, upper 1/3 forearm skin of the volunteers were photographed and recorded. A research medical doctor continually checked and recorded the volunteers' status during SHG/THG observations. The skin conditions of the tested area were evaluated immediately, several hours, 24 h, 3 days, and 1 week after the experiment by a dermatologist doctor. Out of all the volunteers, only one volunteer reported a possible stingy sensation for <1 s during observations; however, the volunteer was not certain about this claim. There were no inflammatory symptoms, skin color change, pigmentation, wound, blister formation, or ulcerations reported in the volunteers' skin. Through this damage evaluation, the noninvasiveness and safety of the SHG/THG system can be further confirmed and according to volunteers' opinions, the procedure was comfortable. This result thus indicates the feasibility of the SHG/THG system for *in vivo* virtual biopsy of skin.

14.3.5.4 Lateral Resolution of *In Vivo* SHG/THG Microscopy

In the *in vivo* observation of the human skin, both the lateral resolution of THG modality at the right skin surface and the lateral resolution of SHG modality deep in the reticular dermis have been measured. Right at the surface of the skin, the lateral resolution of THG microscopy can be analyzed from the multilayer structure of the stratum corneum (Figure 14.22a). As shown in example Figure 14.22b, the THG brightness is plotted versus distance along the yellow line (arrow in Figure 14.22a). The resolution is defined as the FWHM of the fitted Gaussian curves and the lateral resolution measured in this case is 0.49 and 0.56 μm. On the other hand, since THG is too weak for measuring the lateral resolution deep in the dermis, the lateral resolution of SHG is instead analyzed from the collagen fibers and the corresponding lateral resolution of THG microscopy can be obtained. Figure 14.22c shows an example of an SHG image at 200 μm deep, the SHG brightness is plotted in Figure 14.22d) versus distance along the yellow line (arrow in Figure 14.22c). In this case, the lateral resolution of SHG microscopy is measured to be 0.66 μm and the corresponding lateral resolution of THG microscopy is 0.54 μm. Measuring the resolution in the SHG and THG images of different volunteers' skin with the same protocol, the mean lateral resolution of THG microscopy is 0.45 ± 0.05 μm right at the skin surface and 0.75 ± 0.1 μm at a 300 μm depth (Figure 14.22e). Compared with the previous image resolution analysis on fixed human skin (Table 14.3) (Tsai et al. 2006), *in vivo* SGH/THG imaging shows a much lower resolution degradation deep in the live tissue, indicating a much reduced point-spread function aberration (Gu et al. 2000) of the 1230 nm excitation light in the live tissues versus fixed tissues.

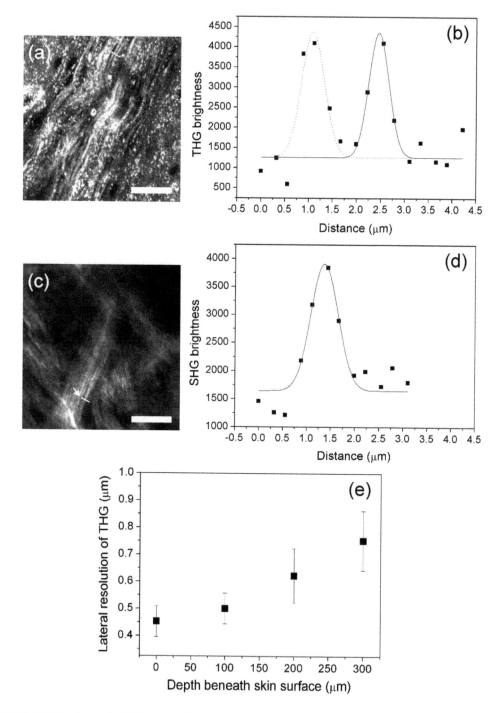

FIGURE 14.22 *In vivo* (a) THG image of stratum corneum at right skin surface and (c) SHG image of reticular dermis at 200-μm deep. The THG and SHG brightness versus distance along the yellow lines (arrows) were plotted in (b) and (d), respectively, and fitted with Gaussian curves. SHG and THG are represented by green and purple pseudocolors. Scale bar: 50 μm.

TABLE 14.3 Lateral Resolution of THG Microscopy in Live Human Skin and Fixed Human Skin

Live Human Skin (NA 1.2)			Fixed Human Skin (NA 0.9)		
	Resolution (nm)			Resolution (nm)	
Depth (μm)	SHG	THG	Depth (μm)	SHG	THG
0	554.5	453.7	75	554	453
100	613.4	500.8	150	608	501
200	762.9	622.8	225	913	752
300	921.7	752.5	300	1287	1051
Theoretical	369	307.5	Theoretical	500	401

Source: Adapted from Tsai, TH et al. 2006, Optical signal degradation study in fixed human skin using confocal micros-copy and higher-harmonic optical microscopy, *Opt. Express*, 14(2), 749–758. With permission of Optical Society of America.

14.4 Conclusion

In this chapter, the combined SHG/THG microscopy, the so-called higher-harmonic generation microscopy or multiharmonic generation microscopy, has been introduced, including the princi-ple, system setup, and biomedical applications. Instead of using a Ti:S laser for combined SHG/2PF microscopy, a Cr:F laser with a wavelength located within 1200–1300 nm is used for combined SHG/THG microscopy. Based on the Cr:F excitation, the wavelength of THG can be shifted into the visible range to facilitate the efficient detection of THG signals. The appropriate wavelength of Cr:F excitation (1230 nm) can reduce both the scattering and absorption in the bio-tissues, which can help to increase the imaging penetrability and reduce the photodamages. Different from 2PF microscopy, only virtual-level transition is involved in THG microscopy. Since both the SHG and THG processes obey the energy conservation and no energy is deposited in the interacted tissues, the combined SHG/THG microscopy can be said to be noninvasive. Utilizing the different characteristics of SHG and THG microscopy, vari-ous image contrasts can be provided by SHG and THG microscopy, respectively, to reveal lots of valu-able information in the bio-tissues. Based on the noninvasiveness nature of the combined SHG/THG microscopy, this imaging modality is very suitable for *in vivo* and clinical investigation, especially the long-term observation. So far, the combined SHG/THG microscopy has been applied to the *ex vivo* and *in vivo* investigation of different standard animal models such as zebrafish and mouse. Furthermore, this imaging modality is also used to perform the clinical diagnosis of human tissues like human skin and human oral mucosa. Owing to the interface sensitivity that THG microscopy possesses, THG microscopy can be used as a generally purposed microscopy to provide histological information of the tissues. In the combined SHG/THG microscopy, THG microscopy plays a similar role to 2PF micros-copy to localize the SHG signals and identify the contrast sources of the SHG signals. However, com-pared with 2PF microscopy, no issues about photodamages and photobleaching have to be taken into consideration in THG microscopy even when the illumination power is around 100 mW. Thus, the combined SHG/THG microscopy is noninvasive and much desirable for clinical applications. In the latter parts of this chapter, some results on the clinical trials of SHG/THG imaging of human skin have been discussed. Making use of the strong SHG contrasts provided by collagen fibers in the dermis and the THG contrasts from cellular structures, the combined SHG/THG microscopy can help to reveal both the connective tissue distribution in the dermis and the cellular morphology in the epidermis. In addition, since THG contrasts can be enhanced by melanin and hemoglobin in the skin through real-level resonance, THG microscopy can provide additional information on the pigmentation dis-tribution and blood flow. All previous studies thus show that SHG and THG microscopy are with the capability to reveal the diagnostic characteristics of various skin diseases such as nevus, melanoma, and BCC. With the diagnostic significance, its noninvasiveness nature, >300-μm imaging penetrabil-ity, and submicron spatial resolution, the combined SHG/THG microscopy is an undoubted desirable tool for clinical diagnosis. To further expand the capability of this imaging modality for disease diagnosis, a combined SHG/THG endoscopy is worth being developed in the future for investigation of organs inside the body. Moreover, for the practical usage in clinical diagnosis, a reliable data base for disease diagnosis has to be first established and robust systems for both imaging and data analysis are also required.

References

Agero, ALC, Busam, KJ, Benvenuto-Andrade, C, Scope, A, Gill, M, Marghoob, AA, Gonzalez, S, and Halpern, AC 2006, Reflectance confocal microscopy of pigmented basal cell carcinoma, *J. Am. Acad. Dermatol.*, 54(4), 638–643.

Anderson, RR and Parish, JA 1981, The optics of human skin, *J. Invest. Dermatol.*, 77(1), 13–19.

Barad, Y, Eisenberg, H, Horowitz, M, and Silberberg, Y 1997, Nonlinear scanning laser microscopy by third harmonic generation, *Appl. Phys. Lett.*, 70(8), 922–924.

Born, M and Wolf, E 1999, *Principles of Optics*, 7th Edition, Cambridge University Press, Cambridge, UK.

Boryczko, K, Dzwinel, W, and Yuen, DA 2003, Dynamical clustering of red blood cells in capillary vessels, *J. Mol. Model*, 9(1), 16–33.

Boyd, RW 1992, *Nonlinear Optics*, Academic Press, San Diego, CA.

Brown, E, McKee, T, diTomaso, E, Pluen, A, Seed B, Boucher, Y, and Jain, RK 2003, Dynamic imaging of collagen and its modulation in tumors *in vivo* using second harmonic generation, *Nat. Med.*, 9, 796–800.

Campagnola, PJ and Loew, LM 2003, Second-harmonic imaging microscopy for visualizing biomolecular arrays in cells, tissues and organisms, *Nat. Biotechnol.*, 21(11), 1356–1360.

Campagnola, PJ, Millard, AC, Terasake, M, Hoppe, PE, Malone, CJ, and Mohler, WA 2002, Three-dimensional high-resolution second-harmonic generation imaging of endogenous structural proteins in biological tissues, *Biophys. J.*, 81, 493–508.

Campagnola, PJ, Wei, MD, Lewis, A, and Loew, LM 1999, High-resolution nonlinear optical imaging of live cells by second harmonic generation, *Biophys. J.*, 77(6), 3341–3349.

Canioni, L, Rivet, S, Sarger, L, Barille, R, Vacher, P, and Viosin, P 2001, Imaging of Ca^{2+} intracellular dynamics with a third-harmonic generation microscope, *Opt. Lett.*, 26(8), 515–517.

Chan, MC, Liu, TM, Tai, SP, and Sun, CK 2005, Compact fiber-delivered Cr:forsterite laser for nonlinear light microscopy, *J. Biomed. Opt.*, 10(5), 054006.

Chang, CF, Yu, CH, and Sun, CK 2010, Multi-photon resonance enhancement of third harmonic generation in human oxyhemoglobin and deoxyhemoglobin, *J. Biophoton.*, 3(10–11), 678–658.

Chen, SY, Chen, SU, Wu, HY, Lee, WJ, Liao, YH, and Sun, CK 2010, *In vivo* virtual biopsy of human skin by using noninvasive higher harmonic generation microscopy, *IEEE J. Select. Topic Quantum. Electron.*, 16(3), 478–492.

Chen, IH, Chu, SW, Sun, CK, Cheng, PC, Lin, and BL 2002, Wavelength dependent damage in biological multi-photon confocal microscopy: A micro-spectroscopic comparison between femtosecond Ti:sapphire and Cr:forsterite laser sources, *Opt. Quantum Electron.*, 34(12), 1251–1266.

Chen, SY, Hsieh, CS, and Chu, SW 2006, Noninvasive harmonics optical microscopy for long-term observation of embryonic nervous system development *in vivo*, *J. Biomed. Opt.*, 11(5), 054022.

Chen, SY, Hsu, CYS, and Sun, CK 2008, Epi-third and second harmonic generation microscopic imaging of abnormal enamel, *Opt. Express*, 16(15), 11670–11679.

Chen, SY, Wu, HY, and Sun, CK 2009a, *In vivo* harmonic generation biopsy of human skin, *J. Biomed. Opt.*, 14(6), 060505.

Chen, SY, Yu, HC, Wang, IJ, and Sun, CK 2009b, Infrared-based third and second harmonic generation imaging of cornea, *J. Biomed. Opt.*, 14(4), 044012.

Chu, SW, Chen, SY, Chern, GW, Tsai, TH, Chen, YC, Lin, BL, and Sun, CK 2004, Studies of $\chi^{(2)}/\chi^{(3)}$ tensors in submicron-scaled bio-tissues by polarization harmonics optical microscopy, *Biophys. J.*, 86(6), 3914–3922.

Chu, SW, Chen, IH, Liu, TM, Cheng, PC, Sun, CK, and Lin, BL 2001, Multimodal nonlinear spectral microscopy based on a femtosecond Cr:forsterite laser, *Opt. Lett.*, 26(23), 1909–1911.

Chu, SW, Chen, IH, Liu, TM, Sun, CK, Lee, SP, Lin, BL, Cheng, PC, Kuo, MX, Lin, DJ, and Liu, HL 2002, Nonlinear bio-photonic crystal effects revealed with multimodal nonlinear microscopy, *J. Micro.*, 208(3), 190–200.

Chu, SW, Chen, SY, Tsai, TH, Liu, TM, Lin, CY, Tsai, HJ, and Sun, CK 2003, *In vivo* developmental biology study using noninvasive multi-harmonic generation microscopy, *Opt. Express*, 11(23), 3093–3099.

Cicchi, R, Kapsokalyvas, D, De Giorgi, V, Maio, V, Van Wiechen, A, Massi, D, Lotti, T, and Pavone, FS 2010, Scoring of collagen organization in healthy and diseased human dermis by multiphoton microscopy, *J. Biophoton.*, 3(1–2), 34–43.

Cicchi, R, Sestini, S, De Giorgi, V, Massi, D, Lotti, T, and Pavone, FS 2008, Nonlinear laser imaging of skin lesions, *J. Biophoton.*, 1(1), 62–73.

Clay, CO, Millard, AC, Schaffer, CB, Aus-der-Au, J, Tsai, PS, Squier, JA, and Kleinfeld, D 2006, Spectroscopy of third-harmonic generation: Evidence for resonances in model compounds and ligated hemoglobin, *J. Opt. Soc. Am. B*, 23(5), 932–950.

Cox, G, Kable, E, Jones, A, Fraser, I, Manconi, F, and Gorrell, MD 2003, 3-dimensional imaging of collagen using second harmonic generation, *J. Struct. Biol.*, 141(1), 53–62.

Debarre, D, Supatto, W, Pena, AM, Fabre, A, Tordjmann, T, Combettes, L, Schanne-Klein, MC, and Beaurepaire, E 2006, Imaging lipid bodies in cells and tissues using third-harmonic generation microscopy, *Nat. Methods*, 3, 47–53.

Dombeck, DA, Kasischke, KA, Vishwasrao, HD, Hyman, BT, and Webb, WW 2003, Uniform polarity microtubule assemblies imaged in native brain tissue by second-harmonic generation microscopy, *Proc. Natl. Acad. Sci. USA*, 100(2), 7081–7086.

Fitzpatrick, TB 1988, The validity and practicality of sun-reaction skin types I through VI, *Arch. Dermatol.*, 124(6), 869–671.

Forman, SB, Ferringer, TC, Peckham, SJ, Dalton, SR, Sasaki, GT, Libow, LF, and Elston, DM 2008, Is superficial spreading melanoma still the most common form of malignant melanoma?, *J. Am. Acad. Dermatol.*, 58(6), 1013–1020.

Fitzpatrick, TB and Breathnach, AS 1963, Das epidermal melanin-Einheit system. Dermatol, *Wochenschr*, 147, 481–489.

Gonzalez, S, Swindells, K, Rajadhyaksha, M, and Torres, A 2003, Changing paradigms in dermatology: Confocal microscopy in clinical and surgical dermatology, *Clin. Dermatol.*, 21(5), 359–369.

Gu, M, Gan X, Kisteman, A, and Xu, MG 2000, Comparison of penetration depth between two-photon excitation and single-photon excitation in imaging through turbid tissue media, *Appl. Phys. Lett.*, 77(10), 1551–1553.

Guo, Y, Ho, PP, Tirksliunas, A, Liu, F, and Alfano, RR 1996, Optical harmonic generation from animal tissues by the use of picosecond and femtosecond laser pulses, *Appl. Opt.*, 35(34), 6810–6813.

Guo, Y, Savage, HE, Liu, F, Schantz, SP, Ho, PP, and Alfano, RR 1999, Subsurface tumor progression investigated by noninvasive optical second harmonic tomography, *Proc. Natl. Acad. Sci. USA*, 96, 10854–10856.

Haus, HA 1984, *Waves and Fields in Optoelectronics*, Prentice-Hall Inc., Englewood Cliffs, New Jersey.

Havell, TC, Hillman, D, and Lessin, LS 2006, Deformability characteristics of sickle cells by microelastimetry, *Am. J. Hematol.*, 4(1), 9–16.

Hsieh, CS, Chen, SU, Lee, YW, Yang, YS, and Sun, CK 2008, Higher harmonic generation microscopy of *in vitro* cultured mammal oocytes and embryos, *Opt. Express*, 16(15), 11574–11588.

Huang, JY, Chen, ZP, and Lewis, A 1989, Second-harmonic generation in purple membrane-poly(vinyl alcohol) films—Probing the dipolar characteristics of the bacteriorhodopsin chromophore in br570 and m412, *J. Phy. Chem.*, 93(8), 3314–3320.

Jalbert, I, Stapleton, F, Papas, E, Sweeney, DF, and Coroneo, M 2003, *In vivo* confocal microscopy of the human cornea, *Br. J. Ophthalmol.*, 87(2), 225–236.

Kimmel, CB, Ballard, WW, Kimmel, SR, Ullmann, B, and Schilling, TF 1995, Stages of embryonic development of the zebrafish, *Dev. Dyn.*, 203(3), 253–310.

Konig, K 2008, Clinical multiphoton tomography, *J. Biophoton.*, 1(1), 13–23.

Konig, K, So, PTC, Mantulin, WW, and Gratton, E 1997, Cellular response to near-infrared femtosecond laser pulses in two-photon microscopes, *Opt. Lett.*, 22(2), 135–136.

Kung, CT, Chuang, CC, Huang, YK, Tsai, HJ, and Sun, CK 2007, *In vivo* continuous observation of vertebrate cardiac valve for congenital heart disease study and drug screening using third harmonic generation microscopy, *Conference on Lasers and Electro-Optics/Quantum Electronics and Laser Science Conference/Conference on Photonic Applications*, Baltimore, MD, paper CTuP4.

Langley, RGB, Fitzpatric, TB, and Sober, A 1998, *Clinical Characteristics. In: Cutaneous Melanoma*, Quality Medical Publishing, St Louis.

Lee, JH, Chen, SY, Chu, SW, Wang, LF, Sun, CK, and Chiang, BL 2009, Noninvasive *in vitro* and *in vivo* assessment of epidermal hyperkeratosis and dermal fibrosis in atopic dermatitis, *J. Biomed. Opt.*, 14(1), 014008.

Li, Z, Joseph, NM, and Easter, SSJR 2000, The morphogenesis of the zebrafish eye, including a fate map of the optic vesicle, *Dev. Dyn.*, 218(1), 175–188.

Liu, TM, Kartner, FX, Fujimoto, JG, and Sun, CK 2005, Multiplying the repetition rate of passive mode-locked femtosecond lasers by an intracavity flat surface with low reflectivity, *Opt. Lett.*, 30(4), 439–441.

Makino, E, Uchida, T, Matsushita, Y, Inaoki, M, and Fujimoto, W 2007, Melanocytic nevi clinically simulating melanoma, *J. Dermatol.*, 34(1), 52–55.

Matts, PJ, Dykes, PJ, and Marks, R 2007, The distribution of melanin in skin determined *in vivo*, *Brit. J. Dermatol.*, 156, 620–628.

Maurice, DM 1984, *The Eye*, Third Edition, Academic Press, London, UK.

McCally, RL and Farrell, RA 1990, *Noninvasive Diagnostic Techniques in Ophthalmology*, Springer-Verlag, New York.

Millard, AC, Wiseman, PW, Fittinghoff, DN, Wilson, KR, Squire, JA, and Muller, M 1999, Third-harmonic generation microscopy by use of a compact, femtosecond fiber laser source, *Appl. Opt.*, 38(36), 7393–7397.

Mohler, W, Millard, AC, and Campagnola, PJ 2003, Second harmonics generation imaging of endogenous structural proteins, *Methods*, 29(1), 97–109.

Moreaux, L, Sandre, O, Blanchard-Desce, M, and Mertz, J 2000a, Membrane imaging by simultaneous second-harmonic and two-photon microscopy, *Opt. Lett.*, 25(5), 320–322.

Moreaux, L, Sandre, O, and Mertz, J 2000b, Membrane imaging by second-harmonic generation microscopy, *J. Opt. Soc. Am. B*, 17(10), 1685–1694.

Muller, M, Squier, J, Wilson, KR, and Brakenhoff, GJ 1998, 3D-microscopy of transparent objects using third-harmonic generation, *J. Microsc*, 191(3), 266–274.

Noguchi, H and Gompper, G 2005, Shape transitions of fluid vesicles and red blood cells in capillary flows, *Proc. Natl. Acad. Sci. USA*, 102(40), 14159–14164.

Nucciotti, V, Stringari, C, Sacconi, L, Vanzi, F, Fusi, L, Linari, M, Piazzesi, G, Lombardi, V, and Pavone, FS 2010, Probing myosin structural conformation *in vivo* by second-harmonic generation microscopy, *Proc. Natl. Acad. Sci. USA*, 107(17), 7763–7768.

Okun, MR 1997, Silhouette symmetry—An unsupportable histologic criterion for distinguishing Spitz nevi and compound nevi from malignant melanoma, *Arc. Pathol. Lab. Med.*, 121(1), 48–53.

Peleg, G, Lewis, A, Linial, M, and Loew, LM 1999, Non-linear optical measurement of membrane potential around single molecules at selected cellular sites, *Proc. Natl. Acad. Sci. USA*, 96(12), 6700–6704.

Plotnikov, SV, Millard, AC, Campagnola, PJ, and Mohler, WA 2006, Characterization of the myosin-based source for second-harmonic generation from muscle sarcomeres, *Biophys. J.*, 90(2), 693–703.

Ramaesh, K, Ramaesh, T, West, JD, and Dhillon, B 2004, Immunolocalisation of leukaemia inhibitory factor in the cornea, *Eye*, 18(10), 1006–1009.

Schins, JM, Schrama, T, Squier, J, Brakenhoff, GJ, and Müller, M 2002, Determination of material properties by use of third-harmonic generation microscopy, *J. Opt. Soc. Am. B*, 19(7), 1627–1634.

Squier, J and Müller, M 2001, High resolution nonlinear microscopy: A review of sources and methods for achieving optimal imaging, *Rev. Sci. Instrum.*, 72(7), 2855–2867.

Squier, JA, Muller, M, Brakenhoff, GJ, and Wilson, KR 1998, Third harmonic generation microscopy, *Opt. Express*, 3(9), 315–324.

Squirrell, JM, Wokosin, DL, White, JG, and Bavister, BD 1999, Long-term two-photon fluorescence imaging of mammalian embryos without compromising viability, *Nat. Biotechnol.*, 17, 763–767.

Sun, CK 2005, Higher harmonic generation microscopy, *Adv. Biochem. Engin./Biotechnol.*, 95, 17–56.

Sun, CK, Chen, CC, Chu, SW, Tsai, TH, Chen, YC, and Lin, BL 2003, Multi-harmonic generation biopsy of skin, *Opt. Lett.*, 28(24), 2488–2490.

Sun, CK, Chu, SW, Chen, SY, Tsai, TH, Liu, TM, Lin, CY, and Tsai, HJ 2004, Higher harmonic generation microscopy for developmental biology, *J. Struct. Bio.*, 147(1), 19–30.

Tai, SP, Lee, WJ, Shieh, DB, Wu, PC, Huang, HY, Yu, CH, and Sun, CK 2006, *In vivo* optical biopsy of hamster oral cavity with epi-third-harmonic-generation microscopy, *Opt. Express*, 14(13), 6178–6187.

Tai, SP, Tsai, TH, Lee, WJ, Shieh, DB, Liao, YH, Huang, HY, Zhang, K, Liu, HL, and Sun, CK 2005, Optical biopsy of fixed human skin with backward-collected optical harmonics signals, *Opt. Express*, 13(20), 8231–8242.

Tai, SP, Yu, CH, Liu, TM, Wen, YC, and Sun, CK 2007, *In vivo* molecular-resonant third harmonic generation microscopy of hemoglobin, *Conference on Lasers and Electro-Optics/Quantum Electronics and Laser Science Conference*, Baltimore, MD, paper CTuF4.

Tsai, MR, Chen, CH, and Sun, CK 2009, Third and second harmonic generation imaging of human articular cartilage, *Proceeding of SPIE*, San Jose, CA, vol. 7183, pp. 71831V-1–71831V.

Tsai, MR, Chen, SY, and Sun, CK 2010a, *In vivo* optical biopsy of human oral cavity with higher-harmonic generation microscopy, *Proceedings of the SPIE*, San Francisco, CA, vol. 7569, pp. 75691Q–75691Q-7.

Tsai, MR, Chiou, YW, Lo, MT, and Sun, CK 2010b, Second harmonic generation imaging of collagen fibers in myocardium for atrial fibrillation diagnosis, *J. Biomed. Opt.*, 15(2), 026002.

Tsai, TH, Tai, SP, Lee, WJ, Huang, HY, Liao, YH, and Sun, CK 2006, Optical signal degradation study in fixed human skin using confocal microscopy and higher-harmonic optical microscopy, *Opt. Express*, 14(2), 749–758.

Tsang, TYF, 1995, Optical third-harmonic generation at interfaces, *Phy. Rev. A*, 52(5), 4116–4125.

Tsukada, K, Sekizuka, E, Oshio, C, and Minamitani, H 2001, Direct measurement of erythrocyte deformability in diabetes mellitus with a transparent microchannel capillary model and high-speed video camera system, *Microvasc. Res.*, 61(3), 231–239.

Wang, BG, Riemann, I, Schubert, H, Schweitzer, D, Konig, K, and Halbhuber, KJ 2007, Multiphoton microscopy for monitoring intratissue femtosecond laser surgery effects, *Lasers Surg. Med.*, 39(6), 527–533.

Yasui, T, Takahashi, Y, Ito, M, Fukushima, S, and Araki, T 2009, *Ex vivo* and *in vivo* second-harmonic-generation imaging of dermal collagen fiber in skin: Comparison of imaging characteristics between mode-locked Cr:forsterite and Ti: sapphire lasers, *Appl. Opt.*, 48(10), D88–D95.

Yelin, D, Oron, D, Korkotian, E, Segal, M, and Silberberg, Y 2002, Third-harmonic microscopy with a titanium-sapphire laser, *Appl. Phys. B*, 74(9), 97–101.

Yelin, D and Silberberg, Y 1999, Laser scanning third-harmonic-generation microscopy in biology, *Opt. Express*, 3(8), 169–175.

Yu, CH, Tai, SP, Kung, CT, Wang, IJ, Yu, HC, Huang, HJ, Lee, WJ, Chan, YF, and Sun, CK 2007, *In vivo* and *ex vivo* imaging of intra-tissue elastic fibers using third-harmonic-generation microscopy, *Opt. Express*, 15(18), 11167–11177.

Zoumi, A, Yeh, A, and Tromberg, BJ 2002, Imaging cells and extracellular matrix *in vivo* by using second-harmonic generation and two-photon excited fluorescence, *Proc. Natl. Acad. Sci. USA*, 99(17), 11014–11019.

SHG Imaging of Collagen and Application to Fibrosis Quantization

Marie-Claire
Schanne-Klein
École Polytechnique
CNRS–Inserm

15.1 Introduction

15.1.1 Collagenous Fibrosis: Biomedical Issues

Collagen is the most abundant protein in the extracellular matrix (ECM), and plays a central role in the formation of fibrillar and microfibrillar networks, basement membranes, as well as other structures of the connective tissue (Hulmes 2002). Many genetically distinct collagen types have been described so far, and show different structures, functions, and distribution in tissues (Ricard-Blum and Ruggiero 2005). The characteristic feature of a collagen molecule is its long triple helical structure. Three polypeptide chains, called α chains and characterized by a (Gly-X-Y) repeated structure, are wrapped around one another in a ropelike right-handed superhelix (Beck and Brodsky 1998). The most abundant family of collagens with more than 90% of the total collagen consists of the fibril-forming collagens: mainly collagen I, II, III, V, and XI, whose helical domains are continuous over typically 1000 amino acids. They are found in a wide variety of tissues such as bone, tendon, skin, ligament, cornea, and internal organs. Once the procollagen molecules are secreted from the cells into the ECM, they are cleaved to collagen molecules which self-assemble into fibrils (diameter: 10–300 nm, length: up to several hundred μm). In tissues such as tendons, collagen I fibrils form bundles or fibers with diameters between 0.5 and 3 μm. Other collagens, such as collagen IV, do not form fibrils, because of numerous interruptions in the helical sequence by non-collagenous domains. Individual collagen IV molecules assemble to form a two-dimensional network in the matrix which is found mainly in basement membranes.

The macromolecular organization of collagens is crucial in the organs' architecture. The quantity and distribution of the various types of collagens result from a balance between synthesis and assembly mechanisms on one hand, and degradation mechanisms on the other, which are regulated by complex signal pathways. In response to various injuries, the three-dimensional (3D) distribution of collagen is modified and fibrillar collagen accumulates in the tissue. This cascade of events is called fibrosis. It alters the structure of affected organs and leads to their functional failure. For instance, renal fibrosis induced by hypertensive inflammatory and mechanical stress is now the first cause of renal insufficiency. However, the relationships between tissue injury, enzymes activity, and tissue remodeling are only poorly understood because of the limitations of conventional imaging techniques. It is therefore crucial to develop new approaches to study such processes taking place at different scales.

In that respect, the most important issues are the visualization of the fibrosis *3D architecture* in intact biological tissues and the determination of *quantitative* indexes of collagen fibrosis. Visualization of the fibrosis network would give insight into the biological mechanisms of fibrosis progression in relationship with other components of the tissue. It necessitates a multimodal approach to visualize simultaneously fibrillar collagens and various proteins of interest or pathological processes, including inflammatory processes. Determination of fibrosis quantitative indexes would enable unambiguous quantization of the role of various enzymes that regulate synthesis/assembly and degradation of collagen in the fibrosis progression and, ultimately, in the fibrosis regression. Such a quantitative imaging method has to be specific to fibrillar collagen that appears to be a better predictor of severe pathological progression than nonfibrillar collagen because it is more resistant to proteolysis.

15.1.2 SHG Imaging of Fibrillar Collagens

In this context, second-harmonic generation (SHG) microscopy appears as a valuable technique since fibrillar collagens exhibit strong endogenous SHG signals. SHG is a nonlinear optical (or multiphoton) process that is complementary to two-photon excited fluorescence (2PEF) and that appears at the harmonic frequency of the laser excitation (that is half the excitation wavelength). Endogenous SHG signals have been observed in a limited number of biological compounds, mainly collagen, skeletal muscle, starch, and so on (Campagnola et al. 2002, Zipfel et al. 2003). SHG signal from collagen is the largest SHG signal in mammals and is specific to fibrillar types of collagen, as verified by comparison with immunochemical labeling (Zoumi et al. 2002, Brown et al. 2003, Strupler et al. 2007).

The reason why SHG is specific to fibrillar collagen is related to the physics of this nonlinear signal. The next section will therefore present the physical origin of SHG endogenous signal in collagen. Readers who are not familiar with nonlinear optics and chemical-physics may just read the summary given as the last subsection. In the third section, we will review SHG imaging of the lung, kidney, and liver fibrotic tissues and discuss the contribution of this technique to fibrosis imaging. The fourth section will be devoted to fibrosis quantization by use of SHG microscopy. Finally, we will discuss the advantages and limitations of SHG microscopy for fibrosis scoring and give some perspectives.

15.2 Physical Origin of SHG Response from Collagen and Specificity to Fibrillar Collagens

The physics of collagen SHG must be considered at two different scales. First, we must identify what chemical entities exhibit a nonlinear response within the collagen triple helix, and second, we have to characterize how the SHG signal builds up at macromolecular scale, that is at the scale of the focal volume in SHG microscopy.

15.2.1 Molecular Origin of Collagen Nonlinear Response

At the molecular scale, SHG is related to the presence of polarizable electrons in a noncentrosymmetric environment, usually between electron donor and electron acceptor chemical groups (Oudar and Chemla 1977). Polarization of these electrons by a strong electric field (such as the one in a multiphoton microscope) results in a nonlinear behavior with components at the second-harmonic frequency since electronic oscillations are favored toward the electron acceptor group. The question is then to identify these so-called "harmonophores" in collagen. Aromatic amino acids have been shown to exhibit a second-order nonlinear response (Duboisset et al. 2010), but they are almost absent from the amino-acid sequence of the collagen triple helix that is mainly composed of glycine, proline, and hydroxyproline. Moreover, SHG signals have been recorded for various types of collagen from many different mammals, independently of the precise amino-acid sequence.

It was therefore proposed that the second-harmonic response of collagen originates in the peptide bond itself (Plotnikov et al. 2006, Tiaho et al. 2007, Han et al. 2008, Deniset-Besseau et al. 2010). This noncentrosymmetrical chemical bond of the peptide backbone indeed exhibits π-electrons delocalized between C=O and N–H groups that behave as slight electron donors and electron acceptors, respectively (see Figure 15.1a) (Levine and Bethea 1976). This assertion is supported by polarization-resolved SHG images of collagenous tissues and polarization-resolved hyper Rayleigh scattering (HRS) measurements of collagen solutions (Plotnikov et al. 2006, Tiaho et al. 2007, Han et al. 2008, Deniset-Besseau et al. 2010). These experiments measure nonlinear responses for excited fields whose polarization is parallel or perpendicular to the fibrils main axis. Their ratio can be related to the direction of the nonlinear electronic oscillation within the harmonophore. They consistently show that this nonlinear electronic oscillation is oriented along the peptide backbone in collagen.

15.2.2 Building the SHG Signal at Macromolecular Scale

SHG is a *coherent* second-order nonlinear process, which means that it builds up as the summation of all the second-harmonic electric fields radiated by all the harmonophores within the focal volume.

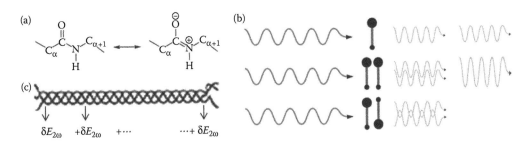

FIGURE 15.1 Physical origin of SHG signal in collagen. (a) Chemical structure of the peptide bond showing delocalization of π-electrons in a noncentrosymmetric environment. (b) Scheme of second-harmonic response from 1 isolated harmonophore, 2 harmonophores with the same direction, and 2 harmonophores with antiparallel directions. In the last case, the 2 harmonic fields have opposite phases and cancel out while the harmonic fields radiated from 2 parallel harmonophores interfere constructively, which results in a doubled harmonic field. SHG intensity that is obtained as the square of the total harmonic field exhibits a quadratic dependence as a function of the number of aligned harmonophores. (c) Scheme of coherent amplification of harmonic response along the rigid and compact collagen triple helix. (Adapted from Strupler, M. et al., 2008. *J. Biomed. Optics* 13:054041. With permission of SPIE.)

A large SHG response is obtained when the harmonophores are aligned in the same direction so that the second-harmonic fields add up in a constructive way as depicted in Figure 15.1b. On the contrary, a centrosymmetric distribution of harmonophores results in destructive interferences and vanishing SHG signal at the macromolecular scale. The SHG intensity is finally obtained as the square of the total SHG field and exhibits a quadratic dependence with the number of aligned harmonophores. SHG is therefore specific to *dense noncentro-symmetric* macromolecular organizations.

Let us now examine how the SHG signal builds up within the collagen triple helix. Since all the peptide bonds are somehow aligned in the same direction, the second-harmonic fields add up mainly in a constructive way. Moreover, the collagen triple helix is a very dense peptide structure. The SHG signal is therefore efficiently amplified along the compact and rigid triple helix (see Figure 15.1c). This mechanism is supported by HRS measurements of the nonlinear response of type I collagen and of a short collagen-like model peptide (Deniset-Besseau et al. 2009). Coherent amplification however saturates in collagen I whose length (300 nm) is close to the optical wavelength (400 nm) because of phase shifts along the triple helix. Model calculations of this effect provide an estimation of the collagen nonlinear response as a function of the length of the triple helical domain, which is the only relevant parameter in that respect (Deniset-Besseau et al. 2009). Advanced quantum chemistry calculations have also been performed recently to test this additive model. They qualitatively reproduce experimental measurements although some theoretical refinements are required (Loison and Simon 2010).

This mechanism of coherent amplification also applies at the macromolecular level. Accordingly, SHG signals have been reported in fibrillar collagens because of the dense and aligned organization of collagen molecules within collagen fibrils (Campagnola et al. 2002, Zoumi et al. 2002, Cox et al. 2003, Zipfel et al. 2003, Strupler et al. 2007). On the contrary, nonfibrillar collagen IV does not exhibit any SHG signal because it is organized as a centrosymmetric network with low density (Strupler et al. 2007). Aligned collagen IV molecules deposited as a thin film do however exhibit an SHG response (Pena et al. 2005). It proves that the vanishing SHG response from collagen IV in tissues is related to its macromolecular organization, not to the precise amino-acid sequence (Strupler et al. 2007). In that respect, SHG imaging is very different from immunochemical techniques.

At higher scale, for a set of fibrils (or a fiber), the SHG signal is a complex process because fibrils may be aligned in opposite directions within the focal volume, so that their radiated harmonic fields cancel out. This effect was demonstrated in rat-tail tendons by use of piezo response force microscopy (Rivard et al. 2011). It results in a decreased SHG signal in the forward direction, while backward-directed SHG signal is less modified because it corresponds to smaller coherence lengths. Nevertheless, forward SHG signal usually exceeds backward SHG signal because phase matching is favored in the forward direction. (It is 10 times larger in Achille tendon, for instance (Legare et al. 2007).) Phase matching along the focal volume is also sensitive to the Gouy phase shift like in third-harmonic microscopy (Barad et al. 1997, Débarre et al. 2005, Williams et al. 2005, Olivier and Beaurepaire 2008). The SHG signal observed in collagenous tissues must therefore be considered as a complex interference pattern from all the fibrils in the focal volume, not as a morphological image of the fibrillar structure (LaComb et al. 2008b, Strupler and Schanne-Klein 2010, Rivard et al. 2011).

15.2.3 Summary

To sum up, the strong SHG response of fibrillar collagen is not related to the presence of a strong harmonophore in the amino-acid sequence but to the tight alignment of many weakly efficient harmonophores that are the peptide bonds. This tight alignment toward the same direction results in an efficient coherent amplification at all the hierarchical levels of the fibrillar collagen organization: single α chain $\Rightarrow$ triple helix $\Rightarrow$ fibril $\Rightarrow$ fiber. Fibrils may however be organized in a complex manner within the focal volume and strong focusing, moreover, complicate phase-matching processes. Consequently, the resulting image is a complex interference process that may not fully reproduce the morphology of the fibrils while highly specific to fibrillar collagens. Note that an SHG signal can also be observed in *aligned liquid* solutions of collagen as reported in collagen liquid crystalline organizations (Deniset-Besseau et al. 2010). While these concentrated solutions at acidic pH do not form fibrils, they show aligned molecular domains with sufficient density to exhibit a significant SHG signal.

Altogether, SHG microscopy is a structural probe of the macromolecular organization of collagen and it is highly specific to fibrillar collagens in tissues. In that respect, it appears to be a relevant method for fibrosis imaging.

15.3 Second-Harmonic Imaging of Tissue Fibrosis

15.3.1 Multiphoton Imaging of Lung, Kidney, and Liver Fibrosis

SHG potential for fibrosis assessment was first mentioned in 2003 in a paper by Cox et al. (2003). SHG imaging and scoring of fibrosis has then been implemented by a few groups and mainly applied to lung, renal, and liver fibrosis (Banavar et al. 2006, Pena et al. 2007, Strupler et al. 2007, 2008, Sun et al. 2008, 2010, Tai et al. 2009b, Gailhouste et al. 2010, He et al. 2010, Raub et al. 2010). It proved relevant to visualize specifically collagen fibrosis in unstained tissues both in animal models and in human biopsies. SHG microscopy has been usually combined with 2PEF microscopy to take advantage of intrinsic fluorescence from various tissue components to visualize the tissue morphology.

Figure 15.2 displays such combined SHG/2PEF images of an intact fibrotic lung from a bleomycin-treated mouse. The SHG signal reveals the accumulation of fibrillar collagen in heterogeneously distributed areas in the alveolar interstitium, which is characteristic of bleomycin-induced fibrosis and of idiopathic pulmonary fibrosis in humans (Pena et al. 2007). Another example of multiphoton imaging of fibrosis is displayed in Figure 15.3, which shows coronal slices of murine control and fibrotic kidneys. In the cortical region, SHG reveals collagen fibrils in the Bowman's capsule and in the tubular interstitium. SHG signal is also observed from fibrillar collagen in the adventitia of arcuate arteries and of arterioles in the interstitium. The 3D distribution of fibrillar collagen in the arterial adventitia and in the tubular interstitium is better visualized in 3D reconstructions obtained from image stacks acquired at increasing depths within the thick tissue (see Figure 15.3d) (Strupler et al. 2008). Finally, SHG images of rat liver fibrosis are displayed in Figure 15.4, which shows morphological changes of collagen distribution during fibrosis: collagen deposition radiates around the portal tract and then extends to form septa that connect and achieve complete bridges with adjacent portal tract and central veins (Sun et al. 2008).

The above examples refer to fibrotic samples from animal models that were mostly used to develop optimized protocols for fibrosis imaging (Strupler et al. 2007, Sun et al. 2008). Nevertheless, human fibrotic tissues have also been imaged by use of SHG microscopy as shown in Figures 15.5 through 15.7. Figure 15.5 displays SHG/2PEF images of normal and fibrotic human lung histological sections: SHG image of idiopathic pulmonary fibrosis highlights marked architectural changes with fibrosis onset and distorted alveolar spaces (Figure 15.5b). It reveals the distribution of collagen mainly in the periphery of fibroblastic foci, with fibroblasts and thinner fibrillar components in the center (Figure 15.5c) (Pena et al. 2007). Human biopsies of fibrotic kidney implants show similar collagen distribution as in murine samples with accumulation of fibrillar collagen in the Bowman's capsule and in the tubular interstitium (see Figure 15.6) (Strupler et al. 2008). The distribution of collagen fibrosis in human liver biopsies is also similar to the one observed in rat tissues with deposits of fibrillar septa connecting portal tracts, although the collagen distribution may differ according to the underlying liver disease (see Figure 15.7) (Gailhouste et al. 2010).

15.3.2 Advantages of SHG Microscopy

The studies reported in the last section highlight many advantages of SHG microscopy compared to conventional techniques for fibrosis assessment. The main advantage is obviously the high specificity of SHG microscopy to fibrillar collagens, as explained in the previous section. The other advantages that often stem from the latter one are detailed in the following.

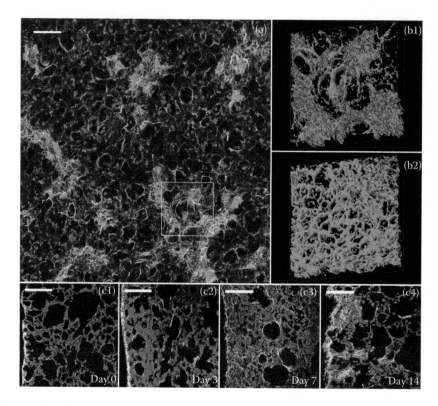

FIGURE 15.2   (See color insert.) SHG/2PEF imaging of lung fibrosis in bleomycin-treated mice. SHG (green: collagen) and 2PEF (red: elastin, macrophages) signals excited with 50 mW excitation power at 860 nm. Yellow color: SHG/2PEF colocalization. (a) Fresh unlabeled fibrotic lung (day 14), at 42 μm under the pleura, evidencing the heterogeneous fibrosis distribution; (b) 3D reconstruction of the SHG signal from the underlined area in (a), within the tissue (b1) and in the pleurae (b2); (c) unstained histological sections embedded in paraffin, from (c1) control, (c2) day 3, (c3) day 7, and (c4) day 14 bleomycin lungs. Bleomycin treatment induces (i) inflammation, which decreases after day 7, as evidenced by the 2PEF signal, and (ii) fibrosis, evidenced by the SHG signal detected as early as day 3. Scale bar: 100 μm. (From Pena, A.-M. et al. Three-dimensional investigation and scoring of extracellular matrix remodeling during lung fibrosis using multiphoton microscopy. *Microsc. Res. Tech.* 2007. 70:162–170. Copyright Wiley-VCH Verlag GmbH & Co. KGaA. Reproduced with permission.)

15.3.2.1 3D Capability

Like other multiphoton microscopies, SHG microscopy offers 3D capability that enables the visualization of the 3D architecture of collagen fibrosis. Other techniques are either restricted to 2D sections (histological or immunochemical labeling) or show limited specificity to fibrillar collagen (reflectance confocal microscopy or optical coherence tomography). SHG microscopy is therefore the only technique that enables highly specific 3D imaging of intact collagen tissues.

15.3.2.2 Highly Contrasted Images/Sensitivity

The high specificity of SHG microscopy to fibrillar collagen results in an excellent signal-to-noise ratio in the SHG images and enables sensitive measurements of the fibrosis distribution. Furthermore, SHG scales *quadratically* with the density of aligned collagen molecules within the focal volume, as explained in the former section (see Figure 15.1). It results in highly contrasted images compared to conventional techniques that all scale *linearly* with the density of (aligned) collagen molecules. This feature was recently illustrated in collagen liquid crystalline samples where both the SHG and the 2PEF signals from collagen were recorded (Deniset-Besseau et al. 2010). The SHG signal intensity unambiguously displayed

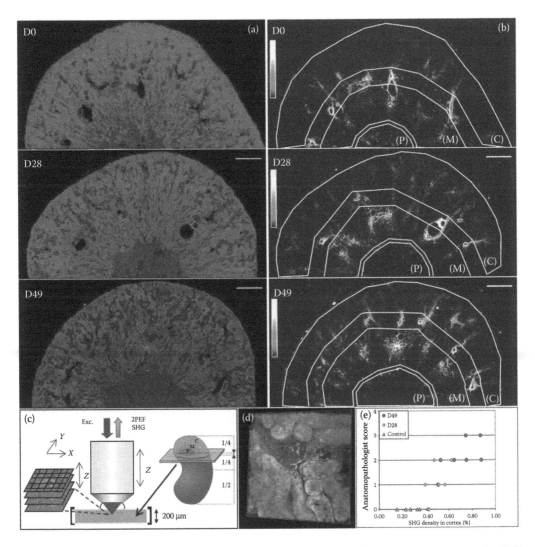

FIGURE 15.3 **(See color insert.)** Multiphoton images of fibrotic murine kidney tissue. (a) SHG/2PEF and (b) segmented SHG images of coronal renal section from a control, 28-days, and 49-days Angiotensin II infused mouse showing renal papilla (P), medulla (M), outer cortex (C), and arcuate arteries along the boundary between cortex and medulla. SHG (green color) reveals collagen fibers in the tubular interstitium and in the arterial adventitia, and 2PEF (red color) underlines tubules. These 4.8×2.4 mm^2 images are obtained by stitching 10×20 laser scanned images (dimension: 270×270 µm^2, pixel size: 0.8×0.8 µm^2) acquired sequentially by moving the kidney sample with a motorized microscope stage. Scale bar: 500 µm. The white lines underline segmentation used for SHG scoring (excluding the artery region and the borders). (c) Scheme of the laser scanning multiphoton microscope showing epidetection of z-stacks of SHG/2PEF images and kidney coronal slicing for multiphoton microscopy. (d) 3D reconstruction showing interstitial fibrosis ($270 \times 270 \times 40$ µm^3 with $0.4 \times 0.4 \times 0.5$ µm^3 voxel size). (e) Automated SHG scoring of cortical fibrosis plotted as a function of a semi-quantitative estimate of interstitial damage for control and hypertensive mice. Both scores are correlated ($\tau = 0.68$, $p < 0.01$). (From Strupler, M. et al. 2008. *J. Biomed. Optics* 13:054041. With permission of SPIE.)

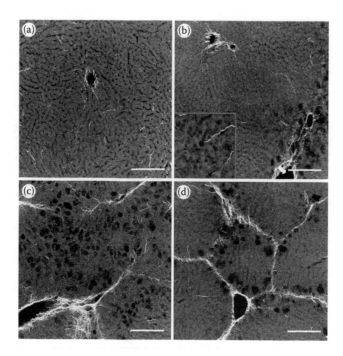

FIGURE 15.4 Morphological changes at different stages of liver fibrosis recorded with SHG and 2PEF microscopies. (a) Collagen fibrils are uniformly distributed throughout the normal liver slice, and no necrosis (dark region) is observed in 2PEF image. (b) Hepatocyte balloon degeneration appears around the central vein (pointed by the white arrow) 3 days after CCl4 injection. Development of necrosis (dark areas) starts at this stage, which can be seen in the magnified view at the bottom left corner. In addition, the amount of distributed collagen fibrils appears to be decreasing as well. (c) Vacuoles caused by hepatocyte necrosis appear massively, and collagen tends to form septa after continuous injection of CCl₄ for 14 days. (d) Bridge fibrosis is formed after injection for 21 days. 2PEF is shown in dark gray, SHG is in gold, and the scale bars shown are 200 μm. (From Sun, W. X. et al. 2008. *J. Biomed. Optics* 13:064010. With permission of SPIE.)

a quadratic behavior as a function of the collagen concentration obtained from the 2PEF signal that is characterized by a linear dependence with the molecular density.

15.3.2.3 Multimodality and Versatility

SHG microscopy is a highly versatile technique that requires no specific tissue preparation. It applies to *in vivo* or *ex vivo* intact tissues, to frozen tissues, to fixed tissues using ethanol, acetone, PFA or any other fixation method and to paraffin-embedded sections. SHG imaging is therefore compatible with any specific protocol aiming at visualizing specific components of the tissue, usually by use of other multiphoton modalities. Indeed, SHG imaging can be easily combined with other multiphoton modalities by taking advantage of their spectral difference: SHG appears at exactly half the excitation wavelength whereas 2PEF is Stokes-shifted and THG appears at the third of the excitation wavelength. Most importantly, SHG microscopy can be combined with 2PEF imaging of fluorescent constructs or immunochemical labels to visualize proteins of interest or specific cells at the same time as the fibrosis distribution within the tissue. This multimodal approach is highly interesting for fundamental studies aiming at deciphering fibrosis mechanisms since it enables tracking of cells that may participate in fibrosis progression or location of proteins that may take in fibrosis signaling pathway. An example is displayed in Figure 15.8 that shows a frozen unfixed section of a fibrotic murine kidney labeled with a histidine-tagged green fluorescent protein (His₆-Xpress-GFP). This fluorescent probe has been shown

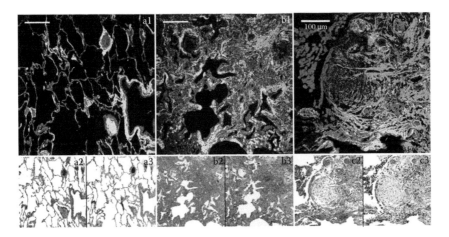

FIGURE 15.5 **(See color insert.)** Normal and fibrotic human lung samples. (Top) SHG/2PEF images and (bottom) transmitted-light images of serial histological sections (scale bar: 200 μm). (Top) Unstained section, with SHG in green and 2PEF in red; (bottom) HPS staining, and Masson's trichrome staining. (a) Control lung showing a bronchovascular axis and surrounding alveolar spaces with thin alveolar walls, as revealed by HPS staining (a2). Masson's trichrome staining (a3) highlights fibrosis in green, which is seen mainly around the vessels and the bronchovascular axis, and correlates to the SHG collagen distribution (a1, green and yellow). (b) Fibrotic lung (idiopathic pulmonary fibrosis) showing marked architectural change with fibrosis and distorted alveolar spaces. (c) Fibroblastic focus composed by myo-fibroblasts (c2), and surrounded by fibrosis (c3, Masson's trichrome). SHG/2PEF image (c1) highlights the collagen distribution mainly in the periphery of the focus, with thinner fibers in its center, and the central accumulation of fibroblasts. (From Pena, A.-M. et al. Three-dimensional investigation and scoring of extracellular matrix remodeling during lung fibrosis using multiphoton microscopy. *Microsc. Res. Tech.* 2007. 70:162–170. Copyright Wiley-VCH Verlag GmbH & Co. KGaA. Reproduced with permission.)

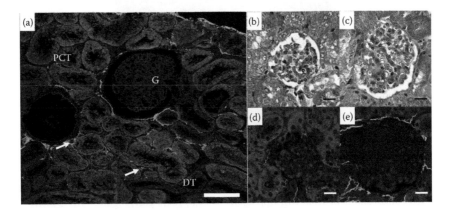

FIGURE 15.6 **(See color insert.)** SHG microscopy versus histological staining of kidney cortex. (a) Multiphoton image of an unstained human renal implant biopsy. SHG (green color) reveals collagen fibers in the Bowman capsule and the tubular interstitium. Endogenous 2PEF (red color) underlines tubules, glomeruli, and arterioles. PCT, convoluted proximal tubules; DT, distal tubules; G, glomerulus; white arrows, arterioles. (b,c) Histological versus (d,e) multiphoton images (same colors as for a) of serial renal sections of control mouse (b,d) and hypertensive mouse infused with AngII for 28 days (c,e). Note that hypertension promotes the accumulation of nonfibrillar ECM within the glomerular tuft stained with Masson's trichrome (c) (SHG silent in (e)) and the accumulation of fibrillar collagen in the Bowman capsule and the surrounding interstitium, similarly to the observation in the human biopsy in (a). Scale bars: 100 μm (a), 20 μm (b,c,d,e). (Simplified from Strupler, M. et al. 2008. *J. Biomed. Optics* 13:054041. With permission of SPIE.)

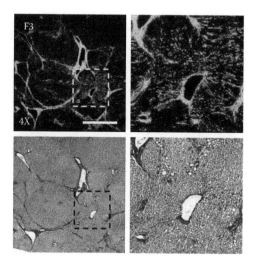

FIGURE 15.7 SHG imaging of human liver fibrosis. Comparison between SHG imaging (top) and transmitted-light microscopy after Sirius red staining (bottom) of human liver fibrosis (F3-Metavir biopsy). Right pictures show high-magnification imaging performed in the area delimited by the dotted square. Samples were 5 μm thick. Laser excitation: 100 mW at 810 nm wavelength; scale bar: 1 mm. (Reprinted from *J. Hepatol.*, 52, Gailhouste, L. et al., Fibrillar collagen scoring by second harmonic microscopy: A new tool in the assessment of liver fibrosis, 398–406, Copyright 2010, with permission from Elsevier.)

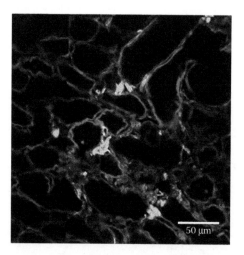

FIGURE 15.8 Simultaneous imaging of TG2 activity and fibrillar collagen deposition. SHG imaging (light grey) reveals collagen fibrosis and 2PEF from His$_6$-Xpress-GFP labeling (dark grey) highlights TG2 activity in the kidney cortex of a fibrotic mice (28-day AngII infusion). Laser excitation: 20 mW at 800 nm. Scale bar: 50 μm. (From Strupler, M. et al., 2008. *J. Biomed. Optics* 13:054041. With permission of SPIE.)

to be a good substrate for tissular transglutaminase (TG2), an enzyme that catalyzes crosslinking of collagen molecules within fibrils (Furutani et al. 2001). Combined SHG/2PEF imaging then enables the compared localization of the TG2 activity and of the fibrosis extent within the kidney cortex.

15.3.2.4 Fibrosis/Inflammation

An interesting application of the multimodal approach described above concerns the simultaneous visualization of fibrosis and inflammation by use of combined SHG/2PEF microscopy of unstained tissues. Indeed, inflammation has been shown to correspond to an increase of the endogenous fluorescence of the tissue. It is presumably due to a strong metabolic activity of the cells involved in the inflammatory response. An example is displayed in Figure 15.2a: we observe many circular structures that exhibit a strong 2PEF signal and are less present in control lungs. These round cells have been shown to correspond to macrophages using immunochemical labeling (Pena et al. 2007). Combined SHG/2PEF microscopy therefore enables the discrimination of two major pathological processes, fibrosis, and inflammation that usually takes place simultaneously or sequentially in fibrotic pathologies. It is exemplified in Figure 15.2c that shows an unstained section of murine lungs at 0, 3, 7, and 14 days after bleomycin-injection. In that animal model, inflammation peaks at days 3–7, before fibrosis that is detectable as early as day 3, and increases until day 14 (Pena et al. 2007).

15.3.2.5 Fibrosis/Sclerosis

As already noted above, SHG is highly specific to fibrillar collagens and gives no signal for the other components of the ECM, including nonfibrillar collagens. As a consequence, SHG microscopy discriminates between *fibrosis* that is precisely defined as the accumulation of fibrillar collagen, and *sclerosis* that is the accumulation of nonfibrillar collagens and other matrix compounds. Both processes are involved in fibrotic pathologies, but fibrosis onset is a crucial step in the pathology progression. Indeed, fibrillar collagen is more resistant to proteolysis than nonfibrillar matrix and fibrosis regression is quite unlikely. Fibrosis therefore appears to be a better predictor of severe pathological progression than sclerosis. For instance, interstitial fibrosis is considered as a strong indication for nephropathy progression to end-stage renal disease (Nath 1992, Nicholson et al. 1996) although it shows poor functional outcomes. Specific visualization and quantization of fibrosis is therefore of great interest for pathologists to improve their diagnosis and to adjust the patient treatment. It is also crucial for fundamental studies to better characterize the progression of fibrotic pathologies.

To summarize this section, SHG microscopy offers many advantages for fibrosis imaging, mainly specificity and sensitivity to fibrillar collagen compared to other components of the matrix. It is therefore suitable for quantitative approaches as shown in the next section.

15.4. Fibrosis Quantitative Scoring Using SHG Microscopy

15.4.1 Issues

The development of quantitative fibrosis indexes is a crucial issue for pathologists since conventional techniques do not provide reliable and reproducible indexes. For instance, the interstitial fibrosis index in the Banff classification that is used to assess renal allograft pathologies (Racusen et al. 1999, Solez et al. 2007) shows poor reproducibility between different pathologists and embarrasses comparison of clinical data from different centers (Marcussen et al. 1995, Furness et al. 2001, Gough et al. 2002, Seron et al. 2002). Fibrosis scoring in liver affected by chronic hepatitis C also shows reproducibility limitations (Bedossa et al. 1994).

The poor reproducibility and reliability of conventional techniques originate from two different limitations. Conventional fibrosis scoring is based on semi-quantitative analysis of stained histological biopsies by trained pathologists. The first limitation is then related to the imaging technique itself, and the second one is related to the image analysis. Regarding the imaging technique, the drawback of histological staining is the lack of specificity to fibrillar collagen. Masson's trichrome, which is used for instance in the Banff score of kidney allograft pathology, reveals in blue-green color all the components of the ECM. Picrosirius red, which is recommended for instance for the METAVIR score of liver fibrosis in hepatitis C, stains all collagen types and cannot discriminate between fibrillar and nonfibrillar collagens (unless the biopsy is visualized by use of polarized microscopy which is not the usual technique in clinical centers, see Section 5.1). Note that immunochemical labeling is a highly specific technique that could be used to improve specificity of fibrosis scoring. However, this technique is not reproducible because it is quite sensitive to many parameters affecting the yield of the immunochemical reaction (accessibility of the epitope, tissue preservation, section thickness, etc.). Generally speaking, any staining procedure, including histological staining, strongly limits the reproducibility of fibrosis scoring because of issues such as dye preservation, section thickness, and so on. In that respect, SHG microscopy is the only specific and reproducible technique for fibrosis scoring because it applies to unstained tissues.

The second limitation regards the image analysis. To improve the reproducibility of fibrosis scoring between different clinical or research centers, image analysis must be automated and provide quantitative scores. There is a lot of activity in that field to develop automated methods based on image analysis to quantify fibrous tissue amount (Pilette et al. 1998, Grimm et al. 2003, Pape et al. 2003, Sund et al. 2004, Friedenberg et al. 2005, Matalka et al. 2006, Goodman et al. 2007, Servais et al. 2007, 2009). Morphometric analysis provides continuous indexes of the fibrosis extent (the ratio of fibrosis area to the total area of the imaged section) instead of the conventional classification in a few grades (ci0 to ci3 in the Banff classification of kidney allograft pathology, F0–F4 in Metavir score of liver hepatitis C). The main difficulty however stems from the variability in the colorations that requires advanced color segmentation. On the contrary, SHG microscopy provides gray-level images of the fibrosis that can be analyzed in a more direct way. Basic thresholding of the SHG image and calculation of the mask area are sufficient to obtain a score of the fibrosis extent. More advanced image processing is however useful to increase the reproducibility and sensitivity of SHG scoring, as presented below. Note however that neither SHG microscopy nor conventional techniques provide properly quantitative fibrosis scores. All these techniques assess the extent of fibrosis but do not quantify the number of collagen molecules accumulating in the tissue. Indeed, the SHG signal depends in a complex way on the distribution and density of collagen in the focal volume as explained in the former sections. Proper quantization of collagen accumulation is a challenging issue that requires complementary experimental and theoretical work (see perspectives below). Consequently, phenomenological approaches have been preferred yet and SHG in practice quantifies the extent of the fibrillar collagen network within the tissue.

15.4.2 Experimental Setup and Protocols

Fibrosis quantitative imaging was first proposed using custom-built multiphoton microscopes. It can also be implemented in recent commercial setups with sensitive nondescanned detection channels if fibrosis detection is optimized as follows:

- The laser excitation has to be circularly polarized to obtain a homogeneous SHG image from fibrils with various orientations within the focal plane. SHG is indeed described by a third-rank tensor and is sensitive to the orientation of the excitation electric field relative to the fibrils orientation. The way to achieve a circular polarization is to put a quarter waveplate in the laser excitation path. It is better to use an achromatic waveplate and to put it at the back pupil of the objective lens. Due to the ellipticity introduced by the laser scanning components, the achieved polarization will not be perfectly circular, unless ellipticity is corrected by another quarter waveplate (Gusachenko et al. 2010). However, it is sufficient to mitigate orientation effects and to image all the fibrils within the focal plane with similar efficiency. Note that advanced polarization shaping of the laser excitation may be used to image out-of-plane fibrils (Yew and Sheppard 2007, Yoshiki et al. 2007).
- SHG signal from thick fibrotic tissues can be detected either in the forward direction or in the backward direction. Backward detection takes advantage of scattering processes that partly redirect forward-directed SHG signal in the backward direction. It is more efficient using an objective with a large field of view to better collect scattered signals (Beaurepaire and Mertz 2002). SHG images of thin sections are better recorded in the forward direction because SHG signals are more important in the forward direction and scattering is negligible in thin samples.
- The choice of the objective is an important issue since it must combine a large field of view to visualize significant regions of the fibrotic tissue and a good resolution to detect thin collagen

fibrils. Water immersion objectives with 0.6–1.0 numerical aperture and 10–40× magnification are a good choice. Since fibrotic processes are usually heterogeneous, it is worth mapping larger sample areas by using a motorized stage to scan the tissue in the focal plane. Subsequent stitching of the images provides centimeter-size images with micrometer resolution of a full human biopsy or animal small organ.

- Quantitative imaging requires using always the same experimental conditions for the sake of reproducibility. In practice, it pertains to the sample preparation and to the excitation power. Quantitative imaging may be performed indifferently in fresh, fixed, or frozen samples; however, since different sample preparations result in different tissue properties, all the samples must be prepared the same way to allow for reliable comparison. Images must then be recorded using the same setup with the same excitation wavelength, the same filters, the same detector settings, and the same excitation power. In these conditions, there is no need for any reference. However, slight misalignments of the setup may induce variations of the laser pulse profile or of the focal volume and affect the SHG response. A reference may then be used for long-term experiments or for comparison between similar setups in different laboratories. The most convenient reference sample is a fixed biopsy of a collagenous tissue (for instance, unstained paraffin-embedded skin dermis).

- Simultaneous recording of the 2PEF image is strongly recommended to visualize the tissue morphology by use of endogenous fluorescence signals. Segmentation of relevant regions of interest may then be implemented to selectively quantify the fibrosis in these regions. For instance, kidney fibrosis has to be scored in the cortex where interstitial fibrosis takes place. We therefore developed a semi-automated segmentation algorithm based on the 2PEF image to extract the cortical region from our murine coronal slices (see Figure 15.3). Briefly, we manually outlined the capsule, the papilla, and the series of arcuate arteries and fitted these curves as ellipses. The ellipses were then dilated or contracted to eliminate the frontiers and delimitate the cortical region without any contribution from the arcuate arteries, the medulla region, and the papilla region (Strupler et al. 2008).

15.4.3 Image Processing to Quantify Fibrosis Extent

Algorithms have been developed to automate SHG image processing and optimize the accuracy of fibrosis quantization. Figure 15.9 displays the algorithm developed by Strupler et al. (2007) for renal fibrosis and highlights the most important steps for reliable fibrosis quantization. Note that it can be implemented using ImageJ software (custom-written macros) or using MATLAB® for large stitched images.

15.4.3.1 Correction for Nonhomogeneous Illumination of the Sample

The laser excitation is usually slightly more important in the center of the field of view of the microscope than at the edges. Strictly speaking, correction for this vignetting effect is not necessary since it is the same for any sample. However, it is preferred for a better outcome when stitching several images or reconstructing 3D volumes. For that purpose, the multiphoton excitation profile is obtained from parabolic 2D fitting of the 2PEF image of a fluorescent plastic slide, and it is used to divide the SHG and 2PEF images. Note that the laser excitation also exhibits depth attenuation due to absorption, scattering, and aberrations in the tissue. We hypothesize that this effect is similar for control and fibrotic sample and do not impede quantitative comparisons. It is usually not corrected for because the signal-to-noise ratio and the resolution also deteriorate when depth increases and cannot be satisfactorily restored by image processing.

15.4.3.2 Background Correction

Background in the SHG images is quite small once the laser excitation is properly filtered out using laser blocking emission filters and the SHG is properly selected using around 10 nm-wide bandpass filters. The small background is then mainly due to the blue-edge component of the fluorescence partly passing

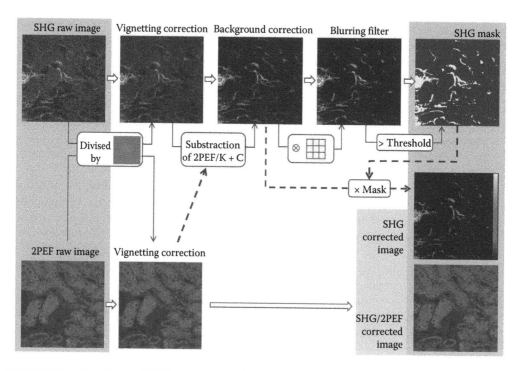

FIGURE 15.9 Algorithm for SHG data processing showing corrections for vignetting and background, filtering to enhance fibrillar structures, and application of a threshold. We obtain a mask of the fibrosis extent, an SHG corrected image and a combined SHG/2PEF image. (From Strupler, M. et al., 2008. *J. Biomed. Optics* 13:054041. With permission of SPIE.)

in the SHG detection channel (Strupler et al. 2007). It can be corrected for by subtracting a few percent of the 2PEF image to the SHG image. The quantity of the 2PEF image to be subtracted can be determined either from the spectral response of the detection channels compared to the fluorescence spectrum of the tissue, or from image analysis in a region with no collagen fibrils (Strupler et al. 2007). Optical noise due to ambient light is also subtracted if necessary.

15.4.3.3 Enhancement of Fibrils' Contrast and Application of a Threshold

After the latter corrections, the SHG image displays fibrillar structures spread out over a salt and pepper background (due to shot noise when using photon-counting photomultiplier tubes). The surface area of this nonsignificant background may not be negligible compared to the SHG signal from sparse collagen fibrils in a low-grade fibrosis. It is better to be filtered out using a blurring filter (3×3 in Figure 15.9) before application of a threshold. This procedure enhances the contrast from fibrillar structures that contain low spatial frequencies compared to high-frequency background (Strupler et al. 2007). Thresholding of the blurred image then yields a so-called SHG mask and multiplication of this mask to the SHG image yields a gray-level SHG corrected image with no background. The latter image is finally combined to the 2PEF corrected image to obtain a corrected multimodal image of the fibrotic tissue (see Figure 15.9).

15.4.3.4 Fibrosis Scoring

Quantization of the fibrosis extent is obtained by calculating the surface density of the SHG mask (D score). A second score is obtained as the average signal in the corrected SHG image (S score). The ratio of the signal-to-density scores S/D yields a third score that corresponds to the average SHG signal in fibrillar structures (SF score). Since fiber bundles show larger signals than thin fibrils, the latter score probes the onset of denser or larger fibers (Strupler et al. 2007). Note that this image processing may be applied to *z*-stacks of 2D images to obtain volume scores.

15.4.4 Biomedical Applications

Fibrosis quantization by means of SHG microscopy was first developed in animal models (Pena et al. 2007, Strupler et al. 2008, Sun et al. 2008, Tai et al. 2009a, He et al. 2010, Raub et al. 2010), and more recently applied to human biopsies (Gailhouste et al. 2010, Sun et al. 2010). We illustrate here the potential of this technique with murine kidney fibrosis.

We studied a murine model of hypertensive fibrosis by subcutaneous injection of Angiotensin II (AngII) for 4 or 7 weeks (Strupler et al. 2007, 2008). Physiological and biochemical parameters, in particular systolic blood pressure, were monitored to verify the progression of hypertension and fibrosis. Kidneys were then harvested and prepared either for histology: a half kidney was fixed in formalin, embedded in paraffin, sectioned, and stained with Masson's trichrome, or for SHG imaging: the other half kidney was fixed in PFA and 200-μm thick coronal slices were cut using a vibrating microtome and imaged without any staining (see Figure 15.3c).

Since fibrosis is heterogeneously distributed, we used a motorized stage to scan the kidney tissue and image the cortical region in our sample (typically 5×3 mm^2) by stitching many SHG/2PEF images acquired with a 20×, 0.9 NA objective (512×512 μm^2 field of view). Typical results are displayed in Figure 15.3. SHG images reveal the progression of fibrosis in AngII-infused mice, whereas endogenous 2PEF signals enable the visualization of the tissue morphology. We then took advantage of the 2PEF image to develop a segmentation algorithm based on the kidney morphology and score the fibrosis only in the cortical region that is relevant for biomedical reasons (see Figure 15.3b). The SHG density scores are displayed in Figure 15.3e for all the mice under study. They are compared to anapathologist semi-quantitative scoring of tubulointerstitial fibrosis in Masson's trichrome sections of the same kidney (Spurney et al. 1992). Both scoring methods show a good agreement as expected. However, SHG imaging advantageously provides continuous scoring with better sensitivity and can distinguish different pathological grades at early stages (see the range of SHG scores in the 0 grade for anapathologist score).

We took advantage of this sensitivity to quantify the role of tissular transglutaminase (TG2) in the fibrosis progression. This collagen cross-linking enzyme is expected to promote fibrosis through collagen assembly. Accordingly, we observed that TG2-deficient mice exhibited significant lower interstitial fibrosis than wild type mice, whereas they showed similar hypertension progression. However, we observed no colocalization between TG2 activity and interstitial fibrosis (see Figure 15.8).

Our SHG images also proved efficient to look at the 3D distribution of collagen fibrosis within the kidney tissue as exemplified in Figure 15.3d. Interestingly, we observed a continuity between perivascular, periglomerular, and tubulointerstitial fibrils (within our optical resolution) that may suggest common mechanisms of progression.

15.5 Discussion: Advantages and Limitations of SHG Microscopy for Fibrosis Scoring

15.5.1 Comparison to Other Imaging Techniques

We have demonstrated in the former sections that SHG microscopy is a valuable tool for fibrosis imaging and quantization. However, it is a complex and expensive method compared to histological techniques and it is worth discussing the crucial advantages of this technique.

The main practical drawback of histological and immunochemical techniques stems from the staining/labeling procedure that limits the reproducibility and restricts these techniques to thin 2D sections. The most effective technique among histological and immunochemical techniques appears to be Picrosirius Red staining when visualized with circularly polarized light microscopy (Junqueira et al.

TABLE 15.1   Comparison of Different Techniques Used for Fibrosis Quantization

	Masson's Trichrome	Picrosirius Red	Picrosirius ↻polarization	Immunochemical Labeling	Confocal Reflectance	OCT	SHG
Specificity to fibrillar collagen	⊖⊖⊖	⊖	⊕⊕	⊕⊕⊕	⊖⊖	⊖⊖	⊕⊕⊕
3D capability	No	No	No	No	⊕⊕	⊕⊕	⊕⊕⊕
Multimodality	⊖	⊖	⊖⊖	⊕	⊕	⊖⊖	⊕⊕⊕
Reproducibility	⊖	⊖	⊖	⊖⊖	⊕⊕⊕	⊕⊕⊕	⊕⊕⊕
Cost-Complexity	⊕⊕⊕	⊕⊕⊕	⊕	⊕	⊖⊖	⊕	⊖⊖⊖

1979, Whittaker et al. 1994). Sirius Red dye indeed aligns along the collagen molecules and enhances the fibrillar collagen birefringence that is detected in polarized light microscopy. The specific advantage of using circular light is the same as in SHG microscopy: it enables homogeneous imaging of fibrils with different orientations in the focal plane. In practice, this technique is an excellent choice when quantifying thin sections from human biopsies as a trade-off between cost and efficiency. SHG microscopy is however a better choice when imaging the 3D distribution of fibrosis in thick tissues and comparing it to the localization of other components of interest. For this purpose, SHG microscopy takes advantage of its 3D capability and multimodality. Confocal reflectance microscopy and optical coherence tomography (OCT) offer the same advantages but lack specificity to fibrillar collagen.

The sensitivity of SHG microscopy and of other techniques has not been fully characterized since it is a complex task. SHG yields the unique advantage of highly contrasted images because of its coherent nonlinear nature (see Section 2.2). It therefore favorably compares to confocal reflectance microscopy or OCT. A rigorous characterization of SHG sensitivity would however require determining the minimal diameter of fibrils that are detected in SHG microscopy. It has not been reported yet to the best of my knowledge. Note that the sensitivity of histological techniques is also poorly characterized.

All these considerations are summed up in Table 15.1. A more rigorous and complete comparison of SHG with other techniques would require using the same standard of image processing in order to distinguish the advantages of the technique itself and the ones due to image processing. Since image processing of histological sections is usually quite basic, the comparison to SHG is somewhat biased and histology should benefit from advanced segmentation and quantization techniques. For instance, automatic quantization of Masson's trichrome-stained renal biopsies based on a colorimetric segmentation algorithm has been reported recently to significantly improve the diagnosis of long-term allograft disease (Servais et al. 2007). In liver pathology, this approach also allowed to detect beneficial effect of antiviral treatments in chronic viral hepatitis that was otherwise not detected with semiquantitative standard evaluation of liver fibrosis (Goodman et al. 2009). I however anticipate that SHG imaging will prove to be more sensitive and reproducible than all other techniques.

15.5.2 Perspectives, Possible Improvements

15.5.2.1 True Quantization

The main limitation of fibrosis quantization by means of SHG microscopy is its inability to retrieve the quantity and 3D distribution of collagen molecules within the focal volume. Actual SHG imaging of fibrotic pathologies quantifies the extent of fibrosis, but it does not make use of the SHG signal intensity. The SHG signal is obtained as the coherent summation of the second-harmonic fields radiated by all the harmonophores within the focal volume. Truly quantitative measurements would require to know the second-order response of one single collagen molecule, or better of the harmonophores (the peptide bonds) within this molecule, and to solve the inverse problem of the second-harmonic fields coherent summation over the 3D distribution of collagen molecules within the fibrillar macro-organization.

The first issue was recently addressed by measuring the second-order hyperpolarizability of the collagen triple helix by use of hyper Rayleigh experiments (Deniset-Besseau et al. 2009). It was shown to be $\beta = 1.25 \times 10^{-27}$ esu for type I collagen from rat tail (relative to the water response: 0.56×10^{-27} esu). Moreover, the response for any collagen type was obtained as a function of the length of the triple helical domain, which was shown to be the only relevant parameter in that issue. For that purpose, we developed model calculation of the coherent summation of the response of all the peptides' bonds tightly aligned along the triple helix (Deniset-Besseau et al. 2009).

The second issue is a highly complicated task and requires more experimental data and more theoretical work. The second-order coherent response of a given distribution of collagen molecules may be calculated using similar approaches to the ones developed for membrane dyes (Moreaux et al. 2000), single fibrils (Williams et al. 2005), or an assembly of fibrils (LaComb et al. 2008b, Strupler and Schanne-Klein 2010). However, the inverse problem cannot be solved unambiguously without complementary data, for instance measuring the ratio of forward- to backward-SHG signals or varying the collection numerical aperture (Williams et al. 2005, Chu et al. 2007, LaComb et al. 2008a, Rivard et al. 2011). Polarization-resolved SHG microscopy may also help to get an insight to the 3D distribution of fibrillar collagen within the focal volume (Roth and Freund 1981, Stoller et al. 2002, 2003, Williams et al. 2005, Erikson et al. 2007, Han et al. 2008) provided a careful analysis of polarization distortions in collagenous tissues (Mansfield et al. 2008, Nadiarnykh and Campagnola 2009, Aït-Belkacem et al. 2010, Gusachenko et al. 2010).

15.5.2.2 Improvement of Image Analysis

SHG quantization of fibrosis extent is based on filtering and thresholding algorithms. These algorithms may be improved to better retrieve significant signals. For instance, Otsu threshold segmentation and erosion and dilation removal of grainy noise were used to quantify liver fibrosis (Sun et al. 2008). Other algorithms have been proposed in the context of collagen quantization in engineered tissues. For instance, adaptative thresholding procedure followed by zeroing of non-interconnected pixels was used to quantify the density of collagen matrices (Bayan et al. 2009). Complementary information may also be obtained by processing the images to obtain orientation indexes of the collagen fibrils as developed for engineered tissues or cornea (Raub et al. 2008, Bayan et al. 2009, Matteini et al. 2009, Bowles et al. 2010). 2D Fourier transform or Hough transform algorithms were used to map the average orientation and the disorder (entropy) of the collagen fibrillar network. Texture analysis was also proposed in cartilage SHG images (Werkmeister et al. 2010).

However, all these approaches are restricted to sequential 2D image processing and subsequent averaging on z-stacks to obtain volume indexes. It would be of great interest to develop direct 3D image processing to better retrieve the 3D fibrillar distribution. For that purpose, a 3D morphological analysis was recently proposed and successfully discriminated different fibrillar organizations in collagen matrices (Altendorf and Jeulin 2009, Altendorf et al. 2012). Similar approaches should be developed for describing the millimeter-scale architecture of fibrosis. For instance, kidney fibrosis in murine models was shown to exhibit a radial interconnected distribution through the cortical region with striking continuity between perivascular, periglomerular, and peritubular fibrosis (Strupler et al. 2008). Quantitative analysis of this distribution would be of great interest for monitoring fibrosis progression and should benefit from 3D morphological approaches.

15.5.2.3 Applicability for *In Vivo* Diagnosis

Application of SHG microscopy to *in vivo* imaging and fibrosis diagnosis would be very interesting. There are two issues in that respect. First, one has to develop fibered SHG microscopes in order to better access the fibrotic organs. It requires complex developments, and work is currently under progress to achieve such advanced endoscopes. The second issue is regarding the penetration depth of SHG imaging. Typical penetration depth in highly dense and scattering tissues such as kidney, liver, or skin is a few hundreds of micrometers. It could be further improved using adaptative optics (Jesacher et al.

2009, Olivier et al. 2009) or optical clearing (see Chapter 8), but SHG imaging will never access the full volume of large organs. This technique is rather probing the "surface" of the organ, and it is important to determine whether this *surface* information reflects the fibrosis progression in the *volume* of the organ.

This issue has been addressed in different studies. SHG scoring of lung fibrosis in the bleomycine murine model was compared in the subpleural region and in the parenchyma (Pena et al. 2007). Both regions showed similar results. A significant superficial fibrosis was also observed in intact fibrotic murine kidney (Strupler 2008). Most importantly, a complete study was recently dedicated to this issue in a rat model of liver fibrosis (He et al. 2010). SHG fibrosis scores on the peripheral liver surface and in the central area of the liver lobes were shown to be strongly correlated. It proved that surface SHG scoring is a reliable method for monitoring liver fibrosis.

15.6 Conclusion

All these considerations illustrate the potential of SHG microscopy for fibrosis 3D imaging and scoring. This method should prove useful for deciphering biological mechanisms of fibrosis progression and proposing new therapeutic approaches. It should also enable pharmacological studies of drugs aiming at limiting fibrosis progression and ultimately at inducing fibrosis regression. Application to diagnosis of human biopsies should improve the reproducibility of multicenter studies and consequently, the accuracy of clinical trial evaluations.

This technique is more generally applicable to any process of tissue remodeling, that is to a variety of pathologies. It is also a relevant method for the evaluation of biomimetic collagenous matrices that may serve as 3D scaffolds for cellular culture or as tissue substitutes to be grafted as implants. Accordingly, similar quantization methods have been developed for assessing the remodeling of these 3D matrices (Raub et al. 2007, Bayan et al. 2009, Rice et al. 2010, Pena et al. 2010).

SHG microscopy could be further improved to better access the 3D organization of collagen fibrils, which requires further fundamental studies as well as the implementation of complementary modalities and new 3D image processing approaches. SHG microscopy will then fully establish as the leading technique for 3D quantitative imaging of tissue remodeling.

Acknowledgments

The author gratefully acknowledges M. Strupler, A.-M. Pena, and A. Deniset-Besseau for their invaluable contributions to this work and for stimulating discussions. The author also thanks P.-L. Tharaux, P. Bedossa, F. Hache, and G. Latour for critical reading of the manuscript. Most of the work presented here was carried out thanks to the close collaborations with E. Beaurepaire (LOB, Ecole Polytechnique, Palaiseau, France), P.-L. Tharaux (Inserm U970, PARCC, Paris, France), B. Crestani, A. Fabre, J. Marchall-Somme (Inserm U700, Hôpital Bichat, Paris, France), and P.-F. Brevet (Lasim, CNRS-Université Lyon I, France). Finally, the author thanks V. Meas-Yedid and C. Olivo-Marin (Institut Pasteur, Paris, France), A. Servais and E. Thervet (Hôpital Necker, Paris, France), E. Decencière, H. Altendorf and D. Jeulin (Mines-ParisTech, Fontainebleau, France), and G. Mosser (UPMC-CNRS, Paris) for stimulating discussions.

References

Aït-Belkacem, D., A. Gasecka, F. Munhoz, S. Brustlein, and S. Brasselet. 2010. Influence of birefringence on polarization resolved nonlinear microscopy and collagen SHG structural imaging. *Opt. Express* 18:14859–14870.

Altendorf, H. and D. Jeulin. 2009. 3D directional mathematical morphology for analysis of fiber orientations. *Image Anal. Stereol.* 28: 143–153.

Altendorf, H. et al. 2012. 3D morphological analysis of collagen fibrils imaged using second harmonic generation. *J. Microscopy* 247:161–175.

Banavar, M., E. P. W. Kable, F. Braet, X. M. Wang, M. D. Gorrell, and G. Cox. 2006. Detection of collagen by second harmonic microscopy as a diagnostic tool for liver fibrosis. *Proc. SPIE* 6089:60891B.

Barad, Y., H. Eisenberg, M. Horowitz, and Y. Silberberg. 1997. Nonlinear scanning laser microscopy by third harmonic generation. *Appl. Phys. Lett.* 70:922–924.

Bayan, C., J. M. Levitt, E. Miller, D. Kaplan, and I. Georgakoudi. 2009. Fully automated, quantitative, non-invasive assessment of collagen fiber content and organization in thick collagen gels. *J. Appl. Phys.* 105:102042.

Beaurepaire, E. and J. Mertz. 2002. Epifluorescence collection in two-photon microscopy. *Appl. Opt.* 41:5376–5382.

Beck, K. and B. Brodsky. 1998. Supercoiled protein motifs: The collagen triple-helix and the alpha-helical coiled coil. *J. Struct. Biol.* 122:17–29.

Bedossa, P., P. Bioulacsage, P. Callard, M. Chevallier, C. Degott, Y. Deugnier, M. Fabre et al. 1994. Intraobserver and interobserver variations in liver-biopsy interpretation in patients with chronic hepatitis-C. *Hepatology* 20:15–20.

Bowles, R. D., R. M. Williams, W. R. Zipfel, and L. J. Bonassar. 2010. Self-assembly of aligned tissue-engineered annulus fibrosus and intervertebral disc composite via collagen gel contraction. *Tissue Eng. A* 16:1339–1348.

Brown, E., T. McKee, E. diTomaso, A. Pluen, B. Seed, Y. Boucher, and R. K. Jain. 2003. Dynamic imaging of collagen and its modulation in tumors in vivo using second-harmonic generation. *Nat. Med.* 9:796–800.

Campagnola, P. J., A. C. Millard, M. Terasaki, P. E. Hoppe, C. J. Malone, and W. A. Mohler. 2002. Three-dimensional high-resolution second-harmonic generation imaging of endogenous structural proteins in biological tissues. *Biophys. J.* 82:493–508.

Chu, S.-W., S.-P. Tai, M.-C. Chan, C.-K. Sun, I.-C. Hsiao, C.-H. Lin, Y.-C. Chen, and B.-L. Lin. 2007. Thickness dependence of optical second harmonic generation in collagen fibrils. *Opt. Express* 15:12005–12010.

Cox, G., E. Kable, A. Jones, I. Fraser, K. Marconi, and M. D. Gorrell. 2003. 3-dimensional imaging of collagen using second harmonic generation. *J. Struct. Biol.* 141:53–62.

Débarre, D., W. Supatto, and E. Beaurepaire. 2005. Structure sensitivity in third-harmonic generation microscopy. *Opt. Lett.* 30:2134–2136.

Deniset-Besseau, A., J. Duboisset, E. Benichou, F. Hache, P.-F. Brevet, and M.-C. Schanne-Klein. 2009. Measurement of the second order hyperpolarizability of the collagen triple helix and determination of its physical origin. *J. Phys. Chem. B* 113:13437–13445.

Deniset-Besseau, A., P. De Sa Peixoto, G. Mosser, and M.-C. Schanne-Klein. 2010. Nonlinear optical imaging of lyotropic cholesteric liquid crystals. *Opt. Express* 18:1113–1121.

Duboisset, J., G. Matar, I. Russier-Antoine, E. Benichou, G. Bachelier, C. Jonin, D. Ficheux, F. Besson, and P. F. Brevet. 2010. First hyperpolarizability of the natural aromatic amino acids tryptophan, tyrosine, and phenylalanine and the tripeptide lysine-tryptophan-lysine determined by hyper-Rayleigh scattering. *J. Phys. Chem. B* 114:13861–13865.

Erikson, A., J. Örtegren, T. Hompland, C. D. L. Davies, and M. Lindgren. 2007. Quantification of the second-order nonlinear susceptibility of collagen I using a laser scanning microscope. *J. Biomed. Opt.* 12:044002.

Friedenberg, M. A., L. Miller, C. Y. Chung, F. Fleszler, F. L. Banson, R. Thomas, K. P. Swartz, and F. K. Friedenberg. 2005. Simplified method of hepatic fibrosis quantification: Design of a new morphometric analysis application. *Liver Int.* 25:1156–1161.

Furness, P. N., N. Taub, and C. Project. 2001. International variation in the interpretation of renal transplant biopsies: Report of the CERTPAP Project. *Kidney Int.* 60:1998–2012.

Furutani, Y., A. Kato, M. Notoya, M. A. Ghoneim, and S. Hirose. 2001. A simple assay and histochemical localization of transglutaminase activity using a derivative of green fluorescent protein as substrate. *J. Histochem. Cytochem.* 49:247–258.

Gailhouste, L., Y. Le Grand, C. Odin, D. Guyader, B. Turlin, F. Ezan, Y. Desille et al. 2010. Fibrillar collagen scoring by second harmonic microscopy: A new tool in the assessment of liver fibrosis. *J. Hepatol.* 52:398–406.

Goodman, Z. D., R. L. Becker, P. J. Pockros, and N. H. Afdhal. 2007. Progression of fibrosis in advanced chronic hepatitis C: Evaluation by morphometric image analysis. *Hepatology* 45:886–894.

Goodman, Z. D., A. M. Stoddard, H. L. Bonkovsky, R. J. Fontana, M. G. Ghany, T. R. Morgan, E. C. Wright et al. 2009. Fibrosis progression in chronic hepatitis C: Morphometric image analysis in the HALT-C Trial. *Hepatology* 50:1738–1749.

Gough, J., D. Rush, J. Jeffery, P. Nickerson, R. McKenna, K. Solez, and K. Trpkov. 2002. Reproducibility of the Banff schema in reporting protocol biopsies of stable renal allografts. *Nephrol. Dial. Transplant.* 17:1081–1084.

Grimm, P. C., P. Nickerson, J. Gough, R. McKenna, E. Stern, and D. N. Rush. 2003. Computerized image analysis of Sirius Red-stained renal allograft biopsies as a surrogate marker to predict long-term allograft function. *J. Am. Soc. Nephrol.* 14:1662–1668.

Gusachenko, I., G. Latour, and M.-C. Schanne-Klein. 2010. Polarization-resolved second harmonic microscopy in anisotropic thick tissues. *Opt. Express* 18:19339–19352.

Han, X., R. M. Burke, M. L. Zettel, P. Tang, and E. B. Brown. 2008. Second harmonic properties of tumor collagen: Determining the structural relationship between reactive stroma and healthy stroma. *Opt. Express* 16:1846–1859.

He, Y., C. H. Kang, S. Xu, X. Tuo, S. Trasti, D. C. S. Tai, A. M. Raja et al. 2010. Towards surface quantification of liver fibrosis progression. *J. Biomed. Optics* 15:056007.

Hulmes, D. J. S. 2002. Building collagen molecules, fibrils, and suprafibrillar structures. *J. Struct. Biol.* 137:2–10.

Jesacher, A., A. Thayil, K. Grieve, D. Debarre, T. Watanabe, T. Wilson, S. Srinivas, and M. Booth. 2009. Adaptive harmonic generation microscopy of mammalian embryos. *Opt. Lett.* 34:3154–3156.

Junqueira, L. C. U., G. Bignolas, and R. R. Brentani. 1979. Picrosirius staining plus polarization microscopy, a specific method for collagen detection in tissue sections. *Histochem. J.* 11:447–155.

LaComb, R., O. Nadiarnykh, S. Carey, and P. J. Campagnola. 2008a. Quantitative second harmonic generation imaging and modeling of the optical clearing mechanism in striated muscle and tendon. *J. Biomed. Opt.* 13:021109-1–021109-11.

LaComb, R., O. Nadiarnykh, S. S. Townsend, and P. J. Campagnola. 2008b. Phase matching considerations in second harmonic generation from tissues: Effects on emission directionality, conversion efficiency and observed morphology. *Opt. Commun.* 281:1823–1832.

Legare, F., C. Pfeffer, and B. R. Olsen. 2007. The role of backscattering in SHG tissue imaging. *Biophys. J.* 93:1312–1320.

Levine, B. F. and C. G. Bethea. 1976. 2nd Order hyperpolarizability of a polypeptide alpha-helix–poly-gamma-benzyl-l-glutamate. *J. Chem. Phys.* 65:1989–1993.

Loison, C. and D. Simon. 2010. Additive model for the second harmonic generation hyperpolarizability applied to a collagen-mimicking peptide (Pro-Pro-Gly)(10). *J. Phys. Chem. A* 114:7769–7779.

Mansfield, J. C., C. P. Winlove, J. Moger, and S. J. Matcher. 2008. Collagen fiber arrangement in normal and diseased cartilage studied by polarization sensitive nonlinear microscopy. *J. Biomed. Opt.* 13:044020-1–044020-13.

Marcussen, N., T. S. Olsen, H. Benediktsson, L. Racusen, and K. Solez. 1995. Reproducibility of the Banff classification of renal-allograft pathology–interobserver and intraobserver variation. *Transplantation* 60:1083–1089.

Matalka, I. I., O. M. Al-Jarrah, and T. M. Manasrah. 2006. Quantitative assessment of liver fibrosis: A novel automated image analysis method. *Liver Int.* 26:1054–1064.

Matteini, P., F. Ratto, F. Rossi, R. Cicchi, C. Stringari, D. Kapsokalyvas, F. S. Pavone, and R. Pini. 2009. Photothermally-induced disordered patterns of corneal collagen revealed by SHG imaging. *Opt. Express* 17:4868–4878.

Moreaux, L., O. Sandre, and J. Mertz. 2000. Membrane imaging by second-harmonic generation microscopy. *J. Opt. Soc. Am. B* 17:1685–1694.

Nadiarnykh, O. and P. J. Campagnola. 2009. Retention of polarization signatures in SHG microscopy of scattering tissues through optical clearing. *Opt. Express* 17:5794–5806.

Nath, K. A. 1992. Tubulointerstitial changes as a major determinant in the progression of renal damage. *Am. J. Kidney Dis.* 20:1–17.

Nicholson, M. L., T. A. McCulloch, S. J. Harper, T. J. Wheatley, C. M. Edwards, J. Feehally, and P. N. Furness. 1996. Early measurement of interstitial fibrosis predicts long-term renal function and graft survival in renal transplantation. *Br. J. Surg.* 83:1082–1085.

Olivier, N. and E. Beaurepaire. 2008. Third-harmonic generation microscopy with focus-engineered beams: A numerical study. *Opt. Express* 16:14703–14715.

Olivier, N., D. Debarre, and E. Beaurepaire. 2009. Dynamic aberration correction for multiharmonic microscopy. *Opt. Lett.* 34:3145–3147.

Oudar, J. L. and D. S. Chemla. 1977. Hyperpolarizabilities of nitroanilines and their relations to excited-state dipole-moment. *J. Chem. Phys.* 66:2664–2668.

Pape, L., T. Henne, G. Offner, J. Strehlau, J. H. H. Ehrich, M. Mengel, and P. C. Grimm. 2003. Computer-assisted quantification of fibrosis in chronic allograft nephropathy by picosirius red-staining: A new tool for predicting long-term graft function. *Transplantation* 76:955–958.

Pena, A.-M., T. Boulesteix, T. Dartigalongue, and M.-C. Schanne-Klein. 2005. Chiroptical effects in the second harmonic signal of collagens I and IV. *J. Am. Chem. Soc.* 127:10314–10322.

Pena, A.-M., A. Fabre, D. Débarre, J. Marchal-Somme, B. Crestani, J.-L. Martin, E. Beaurepaire, and M.-C. Schanne-Klein. 2007. Three-dimensional investigation and scoring of extracellular matrix remodeling during lung fibrosis using multiphoton microscopy. *Microsc. Res. Tech.* 70:162–170.

Pena, A.-M., D. Fagot, C. Olive, J.-F. Michelet, J.-B. Galey, F. Leroy, E. Beaurepaire, J.-L. Martin, A. Colonna, and M.-C. Schanne-Klein. 2010. Multiphoton microscopy of engineered dermal substitutes: Assessment of 3D collagen matrix remodeling induced by fibroblasts contraction. *J. Biomed. Opt.* 15:056018.

Pilette, C., M. C. Rousselet, P. Bedossa, D. Chappard, F. Oberti, H. Rifflet, M. Y. Maiga, Y. Gallois, and P. Cales. 1998. Histopathological evaluation of liver fibrosis: Quantitative image analysis vs semi-quantitative scores—Comparison with serum markers. *J. Hepatol.* 28:439–446.

Plotnikov, S. V., A. C. Millard, P. J. Campagnola, and W. A. Mohler. 2006. Characterization of the myosin-based source for second-harmonic generation from muscle sarcomeres. *Biophys. J.* 90:693–703.

Racusen, L. C., K. Solez, R. B. Colvin, S. M. Bonsib, M. C. Castro, B. Cavallo, B. P. Croker et al. 1999. The Banff 97 working classification of renal allograft pathology. *Kidney Int.* 55:713–723.

Raub, C. B., V. Suresh, T. Krasieva, J. Lyubovitsky, J. D. Mih, A. J. Putnam, B. J. Tromberg, and S. C. George. 2007. Noninvasive assessment of collagen gel microstructure and mechanics using multiphoton microscopy. *Biophys. J.* 92:015004-1–015004-9.

Raub, C. B., J. Unruh, V. Suresh, T. Krasieva, T. Lindmo, E. Gratton, B. J. Tromberg, and S. C. George. 2008. Image correlation spectroscopy of multiphoton images correlates with collagen mechanical properties. *Biophys. J.* 94:2361.

Raub, C. B., S. Mahon, N. Narula, B. J. Tromberg, M. Brenner, and S. C. George. 2010. Linking optics and mechanics in an *in vivo* model of airway fibrosis and epithelial injury. *J. Biomed. Opt.* 15:-.

Ricard-Blum, S. and F. Ruggiero. 2005. The collagen superfamily: From the extracellular matrix to the cell membrane. *Pathol. Biol.* 53:430–442.

Rice, W. L., D. L. Kaplan, and I. Georgakoudi. 2010. Two-photon microscopy for non-invasive, quantitative monitoring of stem cell differentiation. *PlosOne* 5:e10075.

Rivard, M., M. Laliberté, A. Bertrand-Grenier, C. Harnagea, C. P. Pfeffer, M. Vallières, Y. St-Pierre, A. Pignolet, M. A. E. Khakani, and F. Légaré. 2011. The structural origin of second harmonic generation in fascia *Biomed. Opt. Express* 2:26.

Roth, S. and I. Freund. 1981. Optical second-harmonic scattering in rat-tail tendon. *Biopolymers* 20:1271–1290.

Seron, D., F. Moreso, X. Fulladosa, M. Hueso, M. Carrera, and J. M. Grinyo. 2002. Reliability of chronic allograft nephropathy diagnosis in sequential protocol biopsies. *Kidney Int.* 61:727–733.

Servais, A., V. Meas-Yedid, M. Buchler, E. Morelon, J. C. Olivo-Marin, Y. Lebranchu, C. Legendre, and E. Thervet. 2007. Quantification of interstitial fibrosis by image analysis on routine renal biopsy in patients receiving cyclosporine. *Transplantation* 84:1595–1601.

Servais, A., V. Meas-Yedid, O. Toupance, Y. Lebranchu, A. Thierry, B. Moulin, I. Etienne et al. 2009. Interstitial fibrosis quantification in renal transplant recipients randomized to continue cyclosporine or convert to sirolimus. *Am. J. Transplant.* 9:2552–2560.

Solez, K., R. B. Colvin, L. C. Racusen, B. Sis, P. F. Halloran, P. E. Birk, P. M. Campbell et al. 2007. Banff '05 Meeting report: Differential diagnosis of chronic allograft injury and elimination of chronic allograft nephropathy ('CAN'). *Am. J. Transplant.* 7:518–526.

Spurney, R. F., P. Y. Fan, P. Ruiz, F. Sanfilippo, D. S. Pisetsky, and T. M. Coffman. 1992. Thromboxane receptor blockade reduces renal injury in murine lupus nephritis. *Kidney Int.* 41:973–982.

Stoller, P., K. M. Reiser, P. M. Celliers, and A. M. Rubenchik. 2002. Polarization-modulated second harmonic generation in collagen. *Biophys J.* 82:3330–3342.

Stoller, P., P. M. Celliers, K. M. Reiser, and A. M. Rubenchik. 2003. Quantitative second-harmonic generation microscopy in collagen. *Appl. Opt.* 42:5209–5219.

Strupler, M., A.-M. Pena, M. Hernest, P.-L. Tharaux, J.-L. Martin, E. Beaurepaire, and M.-C. Schanne-Klein. 2007. Second harmonic imaging and scoring of collagen in fibrotic tissues. *Opt. Express* 15:4054–4065.

Strupler, M. 2008. Imagerie du collagène par microscopie multiphotonique. Application aux fibroses rénales. PhD thesis, Ecole Polytechnique, Palaiseau.

Strupler, M., M. Hernest, C. Fligny, J.-L. Martin, P.-L. Tharaux, and M.-C. Schanne-Klein. 2008. Second harmonic microscopy to quantify renal interstitial fibrosis and arterial remodeling. *J. Biomed. Optics* 13:054041.

Strupler, M. and M.-C. Schanne-Klein. 2010. Simulating second harmonic generation from tendon: Do we see fibrils? *Biomedical Optics, OSA Technical Digest*, Paper BTuD83.

Sun, T. L., Y. A. Liu, M. C. Sung, H. C. Chen, C. H. Yang, V. Hovhannisyan, W. C. Lin et al. 2010. Ex vivo imaging and quantification of liver fibrosis using second-harmonic generation microscopy. *J. Biomed. Opt.* 15:036002.

Sun, W. X., S. Chang, D. C. S. Tai, N. Tan, G. F. Xiao, H. H. Tang, and H. Yu. 2008. Nonlinear optical microscopy: Use of second harmonic generation and two-photon microscopy for automated quantitative liver fibrosis studies. *J. Biomed. Optics* 13:064010.

Sund, S., P. Grimm, A. V. Reisaeter, and T. Hovig. 2004. Computerized image analysis vs semiquantitative scoring in evaluation of kidney allograft fibrosis and prognosis. *Nephrol. Dial. Transplant.* 19:2838–2845.

Tai, D., N. Tan, C. H. Kang, C. L. Cheng, S. M. Chia, A. Wee, and H. Yu. 2009a. Fibro-C-Index—A standardized quantification of liver fibrosis using second harmonic generation and two-photon microscopy. *Hepatology* 50:815a–816a.

Tai, D. C. S., N. Tan, S. Xu, C. H. Kang, S. M. Chia, C. L. Cheng, A. Wee et al. 2009b. Fibro-C-Index: Comprehensive, morphology-based quantification of liver fibrosis using second harmonic generation and two-photon microscopy. *J. Biomed. Optics* 14:044013.

Tiaho, F., G. Recher, and D. Rouède. 2007. Estimation of helical angle of myosin and collagen by second harmonic generation imaging microscopy *Opt. Express* 15:12286–12295.

Werkmeister, E., N. de Isla, P. Netter, J. F. Stoltz, and D. Dumas. 2010. Collagenous extracellular matrix of cartilage submitted to mechanical forces studied by second harmonic generation microscopy. *Photochem. Photobiol.* 86:302–310.

Whittaker, P., R. A. Kloner, D. R. Boughner, and J. G. Pickering. 1994. Quantitative assessment of myocardial collagen with picrosirius red staining and circularly polarized light. *Basic Res. Cardiol.* 89:397–410.

Williams, R. M., W. R. Zipfel, and W. W. Webb. 2005. Interpreting second-harmonic generation images of collagen fibrils. *Biophys. J.* 88:1377–1386.

Yew, E. Y. S. and C. J. R. Sheppard. 2007. Second harmonic generation polarization microscopy with tightly focused linearly and radially polarized beams. *Opt. Commun.* 275:453–457.

Yoshiki, K., K. Ryosuke, M. Hashimoto, T. Araki, and N. Hashimoto. 2007. Second-harmonic-generation microscope using eight-segment polarization-mode converter to observe three-dimensional molecular orientation. *Opt. Lett.* 32:1680–1682.

Zipfel, W. R., R. M. Williams, R. Christie, A. Y. Nikitin, B. T. Hyman, and W. W. Webb. 2003. Live tissue intrinsic emission microscopy using multiphoton-excited native fluorescence and second harmonic generation. *Proc. Natl. Acad. Sci. USA* 100:7075–7080.

Zoumi, A., A. Yeh, and B. J. Tromberg. 2002. Imaging cells and extracellular matrix in vivo by using second-harmonic generation and two-photon excited fluorescence. *Proc. Natl. Acad. Sci. USA* 99:11014–11019.

16

Quantitative Approaches for Studying the Role of Collagen in Breast Cancer Invasion and Progression

Caroline A. Schneider
The Alliance of Crop, Soil, and Environmental Science Societies (ACSESS)

Carolyn A. Pehlke
University of New Mexico

Karissa Tilbury
University of Wisconsin—Madison

Ruth Sullivan
University of Wisconsin—Madison

Kevin W. Eliceiri
University of Wisconsin—Madison

Patricia J. Keely
University of Wisconsin—Madison

16.1 Collagen and Breast Cancer

16.1.1 Breast Cancer and Tissue Density

About one in eight women in the United States will develop invasive breast cancer in her lifetime, according to data compiled by the American Cancer Society (http://www.cancer.org/cancer/breastcancer/overviewguide/breast-cancer-overview-key-statistics). Although death rates from breast cancer have been decreasing since 1990, breast cancer is still the second leading cause of cancer deaths in American women after lung cancer. In 2011, there were more than 2.6 million breast cancer survivors in the United States. Significant risk factors for breast cancer include family history, age, and genetic mutations (i.e., BRCA1 and BRCA2).

Another significant and emerging risk factor associated with the development of breast cancer is mammographic density [1]. The appearance of breast tissue upon mammography reflects variation in tissue composition. Dark regions indicate fat cells, while lighter areas signify denser tissue made up of epithelial cells and stroma [2,3]. The portion of tissue made up of denser regions is characterized as the percent mammographic density (PMD).

High PMD has been shown to be strongly associated with breast cancer risk [4]. More than 50 studies over the past 30 years have investigated this association. The consistency of the correlation between high PMD and breast cancer places it among risk factors such as age or presence of atypia upon biopsy [3]. A meta-analysis performed in 2006 reviewed data from 42 studies and supported the association between PMD and breast cancer risk [4].

The risk associated with high PMD is significant and robust among many age and ethnic groups and is independent of which breast was used for estimation of PMD [3,5]. Most studies place the increased risk of a PMD more than 75% at four to six times the risk of a PMD <10% [2,6]. This correlation is density-dependent with increasing PMD related to increased risk. Additionally, the risk associated with PMD persists for 8–10 years from the time of the mammogram [4]. While PMD is associated with age and body mass index (BMI), it represents an independent risk factor even after adjustment for these and other risk factors.

Early in the study of PMD and breast cancer risk, much of the correlation was thought to be due to masking; many researchers believed that cancers were difficult to detect in patients with high PMD due to the similar appearance of cancers and high-density tissue. However, studies that have shown an increased risk of breast cancer related to high PMD up to 10 years after the initial screening, which challenges the masking hypothesis and suggests that high PMD has a distinct and real correlation with breast cancer risk [3]. It remains true, however, that women with high PMD have two disadvantages— the higher risk associated with PMD as well as the difficulty of imaging cancers in highly dense breast tissue, which can delay detection of breast cancers [4].

Clinically, the association between PMD and breast cancer has not been fully utilized. The most widely used method of predicting breast cancer risk in the clinic is currently the Gail model. This model uses six risk factors—age, age at menarche, age at first live birth, number of first-degree relatives with breast cancer, number of biopsies, and presence of atypia on biopsy [3]. However, PMD is more strongly correlated with breast cancer risk than any of the factors in the Gail model. Current efforts are underway to incorporate additional risk factors, including PMD, into clinical risk prediction models.

In addition to predicting breast cancer risk, PMD may also be a useful marker of therapy efficacy. PMD can change in response to hormone therapy (increase), menopause (decrease), and tamoxifen therapy (decrease). While a causal connection has not been shown between tamoxifen treatment, reduced PMD and reduced breast cancer risk, it may be possible to use PMD as a marker of therapy efficacy, thus assessing treatments early and easily [1]. This might be especially important for high-risk patients who take tamoxifen prophylactically and do not have a cancer lesion to monitor. PMD could also be used to determine how often a patient should receive mammograms, how likely the mammograms might be to miss early disease, or if other tests should be used to estimate breast cancer risk.

Before PMD is used routinely in a clinical setting, improvements in and standardization of imaging must be established. Radiologists often use subjective tools that lack automation or quantification, making comparisons between mammographic images difficult. New methods are being developed including semi-automated or user-assisted procedures. In particular, the development of a standardized classification system, the Breast Imaging Reporting and Data System (BI-RADS) of the American College of Radiology (ACR) allows comparison of clinical findings across users and treatment sites. The BI-RADS assessment includes numerical description of mammographic density from ACR 1 (fatty) to ACR 4 (extremely dense). (For a recent review, see [7]) Additionally, other imaging modalities could be used to better understand and quantify PMD such as ultrasound, magnetic resonance imaging, or full-field digital mammography (FFDM) [3,5]. These techniques will provide more complete and standardized measurements that will allow for more accurate predictions of breast cancer risk and more useful suggestions for treatment options.

Because of the link between PMD and breast cancer risk, it is important to study the mechanisms underlying the correlation. Mechanistic understanding will both improve the ability to use PMD as a risk predictor, as well as improve the choice of appropriate therapies that take into account the effect of the connective tissue environment on breast cancer progression.

Mammographic density is most strongly linked to an increase in stromal collagen [8], but a direct causal link between density and breast cancer formation has not been established in humans. A functional role for increased collagen in breast cancer progression is suggested by a transgenic mouse model of higher collagen density, where increased stromal collagen increases tumor formation and produces more invasive cancers [9]. The effect is at least in part a direct effect of the increased extracellular matrix

(ECM) on expression of genes linked to cancer phenotypes: cell proliferation and invasion [10]. The role of the ECM may be local, as fibrotic foci containing higher levels of collagen may facilitate metastasis and have been correlated with poor prognosis in breast cancers [11].

16.1.2 Collagen I and Breast Cancer

The ECM is a complex meshwork of proteins (e.g., proteoglycans, collagens, laminin, fibronectin, entactin) and polysaccharides (e.g., glycosaminoglycans) secreted by cells into the spaces between them. Components of the ECM can bind to one another and to adhesion receptors on the cell surface, providing both physical scaffolding to support tissues and biochemical cues to regulate cellular behavior [12]. In the mammary gland, the ECM acts as a signal integrator mediating complex cellular functions [13].

Collagen is a major component of the ECM within the breast tissue and recently has become an area of research focus in understanding breast cancer development and risk. The fundamental unit of collagen is tropocollagen, a long (300 nm) triple-helix protein structure. Cleavage of the pro-peptide termini allows collagen to self-assemble to form collagen fibrils with diameters of 0.5–3 µm. There are 28 different types of collagen in the body, most made up of triple-helical monomeric units that organize into supramolecular structures [14,15]. In particular, collagen I is the most abundant collagen, and is known to have a significant effect on mammary morphogenesis and tumorigenesis [12,16–19].

Collagen regulates cell and tissue function through physical scaffolding, as a biochemical ligand and through its mechanical properties [11]. Changes in the structure and function of ECM and collagen have been found to have profound effects on cell behavior and tumor formation, such as enhancing tumorigenesis through regulating interactions between epithelial cells and the stroma in the breast [6]. Changes in collagen structure can lead to increased rigidity of the ECM, resulting in altered signaling, differentiation, proliferation, and migration of cells. Increased matrix stiffness activates several signaling pathways, including the Ras–MAPK pathway, which promotes growth of epithelial cells [10,20].

The interaction between cells and the ECM plays an important role in maintaining normal cell behavior [13,21], and the disruption of this interaction is one of the hallmarks of the transition from normal tissue to malignancy [22,23]. It is still not clear if the breakdown of structure precedes or follows the invasive transition, but the overexpression of an ECM remodeling enzyme, matrix metalloproteinase 3 (MMP-3), has been shown to stimulate the establishment of a reactive stroma, which is characterized by increased deposition of collagen I prior to tumor formation [24]. MMP-3 has also been shown to promote reactive stroma and fibrosis, leading to carcinogenesis in murine mammary glands [25]. The increased deposition of collagen I, characteristic of a reactive stroma, is just one of several specific collagen I structural changes found in the presence of mammary tumors.

Covalent crosslinking of collagen is a normal part of collagen matrix formation but is often increased in tumor environments. Crosslinking of collagen fibrils is catalyzed by enzymes such as lysyl oxidase (LOX). LOX expression can be induced by growth factor signaling as well as hypoxic conditions in tumor cells [26]. Increased expression of LOX correlates with tumor progression and breast cancer risk, thus highlighting the role of collagen reorganization in breast cancer [11,27]. Crosslinking of collagen fibers increases integrin signaling, which has been shown to promote tumor formation [9].

16.1.3 Collagen as a Cancer Biomarker: Tumor Associated Collagen Signatures

In addition to the amount and cross-linking of collagen, changes in the organization and orientation of collagen fibers have significant effects on cell migration. In normal breast tissue, collagen appears as curly fibers, while collagen associated with tumors often appears thicker and more linear [11]. Provenzano et al. [17] demonstrated a set of changes that accompany tumor progression, termed tumor associated collagen signatures (TACS) (Figure 16.1).

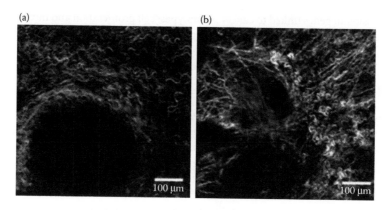

FIGURE 16.1   SHG images of normal murine mammary tissue (a) and murine mammary tissue surrounding a tumor (b).

TABLE 16.1   Descriptions of Tumor-Related Collagen I Structures

Category 1 TACS	Category 2 TACS	Category 3 TACS
• Region of dense collagen I surrounding tumor • Collagen fibers have no specific alignment • Tumors may be pre-palpable	• Collagen I fibers wrapped around tumor • Fibers appear stretched across a relatively smooth tumor boundary • Tangential orientation of fibers (approximately 0° to tumor boundary) predominant	• Collagen I fibers aligned normal to tumor boundary • Tumor boundary is of an irregular shape in these regions • Indicative of local invasion through collective epithelial cell migration • Fibers aligned in the direction of cell invasion (approximately 90° to tumor boundary)

Source: Content taken from Boyd NF et al. *N Engl J Med* 2007, 356(3):227–236.

TACS fall roughly into three categories (see Table 16.1 and Figure 16.2) [17]. Importantly, these collagen changes have prognostic significance: the presence of TACS from category 3, TACS-3, is associated with a significantly increased risk of relapse or death from breast cancer in patients [28].

16.1.4 Collagen as a Therapeutic Target

Not only is collagen a potential clinical biomarker for breast cancer; an important future direction is to use collagen changes as a therapeutic target. Knowing which signaling pathways are activated allows pharmacologic targets to be exploited. Moreover, drugs that target LOX, the crosslinking enzyme, can decrease ECM stiffness, providing a possible treatment option [9,11]. The tumor microenvironment and collagen content can also affect responses to therapy. Collagen has been shown to increase chemo-resistance through interactions with integrins and inhibition of drug delivery due to increased density. Drug delivery can be impaired by collagen concentration and ECM stiffness in several ways, including increasing interstitial fluid pressure, impeding the movement of large drugs or binding and sequestering of drugs [8].

Despite the known role of the stroma and breast density in breast cancer and the growing evidence of collagen as a direct player in breast cancer invasion and progression, there are few, if any, clinical practices that directly harness this information. This is due in part to a need for more knowledge about the specific role that collagen plays in breast tumor progression. Perhaps the greatest need is technologies that not only help researchers unravel the role of collagen, but that can also help clinicians exploit collagen-related changes for clinical treatment. While collagen is the most abundant protein in the body and its role as the glue of the body has been known for generations, it is only in the last few decades that collagen has become

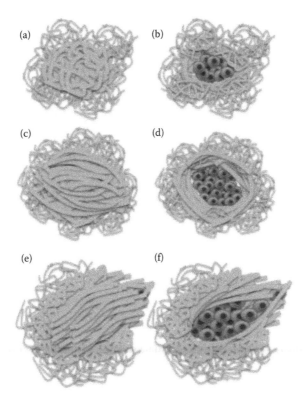

FIGURE 16.2 Graphical representations of tumor associated collagen I structures as described in Table 16.1. (a–b) Category 1, (c–d) Category 2, (e–f) Category 3.

a research target. Likewise, it is only in the last few years that collagen has emerged as a possible clinical target. The challenges of considering collagen as a clinical target are largely due to the properties of collagen: it is intricate in design, dynamic, abundant, needed for normal function and interwoven with a network of other proteins, cellular components, and processes. Even more challenging is that its properties are regulated by other constituents in the ECM and by the physical and chemical conditions under which it is assembled. Thus, collagen polymerized *in vitro* is composed of largely uniform, thin, short fibers and differs from that deposited *in vivo*, which is characterized by a variety of fiber thicknesses and lengths. To truly understand collagen in its *in vivo* functions, whether its beneficial role as a structural protein or a hijacked player in breast cancer invasion and progression, it is necessary to have a method that can monitor all aspects of collagen in the most physiologically relevant and realistic microenvironment possible.

Accomplishing this goal of monitoring and studying the role of collagen in human breast cancer necessitates a very specific type of imaging modality. An effective imaging tool for studying collagen-affiliated processes in breast cancer ideally must: (1) have sufficient resolution and specificity for detecting collagen, normal cells, and cancer cells; (2) be able to detect intrinsic and extrinsic labels for cellular components and cells; (3) have the ability to track over time; and (4) be able to noninvasively image deep into intact tissue.

The quest for early detection and treatment to improve survival rates of cancer is the driving force behind the development of accurate optical biomarkers. Changes in cellular behavior, phenotype, physicochemical properties or alterations to the ECM are all detectable by optical imaging modalities. When quantified in research settings, optical biomarkers have shown potential for earlier detection of carcinoma lesions. Optical biomarkers are innate to the tissue, providing a method to observe natural cellular activities and to correlate underlying biological changes to the marker.

16.2 Imaging Methods

16.2.1 Nonlinear Imaging Techniques for Collagen Characterization

Nonlinear optical imaging techniques, such as multiphoton laser scanning microscopy (MPLSM), [29] are particularly useful for studying the changes taking place at the tumor–stroma boundary [30]. MPLSM is an optical sectioning technique that can simultaneously produce multiphoton excitation (MPE) and second-harmonic generation (SHG). MPE occurs when two or more low-energy photons excite a fluorophore, which then emits a single photon with higher energy than the individual incident photons [29,31] (Figure 16.3). This method restricts fluorophore excitation to the plane of focus (i.e., optical sectioning) and reduces phototoxic effects [30,32] while increasing the effective imaging depth in comparison to conventional confocal microscopy [33].

The noncentrosymmetric crystalline structure of fibrillar collagen I makes it an ideal candidate for imaging via SHG microscopy [34,35]. SHG is a coherent, nonlinear, second-order polarization resulting from the nonabsorptive interaction between a pulsed laser source and a medium lacking a center of symmetry [30,36]. Other key advantages of SHG involve optical sectioning and lack of photobleaching. Additionally, if both forward and backward SHG is collected, information about the structure of collagen can be inferred.

In addition to the collection of SHG, MPLSM is also useful for imaging a number of endogenous (ex: NADH, FAD) and exogenous (ex: GFP, phalloidin) fluorophores. Imaging of fluorophores can be used to determine the metabolic state of a cell or as markers for a range of molecular events within a cell. This information can then be combined with SHG data to create a complete picture of collagen structure and the states of associated cells.

Using SHG, collagen has been exploited as an optical biomarker in research. Being a fundamental component of the cellular microenvironment, collagen has been visualized to study normal and abnormal cellular behavior. As detailed above, Provenzano et al. (2006) [17] used this approach to describe tumor associated collagen signatures (TACS) and the corresponding changes in collagen as breast cancer progresses. More recently, collagen order and ovarian cancer were investigated with forward and backward scatter collection of SHG [37] to further work done by Kirkpatrick et al. [38].

16.2.2 SHG versus Gold Standard Histology Approaches

In addition to SHG, collagen can be visualized by staining with picrosirius red, a histology stain popularized in the 1980s. Initially Sirius red served as a substitute for acid fuchsin in Van Gieson's trichrome to preserve the sample's colors for longer periods of time, up to months [39]. Trichromes, unlike

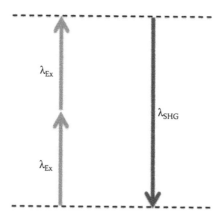

FIGURE 16.3 Jablonski diagram showing the two excitation photons and the single SHG emission photon.

picrosirius, bind differentially depending on the tissue properties (physical structures, amino acid composition) and the size of the dye molecules. Due to differential binding, trichromes poorly stain basement membranes, reticulum fibers, and thin collagen fibers resulting in an underestimation of collagen content or lack of structural information [40]. The shortcomings of trichromes should be noted, since in clinical settings trichromes are primarily used. Polarized picrosirius red, however, is a direct dye that is collagen specific, while standard H&E stains are not specific for collagen. Picric acid dye molecules align their long axis (46 Å) parallel to the collagen fibers to enhance the birefringence of collagen in polarized microscopy. The basic sulfonic groups of lysine, hydroxylysine, and arginine in collagen bind well with acidic picrosirius red dye molecules. However, quantification studies of collagen must proceed with caution since the picric acid dye molecule interaction is not simply stoichiometric but depends on the configuration and substituents of dye molecule, ratio of ionic to nonionic sites and tendency to form aggregates [41]. Picrosirius red staining methods have been used to study normal and abnormal intestinal walls and vaginal mucosa in research settings [42,43].

Further investigation into picrosirius red and direct comparison with SHG in its ability to detect collagen signatures is a vital component of the potential incorporation of collagen signatures in diagnostic environments. If picrosirius red is validated, it can provide a cheap and simple method for pathologists to incorporate collagen signature analysis into their workflow as the importance of collagen in disease progression is validated and introduced into the clinical setting.

Picrosirius red and SHG both demonstrate the ability to distinguish similar patterns of collagen deposition in ECM in various stromal components. In Figure 16.4, the imaging techniques are compared inside the stroma of a rather developed mouse mammary tumor. Straightened collagen is visually detectable with both techniques. Figure 16.5 further illustrates the techniques' ability to distinguish collagen deposition patterns in early dysplasia. Both picrosirius red and SHG are able to distinguish between the dense collagen in the tunica adventitia surrounding the blood vessels and collagen composing the tumor stroma and boundary. The collagen signal from the tunica adventitia is equivalent in

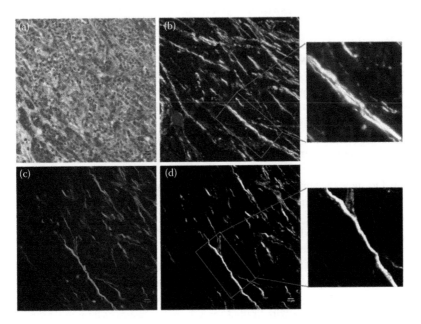

FIGURE 16.4 **(See color insert.)** Collagen in stroma imaged by both polarized picrosirius and SHG images all taken at 40× magnification from Col1a1^{tm1Jae}/+ PyVT (+/+) palpable mammary mouse tissue. (a) H/E, (b) polarized picrosirius red, (c) multiphoton and second harmonic composite, (d) second harmonic. Zoomed in (150%) regions demonstrate differences in collagen fiber detection between techniques. Scale bars = 10 μm.

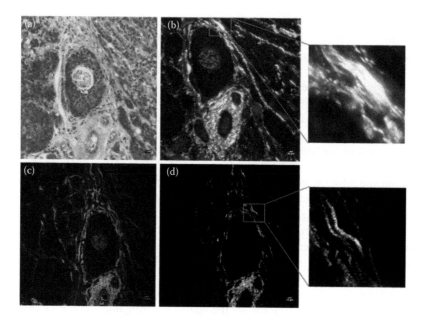

FIGURE 16.5 **(See color insert.)** Collagen in early dysplasia imaged by both polarized picrosirius and SHG images all taken at 40× magnification from Col1a1^{tm1Jae}/+ PyVT (+/+) palpable mammary mouse tissue. (a) H/E, (b) polarized picrosirius red, (c) multiphoton and second-harmonic composite, (d) second harmonic. Zoomed in (150%) regions demonstrate differences in collagen fiber detection between techniques. Scale bars = 10 μm.

both picrosirius red and SHG; however, a visual distinction is apparent in the signal arising from the boundary collagen. Here, the picrosirius red signal appears saturated and fails to provide any indication of individual wavy collagen fibers as visually seen in the SHG image. Also in Figure 16.5, the already controversial interpretation of fiber coloration in picrosirius red is further complicated with alterations in the acquisition settings during picrosirius red collection.

A clear distinction between picrosirius red and SHG is demonstrated in Figure 16.6, where collagen is not readily abundant. Collagen's signal in SHG is much stronger than picrosirius red although it still is visible in both techniques. The low collagen signal in picrosirius red may be explained by its dependency on the picric molecule binding and aligning parallel to the collagen fiber to maximize the birefringence of the dyed collagen fiber. SHG is an intrinsic nonlinear optical effect and is sensitive to both the concentration of collagen (squared) and collagen organization. Due to this fundamental limitation, it is difficult to perform a robust quantitative analysis comparing picrosirius red and SHG despite previous attempts.

We attempted to show the ability of both techniques to decipher small angular changes in collagen fibers in the tumor stroma boundary by selecting five specific fibers in each picrosirius red and SHG image and measuring its angular orientation with respect to the horizontal by hand. Superimposed lines on the collagen were best matched to the overall shape of the collagen fibers to permit relative comparison between the measurements of both techniques (Figure 16.6). Direct comparison between the angular measurements is impossible due to the following limitations: (1) the image frame of the picrosirius red and SHG have a rotation; and (2) the areas of interest are different pathological 5 μm sections. Relative to each other, there is no quantitative difference shown between average angular measurements or standard deviation measurements. Therefore, with limited analysis tools, it seems as if picrosirius red and SHG are equivalent in their ability to detect collagen signals in areas of abundant collagen. However, we anticipate seeing a difference in angular orientation sensitivity with more robust analysis tools.

Upon further digital zoom, as shown in panels (a) and (b) of Figure 16.5, we note that the SHG images show finer details in the structure of collagen. The waviness of the collagen is still intact in the

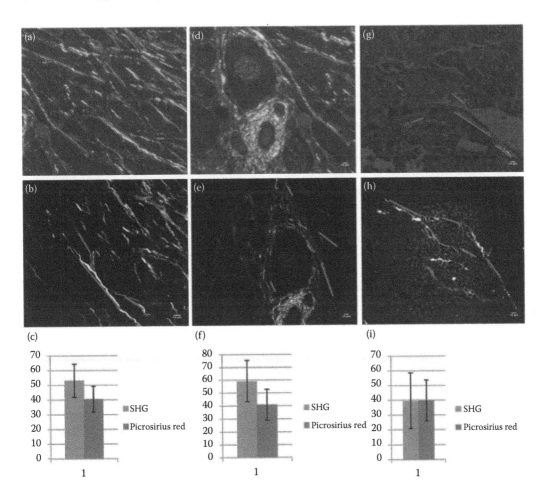

FIGURE 16.6 (See color insert.) Highlighted fiber orientation relative to the horizontal was measured in Gimp, a freely available imaging processing tool. The graph shows average angle orientation with error bars indicating (+/–) standard deviations of the 5 angle measurements per sample of both the SHG and picrosirius techniques. Scale bars = 10 μm.

SHG image but is lost in the picrosirius red image. To quantify this further, we developed a curvelet transform approach (described below) that can quantify the straightness/waviness of collagen fibers. With increased collagen waviness, we would expect the standard deviation of the angular orientation to increase, which is what we observe (data not shown).

We assume that incorporation of the picrosirius red technique into clinical settings may be more practical than SHG since the picrosirius red technique only alters the current slide preparation protocol and can utilize readily available hardware. The picrosirius red technique used here utilized a standard pathology microscope and polarizers and demonstrates the ability to detect collagen and its various signatures. However, if more robust quantitative measurements are desired, additional work with both picrosirius red and SHG measurements is necessary. Further analysis with picrosirius red would benefit from use of multiple polarization orientations of the light to capture all of the picric-dyed birefringent collagen fibers, not simply the fibers in the orientation selected by the viewer, which is possible with a pol-scope. Also, additional information is easily attainable from SHG. Using circularly polarized light instead of linearly polarized light removes the orientation dependence of the collagen signal [44]. In conjunction with circularly polarized light, both the forward and backward signals can be detected to provide a direct read-out of quantitative collagen structure. Future studies may incorporate both the pol-scope and the changes mentioned in SHG collagen signal collection, providing more rigorous comparison between the techniques.

Collagen and its structure have been the focus of much research involving epithelial cell-based cancers using SHG. However to date, no study compared the picrosirius red to SHG. Fundamentally, picrosirius red has inferior resolution than SHG due to its dependence on the picric acid molecules binding to the collagen fibers as well as the light source resolution limitations. Also picrosirius red requires fixation, limiting it to one optical image plane, whereas SHG signal is innate to collagen and permits optical sectioning, thus providing a 3D reconstruction of the microenvironment. Besides fundamental differences between the techniques, it is quite difficult to obtain picrosirius red and SHG images from the same location within the biopsy due to pathological sectioning used to compare the techniques. Despite these limitations, picrosirius and other types of contrast imaging should be explored and compared with second-harmonic imaging to potentially yield new information and perhaps provide more clinically accessible techniques.

Both picrosirius and SHG approaches are capable of visualizing wavy, straightened, and dense collagen deposition in the tumor microenvironment. SHG seems to be more sensitive in less abundant regions of collagen and displays a higher sensitivity to changes in waviness than picrosirius red. However, picrosirius red does show changes within collagen structures well enough to have potential clinical use since it is much simpler, less expensive, and less disruptive to the current workflow in the clinic than SHG.

16.3 Quantifying SHG Data

16.3.1 The Need for Improved Computational Methods

Collagens are the most abundant proteins in mammalian tissues and the major protein constituents of the ECM, which principally maintains shape and structural integrity for cells and tissues, and plays an important role in wound healing, tissue repair, and morphogenesis. Advanced imaging techniques have increased our ability to study this complex relationship [29–31,36,45] and have led to the discovery of some important phenomena [16,17,28,46,47]. Despite the sophisticated imaging techniques available to examine the relationship between changes in collagen I and disease state—dermal wounds [48–50], breast cancer [16,17,28,51,52], ovarian cancer [37], prostate cancer [53], asthma [54,55]—there is still a dearth of robust computational methods for characterizing these changes. The development of new computational analysis techniques will assist researchers in fully exploiting the potential of information-rich imaging data. The techniques described below have the common goals of improving image data analysis for biologists and developing techniques that will aid researchers in the exploration of stromal collagen I as a biomarker for disease.

16.3.2 Digital Signal Processing for Collagen Analysis

The images resulting from these microscopy techniques can be quantitatively analyzed using methods based on digital signal processing. A signal can be defined as a description of the way that the behavior of one process or parameter depends on another [47]. This relationship is typically represented as a mathematical function of one or more independent variables [56]. A one-dimensional signal, such as an audio signal, consists of an independent variable, which commonly represents time, distance, and so on, and a dependent variable, which is a function of the independent variable (Figure 16.7). If the independent variable occurs over a range of continuous values, then the signal is said to be a continuous signal. If the independent variable occurs over a range of discrete values, then the signal is a discrete or digital signal. For a digital signal, the independent variable describes the sampling of the process and the dependent variable is the value of each sample [47].

The independent variable can also be said to define the domain of a signal. A signal for which the independent variable represents time is a time-domain signal. Likewise, a signal for which the independent

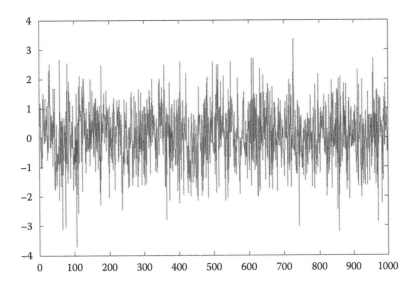

FIGURE 16.7 Example of a 1D digital signal. The *x*-axis describes the sampling of the process and the *y*-axis is the value of each sample.

variable represents distance or frequency is known as a spatial-domain or frequency-domain signal, respectively. A fundamental area of digital signal processing deals with mathematical techniques for converting signals from one domain to another. The Fourier transform is a well-known example of these techniques and is used to convert time-domain signals into frequency-domain signals.

An image, such as might be acquired using the advanced imaging techniques described previously, can be thought of as a matrix of values, with the rows and columns represented by the independent variables *x* and *y* (denoting the distance from the origin) and the dependent variable, $f(x,y)$, representing the value of the matrix at a given location. Each one of these matrix locations is known as a pixel and, in a grayscale image with 8 bits per pixel, the value of each pixel can range from 0 (black) to 255 (white) (Figure 16.8). This number is related to the amount of energy reflected back from or passing through the corresponding location on the surface of the object being imaged at acquisition [47]. The value of a pixel is commonly referred to as the pixel's intensity. The spatial-domain information of an image is in the form of edges—transitions between areas of high intensity and areas of low intensity.

An image may also be thought of as being composed of a complex combination of 2D signal components. For example, Figure 16.9 shows how several 2D sinusoids may be combined to form a new image.

Digital image processing is generally divided into four categories: image coding, image enhancement, image restoration, and image feature extraction [57]. Image coding refers to processes related to image storage and transportation. Image enhancement improves human visual perception of an image by altering image features. Correction of noise, blurring, or distortion is known as image restoration. Finally, image feature extraction is the transformation of one image into another image from which quantitative measurements may be taken [57]. Feature extraction and subsequent analysis of those features are the goals of the projects described here.

Decomposing an image into its component waveforms is a method for simplifying a complex problem and making the task of feature extraction more manageable. Image decomposition can aid in applications such as geometric feature extraction, image filtering, image reconstruction, and image compression. One of the best known and most used image decomposition methods is the Fourier transform, which decomposes a signal (1D or 2D) into its component sinusoids. As mentioned earlier, the Fourier transform is a time- or space-to-frequency transform (i.e., the input is a spatial- or time-domain signal,

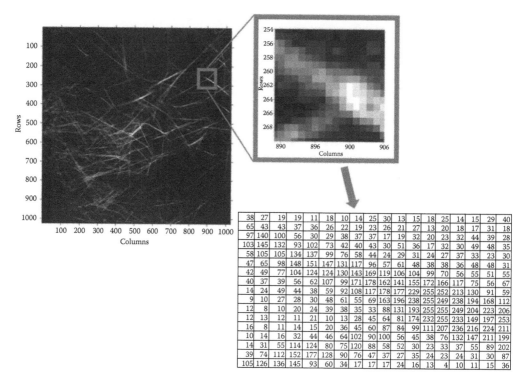

FIGURE 16.8 Digital image structure. A digital image is a matrix of values representing the intensity at each matrix location, or pixel.

and the output is the frequency-domain equivalent). The frequency domain contains exactly the same information as the spatial (or time) domain but in a different form. In the case of the Fourier transform, the frequency domain describes the amplitudes of the sine and cosine waves of which the original signal (image) is composed. These sine and cosine waves are known as the basis functions of the Fourier transform [47].

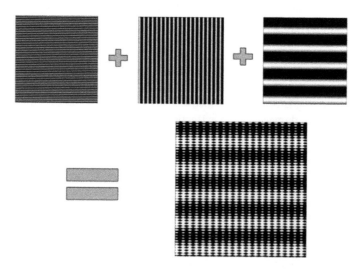

FIGURE 16.9 The combination of different sinusoids yields a new pattern. The more unique sinusoids combined, the more complex the resultant image will be.

16.3.3 The Curvelet Transform

There are many other image decomposition methods, or image transforms, which operate similarly to the Fourier transform but with a different set of basis functions. While the Fourier transform allows a global analysis, the wavelet transform allows a local spatial analysis of image data. A further transform, the curvelet transform [58], derives from wavelets, with the added ability to capture orientation in addition to scale.

The curvelet transform is a nonadaptive, multi-scale transform capable of representing objects in a sparse manner while retaining information on the object's scale, location, and orientation [59]. While curvelets have been used in mammogram-based breast cancer diagnosis [60], we present below, to our knowledge, the first application of the curvelet transform to the problem of collagen alignment in cell resolution image data, particularly in SHG images.

The fundamental advantage of the curvelet transform for collagen alignment analysis is the ability of the transform to retain orientation information from the image. As shown in Figure 16.10, the curvelet transform detects both the scale and orientation of the edges. This results in the ability to examine all prominent edges at a particular orientation and a particular scale (varying only on location of the fixed scale and orientation curvelet in the image). When applied to the collagen alignment analysis problem, the curvelet transform becomes a powerful tool for detecting the presence of filamentous structures and their location, scale, and orientation. By obtaining accurate quantitative data regarding collagen amount, morphology, and organization/orientation, biologically relevant data can be derived. To make use of the curvelet transform to capture information about the spatial organization of collagen, we have developed CurveAlign analysis software.

CurveAlign is not intended to precisely follow individual fibers, but rather to determine the overall trend in fiber alignment in an image. The measured angles can be binned according to the collagen fiber structures in diseases such as cardiac disease [61]. As well, CurveAlign can detect trends in other filamentous structures such as microtubules [62]. In breast cancer invasion and progression studies, CurveAlign was used to quantitatively assess the effect of cancer cells on the remodeling of collagen during cancer onset and progression, as well as in human pathological samples (Figure 16.11).

The integration of these concepts—the influence of stromal collagen I on the tumor microenvironment, nonlinear optical imaging techniques, and digital image processing—forms the basis for a toolkit for using quantitative SHG to understand the role of collagen. While our work to date has focused on breast cancer, these digital image analysis techniques can be generalized, some in the spatial domain

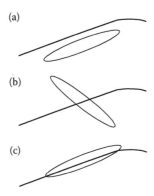

FIGURE 16.10 Graphical representation of curvelets. (a) A curvelet whose length-wise support does not intersect a discontinuity. The curvelet coefficient magnitude will be zero. (b) A curvelet whose length-wise support intersects with a discontinuity, but not at a critical angle. The curvelet coefficient magnitude will be close to zero. (c) A curvelet whose length-wise support intersects with a discontinuity and is tangent to that discontinuity. The curvelet coefficient magnitude will be much larger than zero.

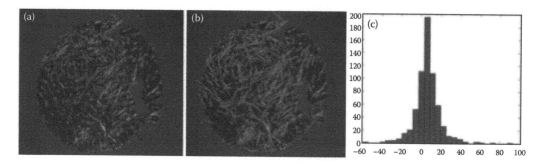

FIGURE 16.11   **(See color insert.)** Use of CurveAlign to quantify collagen orientation in a human breast cancer specimen. An SHG image of a histopathological section (a) was analyzed by CurveAlign (b) and exported as a histogram (c).

and some in the frequency domain, to SHG imaging data sets in an effort to better understand the behavior and influence of collagen I in various disease states.

16.3.4 The Future of Collagen Imaging by SHG

In this chapter, we have explored the changes that occur in collagen surrounding mammary tumors and the robust means to image those changes by SHG. Although collagen has not been previously exploited as a biomarker, the finding that collagen alignment predicts outcome for breast cancer patients suggests that collagen could be used for prognostic value. The use of SHG has many advantages over classic histology stains including picrosirius red. Because SHG requires no stain, it can be used with live or unfixed tissue and can be exploited to understand tissue organization in three-dimensional space. The application of image analysis and quantification of collagen features has provided robust, nonsubjective means to assess the changes in collagen deposition, structure, and alignment that occur around tumors. Thus, with these tools in place, we as a research community are now poised to bring imaging of collagen by SHG into the clinical setting. The ability to image collagen in live tissue may allow a better determination of tumor margins during surgery or a rapid means to determine if a tumor is invasive upon biopsy prior to more extensive pathology processing. Quantification of collagen in biopsy samples may provide prognostic information that will help physicians choose the most effective course of action in treatment.

References

1. Brower V: Breast density gains acceptance as breast cancer risk factor. *J Natl Cancer Inst*, 102(6): 374–375.
2. Boyd NF, Guo H, Martin LJ, Sun L, Stone J, Fishell E, Jong RA, Hislop G, Chiarelli A, Minkin S et al. Mammographic density and the risk and detection of breast cancer. *N Engl J Med* 2007, 356(3):227–236.
3. Vachon CM, van Gils CH, Sellers TA, Ghosh K, Pruthi S, Brandt KR, Pankratz VS: Mammographic density, breast cancer risk and risk prediction. *Breast Cancer Res* 2007, 9(6):217.
4. Boyd NF, Martin LJ, Yaffe MJ, Minkin S: Mammographic density and breast cancer risk: Current understanding and future prospects. *Breast Cancer Res* 2011, 13(6):223.
5. Vachon CM, Brandt KR, Ghosh K, Scott CG, Maloney SD, Carston MJ, Pankratz VS, Sellers TA: Mammographic breast density as a general marker of breast cancer risk. *Cancer Epidemiol Biomarkers Prev* 2007, 16(1):43–49.
6. Maskarinec G, Woolcott CG, Kolonel LN: Mammographic density as a predictor of breast cancer outcome. *Future Oncol* 2010, 6(3):351–354.

7. Obenauer S, Hermann KP, Grabbe E: Applications and literature review of the BI-RADS classification. *Eur Radiol* 2005, 15(5):1027–1036.
8. Alowami S, Troup S, Al-Haddad S, Kirkpatrick I, Watson PH: Mammographic density is related to stroma and stromal proteoglycan expression. *Breast Cancer Res* 2003, 5(5):R129–R135.
9. Brower V: Homing in on mechanisms linking breast density to breast cancer risk. *J Natl Cancer Inst* 2010, 102(12):843–845.
10. Provenzano PP, Inman DR, Eliceiri KW, Keely PJ: Matrix density-induced mechanoregulation of breast cell phenotype, signaling and gene expression through a FAK-ERK linkage. *Oncogene* 2009, 28(49):4326–4343.
11. Egeblad M, Rasch MG, Weaver VM: Dynamic interplay between the collagen scaffold and tumor evolution. *Curr Opin Cell Biol* 2010, 22(5):697–706.
12. Ghajar CM, Bissell MJ: Extracellular matrix control of mammary gland morphogenesis and tumorigenesis: Insights from imaging. *Histochem Cell Biol* 2008, 130:1105–1118.
13. Roskelley CD, Srebrow A, Bissell MJ: A hierarchy of ECM-mediated signalling regulates tissue-specific gene expression. *Curr Opin Cell Biol* 1995, 7(5):736–747.
14. Plant AL, Bhadriraju K, Spurlin TA, Elliott JT: Cell response to matrix mechanics: Focus on collagen. *Biochim Biophys Acta* 2009, 1793(5):893–902.
15. Shoulders MD, Raines RT: Collagen structure and stability. *Annu Rev Biochem* 2009, 78:929–958.
16. Provenzano PP, Inman DR, Eliceiri KW, Knittel JG, Yan L, Rueden CT, White JG, Keely PJ: Collagen density promotes mammary tumor initiation and progression. *BMC Med* 2008, 6:11.
17. Provenzano PP, Eliceiri KW, Campbell JM, Inman DR, White JG, Keely PJ: Collagen reorganization at the tumor–stromal interface facilitates local invasion. *BMC Med* 2006, 4(1):38.
18. Keely PJ, Wu JE, Santoro SA: The spatial and temporal expression of the [alpha]2[beta]1 integrin and its ligands, collagen I, collagen IV, and laminin, suggest important roles in mouse mammary morphogenesis. *Differentiation* 1995, 59(1):1–13.
19. Simian M, Hirai Y, Navre M, Werb Z, Lochter A, Bissell MJ: The interplay of matrix metalloproteinases, morphogens and growth factors is necessary for branching of mammary epithelial cells. *Development* 2001, 128(16):3117–3131.
20. Paszek MJ, Zahir N, Johnson KR, Lakins JN, Rozenberg GI, Gefen A, Reinhart-King CA, Margulies SS, Dembo M, Boettiger D et al.: Tensional homeostasis and the malignant phenotype. *Cancer Cell* 2005, 8(3):241–254.
21. Keely P, Fong A, Zutter M, Santoro S: Alteration of collagen-dependent adhesion, motility, and morphogenesis by the expression of antisense alpha 2 integrin mRNA in mammary cells. *J Cell Sci* 1995, 108(2):595–607.
22. Burstein H, Polyak K, Wong J, Lester S, Kaelin C: Ductal carcinoma *in situ* of the breast. *NEJ Med* 2004, 350(14):1430–1441.
23. Butcher DT, Alliston T, Weaver VM: A tense situation: Forcing tumour progression. *Nat Rev Cancer* 2009, 9(2):108–122.
24. Thomasset N, Lochter A, Sympson CJ, Lund LR, Williams DR, Behrendtsen O, Werb Z, Bissell MJ: Expression of autoactivated stromelysin-1 in mammary glands of transgenic mice leads to a reactive stroma during early development. *Am J Pathol* 1998, 153(2):457–467.
25. Sternlicht MD, Lochter A, Sympson CJ, Huey B, Rougier J-P, Gray JW, Pinkel D, Bissell MJ, Werb Z: The stromal proteinase MMP3/stromelysin-1 promotes mammary carcinogenesis. *Cell* 1999, 98(2):137–146.
26. Erler JT, Bennewith KL, Cox TR, Lang G, Bird D, Koong A, Le QT, Giaccia AJ: Hypoxia-induced lysyl oxidase is a critical mediator of bone marrow cell recruitment to form the premetastatic niche. *Cancer Cell* 2009, 15(1):35–44.
27. Wolf K, Alexander S, Schacht V, Coussens LM, von Andrian UH, van Rheenen J, Deryugina E, Friedl P: Collagen-based cell migration models *in vitro* and *in vivo*. *Semin Cell Dev Biol* 2009, 20(8):931–941.

28. Conklin MW, Eickhoff JC, Riching KM, Pehlke CA, Eliceiri KW, Provenzano PP, Friedl A, Keely PJ: Aligned collagen is a prognostic signature for survival in human breast carcinoma. *Am J Pathol* 2011, 178(3):1221–1232.

29. Denk W, Strickler JH, Webb WW: Two-photon laser scanning fluorescence microscopy. *Science* 1990, 2484951:73–76.

30. Provenzano PP, Eliceiri KW, Yan L, Ada-Nguema A, Conklin MW, Inman DR, Keely PJ: Nonlinear optical imaging of cellular processes in breast cancer. *Microsc Microanal* 2008, 14(6):532–548.

31. Diaspro A, Sheppard CJR: Two-Photon excitation fluorescence microscopy. In: *Confocal and Two-Photon Microscopy: Foundations, Applications, and Advances.* Edited by Diaspro A. New York: Wiley-Liss, Inc.; 2002:39–73.

32. Squirrell JM, Wokosin DL, White JG, Bavister BD: Long-term two-photon fluorescence imaging of mammalian embryos without compromising viability. *Nat Biotechnol* 1999, 17(8):763–767.

33. Centonze VE, White JG: Multiphoton excitation provides optical sections from deeper within scattering specimens than confocal imaging. *Biophys J* 1998, 75(4):2015–2024.

34. Campagnola PJ, Clark HA, Mohler WA, Lewis A, Loew LM: Second-harmonic imaging microscopy of living cells. *J Biomed Opt* 2001, 6(3):277–286.

35. Plotnikov SV, Kenny AM, Walsh SJ, Zubrowski B, Joseph C, Scranton VL, Kuchel GA, Dauser D, Xu M, Pilbeam CC et al.: Measurement of muscle disease by quantitative second-harmonic generation imaging. *J Biomed Opt* 2008, 13(4):044018.

36. Mohler W, Millard AC, Campagnola PJ: Second harmonic generation imaging of endogenous structural proteins. *Methods* 2003, 29(1):97–109.

37. Nadiarnykh O, LaComb RB, Brewer MA, Campagnola PJ: Alterations of the extracellular matrix in ovarian cancer studied by second harmonic generation imaging microscopy. *BMC Cancer* 2010, 10:94.

38. Kirkpatrick ND, Brewer MA, Utzinger U: Endogenous optical biomarkers of ovarian cancer evaluated with multiphoton microscopy. *Cancer Epidemiol Biomarkers Prev* 2007, 16(10):2048–2057.

39. Puchtler H, Sweat F: Histochemical specifity of staining methods for connective tissue fibers: Resorcin-fuchsin and Van Gieson's Picro-Fuchsin. *Z Zellforch Microsk Anat Histochem* 1964, 79:24–34.

40. Sweat F, Meloan SN, Puchtler H: A modified one-step trichrome stain for demonstration of fine connective tissue fibers. *Stain Technol* 1968, 43(4):227–231.

41. Puchtler H, Sweat F, Gropp S: An investigation into the relation between structure and fluorescence of azo dyes. *J R Microsc Soc* 1967, 87(3):309–328.

42. Rabau MY, Dayan D: Polarization microscopy of picrosirius red stained sections: A useful method for qualitative evaluation of intestinal wall collagen. *Histol Histopathol* 1994, 9(3):525–528.

43. Borges LF, Gutierrez PS, Marana HR, Taboga SR: Picrosirius-polarization staining method as an efficient histopathological tool for collagenolysis detection in vesical prolapse lesions. *Micron* 2007, 38(6):580–583.

44. Campagnola PJ, Loew LM: Second-harmonic imaging microscopy for visualizing biomolecular arrays in cells, tissues and organisms. *Nat Biotechnol* 2003, 21(11):1356–1360.

45. Stoller P, Kim BM, Rubenchik AM, Reiser KM, Da Silva LB: Polarization-dependent optical second-harmonic imaging of a rat-tail tendon. *J Biomed Opt* 2002, 7(2):205–214.

46. Provenzano PP, Inman DR, Eliceiri KW, Trier SM, Keely PJ: Contact guidance mediated three-dimensional cell migration is regulated by Rho/ROCK-dependent matrix reorganization. *Biophys J* 2008, 95(11):5374–5384.

47. Smith SW: *The Scientist and Engineer's Guide to Digital Signal Processing.* San Diego, CA: California Technical Publishing; 1997.

48. Torkian BA, Yeh AT, Engel R, Sun C-H, Tromberg BJ, Wong BJF: Modeling aberrant wound healing using tissue-engineered skin constructs and multiphoton microscopy. *Arch Facial Plast Surg* 2004, 6(3):180–187.

49. Yeh AT, Kao B, Jung WG, Chen Z, Nelson JS, Tromberg BJ: Imaging wound healing using optical coherence tomography and multiphoton microscopy in an *in vitro* skin-equivalent tissue model. *J Biomed Opt* 2004, 9(2):248–253.

50. Zhuo S, Chen J, Jiang X, Cheng X, Xie S: Visualizing extracellular matrix and sensing fibroblasts metabolism in human dermis by nonlinear spectral imaging. *Skin Res Technol* 2007, 13(4):406–411.

51. Pollard JW: Macrophages define the invasive microenvironment in breast cancer. *J Leukocyte Biol* 2008, 84(3):623–630.

52. Round AR et al.: A preliminary study of breast cancer diagnosis using laboratory based small angle x-ray scattering. *Phys Med Biol* 2005, 50(17):4159.

53. Tuxhorn JA, Ayala GE, Smith MJ, Smith VC, Dang TD, Rowley DR: Reactive stroma in human prostate cancer. *Clin Cancer Res* 2002, 8(9):2912–2923.

54. Araujo BB, Dolhnikoff M, Silva LFF, Elliot J, Lindeman JHN, Ferreira DS, Mulder A, Gomes HAP, Fernezlian SM, James A et al.: Extracellular matrix components and regulators in the airway smooth muscle in asthma. *Eur Resp J* 2008, 32(1):61–69.

55. Bergeron C, Al-Ramli W, Hamid Q: Remodeling in Asthma. *Proc Am Thorac Soc* 2009, 6(3):301–305.

56. Prince JL, Links JM: *Medical Imaging Signals and Systems*. Upper Saddle River, NJ: Pearson Education, Inc; 2006.

57. vander Heijden F: *Image Based Measurement Systems: Object Recognition and Parameter Estimation*. New York: Wiley; 1995.

58. Candès E, D. Donoho: Curvelets—A surprisingly effective nonadaptive representation for objects with edges. In: *Curves and Surface Fitting: Saint-Malo 1999*. Edited by Cohen A, Rabut C, Schumaker L. Saint-Malo, France: Vanderbilt University Press; 1999:105–120.

59. Candés E: What is a curvelet? *Notices Am Math Soc* 2003, 50(11):1402–1403.

60. Eltoukhy MM, Faye I, Samir BB: Breast cancer diagnosis in digital mammogram using multiscale curvelet transform. *Comput Med Imaging Graphics* 2010, 34(4):269–276.

61. Kouris NA, Squirrell JM, Jung JP, Pehlke CA, Hacker T, Eliceiri KW, Ogle BM: A nondenatured, noncrosslinked collagen matrix to deliver stem cells to the heart. *Regen Med* 2011, 6(5):569–582.

62. Heck JN, Garcia-Mendoza MG, Ponik SM, Pehlke CA, Inman DR, Eliceiri KW, Keely PJ: Microtubules regulate Gef-H1 in response to extracellular matrix stiffness. *Mol Biol Cell* 2012, 23(13):2583–2592.

17

SHG in Tumors: Scattering and Polarization

Seth W. Perry
University of Rochester

Xiaoxing Han
University of Rochester

Edward B. Brown
University of Rochester

Tumor physiology, biochemistry, structural organization, and cellular composition are believed to make integrated contributions to cancer pathology. The tumor stroma for one includes the basement membrane, the extracellular matrix (ECM), and nonmalignant cells in the tumor, and plays significant roles in tumor growth and metastasis [1,2]. The biochemical and phenotypic changes in stromal cells surrounding malignant tumor cells, including immune cells and connective tissue cells such as fibroblasts (a principal collagen-producing cell type), modify the synthesis and breakdown of key ECM components, resulting in a "reactive stroma" characteristic of neoplastic transformation and thus contributing to tumor progression and metastasis [1–4]. This key role that the "reactive stroma" plays in cancer pathogenesis has led to significant interest in understanding the particular characteristics of reactive stroma that may provide prognostic value for predicting cancer pathogenesis or its outcome [5–8]. Since collagen is a key component of the ECM and is also believed to play important roles in cancer, recent years have seen a burgeoning interest in monitoring the second-harmonic generation (SHG) signal from fibrillar collagen in tumor stroma, with the idea that it may provide diagnostic or predictive value in comparison of normal versus malignant tissue [9–16, and other references herein].

Indeed, SHG signal is proportional to the total collagen content and changes dynamically with tumor growth and biochemical modification of the ECM [9], and several studies have demonstrated relative changes in the overall SHG levels or intensity in comparison of control versus cancerous tissue from both experimental animal models and humans [17–23, and reviewed in Ref. [24] and elsewhere in the book]. However, beyond gross changes in SHG intensity, SHG's coherent properties allow us to extract other information about collagen's molecular structure from the SHG signal that is also relevant to tumor pathology. Specifically, the ratio of forward-scattered (*F*) to backward-scattered (*B*) SHG signal (*F/B* ratio) is sensitive to the axial length scale over which collagen scatterers are ordered for SHG [1,25–32]. Monitoring these fibril characteristics can provide important insights into cancer pathophysiology, since the reactive tumor stroma is characterized by altered collagen production and degradation [2–4], and the changes in collagen fiber diameter or shape have been linked to human cancer [33–35]. (Collagen fibers are larger diameter bundles of collagen fibrils.)

The measurements of SHG *polarization,* on the other hand, can provide several pieces of information regarding secondary and higher-order organization of collagen molecules and fibrils. First,
collagen SHG polarization may be used to indicate the angular orientation of the helix structure of
individual collagen molecules (i.e., the collagen pitch angle or helical angle) [36,37]. Polarization-
modulated SHG of collagen can also be employed to assess the angular distribution of collagen fibrils
[38–40]. These properties of collagen are significant for cancer research because an increasing number of studies support the idea that the changes in collagen's secondary and higher-order molecular
structure may predict the development or outcome of cancer. For example, the axial period of collagen
fibrils, which is in part a function of collagen's helical angle and higher molecular organization [41], is
demonstrably different in human breast cancer versus control tissue [42]. In addition, the proportions
of matrix molecules such as collagen I, III, fibrin, and fibronectin are frequently altered in reactive
tumor stroma [2,43,44]. The changes in the molecular collagen I:collagen III ratio in particular are
likely to alter several higher-order molecular characteristics of collagen fibrils [41,45], which may in
turn affect tumorigenesis or prognosis.

As such, in this chapter we highlight several optical methods and publications that have utilized the
coherent properties of SHG, specifically the scattering directionality and polarization, to investigate
collagen properties in tumor stroma.

17.1 Scattering Properties of SHG in Tumors

17.1.1 Factors Affecting the *F/B* Signal

The SHG signal propagates both forward and backwards relative to the excitation laser axis. The *F/B*
ratio from a distribution of scatterers, such as triple helices bundled into collagen fibrils, equals 1 for
small distributions of scatterers and increases as the size scale of the distribution surpasses the SHG
wavelength [1,25,46,47]. Therefore, the *F/B* ratio is closely related to the optically apparent thickness of the fibril along the laser axis and within the focal volume [1,25–32], but is also influenced
by the excitation wavelength (as noted above), collagen packing density and order [47,48], and ionic
strength of the collagen's environment [27]. Briefly, the *F/B* ratio has been shown to be sensitive to
both collagen fibril diameter [31] and fibril orientation (angle) relative to the laser beam [32] (Figure
17.1), two factors that effectively determine the "optically apparent" fibril thickness in the z-axis.
Moreover, the *measured F/B* SHG ratio arises from the initial forward and backward SHG directionalities, as well as subsequent scattering of these signals (e.g., initially forward-directed SHG can scatter to become backward-directed SHG), the latter of which is principally a function of tissue density
[46]. Similarly, biologic tissues will differentially attenuate the F and B SHG signals and thus, the *F/B*
ratio is dependent on the tissue thickness, and in particular, the axial (z-axis) location of the focal
volume relative to the overall tissue depth [28,29,47–49]. Finally, reduction in ionic strength of the
collagen's environment dramatically increases the observed *F/B* (higher F and lower B), an effect that
is postulated to arise from swelling of the SHG-emitting collagen structures under hypo-osmotic
conditions [27].

As discussed elsewhere [28,46,48], we can also intuitively see that the contribution of some of these
factors (e.g., tissue depth and density) means that in the biologic tissue, the *final detected* (i.e., measured) *F/B* ratio generally differs from the scatterers' *initial emitted* or *original F/B* SHG directionality,
since subsequent scattering and absorption of the *initial* SHG signal with tissue depth and density will
alter the final measured SHG *F/B*. With calibration approaches, the *original* initial emitted *F/B* can be
inferred from the measured *F/B* [1,27], but it can only effectively be measured directly when the tissue
thickness is <1 mean free path (MFP) [28] (see discussions below for further details). Thus, the *detected
F/B* is dependent on both collagen structure *and* other tissue characteristics, whereas the *initial emitted F/B* is primarily dependent on collagen structural characteristics. Herein, we will use F_d/B_d when
referring to the *final detected F/B* ratio measured in tissues, and F_i/B_i when referencing studies that have

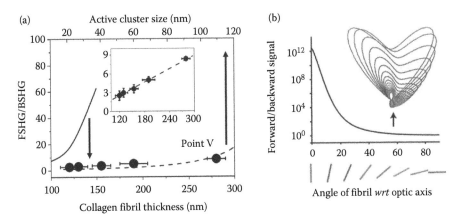

FIGURE 17.1 Collagen *F/B* SHG ratio varies with fibril diameter and orientation in the *z*-axis. Collagen (a) fibril diameter and (b) orientation with respect to the *z*-axis determine the effective fibril diameter in the optical axis, which in turn affects the *F/B* ratio, with the *F/B* ratio increasing with increasing effective fibril diameter. However, the relationship between fibril diameter (as determined by atomic force microscopy) and *F/B* is complex, as detailed in Ref. [31]. In (a), solid circles are measured as *F/B* ratio relative to collagen fibril thickness. If all the collagen molecules act as active SHG scatterers, the calculated *F/B* ratio is plotted as the solid line. The dashed line is *F/B* calculation based on active cluster sizes, indicating that only a fraction of the collagen molecules inside a fibril produces SHG effectively. (a, inset) An empirical linear relation on *F/B* ratio over collagen thickness is obtained. Panel (b) represents *F/B* SHG as a function of fibril angle to the optical axis. (b, inset) The expected SHG intensity profiles from variously oriented rods. (Panel (a) and its legend are reprinted from Chu, S.W. et al. 2007. Thickness dependence of optical second harmonic generation in collagen fibrils. *Opt Express*, 15:12005–12010, Copyright 2007. With permission of Optical Society of America; Panel (b) and its legend are reprinted with permission from Zipfel, W.R. et al. Live tissue intrinsic emission microscopy using multiphoton-excited native fluorescence and second harmonic generation. *Proc Natl Acad Sci USA*, 100:7075–7080, Copyright 2003, National Academy of Sciences, USA.)

endeavored to infer the *initial* emitted *F/B* ratio (our F_i/B_i here, to describe the initial emitted *F/B* SHG directionality is equivalent to the F_{SHG}/B_{SHG} designation used in Chapter 6).

In this manner, with consideration to the other factors described above, the *F/B* ratio can provide some indication of the overall spatial distribution and orientation of collagen fibrils, which as we suggest above may have diagnostic value in differentiating normal versus cancerous tissue. In this section, we will review several literature reports that have specifically analyzed *F/B* ratios in cancer tissues as a means of distinguishing tumor stroma from healthy stroma, after which we will highlight several technical instrumentation advances intended to facilitate the collection of *F/B* data from thicker tissue sections.

17.1.2 *F/B* Measurements in Tumor Tissue

Numerous reports have investigated differences in tissue collagen SHG signal in cancerous versus normal tissue (see [25] and references therein), but to our knowledge, only a few reports have looked specifically at collagen *F/B* ratios in cancer tissue. In one report using mouse tumor models, Han et al. found morphologic differences in collagen organization between healthy and tumorous murine mammary tissue, with healthy mammary tissue displaying more delicate and evenly distributed collagen fibers compared to the thicker bands of collagen observed in tumor-implanted murine mammary fat pads [1]. However, they found no statistically significant differences in collagen F_i/B_i ratios between the normal and tumor-implanted murine mammary fat pad tissues. In this study, the unknown collagen F_i/B_i ratio was indirectly measured by normalizing the detected collagen *F/B* to the detected *F/B* from fluorescent beads in the same image plane because the F_i/B_i of the beads was known from a separate measurement [1,27].

Since altered collagen production and degradation are hallmarks of reactive stroma [2–4], measurable differences in collagen F_i/B_i ratios might be expected in tumor versus healthy tissue. Thus, these results might seem somewhat surprising, until we consider the finding that F_i/B_i ratios also did not vary with apparent fiber diameter [1] (also see the comparative discussion of F_i/B_i and F_d/B_d ratios in cancer tissue, in this section below). A uniform F_i/B_i despite a changing fiber diameter suggests that the SHG-generating collagen features maintain a stable axial length/λ_{SHG} ratio despite variations in the overall fiber thickness. At least two possible models may account for how this could occur: (1) Tumor collagen fibrils might only be appropriately ordered for strong SHG emission within an exterior "shell" region of relatively constant thickness (and thus relatively uniform F_i/B_i), with an SHG-deficient center region that varies to affect the overall fibril diameter (as was found in the rat-tail model [27]), or (2) tumor collagen fibrils produce strong SHG emission throughout, and are a collection of similar-diameter (i.e., uniform F_i/B_i), rod-like collagen "building blocks." In each case, the uniform F_i/B_i fibrils assemble in varying numbers to determine visible collagen fiber thickness [1]. In the Han study, further analysis of the F_i/B_i ratios (that averaged ~34 for all tissues), along with confirmatory electron microscopy, found that both healthy and tumorous murine mammary tissue were composed of core ~70 nm diameter collagen rods (i.e., apparent "building blocks") whose diameter remained constant, but which aggregated to form larger diameter fibrils, as per the second model above [1]. These results are interesting because they inform us that at least in this particular animal cancer model, some base level of collagen organizational structure as measured by F_i/B_i does not change significantly between normal and cancerous tissues, despite clearly visible higher-order changes in collagen morphology. Additional studies will help determine whether differences in collagen F_i/B_i can be detected in normal versus tumor tissue, across a wider range of animal and human cancer models.

In another study, Nadiarnykh et al. [48] found that away from the surface epithelium, human ovarian cancer tissue contained fewer cells, and more dense and more regularly patterned collagen, as compared to the control ovarian tissue. These morphologic differences manifested as several quantitative changes related to scattering properties of collagen SHG, as follows. First, the increased density in cancer tissue was reflected, as expected, in higher scattering coefficients (μ_s) (as calculated by Monte Carlo simulations), which is the inverse of the MFP, or the distance a photon will travel before undergoing a direction-changing scattering collision [48]. The readers can refer to Refs. [28,48] for further details on how these μ_s values were obtained. Second, F_d/B_d ratios were decreased in human ovarian cancer tissue compared to normal controls, which reflected decreased F_d SHG in the cancer tissues, consistent with a *higher* μ_s (and density) in this population such that initially forward-directed (F_i) SHG photons were more likely to be scattered in different directions. Moreover, as expected for *detected F/B*, F_d/B_d ratios for each condition increased with increasing tissue depth [48], which photon diffusion theory informs is consistent with less scattering/redirection of the F_i SHG as the distance from the focal point to the forward tissue boundary becomes smaller, thus increasing F_d SHG (and decreasing B_d) and therefore increasing F_d/B_d with depth into the tissue [48,50]. Overall, the authors found these results support the idea that collagen's structural and organizational changes in cancer likely arise from new synthesis of collagen, rather than from reorganization of the existing collagen pools.

17.1.3 Technical Advances for Measuring *F/B*

The reports described above have advanced methods for determining *F/B* SHG in malignant tissues, to evaluate the significance of these parameters as they predict collagen structure and organization in cancer diseased versus undiseased states. A series of studies, some of which were discussed above, have established elegant methods for determining F_d/B_d in biologic tissues as influenced by tissue depth and scattering properties [28,29,47–49,51], as well as by calibration methods to extract F_i/B_i from measured F_d/B_d [1,27]. These studies were enabled by an abundance of prior work that has laid the foundation for deciphering the meaning of *F/B* ratios as they pertains to collagen's organizational structure

[31,47,49,52,53]. In another recent study, Ajeti et al. [54] found that in mixed collagen V/collagen I gels, the F_d/B_d ratio was effectively equivalent in 0% and 5% collagen V gels, but was significantly lower in 20% collagen V gels. These results are significant because collagen V interacts with collagen I to influence fibril diameter distribution [55] (0% and 5% collagen V gels have similar fibril diameter distributions [55], hence no difference in F/B above), and collagen V is found upregulated in human breast carcinoma [56]. Therefore, the ability to distinguish changes in proportions of collagen I/V (or other) isoptypes by collagen F_d/B_d SHG may provide a useful prognostic tool for understanding and treating breast cancer.

Other groups have focused on developing novel instrumentation for obtaining F/B data from human or other biologic samples. The most direct and typical means of obtaining F/B measurements involves placing an objective lens on both sides of an excised and sectioned tissue specimen. However, this two-lens approach is not feasible for very thick biopsy specimens, or for *in vivo* clinical applications such as endoscopy.

To overcome these limitations, Han et al. [26] recently described a system that can capture collagen F/B SHG from intact specimens *in vivo*, using a single-objective lens. This system is illustrated in Figure 17.2a. Briefly, this system relies on the properties that at shallow imaging depths, the initially backward-directed SHG (B_i) will exit the image from the focal volume with minimal subsequent scatter, whereas the initially forward-propagating SHG (F_i) will subsequently scatter such that a fraction (as much as ~20%, see [57]) of this signal ultimately exits in the backward direction (and therefore can be captured with the same objective lens and detector as the initially backward-directed SHG, i.e., B_i) (Figure 17.2b). This results in a Gaussian distribution of the B_i signal in an image plane, and a subsequently backscattered F_i signal whose photon intensity distribution in an image plane is constant and does not vary significantly with position over an approximately 50 μm radius from the focal volume [26] (Figure 17.3):

$$I_{SHG}(r) = B\exp\left[-2\left(\frac{r}{\omega}\right)^2\right] + FC \tag{17.1}$$

As defined in Ref. [26], I_{SHG} is the total SHG signal intensity distribution on the object plane, r is the radius from the focal point on the image plane, ω is the e^{-2} Gaussian spot size on the image plane of the direct backward-propagating SHG (B_i), and F and B are the absolute intensities of the F- and B-propagating SHG signals. The C parameter relates the initially forward-propagating SHG (F_i) signal intensity to the average intensity of the uniform distribution of (now backscattered) SHG light that reaches the object plane, and is a function of the scattering and absorption properties of the tissue [26]. This equation can also be written as

$$I_{SHG}(r) = B\left[\exp\left[-2\left(\frac{r}{\omega}\right)^2\right] + \frac{F}{B}C\right] \tag{17.2}$$

where F/B is the collagen fiber SHG F/B ratio [26] (i.e., F_i/B_i, as we define it here). When a series of collagen fiber SHG images are generated through a series of different sized confocal pinholes, each image pixel represents an integration of the total SHG signal over the pinhole area that can be expressed as follows:

$$I_{pixel} \propto \int_0^{2\pi} d\theta \int_0^R \left\{\exp\left[-2\left(\frac{r}{\omega}\right)^2\right] + \left(\frac{F}{B}\right)C\right\} r\,dr \tag{17.3}$$

where R is the pinhole size with respect to the direct backward-propagating SHG (B_i) Gaussian spot size, that is, $R = r_{pinhole}/\omega$ [26]. By normalizing pixel intensities at the various pinhole sizes to the maximum

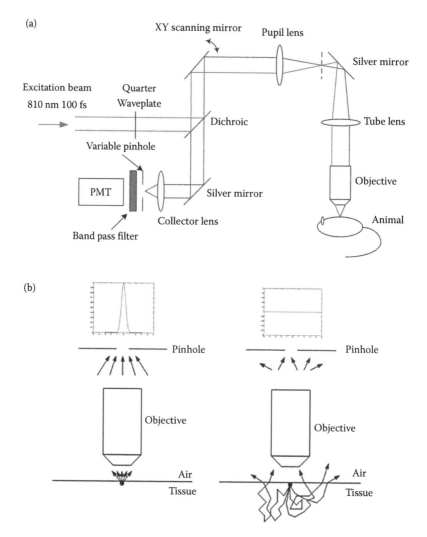

FIGURE 17.2 Collagen *F/B* SHG measured with a single objective. Using the system illustrated in panel (a), both the forward- and backward-directed SHG signals can be captured with a single objective, and *F/B* ratios can be calculated via serial scans of the same region of interest (ROI) through different sized pinholes. This is possible because as illustrated in panel (b), at shallow imaging depths, the backward-propagating SHG signal exits the two-photon focal volume with minimal subsequent scatter and will produce a sharp peak in an image plane, which will pass through a confocal pinhole (b, left image). On the other hand, a fraction of the initially forward-propagating SHG signal will ultimately backscatter toward the objective lens, passing through the object plane at multiple locations, and resulting in a weak diffuse signal at an image plane. By repeatedly imaging the sample through a series of different sized pinholes, the shape of this total SHG distribution can be measured and, with suitable calibration, the underlying *F/B* ratio can be determined, as described [26]. (Reprinted from Han, X. and E. Brown. 2010. Measurement of the ratio of forward-propagating to back-propagating second harmonic signal using a single objective. *Opt Express*, 18:10538–10550, Copyright 2010. With permission of Optical Society of America.)

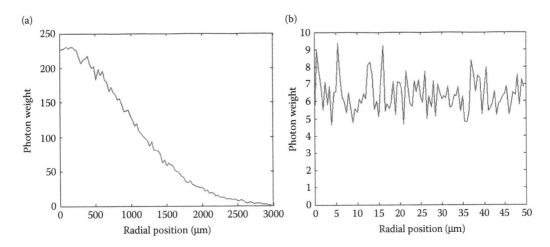

FIGURE 17.3 Spatial distribution of forward-propagating and subsequently backscattered SHG photons reaching the object plane. Monte Carlo simulation was used to model the scattering of initially forward-propagating (F_i) and subsequently backscattered SHG photons in the tissue, to predict the radial distribution of these photons that exit the tissue interface toward the objective lens. (a) Steady-state radial distribution of the backscattered SHG photons over a large range of radial position (3 mm) from the initial emission point (i.e., focal volume), demonstrating that over a large length scale, backscattered SHG photon intensity decays exponentially with the distance from the emission point. (b) However, the same distribution over a short radial range demonstrates that within ~50 μm from the emission point, initially forward propagating then backscattered SHG photon intensity remains constant with the distance from the emission point. (Reprinted from Han, X. and E. Brown. 2010. Measurement of the ratio of forward-propagating to back-propagating second harmonic signal using a single objective. *Opt Express*, 18:10538–10550, Copyright 2010. With permission of Optical Society of America.)

pixel intensity at the largest pinhole size, then the relative pixel intensity as a function of the relative pinhole size R can be expressed as

$$I_{rel}(R) = \frac{\int_0^{2\pi} d\theta \int_0^R \left\{\exp[-2(r/\omega)^2] + (F/B)C\right\} r\, dr}{\int_0^{2\pi} d\theta \int_0^{R_{max}} \left\{\exp[-2(r/\omega)^2] + (F/B)C\right\} r\, dr} \tag{17.4}$$

where R_{max} is the largest pinhole size [26]. This expression allows relative pixel intensity versus pinhole size to be plotted and fitted to generate $(F/B)/C$. To determine F/B from these data, we must eliminate C, that is, the fraction of the signal that originally propagates forward but is then backscattered by the tissue to reach the pinhole plane. This will vary with the scattering properties of the underlying tissue. It can be eliminated by including in the sample object plane a fluorescent reference with a known F/B emission ratio, whose C value will be the same (i.e., forward-propagating fluorescent signal emitted from the reference object in the sample plane will be subject to the same backscattering in the tissue, as the initially forward-propagating SHG signal). Fluorescent polystyrene beads, for which the F/B fluorescence emission ratio has been previously measured in buffer alone, are then sprinkled on the sample surface to serve as this reference standard. Thus, by taking serial images of the collagenous tissue plus beads through a series of different sized confocal pinholes (Figure 17.4), plotting the epidetected total collagen SHG intensity of a pixel or small region of interest, as well as the total bead fluorescence intensity versus pinhole size (Figure 17.5), we can fit to Equation 17.4 to obtain $(F/B)/C$ for both the collagen fibers of interest and the calibration beads [26]. Figure 17.6 further illustrates how the relative pixel

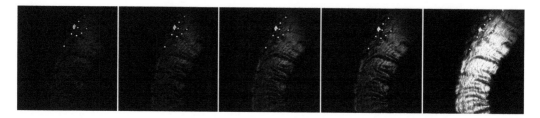

FIGURE 17.4 Serial pinhole *in vivo* SHG images of collagen fibers in an intact rat tail. Images 1 through 5 are SHG images of the same ROI when the size of the pinhole was varied from 60, 100, 150, 200, to 7000 μm, respectively. The bright spots are blue fluorescent polystyrene beads for calibration. The images are 600 μm across. Research was conducted in compliance with the guidelines of the Institutional Animal Care and Use Committee (IACUC). (Reprinted from Han, X. and E. Brown. 2010. Measurement of the ratio of forward-propagating to back-propagating second harmonic signal using a single objective. *Opt Express*, 18:10538–10550, Copyright 2010. With permission of Optical Society of America.)

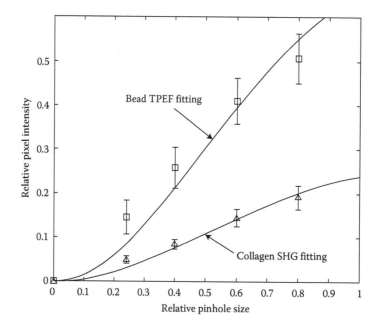

FIGURE 17.5 Determining collagen SHG F_i/B_i from collagen SHG and bead pixel intensities, by the single-objective method. The epidetected total collagen SHG intensity (triangles), or total bead two-photon excitation fluorescence (TPEF) intensity (squares) versus pinhole size, fit to the model given by Equation 17.4 (lines). For each set of images, these signal intensities were averaged from five ROIs around collagen fibers and three ROIs around the calibration beads, respectively, and were normalized such that the maximum intensity with the largest pinhole was set to one. The data plotted represent five image sets, that is, the mean SHG of 25 collagen ROIs, and the mean TPEF of 15 bead ROIs. Fitting the plotted data to Equation 17.4 as shown provides the *measured* collagen SHG F/B and bead TPEF F/B values, then the true (emitted) collagen SHG F/B value (i.e., F_i/B_i) is calculated by Equation 17.5. The horizontal axis is pinhole size in units of backward propagating SHG spot size on the pinhole plane, that is, fraction of ω. The vertical axis is normalized SHG intensity. Note that the fitting curves identically equal 1 when the pinhole size reaches $R = 28$, whose data point is not shown, and that the intensity information for $R = 28$ provides the normalization value and hence is included in the overall fit. Research was conducted in compliance with the guidelines of the Institutional Animal Care and Use Committee (IACUC). (Reprinted from Han, X. and E. Brown. 2010. Measurement of the ratio of forward-propagating to back-propagating second harmonic signal using a single objective. *Opt Express*, 18:10538–10550, Copyright 2010. With permission of Optical Society of America.)

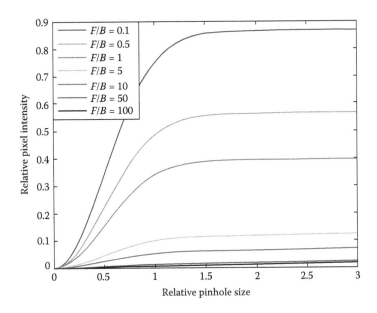

FIGURE 17.6 Relationship between pixel intensity, pinhole size, and *F/B* ratios, by the single-objective method. The relative SHG intensity versus relative pinhole size curves at different collagen SHG *F/B* ratios were based upon Equation 17.4 and $C = 0.001$. The relative SHG intensities equal 1 at a pinhole size of $r = R_{max} = 28$. Note the sharp curvature due to the Gaussian distribution of direct backward-propagating SHG (B_i) (most evident in the steep early rise of the curves with lower *F/B*), and the slow and steady curve rise due to the subsequently backscattered forward-propagating SHG (F_i) that produces a diffuse signal that does not vary with radius (most evident in the slow rise of the curves with large values of *F/B*). Also note that given the typical variation in experimental data (e.g., as seen in Figure 17.5 above), when $F/B \gg \sim 5$, the curves may not be statistically distinguishable, and thus, this method may be best suited for samples with collagen fiber *F/B* ratios $< \sim 5$ (see the text for further detail). (Reprinted from Han, X. and E. Brown. 2010. Measurement of the ratio of forward-propagating to back-propagating second harmonic signal using a single objective. *Opt Express*, 18:10538–10550, Copyright 2010. With permission of Optical Society of America.)

intensity will vary with relative pinhole size for different *F/B* ratios, as fit by Equation 17.4. The correction factor *C* can be eliminated and the collagen fiber F_i/B_i SHG ratio can be determined by the equation:

$$\frac{\text{Measured collagen SHG } F/B \text{ ratio}}{\text{Measured beads TPEF } F/B \text{ ratio}} = \frac{\text{real collagen SHG } F/B \text{ ratio}}{\text{real beads TPEF } F/B \text{ ratio}} \tag{17.5}$$

In this manner, this method was utilized to extract collagen F_i/B_i values from intact (not sectioned) rat tail as well as from muscle fascia using the SHG signals captured from a single objective, and these F_i/B_i values were not significantly different from F_i/B_i values for collagen obtained using both forward and backward detectors (i.e., the traditional approach using two objectives, one on either side of the sample) [26].

Overall, this technique demonstrates that by imaging the sample repeatedly through different sized pinholes to measure the shape of the total SHG distribution, and extrapolating the F_i/B_i from the F_d/B_d by a bead calibration method as described above, F_i/B_i measurements can thus be obtained *in vivo* from intact biologic specimens from a single objective. This represents an important advancement that should improve our ability to conveniently obtain *F/B* measurements from biologic tissue under *in vivo* conditions, which may in turn ultimately yield real-time clinical imaging diagnostics for cancer or other diseases.

There are, however, limits to this method. First, assuming typical variation in experimental data, the SHG intensity versus pinhole size calibration curves may not be able to distinguish between F_i/B_i values much above ~5 (note the decreasing separation of the curves in Figure 17.6, as $F/B \gg 5$). Fortunately, F/B ratios (extrapolated to F_i/B_i as appropriate, for comparison) in many biological samples of interest, including cancer tissues, are often <5 [26–29,48,57]. Second, because this model assumes there is no subsequent scattering of initially backscattered SHG B_i, this model is limited to tissue imaging depths <1 MFP, such that significant scattering of B_i will not occur.

17.2 Polarization of SHG in Tumors

17.2.1 Polarization Properties of Collagen SHG

Collagen SHG emission is influenced by the polarization of the incoming excitation photons, together with the overall orientation of the scattering collagen fibril relative to the laser axis, and the pitch angle of the collagen helix (see the Appendix). Consequently, polarization of SHG signals can be exploited to interrogate these tissue properties, as well as related properties such as collagen organization and alignment. In this section, we review the efforts to determine collagen fibril orientation and helix pitch angle, and their application to understand the structure of tumor stroma. Before discussing the application of these SHG polarization-related techniques to tumor tissues (to our knowledge, there have only been a handful of such reports), it is informative to discuss some of the preceding literature in nontumor tissue that has helped to lay the foundation for interpreting polarization effects of SHG as they apply to collagen's structure, organization, and molecular properties.

The work from several groups has helped us to define and understand the relationships between incoming laser polarization and collagen SHG emission intensity [27,36,37,39,40,58–60], and these relationships are in turn principally sensitive to the orientation of SHG-emitting dipoles [60]. As such, SHG polarization anisotropy (PA) involves rotating the polarization plane of the incoming linearly polarized laser light relative to the plane of the SHG emitter, and measuring the resultant SHG intensity (I_{SHG}) over the range of relative polarization states. In general, I_{SHG} minima and maxima occur when laser polarization is perpendicular and parallel to the collagen fiber direction, respectively [61,62]. Of particular importance for imaging clinical samples, to help prevent sample damage that may occur by taking scans at many different polarization states, methods have been advanced to allow for obtaining meaningful PA data from as little as four scans [39], or from continuous scanning techniques [40]. In addition, this dependence of I_{SHG} on laser polarization is impacted by tissue depth [40,51], proportionate to tissue scattering effects that will effectively depolarize the laser emission before it reaches the focal volume [51]. This phenomena should be heeded by investigators as they compare anisotropy measurements at different tissue depths and across tissues with different scattering properties, and indeed has been addressed in part by an elegant "optical clearing" method of tissue preparation that serves to reduce scattering effects in biologic tissues, thus normalizing the polarization dependence at varying tissue depths [51]. Thus, scattering effects relative to tissue thickness and the location of the focal plane are also important considerations when deciding whether to measure the forward or backward SHG signal, both of which have been employed for polarization anisotropy.

For the purposes of our discussions in this chapter, PA relationships with I_{SHG} enable mathematical derivation of several key pieces of information related to collagen's structure and organization. First, we can obtain the angular orientation of an SHG-emitting collagen fibril in the XY plane (see the Appendix, Figure A.1), information that in turn can be used to generate "orientation field maps" reflecting the overall orientation of fibril or fiber ensembles throughout an field of view (FOV) [38,39]. Second, we can determine a related "anisotropy parameter" β, which represents a measure of collagen ordering in the sample, and is calculated by: $\beta = (I_{SHGpar} - I_{SHGperp})/(I_{SHGpar} + 2I_{SHGperp})$, where

β varies between -0.5 and 1, and I_{SHGpar} and $I_{SHGperp}$ correspond to the intensities of the SHG signals polarized parallel and perpendicular to the laser polarization, respectively. A similar PA parameter has also been expressed as PA $= (I_{SHGpar} - I_{SHGperp})/(I_{SHGpar} + I_{SHGperp})$, where PA varies between 0 and 1 [63,64]. In both cases, β or PA $= 0$ corresponds to less ordered collagen, and β or PA $= 1$ corresponds to more ordered collagen [46,51,60,63,64]. Third, we can obtain the helical pitch angle of collagen (see the Appendix, Figure A.2, and description) [36,37,39,65]. Finally, we can obtain the "tilt" angle of individual collagen triple helices relative to the collagen fibril axis (see the Appendix, Figure A.3, and description) [66]. These latter two properties are particularly intriguing because such measurements may enable us to determine by optical imaging approaches, whether fundamental molecular properties of collagen are altered by different disease states, including cancer. Overall, we see that SHG intensity under varied polarization states contains not just information with regard to the orientation of the fibril in the XY plane, but from this, we can also extract information regarding the molecular structure of collagen, which together make this a powerful optical imaging modality for investigating cancer. See the Appendix for mathematical derivations and more detailed discussions of these concepts.

17.2.2 SHG Polarization Measurements in Tumor Tissue

Contrary to the relative abundance of studies on forward and backward SHG in tumors, to our knowledge, only a few studies have carried out polarization-related SHG measurements specifically on tumor tissue. Using PA of collagen SHG to calculate θ (see the Appendix), Han et al. [1] found no differences in θ between normal mouse mammary fat pad, and mouse mammary tumors. On the one hand, perhaps, this is not surprising because one might expect the fundamental aspects of collagen's molecular structure (e.g., helical pitch angle) to remain unchanged even in diseased tissue, with any SHG-identifiable changes between normal and tumor tissue resulting from higher-order changes in collagen's organization. On the other, it would be intriguing if disease states *did* alter the fundamental aspects of collagen's molecular structure such as pitch angle, and thus, further measurements of θ in a wider range of cancer tissues should prove informative.

Other SHG anisotropy studies of normal versus tumor tissues have looked at related polarization parameters. A series of studies from one group found differences in the d_{22} coefficient (a measure of collagen's second-order nonlinear susceptibility) [63,67] and in the PA parameter (PA, as defined above) [63], between human tumor tissues xenografted into mice (i.e., human tumor cell lines injected into mice) and normal mouse tissue taken from the same anatomical region. The authors were able to attribute changes in these parameters to observed changes in collagen amount or organization between the normal and tumor tissues (xenografted human tumors had much more sparse collagen compared to anatomically matched normal mouse tissue), suggesting that these measures may provide a quantifiable means of discriminating collagen organizational changes in cancer.

In a related extensive study of normal versus cancerous human ovarian tissue, Nadiarnykh et al. [48] found a higher anisotropy parameter (β, as defined above) in cancerous versus normal tissue, which they attributed to more organized and aligned collagen fibers in the cancer conditions. This finding was additionally supportive of and consistent with the interpretation of their F/B data, as discussed above.

Curiously, at least one study has found that SHG-calculated θ values may differ in collagen from different species [1], and for example, the above d_{22} coefficient is sensitive to dispersion and thus collagen packing and density [63,67], which in turn is often tissue specific [63] and could conceivably vary across species. In light of these intriguing results, it would be informative to undertake more comprehensive studies as to how these and related anisotropy parameters (and F/B ratios) may vary in collagen taken from different anatomical regions and from different species, thus facilitating translation of the existing data across experimental models, cancer types, and species.

17.3 Summary and Conclusions

In summary, we have endeavored to present a brief yet informative overview of the application of scattering- and polarization-related SHG techniques to investigations of collagen in tumor tissue, as distinct (as possible) from other discussions of SHG and cancer presented elsewhere in the book, and with just enough technical background to be able to understand the meanings of these results as they apply to changes in specific collagen properties with cancer. Together with our colleague's other chapters in this book, we are confident readers will gain a broad perspective and understanding of both the origins of SHG, and how it can be used to advance our understanding, and ultimately we hope treatment of human disease.

Acknowledgments

This work was supported by the Department of Defense Breast Cancer Research Program (DoD BCRP) Era of Hope Scholar Award W81XWH-09-1-0405 (to EBB), DoD BCRP Predoctoral Traineeship Award W81WXH-08-1-0323 (to XH), National Institutes of Health (NIH) Director's New Innovator Award 1DP2 OD006501-01 (to EBB), and NIH Exploratory Developmental Research Grant Award R21DA030256 (to SWP).

Appendix: Theory of Polarization Analysis

An incoming excitation beam with electric field vector $\vec{E}$ will induce polarization $\vec{P}$ in the fibril given by

$$\vec{P} = \chi^{(1)} * \vec{E} + \chi^{(2)} * \vec{E} * \vec{E} + \chi^{(3)} * \vec{E} * \vec{E} * \vec{E} \tag{A.1}$$

In the case of second-harmonic generation from a system with cylindrical symmetry (such as the collagen fibril), this simplifies to

$$\vec{P} = a\hat{s}(\hat{s} \cdot \vec{E})^2 + b\hat{s}(\vec{E} \cdot \vec{E}) + c\vec{E}(\hat{s} \cdot \vec{E}) \tag{A.2}$$

where $\hat{s}$ is the direction of collagen orientation and the coefficients a, b, and c are related to elements of the second-order nonlinear susceptibility tensor $\chi^{(2)}$. If the resultant SHG signal is detected through a polarizer orientated at a direction $\hat{e}$, the detected SHG intensity is $I_e \propto (\vec{P} \cdot \hat{e})^2$. This is given by

$$I_e \propto \left(\vec{P} \cdot \hat{e}\right)^2 \propto \left[a\cos\alpha\cos^2\varphi + b\cos\alpha + c\cos(\alpha - \varphi)\cos\varphi\right]^2 \tag{A.3}$$

where as in Figure A.1, α is the angle between the collagen fibril and the polarizer axis, and φ is the angle between the fibril and excitation beam polarization. If we consider a z-directed excitation beam interacting with a y-aligned fibril, then when $\alpha = 0$

$$I_y \propto [(a + c)\cos^2\varphi + b]^2 \tag{A.4}$$

and when $\alpha = \dfrac{\pi}{2}$

$$I_x \propto \left[\frac{c}{2}\sin 2\varphi\right]^2 \tag{A.5}$$

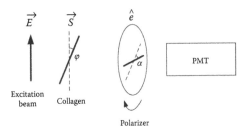

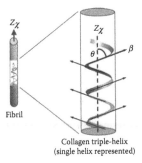

So, we have

$$\frac{I_y}{I_x} = \frac{\left[(a+c)\cos^2\varphi + b\right]^2}{\left[\frac{c}{2}\sin(2\varphi)\right]^2} \tag{A.6}$$

As illustrated in Figure A.2, if the collagen fibril (Figure A.2, *left image*) is considered to be a cylindrically symmetric collection of single-axis scatterers (e.g., a collection of helical turns from collagen single helices, three of which intertwine to form the superhelical collagen triple helix) with a constant polar angle θ (Figure A.2, *right image*) and a random azimuthal angle φ, there are only two independent elements of $\chi^{(2)}$, the nonlinear susceptibility tensor, with the result that

$$
\begin{aligned}
a &= n - 3m \\
c &= 2b = m \\
n &= \chi^{(2)}_{zzz} = N\cos^3\theta\beta \\
m &= \chi^{(2)}_{zxx} = \chi^{(2)}_{xxz} = N/2\,\cos\theta\sin^2\theta\beta
\end{aligned} \tag{A.7}
$$

where β is the hyperpolarizability of these individual single-axis scatterers (B, *right image*), of density N. This simplification signifies that the susceptibility tensor has Kleinman symmetry regardless of whether the excitation energy is significantly off-resonance. In this case, Equation A.6 becomes

$$\tan^2\theta = \frac{2\cos^2\varphi}{\sqrt{\dfrac{I_y}{I_x}\sin(2\varphi) - \sin^2(\varphi)}} \tag{A.8}$$

And a measurement of directional SHG intensities I_y and I_x at a known φ allows the deduction of θ, the polar angle of the SHG scatterers within a given fibril, that is, the helical pitch angle for collagen (B, *right image*). These directional SHG intensities relate to the total SHG intensity measured with a perpendicularly polarized illumination beam (I_p) as [27]

$$I_y(\varphi) = I_p\left[\rho\cos^2\varphi + \sin^2\varphi\right]^2$$
$$I_x(\varphi) = I_p\left[\sin^2\varphi\right]^2$$

where the fitting parameter ρ is used as a measure of the axial polarizing effects of the fibril. Thus

$$\tan^2\theta = 2/\rho$$

Hence, we see that the ρ term of Williams et al. and others [27,38,69] can also give us θ, the polar angle of the SHG scatterers in a given fibril.

In further support, SHG-based estimations of this θ [36,37,39] have correlated well with x-ray diffraction measurements of the pitch angle for a single collagen helix [70,71]. Furthermore, a very recent elegant study suggests that when making SHG-based measurements of θ in collagen, accounting for the distinct pitch angles of both peptide and methylene groups in single-collagen helices will yield θ values that best agree with gold standard x-ray diffraction data, versus approaches that assume that SHG-measured θ values arise principally from peptide contributions alone [65].

We also point out that using slightly different mathematical modeling, the polar angle of the SHG scatterers in a given fibril, rather than providing the helical pitch angle as above, can inform us as to the "tilt" angle of individual collagen triple helices away from the fibril axis (θ_{t-h}, in Figure A.3) (for details, see Ref. [66]).

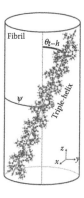

FIGURE A.3 Model 2, θ as triple-helix tilt angle. (Reprinted with permission from Tuer, A.E. et al. 2011. Nonlinear optical properties of type I collagen fibers studied by polarization dependent second harmonic generation microscopy. *J Phys Chem B*. Copyright 2011, American Chemical Society.)

From this additional elegant model, we can also infer that if a sufficient proportion of collagen triple helixes are not axially aligned with the fibril axis, this may impact our ability to obtain accurate helical pitch angle measurements as per Model 1 above, and this may be another factor to account for some of the small discrepancies seen between SHG-based and x-ray diffraction-based measurements of collagen helix pitch angle.

Overall, we see that SHG intensity under varied polarization states contains not just information as regards the orientation of the collagen fibril in the *xy* plane, but we can also extract information regarding the molecular structure of collagen, namely, collagen's helical pitch angle and the angular orientation of collagen triple helices relative to the fibril axis.

The equations and text in this box are reprinted (with some modifications) with permission from Han, X. et al. 2008. Second harmonic properties of tumor collagen: Determining the structural relationship between reactive stroma and healthy stroma. *Opt Express*, 16:1846–1859. Copyright 2008, Optical Society of America. The figures in this box are reprinted with permission, as indicated.

References

1. Han, X., R.M. Burke, M.L. Zettel, P. Tang, and E.B. Brown. 2008. Second harmonic properties of tumor collagen: Determining the structural relationship between reactive stroma and healthy stroma. *Opt Express*, 16:1846–1859.
2. Kalluri, R. and M. Zeisberg. 2006. Fibroblasts in cancer. *Nat Rev Cancer*, 6:392–401.
3. Shekhar, M.P., R. Pauley, and G. Heppner. 2003. Host microenvironment in breast cancer development: Extracellular matrix–stromal cell contribution to neoplastic phenotype of epithelial cells in the breast. *Breast Cancer Res*, 5:130–135.
4. Haslam, S.Z. and T.L. Woodward. 2003. Host microenvironment in breast cancer development: Epithelial-cell–stromal-cell interactions and steroid hormone action in normal and cancerous mammary gland. *Breast Cancer Res*, 5:208–215.
5. Hasebe, T., S. Sasaki, S. Imoto, T. Mukai, T. Yokose, and A. Ochiai. 2002. Prognostic significance of fibrotic focus in invasive ductal carcinoma of the breast: A prospective observational study. *Mod Pathol*, 15:502–516.
6. Jukkola, A., R. Bloigu, K. Holli, H. Joensuu, R. Valavaara, J. Risteli, and G. Blanco. 2001. Postoperative PINP in serum reflects metastatic potential and poor survival in node-positive breast cancer. *Anticancer Res*, 21:2873–2876.
7. Jensen, B.V., J.S. Johansen, T. Skovsgaard, J. Brandt, and B. Teisner. 2002. Extracellular matrix building marked by the N-terminal propeptide of procollagen type I reflect aggressiveness of recurrent breast cancer. *Int J Cancer*, 98:582–589.
8. Keskikuru, R., R. Bloigu, J. Risteli, V. Kataja, and A. Jukkola. 2002. Elevated preoperative serum ICTP is a prognostic factor for overall and disease-free survival in breast cancer. *Oncol Rep*, 9:1323–1327.
9. Brown, E., T. McKee, E. diTomaso, A. Pluen, B. Seed, Y. Boucher, and R.K. Jain. 2003. Dynamic imaging of collagen and its modulation in tumors *in vivo* using second-harmonic generation. *Nat Med*, 9:796–800.
10. Perentes, J.Y., T.D. McKee, C.D. Ley, H. Mathiew, M. Dawson, T.P. Padera, L.L. Munn, R.K. Jain, and Y. Boucher. 2009. *In vivo* imaging of extracellular matrix remodeling by tumor-associated fibroblasts. *Nat Methods*, 6:143–145.
11. Hompland, T., A. Erikson, M. Lindgren, T. Lindmo, and C. de Lange Davies. 2008. Second-harmonic generation in collagen as a potential cancer diagnostic parameter. *J Biomed Opt*, 13:054050.
12. Zipfel, W.R., R.M. Williams, R. Christie, A.Y. Nikitin, B.T. Hyman, and W.W. Webb. 2003. Live tissue intrinsic emission microscopy using multiphoton-excited native fluorescence and second harmonic generation. *Proc Nat Acad Sci USA*, 100:7075–7080.
13. Wilder-Smith, P., T. Krasieva, W.G. Jung, J. Zhang, Z. Chen, K. Osann, and B. Tromberg. 2005. Noninvasive imaging of oral premalignancy and malignancy. *J Biomed Opt*, 10:051601.

14. Lin, S.J., S.H. Jee, C.J. Kuo, R.J. Wu, W.C. Lin, J.S. Chen, Y.H. Liao et al. 2006. Discrimination of basal cell carcinoma from normal dermal stroma by quantitative multiphoton imaging. *Opt Lett*, 31:2756–2758.

15. Provenzano, P.P., K.W. Eliceiri, J.M. Campbell, D.R. Inman, J.G. White, and P.J. Keely. 2006. Collagen reorganization at the tumor–stromal interface facilitates local invasion. *BMC Med*, 4:38.

16. Campagnola, P.J., H.A. Clark, W.A. Mohler, A. Lewis, and L.M. Loew. 2001. Second-harmonic imaging microscopy of living cells. *J Biomed Opt*, 6:277–286.

17. Conklin, M.W., J.C. Eickhoff, K.M. Riching, C.A. Pehlke, K.W. Eliceiri, P.P. Provenzano, A. Friedl, and P.J. Keely. 2011. Aligned collagen is a prognostic signature for survival in human breast carcinoma. *Am J Pathol*, 178:1221–1232.

18. Thrasivoulou, C., G. Virich, T. Krenacs, I. Korom, and D.L. Becker. 2011. Optical delineation of human malignant melanoma using second harmonic imaging of collagen. *Biomed Opt Express*, 2:1282–1295.

19. Wang, W., J.B. Wyckoff, V.C. Frohlich, Y. Oleynikov, S. Huttelmaier, J. Zavadil, L. Cermak et al. 2002. Single cell behavior in metastatic primary mammary tumors correlated with gene expression patterns revealed by molecular profiling. *Cancer Res*, 62:6278–6288.

20. Guo, Y., H.E. Savage, F. Liu, S.P. Schantz, P.P. Ho, and R.R. Alfano. 1999. Subsurface tumor progression investigated by noninvasive optical second harmonic tomography. *Proc Natl Acad Sci USA*, 96:10854–10856.

21. Lin, S.J., S.H. Jee, C.J. Kuo, R.J. Wu, W.C. Lin, J.S. Chen, Y.H. Liao et al. 2006. Discrimination of basal cell carcinoma from normal dermal stroma by quantitative multiphoton imaging. *Opt Lett*, 31:2756–2758.

22. Wilder-Smith, P., T. Krasieva, W.G. Jung, J. Zhang, Z. Chen, K. Osann, and B. Tromberg. 2005. Noninvasive imaging of oral premalignancy and malignancy. *J Biomed Opt*, 10:051601.

23. Provenzano, P.P., K.W. Eliceiri, J.M. Campbell, D.R. Inman, J.G. White, and P.J. Keely. 2006. Collagen reorganization at the tumor–stromal interface facilitates local invasion. *BMC Med*, 4:38.

24. Perry, S.W., R.M. Burke, and E.B. Brown. 2012. Two photon and second harmonic microscopy in clinical and translational cancer research. *Ann Biomed Eng*, 40(2):277–291.

25. Mertz, J. and L. Moreaux. 2001. Second harmonic generation by focused excitation of inhomogeneously distributed scatterers. *Opt Commun*, 196:325–330.

26. Han, X. and E. Brown. 2010. Measurement of the ratio of forward-propagating to back-propagating second harmonic signal using a single objective. *Opt Express*, 18:10538–10550.

27. Williams, R.M., W.R. Zipfel, and W.W. Webb. 2005. Interpreting second-harmonic generation images of collagen I fibrils. *Biophys J*, 88:1377–1386.

28. Lacomb, R., O. Nadiarnykh, and P.J. Campagnola. 2008. Quantitative second harmonic generation imaging of the diseased state osteogenesis imperfecta: Experiment and simulation. *Biophys J*, 94:4504–4514.

29. Nadiarnykh, O., R.B. Lacomb, P.J. Campagnola, and W.A. Mohler. 2007. Coherent and incoherent SHG in fibrillar cellulose matrices. *Opt Express*, 15:3348–3360.

30. Kwan, A.C., D.A. Dombeck, and W.W. Webb. 2008. Polarized microtubule arrays in apical dendrites and axons. *Proc Natl Acad Sci USA*, 105:11370–11375.

31. Chu, S.W., S.P. Tai, M.C. Chan, C.K. Sun, I.C. Hsiao, C.H. Lin, Y.C. Chen, and B.L. Lin. 2007. Thickness dependence of optical second harmonic generation in collagen fibrils. *Opt Express*, 15:12005–12010.

32. Zipfel, W.R., R.M. Williams, R. Christie, A.Y. Nikitin, B.T. Hyman, and W.W. Webb. 2003. Live tissue intrinsic emission microscopy using multiphoton-excited native fluorescence and second harmonic generation. *Proc Natl Acad Sci USA*, 100:7075–7080.

33. Smolle, J., M. Fiebiger, R. Hofmann-Wellenhof, and H. Kerl. 1996. Quantitative morphology of collagen fibers in cutaneous malignant melanoma and melanocytic nevus. *Am J Dermatopathol*, 18:358–363.

34. Taboga, S.R. and C. Vidal Bde. 2003. Collagen fibers in human prostatic lesions: Histochemistry and anisotropies. *J Submicrosc Cytol Pathol*, 35:11–16.

35. Falzon, G., S. Pearson, and R. Murison. 2008. Analysis of collagen fibre shape changes in breast cancer. *Phys Med Biol*, 53:6641–6652.

36. Tiaho, F., G. Recher, and D. Rouede. 2007. Estimation of helical angles of myosin and collagen by second harmonic generation imaging microscopy. *Opt Express, 15*:12286–12295.

37. Plotnikov, S.V., A.C. Millard, P.J. Campagnola, and W.A. Mohler. 2006. Characterization of the myosin-based source for second-harmonic generation from muscle sarcomeres. *Biophys J, 90*:693–703.

38. Stoller, P., K.M. Reiser, P.M. Celliers, and A.M. Rubenchik. 2002. Polarization-modulated second harmonic generation in collagen. *Biophys J, 82*:3330–3342.

39. Odin, C., T. Guilbert, A. Alkilani, O.P. Boryskina, V. Fleury, and Y. Le Grand. 2008. Collagen and myosin characterization by orientation field second harmonic microscopy. *Opt Express, 16*:16151–16165.

40. Stoller, P., B.M. Kim, A.M. Rubenchik, K.M. Reiser, and L.B. Da Silva. 2002. Polarization-dependent optical second-harmonic imaging of a rat-tail tendon. *J Biomed Opt, 7*:205–214.

41. Cameron, G.J., D.E. Cairns, and T.J. Wess. 2007. The variability in type I collagen helical pitch is reflected in the D periodic fibrillar structure. *J Mol Biol, 372*:1097–1107.

42. Fernandez, M., J. Keyrilainen, R. Serimaa, M. Torkkeli, M.L. Karjalainen-Lindsberg, M. Tenhunen, W. Thomlinson, V. Urban, and P. Suortti. 2002. Small-angle x-ray scattering studies of human breast tissue samples. *Phys Med Biol, 47*:577–592.

43. Dvorak, H.F. 1986. Tumors: Wounds that do not heal. Similarities between tumor stroma generation and wound healing. *N Engl J Med, 315*:1650–1659.

44. Kauppila, S., F. Stenback, J. Risteli, A. Jukkola, and L. Risteli. 1998. Aberrant type I and type III collagen gene expression in human breast cancer *in vivo*. *J Pathol, 186*:262–268.

45. Cameron, G.J., I.L. Alberts, J.H. Laing, and T.J. Wess. 2002. Structure of type I and type III heterotypic collagen fibrils: An x-ray diffraction study. *J Struct Biol, 137*:15–22.

46. Campagnola, P. 2011. Second harmonic generation imaging microscopy: Applications to diseases diagnostics. *Anal Chem, 83*:3224–3231.

47. Lacomb, R., O. Nadiarnykh, S.S. Townsend, and P.J. Campagnola. 2008. Phase matching considerations in second harmonic generation from tissues: Effects on emission directionality, conversion efficiency and observed morphology. *Opt Commun, 281*:1823–1832.

48. Nadiarnykh, O., R.B. LaComb, M.A. Brewer, and P.J. Campagnola. 2010. Alterations of the extracellular matrix in ovarian cancer studied by second harmonic generation imaging microscopy. *BMC Cancer, 10*:94.

49. LaComb, R., O. Nadiarnykh, S. Carey, and P.J. Campagnola. 2008. Quantitative second harmonic generation imaging and modeling of the optical clearing mechanism in striated muscle and tendon. *J Biomed Opt, 13*:021109.

50. Wang, L., S.L. Jacques, and L. Zheng. 1995. MCML—Monte Carlo modeling of light transport in multilayered tissues. *Comput Methods Programs Biomed, 47*:131–146.

51. Nadiarnykh, O. and P.J. Campagnola. 2009. Retention of polarization signatures in SHG microscopy of scattering tissues through optical clearing. *Opt Express, 17*:5794–5806.

52. Pfeffer, C.P., B.R. Olsen, and F. Legare. 2007. Second harmonic generation imaging of fascia within thick tissue block. *Opt Express, 15*:7296–7302.

53. Theodossiou, T.A., C. Thrasivoulou, C. Ekwobi, and D.L. Becker. 2006. Second harmonic generation confocal microscopy of collagen type I from rat tendon cryosections. *Biophys J, 91*:4665–4677.

54. Ajeti, V., O. Nadiarnykh, S.M. Ponik, P.J. Keely, K.W. Eliceiri, and P.J. Campagnola. 2011. Structural changes in mixed col I/col V collagen gels probed by SHG microscopy: Implications for probing stromal alterations in human breast cancer. *Biomed Opt Express, 2*:2307–2316.

55. Birk, D.E., J.M. Fitch, J.P. Babiarz, K.J. Doane, and T.F. Linsenmayer. 1990. Collagen fibrillogenesis *in vitro*: Interaction of types I and V collagen regulates fibril diameter. *J Cell Sci, 95(4)*:649–657.

56. Barsky, S.H., C.N. Rao, G.R. Grotendorst, and L.A. Liotta. 1982. Increased content of type V collagen in desmoplasia of human breast carcinoma. *Am J Pathol, 108*:276–283.

57. Legare, F., C. Pfeffer, and B.R. Olsen. 2007. The role of backscattering in SHG tissue imaging. *Biophys J, 93*:1312–1320.

58. Campagnola, P. and C.Y. Dong. 2011. Second harmonic generation microscopy: Principles and applications to disease diagnostics. *Laser Photonics Rev, 5(1)*:13–26.

59. Stoller, P., P.M. Celliers, K.M. Reiser, and A.M. Rubenchik. 2003. Quantitative second-harmonic generation microscopy in collagen. *Appl Opt, 42*:5209–5219.

60. Campagnola, P.J., A.C. Millard, M. Terasaki, P.E. Hoppe, C.J. Malone, and W.A. Mohler. 2002. Three-dimensional high-resolution second-harmonic generation imaging of endogenous structural proteins in biological tissues. *Biophys J, 82*:493–508.

61. Kim, B.M., J. Eichler, K.M. Reiser, A.M. Rubenchik, and L.B. Da Silva. 2000. Collagen structure and nonlinear susceptibility: Effects of heat, glycation, and enzymatic cleavage on second harmonic signal intensity. *Lasers Surg Med, 27*:329–335.

62. Yasui, T., Y. Tohno, and T. Araki. 2004. Determination of collagen fiber orientation in human tissue by use of polarization measurement of molecular second-harmonic-generation light. *Appl Opt, 43*:2861–2867.

63. Hompland, T., A. Erikson, M. Lindgren, T. Lindmo, and C. de Lange Davies. 2008. Second-harmonic generation in collagen as a potential cancer diagnostic parameter. *J Biomed Opt, 13*:054050.

64. Yasui, T., Y. Tohno, and T. Araki. 2004. Characterization of collagen orientation in human dermis by two-dimensional second-harmonic-generation polarimetry. *J Biomed Opt, 9*:259–264.

65. Su, P.J., W.L. Chen, Y.F. Chen, and C.Y. Dong. 2011. Determination of collagen nanostructure from second-order susceptibility tensor analysis. *Biophys J, 100*:2053–2062.

66. Tuer, A.E., S. Krouglov, N. Prent, R. Cisek, D. Sandkuijl, K. Yasufuku, B. Wilson, and V. Barzda. 2011. Nonlinear optical properties of type I collagen fibers studied by polarization dependent second harmonic generation microscopy. *J Phys Chem B, 115(44)*:12759–12769.

67. Erikson, A., J. Ortegren, T. Hompland, C. de Lange Davies, and M. Lindgren. 2007. Quantification of the second-order nonlinear susceptibility of collagen I using a laser scanning microscope. *J Biomed Opt, 12*:044002.

68. Han, X. 2011. Novel techniques for quantitative second harmonic generation microscopy. PhD dissertation, University of Rochester. Ann Arbor: ProQuest/UMI.

69. Freund, I., M. Deutsch, and A. Sprecher. 1986. Connective tissue polarity. Optical second-harmonic microscopy, crossed-beam summation, and small-angle scattering in rat-tail tendon. *Biophys J, 50*:693–712.

70. Beck, K. and B. Brodsky. 1998. Supercoiled protein motifs: The collagen triple-helix and the α-helical coiled coil. *J Struct Biol, 122*:17–29.

71. Bella, J., M. Eaton, B. Brodsky, and H.M. Berman. 1994. Crystal and molecular structure of a collagen-like peptide at 1.9 A resolution. *Science, 266*:75–81.

18

SHG Imaging for Tissue Engineering Applications

Annika Enejder
*Chalmers University of
Technology*

**Christian
Brackmann**
Lund University

18.1 Background

The loss or failure of tissue or an organ is a recurrent and costly health problem worldwide. The aim of tissue engineering is to restore, maintain, or improve the functionality of lost or damaged tissue by the use of cells. This requires detailed characterization of the generated construct and microscopy techniques based on nonlinear optical interactions have become increasingly used for this purpose. This chapter is focused on second-harmonic generation (SHG) microscopy applied as tool in tissue engineering. A short background on tissue engineering and microscopy techniques for tissue characterization is followed by presentations of SHG microscopy applications.

18.1.1 Tissue Engineering Using Biomaterial Scaffolds

Treatment of lost tissue often relies on transplantations, either of donor or of autologous tissue. Both alternatives have limitations; there is for example a limited supply of donor transplants, which also require immunosuppression therapy with possible side effects. Transplanted autologous tissue may lack some of the functions of the original tissue and the procedure may also introduce complications at the donor site. In some cases, artificial substitutes manufactured from nonbiological materials can be used, for example, synthetic polymer blood vessels or joint replacement prostheses. However, these replacements have drawbacks such as risk for infections, limited material durability, and lack of mechanisms for repair, growth, and remodeling. For these reasons, development of advanced artificial tissue constructs with adaptive capabilities is desirable.

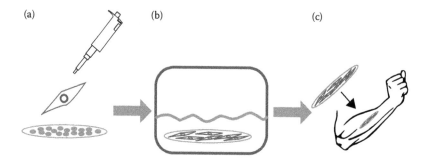

FIGURE 18.1 Schematic showing the principle steps of tissue engineering using a biomaterial scaffold. (a) A porous scaffold material is seeded with cells, (b) maintained in a bioreactor for cell proliferation and ECM production, and (c) eventually the cell-scaffold construct can be implanted at the site requiring tissue replacement.

This requires suitable biomaterials and their development can be described by three generations of materials (Rabkin and Schoen 2002, Schoen 2004). The first generation included industrial materials that had properties compatible with the structure to be replaced as well as *bio-inertness*, that is, induced a minimal response from the host tissue to the implant. This was followed by *bioactive* materials that allowed for controlled interactions with the host tissue, for example, degradable materials for which the interface between implant and host tissue is gradually eliminated. The materials of the third generation are designed for interactions on a molecular level, that is, *functional* materials. This development has been a prerequisite for tissue engineering in which biomaterial scaffolds are implanted at the tissue replacement site.

The scaffold forms a supporting structure to guide cell arrangement and initial tissue development. It can either be seeded with cells for growth and differentiation *in vitro* prior to implantation, as shown in Figure 18.1, or designed to attract endogenous cells *in vivo*. After cell seeding, the scaffold is maintained in a so-called bioreactor (cf. Figure 18.1), providing mechanical support for the scaffold and an environment with nutritional supply for the cells. In the bioreactor, the cells are able to proliferate, produce extracellular matrix (ECM), and eventually the construct can be implanted at the desired site. After implantation, further cell proliferation, cell differentiation, ECM production, and gradual scaffold degradation can take place *in vivo*.

18.1.2 Scaffold Properties and Synthesis

It is crucial but far from trivial to design a tissue engineering scaffold that has suitable structure and provides an appropriate biological environment for the desired type of cells. To achieve this goal, several parameters need to be considered:

- The material structure and morphology should be favorable for cell adhesion and ingrowth. Many scaffolds consist of porous material structures or fiber networks, making a three-dimensional arrangement of cells possible.
- Interconnectivity between pores in the scaffold structure is necessary for transport of nutrition, oxygen, and waste products. In addition, it facilitates tissue vascularization, that is, the build-up of a blood vessel structure.
- The mechanical properties of the scaffold should be similar to those of the tissue or organ intended for regeneration.
- The chemical/biochemical properties of the scaffold should be favorable for cell adhesion, proliferation, differentiation, and ECM production.
- The scaffold material should be biocompatible, that is, no inflammatory or immune response should be induced after implantation.

- With biodegradable scaffolds completely natural tissue and long-term biocompatibility is eventually achieved. However, the degradation rate needs to match the development of the regenerated tissue.

Scaffold materials include decellularized extracellular matrix (Gilbert et al. 2006), synthetic polymers, for example, poly(glycolic) acid (PGA), poly(L-lactic acid) (PLLA), and natural polymers such as collagens, gelatins, fibrin, carbohydrates, peptides, and nucleic acids (Pachence et al. 2007). A number of methods are available for polymer scaffold fabrication, for example, fiber bonding and electrospinning render fiber network structures similar to native tissue matrices. Solvent casting, particulate leaching, and melt molding result in porous structures and the development of rapid prototyping techniques has provided a tool for design of advanced three-dimensional structures. Overviews of polymer scaffold fabrication methods can be found in textbooks, for example, *Principles of Tissue Engineering* (Murphy and Mikos 2007).

Fibrous scaffold materials can also be obtained from bacteria spinning cellulose (Czaja et al. 2007) or from worms spinning silk fibroin (Mandal and Kundu 2010). Another alternative for scaffold synthesis is to use genetically encoded polypeptides, allowing for precise control of protein structure and hence scaffold function (Chow et al. 2008, Sengupta and Heilshorn 2010). Furthermore, scaffold material properties can be tuned by forming composites, for example, between natural and synthetic fiber polymers (Heydarkhan-Hagvall et al. 2008).

In all, an impressive number of advanced technologies are available to produce artificial tissues, and with innovations made within material science, nano- and biotechnologies, as well as stem cell biology, highly sophisticated materials can be expected in the future. This sets high demands on methods to characterize the morphology of intact constructs in three dimensions under native conditions. In addition, it is desirable to be able to monitor their function in and interaction with host tissue *in vivo*.

18.1.3 Microscopy Techniques in Tissue Engineering

A number of microscopy techniques are available for characterization of structure and morphology of the cell-scaffold constructs synthesized in tissue engineering. Electron microscopy remains an invaluable tool for this purpose with the ability to provide high-resolution images of the scaffold architecture, allowing 10 nm-sized structures to be resolved. However, electron microscopy requires extensive sample preparation involving fixation, dehydration, freeze-drying, and material coating (Abeysekera et al. 1993, Czaja et al. 2004), all potentially affecting the properties of soft materials and excluding the possibility of observing the cell-scaffold matrix under native conditions. In addition, it only provides access to the most superficial layer, requiring physical sectioning in order to examine the interior sample structure, which in turn could introduce compression and edge artifacts to the architecture. Atomic force microscopy (AFM) eliminates the need for harsh sample preparation and also has the potential to render images with very fine details resolved. However, an AFM tip has a tendency to stick to soft material, which is a serious practical limitation for characterization of tissue (Shao et al. 1996). In addition, the information obtained is merely of topographical character, giving no access to the morphology below the sample surface. Access to the internal sample structure can be obtained by confocal fluorescence microscopy, which gives an optical sectioning of the sample. However, fluorescence microscopy usually requires the sample to be labeled with a marker molecule that specifically attaches to the structure or cellular component to be imaged. This indirect visualization method via a tracer is dependent on the uptake and expression efficiencies of the fluorophore in the material as well as an often unknown fluorescence yield, which both introduce measurement uncertainties. An additional limitation of fluorescence microscopy is photobleaching, which makes exposure times longer than the order of minutes unrealistic and long-term studies of cell vs. scaffold interaction processes infeasible. In addition, fluorescence labeling may also affect the viability of living matter; for example, it has been shown that it influences and inhibits the formation of nanofibrils during cellulose synthesis by bacteria (Colvin and Witter 1983).

TABLE 18.1 SHG-Active Materials of Interest in
Tissue Engineering Listed with Their Function

Material	Function in Tissue Engineering
Collagen	Scaffold material, ECM component
Cellulose	Scaffold material
Silk	Scaffold material
Myosin	Muscle tissue component

These limitations of established microscopy techniques make nondestructive, label-free methods for three-dimensional monitoring of unmodified cell-scaffold constructs of high interest. SHG microscopy, introduced for the first time in the 1980s (Freund and Deutsch 1986), has in the last decade become a widely used imaging method benefitting from developments in laser and microscope technology (Campagnola et al. 2002, Campagnola and Loew 2003, Zipfel et al. 2003). SHG microscopy permits specific label-free visualization with high spatial resolution in three dimensions probing nonlinear optical properties in structures having noncentrosymmetric molecular arrangement. Although being limited to monitoring molecules of specific symmetry properties (and their assembly), this technique is a valuable characterization method in tissue engineering since a number of materials of interest are SHG-active. Table 18.1 presents these materials and their functions in tissue engineering.

A number of publications have emerged in later years using SHG as well as other nonlinear microscopy methods for tissue characterization and review papers addressing applications to tissue engineering have also been presented (Georgakoudi et al. 2008, Schenke-Layland 2008). In addition, technical aspects of importance for SHG microscopy in tissue have been investigated, such as polarization effects (Matcher 2009) and phase-matching considerations (LaComb et al. 2008). The usage of microscopy techniques based on nonlinear optics is promoted by the development of smaller, more user-friendly, low-cost, short-pulse laser systems (Svedberg et al. 2010, Tang et al. 2009). In particular, combining SHG microscopy with other techniques, such as multi-photon fluorescence, coherent anti-Stokes Raman scattering (CARS) (Cheng 2007), and THG (Yu et al. 2007), valuable insights in the relationship between architecture and function of artificial tissue constructs can be obtained. Thus, SHG microscopy is a valuable instrument within tissue engineering and an increased future use of the technique can be expected. In the following sections, some examples showing the potential of the technique will be presented.

18.2 SHG Microscopy on Collagen in Tissue Engineering

Collagen is a class of natural proteins containing more than 20 identified types with the fibrillar type I being the most abundant and the major component of connective tissue. The primary unit of collagen type I, tropocollagen, consists of three polypeptides arranged in a right-handed triple helix formation, typically 300 nm long and 1 nm wide, stabilized by hydrogen bonds. Tropocollagen units are able to self-organize into arrays and can then form larger structures such as microfibrils, fibrils, and fascicles. The ordering of the material structure makes it highly SHG-active, resulting in strong signals (Campagnola et al. 2002), and the SHG process in collagen type I fibers has been well characterized (Stoller et al. 2002, Williams et al. 2005). SHG microscopy has been compared to established techniques for collagen imaging such as histochemical staining with Masson's Trichrome dye and fluorescence microscopy using Sirius Red dye (Cox et al. 2003). Detailed comparison between corresponding images showed that it was possible to identify finer collagen fibrils by SHG microscopy, not visualized using the conventional methods. Furthermore, Sirius Red to some degree also labeled other, less-crystalline types of collagen and thus showed less specificity to collagen type I than SHG. The van Giesen stain is an alternative method to label and visualize collagen. A comparison between this stain and SHG microscopy of fibrous liver tissue showed good overall agreement with the advantage for SHG in imaging small collagen structures (Brackmann et al. 2010b).

Tissue engineering scaffolds of different structure can be synthesized from collagen, for example, fibrillar matrices (Caves et al. 2010, Zeugolis et al. 2008) or porous sponge-like structures (Glowacki and Mizuno 2007). In addition, collagen is the main component in scaffolds obtained by means of decellularization of native tissue. The material induces a relatively low immunoresponse as an implanted scaffold and a large number of collagen-based medical devices for tissue repair exist (Pachence 1996). Furthermore, collagen incorporates integrin-binding domains in terms of peptide (arginine–glycine–aspartate, RGD, and GFOGER) sequences that promote the development of cell attachment sites. Thus, it is a very attractive scaffold material and has been used for this purpose in a large number of tissue engineering applications. In the following some examples where SHG microscopy has been used for collagen imaging will be presented, either for investigation of collagen scaffolds or for monitoring ECM collagen produced by cells.

18.2.1 Collagen Scaffolds

SHG microscopy has been applied in several studies of collagen scaffold structure, for example, in characterization of spun fiber matrices (Caves et al. 2010, Zeugolis et al. 2008). In addition to studies of the isolated material itself, scaffolds seeded with different types of cells have been investigated. Figure 18.2 shows two SHG images of collagen scaffolds, one acellular (Figure 18.2a) and one seeded with rat aortic smooth muscle cells (Figure 18.2b). Comparing the images, it can be seen that the seeded scaffold shows a higher collagen density due to material remodeling induced by the cells. In particular, this can be seen in the vicinity of the cells, which appear as dark regions in the image and are indicated by arrows in Figure 18.2b.

Collagen scaffolds seeded with fibroblasts and/or keratinocytes have been used as model systems for skin and SHG microscopy has been used to visualize the collagen matrix. In a study of the skin-healing process, the collagen fiber rearrangement induced by fibroblast contractions was investigated (Abraham et al. 2010). Chemically stimulated contraction induced elongation of fibroblasts and accumulation of collagen in their vicinity, whereas in samples with no stimulation cells showed round morphology and no collagen accumulation was observed. Wound-healing mechanisms where tissue damage was induced by laser ablation have also been studied on a similar cell-scaffold system and collagen tissue recovery could be observed 7 days after damage (Yeh et al. 2004). Furthermore, a method for photodynamic therapy of scars has been characterized with experiments, including SHG microscopy, on a collagen/

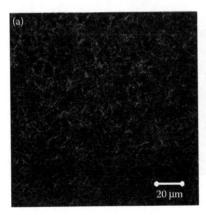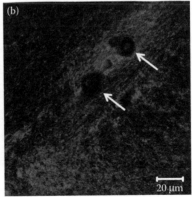

FIGURE 18.2 SHG images of collagen scaffolds. (a) Acellular scaffold, showing a network of randomly oriented fibers. (b) Scaffold seeded with rat aortic smooth muscle cells. A denser fiber structure can be seen compared to the acellular scaffold, which is due to remodeling of the collagen by the cells. This is especially evident around the cells, which appear as dark image regions indicated by arrows. The images have kindly been provided by Mr. Matthew Dalene and Associate Professor Jan Stegemann, University of Michigan.

fibroblast/keratinocyte system (Chiu et al. 2005). Pena et al. have investigated the effect of collagen accumulation around fibroblasts as an effect of skin aging and observed a reduced effect when using an inhibitor agent. Figure 18.3 presents results from their investigation and shows combined SHG (green) and two-photon fluorescence (red) images of the collagen matrix and fibroblasts, respectively. The images in Figure 18.3a–c were measured in a control sample without inhibitor added and an accumulation of collagen around the cells can be observed. The images in Figure 18.3d–f were obtained in a sample initially treated with the inhibitor agent and no collagen accumulation can be observed up to 24 h (Figure 18.3e). At this time, the inhibitor was removed and the image acquired after 48 h (Figure 18.3f) shows that collagen accumulation has been reinitiated.

To mimic the geometry of the intervertebral disc, Bowles et al. synthesized collagen gels seeded with ovine annulus fibrosus cells (Bowles et al. 2010). SHG images measured on collagen gels of different density were compared in terms of fiber alignment and material heterogeneity to identify a suitable construct for a functional tissue-engineered intervertebral disc. In an investigation of tissue tumor invasion, cell invasion has been monitored in synthetic as well as native collagen tissues by combined two-photon fluorescence and SHG microscopy (Wolf et al. 2009). A similar combination of techniques has been used to investigate cell-material invasion for endothelial cells grown on collagen scaffolds (Lee et al. 2009). Their results showed an alignment of extruding cell processes along collagen fibers together with degradation and displacement of the matrix during cell migration. Collagen has also been monitored in a study of chondrocytes seeded on collagen type I/III membranes in order to optimize growth for obtaining suitable implants for knee joints (Martini et al. 2006).

To develop a scaffold promoting spread of Schwann cells, constructs consisting of collagen treated with Growth Factor Reduced Matrigel and seeded with cells have been investigated (Dewitt et al. 2009).

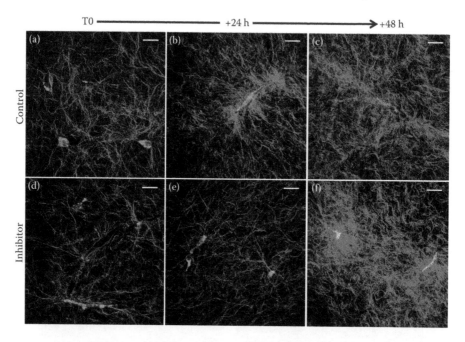

FIGURE 18.3 **(See color insert.)** Combined SHG (green) and two-photon fluorescence (red) images measured on a collagen scaffold seeded with fibroblasts. Images (a)–(c) were obtained in a control sample and an increased accumulation of collagen can be observed around the cells. For a sample where an inhibitor agent has been added initially, images (d) and (e) show no collagen accumulation around the cells. However, after the inhibitor has been removed at 24 h, collagen accumulation is initiated and clearly seen after 48 h in image (f). Scale bars 30 μm. (Reprinted from Pena, A.-M. et al. 2010. *Journal of Biomedical Optics* 15:056018, with the kind permission of the authors.)

The Matrigel treatment was found to promote outgrowth of cells as well as cell survival. The collagen structure was monitored by SHG microscopy and it was observed that increasing amounts of added Matrigel resulted in a less homogenous matrix.

Dumas et al. have presented combined two-photon fluorescence and SHG microscopy with detection utilizing time-correlated single-photon counting to distinguish between the signals. The technique was applied to investigate mesenchymal human stem cells seeded on collagen sponge scaffolds (Dumas et al. 2010). The advance of SHG microscopy as a tool in tissue engineering also requires methods for image analysis. One example of this is an algorithm for analysis of fiber content and orientation in thick collagen gels developed by Bayan et al. and applied to study the influence of fibroblast seeding on the scaffold (Bayan et al. 2009).

18.2.2 Extracellular Matrix Collagen

In addition to being employed as a scaffold material, collagen is a major component of the ECM synthesized by many types of cells. ECM collagen production has been traced using SHG microscopy in a number of investigations, for example, in studies of chondrocytes differentiated from stem cells (Chen et al. 2008, 2010) or in explanted cartilage tissue (Werkmeister et al. 2010). A delayed collagen production (Chen et al. 2008) as well as a more aligned fiber arrangement could be observed for samples under mechanical stress compared to unstressed samples (Chen et al. 2008, Werkmeister et al. 2010). Studies of ECM collagen have also been carried out on mesenchymal stem cells differentiated into osteogenic cells (Pallotta et al. 2009, Rice et al. 2010). In a study of megakaryocytes, Pallotta et al. compared the production of ECM collagen at oxygen levels of 5% and 20% for isolated osteoblasts as well as for co-cultures with hematopoietic stem cells (Pallotta et al. 2009). In agreement with the trend observed for collagen levels monitored via hydroxyproline detection, SHG data showed an earlier development of ECM collagen from osteoblasts in co-culture at the lower oxygen condition. Results from a similar study carried out by Rice et al., comparing osteoblast collagen production at low (5%) and high (20%) levels of oxygen access, also showed an earlier and in total higher production of collagen for cells grown at low oxygen supply. Another example of ECM collagen produced by osteoblasts and monitored by SHG microscopy is presented in Section 18.3.2.2 (Figure 18.9). The technique has also been employed in cardiovascular tissue engineering where König and coworkers studied the influence of pressure on collagen synthesis in engineered heart-valve leaflets (König 2008). Images of tissue exposed to pressures above physiological levels showed signs of disturbed matrix formation and cell growth compared to tissue at normal pressure level. The same author has also presented results from SHG applied to investigate collagen generation by differentiated human stem cells (König 2008). In addition to investigations, *in vitro* SHG microscopy has also been employed for imaging ECM collagen in native tissue such as heart valve leaflets, cornea, and cartilage (Schenke-Layland 2008).

18.3 Biosynthesized Cellulose

Cellulose is the most abundant naturally occurring polymer and a major component of biomass from plants. It can also be produced by bacteria and algae, resulting in a matrix of crosslinked fibers (Ross et al. 1991). The cellulose polymer consists of a linear arrangement of glucose molecules, where parallel polymer chains attach with hydrogen bonds and form the crystalline cellulose structure. The molecular structure of cellulose and a schematic of the synthesis carried out by bacteria of strain *Gluconacetobacter xylinus* are shown in Figure 18.4.

In the cellulose synthesis, glucan chains, extruded from cellulose synthase complexes located on the bacterium surface, are initially assembled into a sub-elementary fibril, then further on into a crystalline microfibril and finally a fiber ribbon assembly, which is a few hundred nanometers in diameter (Brown et al. 1976, Ross et al. 1991). The ordered crystalline arrangement makes the material non-centrosymmetric and it has also proven to be SHG-active (Brown et al. 2003). The first visualizations of microbial cellulose

(a)   (b)

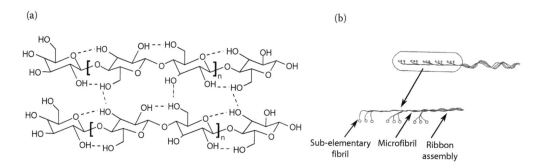

FIGURE 18.4 (a) The molecular structure of cellulose where linear polymer chains containing n glucose molecules are attached by hydrogen bonds (thin dashed lines). (b) Cellulose synthesis by *G. xylinus*. Glucan chains are synthesized into sub-elementary fibrils at specific sites on the bacteria surface. The fundamental fibrils are then combined into microfibrils and further into fiber ribbon assemblies having diameters of typically 200 nm. (The illustration in (b) has been adapted from Hirai, A. and F. Horii 1999. *ICR Annual Report* 6:28–29.)

using SHG microscopy (Brown et al. 2003) have been followed by a more detailed characterization of the forward- and backward-scattered components of the SHG signal from cellulose matrices (Nadiarnykh et al. 2007). Furthermore, the synthesis and structure of cellulose scaffolds intended as blood vessel replacements have been investigated by SHG microscopy (Brackmann et al. 2010a).

18.3.1 SHG Microscopy on Biosynthesized Cellulose

Figure 18.5 shows SHG images measured on cellulose synthesized by *G. xylinus*. The image in Figure 18.5a, measured during the first hour of the synthesis process, is an overlay of SHG and CARS microscopy images showing bacteria, visualized by CARS, in red color and an initial network of a few cellulose fibers in blue. Figure 18.5b shows an SHG image of a developed cellulose fiber network obtained after 7 days of growth, the fiber ribbon assemblies can be clearly distinguished, and a volume representation showing the fiber arrangement in three dimensions is shown in Figure 18.5c. Both cellulose and bacteria were monitored under native conditions without labeling or specific sample preparation for microscopy. For the initially produced cellulose tissue having low density of fibers (cf. Figure 18.5a), the generated SHG signal can be monitored using forward detection geometry, whereas the developed compact material shown in Figures 18.5b and 18.5c requires detection of the backscattered SHG signal in epi-mode. Nadiarnykh et al. have investigated backward scattered SHG from cellulose in detail and identified two signal contributions. Primarily SHG-active structures much smaller than the excitation

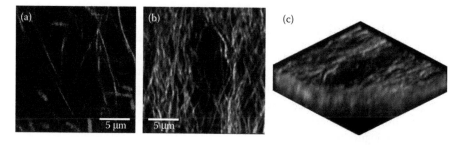

FIGURE 18.5 **(See color insert.)** (a) Overlay SHG/CARS microscopy image showing fibers (blue) and bacteria (red) monitored during the first hour of cellulose production by SHG and CARS microscopy, respectively. (b) SHG microscopy image showing a developed cellulose matrix after 7 days of growth with clearly distinguishable fibers (c) volume representation generated from a stack of SHG images, showing the fiber matrix in three dimensions (volume depth 5 μm).

wavelength generate coherently emitted SHG in forward as well as backward direction. Generally, the combined contributions from individual dipole sources result in destructive interference and suppression of the coherent back-scattered SHG whereas constructive interference is obtained for SHG-emission in the forward direction. However, for sources much smaller than the excitation wavelength, the backscattered SHG is not completely cancelled and a direct backward-emitted signal can still be maintained. An additional incoherent backward-scattered SHG signal contribution is obtained in dense media, such as the developed cellulose fiber matrices, due to multiple scattering of the forward-directed SHG (Nadiarnykh et al. 2007). Similar observations have also been made for SHG generated in collagen (Williams et al. 2005).

Figure 18.6 shows results from SHG microscopy measurements carried out to further characterize the SHG process in biosynthesized cellulose. A comparison between excitation at 800 and 1064 nm is shown in Figure 18.6a and 18.6b, and it can be seen that similar signal strengths are obtained. Cellulose is reported to have electronic resonances in the ultraviolet wavelength regime and the index of refraction for the material has been shown to be similar in the visible and near-infrared wavelength regimes (Nadiarnykh et al. 2007). Thus, similar SHG signals can be expected for the two near-infrared excitation wavelengths. The back-scattered SHG signal is dependent on material bulk density as presented in Figures 18.6c through 18.6e, showing SHG images measured in samples of different density, expressed as weight percentage of the mass of cellulose relative to the total mass of cellulose and absorbed water. A linear relation is obtained between the average SHG signal and cellulose sample density, Figure 18.6f, confirming a larger fraction of backscattered SHG signal for denser samples and that primarily multiple-scattered forward-directed SHG signal is detected.

The SHG signal of individual fibers is dependent on their orientation relative to the laser polarization, as demonstrated in Figures 18.6g and 18.6h measured in the same field of view using vertical and horizontal laser polarization, respectively. Although the large-scale features of the sample appear similar in the images, the fibers aligned with the laser polarization generate the strongest signal contribution to the image. This can be seen comparing Figures 18.6g and 18.6h and also clearly in the images of Figures 18.6i and 18.6j, covering a smaller field of view.

18.3.2 Tissue Engineering Applications of Biosynthesized Cellulose

The use of microbial-derived cellulose in medical industry has mainly been focused on liquid-loaded pads, wound dressings, and other external applications (Czaja et al. 2007). Nevertheless, biosynthesized cellulose has interesting properties in its wet, unmodified state. The high water content of around 99% suggests that the material can be considered a hydrogel, which are known for their favorable biocompatible properties, much due to little protein adsorption. The versatility of the material, allowing it to be manufactured in various sizes and shapes, depending on product requirements, has made it of interest to explore its use as an implant in biomedical applications, such as a bone graft material (Zaborowska et al. 2010), blood vessel substitute (Bodin et al. 2007a, Klemm et al. 2001), cartilage replacement (Bodin et al. 2007c), or as a hydrophilic coating of other biomaterials (Charpentier et al. 2006).

18.3.2.1 Biosynthetic Blood Vessels

The most common vascular graft material in use today is autologous veins and arteries. However, about 10–20% of the patients that need vascular bypass are lacking usable small-diameter (<5 mm) vessels as graft material. Furthermore, purely synthetic grafts are not suitable replacements for small vessels due to problems with occlusion and mechanical mismatch to the native vessel. Thus, there is a strong demand for small replacement blood vessels with mechanical and surface properties similar to those of native ones. The possibility to produce small-diameter cellulose tubes (see photo Figure 18.7a) with designed tissue properties makes the material of high interest for tissue-engineered blood vessels. In addition, the material has promising mechanical properties for use as a blood vessel, such as a high tensile strength, ascribed to the super-molecular structure in which the microfibrils are tightly bound

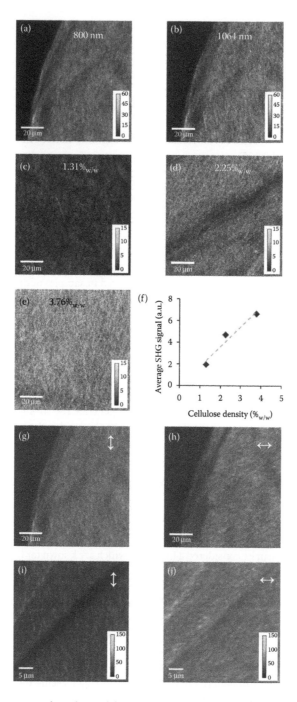

FIGURE 18.6 SHG microscopy data obtained from microbial biosynthesized cellulose. Comparing excitation at 800 nm (a) and 1064 nm (b) shows similar signal strengths for both cases. SHG images measured at different sample densities expressed as %$_{w/w}$ cellulose; (c) 1.31%$_{w/w}$, (d) 2.25%$_{w/w}$, and (e) 3.76%$_{w/w}$. (f) Average SHG-signal versus sample density, an increase in epi-detected SHG signal is obtained with increasing sample density, due to more efficient back-scattering of the signal. SHG images measured in the same fields of view using vertical and horizontal laser polarizations (double arrows). (g) and (h) 100×100 μm images, (i) and (j) 40×40 μm images. The large-scale material structures appear similar, (g) and (h), however, fibers aligned with the laser polarization give the dominant signal contribution, (i) and (j).

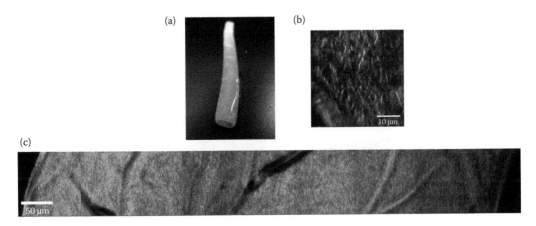

FIGURE 18.7 (a) Photo of biosynthesized cellulose tube. (b) SHG image measured on a cellulose tube and showing the detailed material fiber structure. (c) Radial profile of a cellulose tube obtained from a montage of SHG
images, luminal side to the right. Regions of relatively compact, homogeneous material are interspersed by denser
structures and voids.

by hydrogen bonds (Bodin et al. 2007b). Furthermore, the cellulose has shown good biocompatibility
(Helenius et al. 2006), but the degradability of the material needs to be further evaluated.

Figures 18.7b and c show SHG microscopy images measured on a cellulose tube in its natural state
without special sample preparation for microscopy. Detailed information on the morphology at a fibrillar level can be seen in the close-up of Figure 18.7b whereas the montage of images in Figure 18.7c shows
a full radial profile of the tube presenting the morphology on a larger scale.

Figure 18.8 shows SHG images measured on a cellulose tube that has been implanted in a sheep during 1 year for evaluation of the tube as a blood vessel replacement. The images were measured on the
outer side of the graft in the contact region between cellulose and the collagen of the surrounding native
tissue. Both materials are visible in the image and the SHG signal from collagen (bright image structures to the left) is much stronger than that of the cellulose (darker gray image structures to the right)
allowing them to be distinguished from each other. The magnified view of Figure 18.8b shows the close

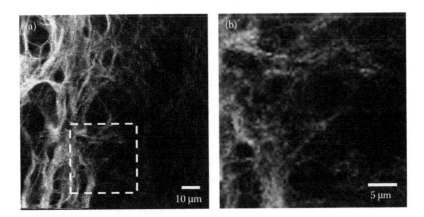

FIGURE 18.8 SHG images measured on a cellulose blood vessel graft, previously implanted in a sheep during
1 year. Both native collagen tissue and cellulose are detected in the image with high (bright structures) and low
(dark gray structures) SHG signals, respectively. (a) 100 × 100 μm overview. (b) Close-up image corresponding to
the region indicated by the white square in (a) showing a detailed view of the integration between the cellulose graft
and native collagen tissue.

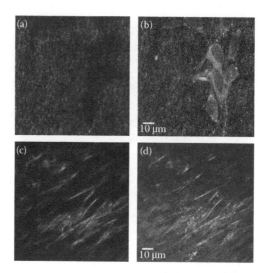

FIGURE 18.9   **(See color insert.)** SHG microscopy images of cellulose scaffolds seeded with osteoblasts to obtain a construct intended as a bone tissue replacement. (a) SHG image of cellulose scaffold measured 4 days after cell seeding (b) overlay with simultaneously measured CARS image (orange color), showing a group of osteoblasts located on the scaffold. (c) SHG image measured 8 days after seeding, when ECM collagen fibers could be detected in regions of high cell density. (d) Corresponding SHG/CARS overlay image showing the collagen fibers located in between the cells.

contact with a mesh of interconnected fibers from both materials, confirming adaptation of the graft *in vivo* after implantation.

18.3.2.2 Biosynthesized Cellulose as a Scaffold for Bone Regeneration

The potential for using biosynthesized cellulose as a bone graft material has also been investigated (Zaborowska et al. 2010). For this purpose, a porous version of the material was manufactured using particle leaching with paraffin wax particles added as porogens during fermentation. Figures 18.9a and 18.9c show SHG images measured on cellulose scaffolds with osteoblasts, 4 and 8 days after cell seeding, respectively. In Figure 18.9a, the SHG signal is obtained from the cellulose material surrounding a group of cells, whose positions are indicated by the dark region in the image. After 8 days of growth when denser confluent cell regions could be observed on the scaffold, SHG images also showed ECM collagen fibers as shown in Figure 18.9c. The osteoblasts can be imaged by CARS microscopy probing their hydrocarbon content via the CH_2 symmetric stretch vibration at wavenumber 2845 cm^{-1}, in this case using combined excitation at 817 and 1064 nm. By means of simultaneous SHG and CARS microscopy, the combined arrangement of cells, cellulose, and collagen can be visualized as shown in the overlay images of Figures 18.9b and 18.9d. In the overlay image in Figure 18.9d, it can be seen that the detected ECM collagen fibers are located in between the cells.

18.4 Silk Fibroin

Silk fibroin is a natural biopolymer spun into fibers by silk worms such as the *Bombyx mori* and *Antherea mylitta*. Properties such as high mechanical strength and low inflammatory effects when implanted *in vivo* (Mandal and Kundu 2010, Murphy and Kaplan 2009) make it of interest as a tissue engineering scaffold material. For this purpose, silk can either be used in its raw form obtained from worm spinning or, more commonly, in regenerated form. A number of tissue engineering applications using silk as scaffold material has been presented in recent years. Some examples are vascular tissue (Enomoto et al. 2010, Zhou et al. 2010), cartilage (Chao et al. 2010), ligament (Liu et al. 2008), bone (MacIntosh et al. 2008, Weska et al. 2009), cornea (Lawrence et al. 2009), and nerve grafts (Yang et al. 2007).

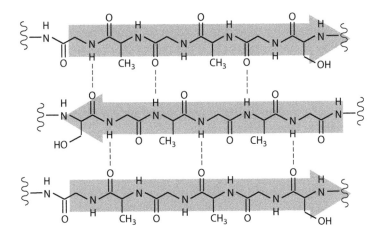

FIGURE 18.10 The molecular structure of Silk II. The antiparallel arrangement of β-sheets, indicated by broad gray arrows, connected via hydrogen bonds (thin dashed lines) makes the material SHG-active.

Silk fibroin can be described by two structural models; Silk I that consists of type II β-turn random coil domains together with structures including alpha helices and Silk II that is assembled of antiparallel β-pleated sheets (cf. Figure 18.10) linked with hydrogen bonds giving the material its high mechanical strength. The SHG signal obtained in silk fibroin assemblies of different structure has been investigated (Rice et al. 2008) and it was identified that the ordered β-sheet arrangement of Silk II makes this type SHG-active (cf. Figure 18.10). Results from their investigation are shown in Figure 18.11, presenting

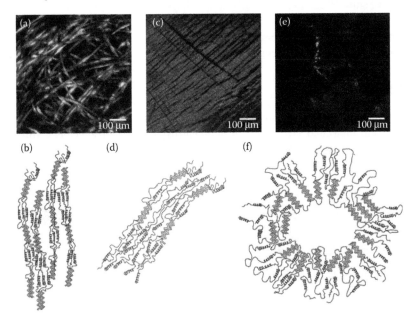

FIGURE 18.11 Characterization of silk fibroin materials of different structures using SHG microscopy. (a) SHG image measured on an array of silk fibers with high degree of β-sheet arrangement (b) resulting in efficient signal generation. Whereas structures with a randomized β-sheet orientation such as aqueous films or gels result in no SHG signal generation, an SHG signal can be detected in compressed and stretched film (c) that has retrieved β-sheet ordering (d). (e) SHG image measured on a porous silk scaffold having local regions of β-sheet alignment around the pores (f), making the pore edges visible in the SHG image. (Adapted from Rice W.L. et al. 2008. *Biomaterials* 29:2015–24. Images kindly provided by Mr. William Rice and Associate Professor Irene Georgakoudi, Tufts University.)

SHG images together with schematic illustrations of secondary structure features such as β-sheet content and alignment. Strong SHG signals were obtained for single silk fibers which exhibit a high degree of β-sheet alignment, Figures 18.11a and 18.11b. When the material is rearranged into an aqueous film or a gel, the β-sheet arrangement is randomized resulting in no SHG signal (data not shown). However, after compression and stretching an aqueous film recaptures some β-sheet ordering, resulting in SHG from the material (Figures 18.11c and 18.11d). For a silk fibroin scaffold, Figures 18.11e and 18.11f, partial β-sheet alignment can be observed locally around pores, Figure 18.11f, and some SHG signal can be detected at the pore edges, as shown in Figure 18.11e.

18.5 Skeletal Muscle Tissue Engineering

Loss of skeletal muscle function can be induced by a number of factors, for example, congenital defects, tumor ablation, prolonged denervation, traumatic injury or myopathies. Tissue engineering does offer methods for reconstruction of lost skeletal muscle function; however, a number of requirements for synthetic muscle tissue must be fulfilled. A parallel arrangement of muscle fibers consisting of myosin/actin filaments as well as acetylcholine receptors must be achieved. In addition, the tissue must be vascularized in order to have efficient transport of oxygen, carbon dioxide, nutrients, and waste products. This is particularly important since relatively large amounts of synthetic tissue may be required. Furthermore, the tissue must be innervated, that is, the muscle fibers must be connected to motor neurons for control. Skeletal muscle tissue harbors its own organ-specific mesenchymal stem cells, the so-called satellite cells, and a number of cell-scaffold constructs using this type of cells have been investigated for synthesis of muscle tissue (Koning et al. 2009, Liao and Zhou 2009). The SHG-active myosin filaments in muscle fibers allow them to be visualized by means of SHG microscopy, as exemplified by the two SHG images shown in Figure 18.12, measured on a mouse diaphragm. Thus, SHG microscopy provides an excellent tool for characterization of skeletal muscle tissue (Nucciotti et al. 2010, Plotnikov et al. 2008, 2006), for example by measuring sarcomere lengths to gauge function, and generally a valuable instrument for this branch of tissue engineering.

18.6 Summary

SHG microscopy has developed into a powerful imaging tool for use within tissue engineering. Although restricted to certain molecular structures, of non-centrosymmetric arrangement, this includes the important fibrillar structural protein collagen type I, the major component of native extracellular

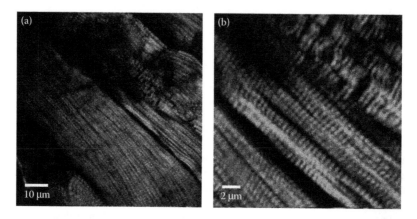

FIGURE 18.12 SHG images measured on a mouse diaphragm showing the skeletal muscle tissue structure with fibers (a) and myo-filaments (b).

matrix and a widely used scaffold material. In addition, two other promising scaffold materials are SHG-active, namely biosynthesized cellulose and silk fibroin. Finally, one of the major components of skeletal muscle tissue, myosin, can also be selectively visualized using SHG microscopy. These imaging abilities are sufficient to provide relevant information on a multitude of tissue engineering constructs, as confirmed by the increasing number of publications where SHG microscopy has been applied for this purpose. This growing trend can be expected to continue since recent developments in laser technology promote the spread and future establishment of the SHG technique within tissue engineering as well as other fields of chemistry and biology.

Acknowledgments

The authors would like to gratefully acknowledge the following persons for their contributions to this chapter; Mr. Matthew Dalene and Associate Professor Jan Stegemann, University of Michigan (Figure 18.2); Ms. Ana-Maria Pena, LOréal Research and Innovation (Figure 18.3); Mr. Willam Rice and Associate Professor Irene Georgakoudi, Tufts University (Figure 18.11). The long-term fruitful collaboration with the group of Professor Paul Gatenholm, Polymer Science, Chalmers University of Technology, Sweden, on the studies of biosynthesized cellulose is gratefully acknowledged by the authors.

References

Abeysekera, R. M., A. W. Robards, A. B. Hodgson, and D. M. Goodall 1993. Improved visualization of folded collagen alpha-chains by ultra-rapid freezing. *International Journal of Biological Macromolecules* 15:313–15.

Abraham, T., J. Carthy, and B. McManus 2010. Collagen matrix remodeling in 3-dimensional cellular space resolved using second harmonic generation and multiphoton excitation fluorescence. *Journal of Structural Biology* 169:36–44.

Bayan, C., J. M. Levitt, E. Miller, D. Kaplan, and I. Georgakoudi 2009. Fully automated, quantitative, noninvasive assessment of collagen fiber content and organization in thick collagen gels. *Journal of Applied Physics* 105:102042-1–102042-11.

Bodin, A., H. Bäckdahl, B. Risberg, and P. Gatenholm 2007a. Nano cellulose as a scaffold for tissue engineered blood vessels. *Tissue Engineering* 13:885.

Bodin, A., H. Bäckdahl, H. Fink et al. 2007b. Influence of cultivation conditions on mechanical and morphological properties of bacterial cellulose tubes. *Biotechnology and Bioengineering* 97:425–34.

Bodin, A., S. Concaro, M. Brittberg and P. Gatenholm 2007c. Bacterial cellulose as a potential meniscus implant. *Journal of Tissue Engineering and Regenerative Medicine* 1:406–08.

Bowles, R. D., R. M. Williams, W. R. Zipfel, and L. J. Bonassar 2010. Self-assembly of aligned tissue-engineered annulus fibrosus and intervertebral disc composite via collagen gel contraction. *Tissue Engineering Part A* 16:1339–48.

Brackmann, C., A. Bodin, M. Åkeson, P. Gatenholm, and A. Enejder 2010a. Visualization of the cellulose biosynthesis and cell integration into cellulose scaffolds. *Biomacromolecules* 11:542–48.

Brackmann, C., B. Gabrielsson, F. Svedberg et al. 2010b. Nonlinear microscopy of lipid storage and fibrosis in muscle and liver tissues of mice fed high-fat diets. *Journal of Biomedical Optics* 15:066008.

Brown, R. M., A. C. Millard, and P. J. Campagnola 2003. Macromolecular structure of cellulose studied by second-harmonic generation imaging microscopy. *Optics Letters* 28:2207–09.

Brown, R. M., J. H. M. Willison, and C. L. Richardson 1976. Cellulose biosynthesis in *Acetobacter xylinum*: Visualization of site of synthesis and direct measurements of the *in vivo* process. *Proceedings of the National Academy of Sciences of the United States of America* 73:4565–69.

Campagnola, P. J. and L. M. Loew 2003. Second-harmonic imaging microscopy for visualizing biomolecular arrays in cells, tissues and organisms. *Nature Biotechnology* 21:1356–60.

Campagnola, P. J., A. C. Millard, M. Terasaki et al. 2002. Three-dimensional high-resolution second-harmonic generation imaging of endogenous structural proteins in biological tissues. *Biophysical Journal* 82:493–508.

Caves, J. M., V. A. Kumar, J. Wen et al. 2010. Fibrillogenesis in continuously spun synthetic collagen fiber. *Journal of Biomedical Materials Research Part B-Applied Biomaterials* 93B:24–38.

Chao, P. H. G., S. Yodmuang, X. Q. Wang et al. 2010. Silk hydrogel for cartilage tissue engineering. *Journal of Biomedical Materials Research Part B-Applied Biomaterials* 95B:84–90.

Charpentier, P. A., A. Maguire, and W. K. Wan 2006. Surface modification of polyester to produce a bacterial cellulose-based vascular prosthetic device. *Applied Surface Science* 252:6360–67.

Chen, W. L., C. C. Chang, L. L. Chiou et al. 2008. Monitoring the effect of mechanical stress on mesenchymal stem cell collagen production by multiphoton microscopy—art. no. 685806. *Optics in Tissue Engineering and Regenerative Medicine* II 6858:685806.

Chen, W. L., C. H. Huang, L. L. Chiou et al. 2010. Multiphoton imaging and quantitative analysis of collagen production by chondrogenic human mesenchymal stem cells cultured in chitosan scaffold. *Tissue Engineering Part C-Methods* 16:913–20.

Cheng, J.-X. 2007. Coherent anti-Stokes Raman scattering microscopy. *Applied Spectroscopy* 61:197–208.

Chiu, L. L., C. H. Sun, A. T. Yeh et al. 2005. Photodynamic therapy on keloid fibroblasts in tissue-engineered keratinocyte-fibroblast co-culture. *Lasers in Surgery and Medicine* 37:231–44.

Chow, D., M. L. Nunalee, D. W. Lim, A. J. Simnick, and A. Chilkoti 2008. Peptide-based biopolymers in biomedicine and biotechnology. *Materials Science and Engineering R* 62:125–55.

Colvin, J. R. and D. E. Witter 1983. Congo Red and Calcofluor White inhibition of *Acetobacter xylinum* cell-growth and of bacterial cellulose microfibril formation—Isolation and properties of a transient, extracellular glucan related to cellulose. *Protoplasma* 116:34–40.

Cox, G., E. Kable, A. Jones et al. 2003. 3-dimensional imaging of collagen using second harmonic generation. *Journal of Structural Biology* 141:53–62.

Czaja, W., D. Romanovicz, and R. M. Brown 2004. Structural investigations of microbial cellulose produced in stationary and agitated culture. *Cellulose* 11:403–11.

Czaja, W. K., D. J. Young, M. Kawecki, and R. M. Brown 2007. The future prospects of microbial cellulose in biomedical applications. *Biomacromolecules* 8:1–12.

Dewitt, D. D., S. N. Kaszuba, D. M. Thompson, and J. P. Stegemann 2009. Collagen I-Matrigel scaffolds for enhanced Schwann cell survival and control of three-dimensional cell morphology. *Tissue Engineering Part A* 15:2785–93.

Dumas, D., C. Henrionnet, S. Hupont et al. 2010. Innovative TCSPC-SHG microscopy imaging to monitor matrix collagen neo-synthetized in bioscaffolds. *Bio-Medical Materials and Engineering* 20:183–88.

Enomoto, S., M. Sumi, K. Kajimoto et al. 2010. Long-term patency of small-diameter vascular graft made from fibroin, a silk-based biodegradable material. *Journal of Vascular Surgery* 51:155–64.

Freund, I. and M. Deutsch 1986. Second-harmonic microscopy of biological tissue. *Optics Letters* 11:94–96.

Georgakoudi, I., W. L. Rice, M. Hronik-Tupaj, and D. L. Kaplan 2008. Optical spectroscopy and imaging for the noninvasive evaluation of engineered tissues. *Tissue Engineering* 14:321–40.

Gilbert, T. W., T. L. Sellaro, and S. F. Badylak 2006. Decellularization of tissues and organs. *Biomaterials* 27:3675–83.

Glowacki, J. and S. Mizuno 2007. Collagen scaffolds for tissue engineering. *Biopolymers* 89:338–44.

Helenius, G., H. Bäckdahl, A. Bodin et al. 2006. *In vivo* biocompatibility of bacterial cellulose. *Journal of Biomedical Materials Research Part A* 76A:431–38.

Heydarkhan-Hagvall, S., K. Schenke-Layland, A. P. Dhanasopon et al. 2008. Three-dimensional electrospun ECM-based hybrid scaffolds for cardiovascular tissue engineering. *Biomaterials* 29:2907–14.

Hirai, A. and F. Horii 1999. Cellulose assemblies produced by *Acetobacter xylinum*. *ICR Annual Report* 6:28–29.

Klemm, D., D. Schumann, U. Udhardt, and S. Marsch 2001. Bacterial synthesized cellulose—Artificial blood vessels for microsurgery. *Progress in Polymer Science* 26:1561–603.

König, K. 2008. Multiphoton tomography for tissue engineering—art. no. 68580C. *Optics in Tissue Engineering and Regenerative Medicine* II 6858:68580C.

Koning, M., M. C. Harmsen, M. J. A. van Luyn, and P. M. N. Werker 2009. Current opportunities and challenges in skeletal muscle tissue engineering. *Journal of Tissue Engineering and Regenerative Medicine* 3:407–15.

LaComb, R., O. Nadiarnykh, S. S. Townsend, and P. J. Campagnola 2008. Phase matching considerations in second harmonic generation from tissues: Effects on emission directionality, conversion efficiency and observed morphology. *Optics Communications* 281:1823–32.

Lawrence, B. D., J. K. Marchant, M. A. Pindrus, F. G. Omenetto, and D. L. Kaplan 2009. Silk film biomaterials for cornea tissue engineering. *Biomaterials* 30:1299–308.

Lee, P. F., A. T. Yeh, and K. J. Bayless 2009. Nonlinear optical microscopy reveals invading endothelial cells anisotropically alter three-dimensional collagen matrices. *Experimental Cell Research* 315:396–410.

Liao, H. and G. Q. Zhou 2009. Development and progress of engineering of skeletal muscle tissue. *Tissue Engineering Part B-Reviews* 15:319–31.

Liu, H., H. Fan, E. J. W. Wong, S. L. Toh, and J. C. H. Goh 2008. Silk-based scaffold for ligament tissue engineering. *14th Nordic-Baltic Conference on Biomedical Engineering and Medical Physics* 20:34–37.

MacIntosh, A. C., V. R. Kearns, A. Crawford, and P. V. Hatton 2008. Skeletal tissue engineering using silk biomaterials. *Journal of Tissue Engineering and Regenerative Medicine* 2:71–80.

Mandal, B. B. and S. C. Kundu 2010. Biospinning by silkworms: Silk fiber matrices for tissue engineering applications. *Acta Biomaterialia* 6:360–71.

Martini, J., K. Tonsing, M. Dickob et al. 2006. 2-Photon laser scanning microscopy on native cartilage and collagen-membranes for tissue-engineering—art. no. 60891N. *Multiphoton Microscopy in the Biomedical Sciences VI* 6089:60891N.

Matcher, S. J. 2009. A review of some recent developments in polarization-sensitive optical imaging techniques for the study of articular cartilage. *Journal of Applied Physics* 105:102041-1–102041-11.

Murphy, A. R. and D. L. Kaplan 2009. Biomedical applications of chemically-modified silk fibroin. *Journal of Materials Chemistry* 19:6443–50.

Murphy, M. B. and A. G. Mikos 2007. Polymer scaffold fabrication. In: *Principles of Tissue Engineering*, eds. R. Lanza, R. Langer, and J. Vacanti, pp. 309–21. San Diego: Elsevier Academic Press.

Nadiarnykh, O., R. LaComb, P. J. Campagnola, and W. A. Mohler 2007. Coherent and incoherent SHG in fibrillar cellulose matrices. *Optics Express* 15:3348–60.

Nucciotti, V., C. Stringari, L. Sacconi et al. 2010. Probing myosin structural conformation *in vivo* by second-harmonic generation microscopy. *Proceedings of the National Academy of Sciences of the United States of America* 107:7763–68.

Pachence, J. M. 1996. Collagen-based devices for soft tissue repair. *Journal of Biomedical Materials Research* 33:35–40.

Pachence, J. M., M. P. Bohrer, and J. Kohn 2007. Biodegradable polymers. In *Principles of Tissue Engineering*, eds. R. Lanza, R. Langer and J. Vacanti, pp. 323–40. San Diego: Elsevier Academic Press.

Pallotta, I., M. Lovett, W. Rice, D. L. Kaplan, and A. Balduini 2009. Bone marrow osteoblastic niche: A new model to study physiological regulation of megakaryopoiesis. *Plos One* 4:e8359.

Pena, A.-M., D. Fagot, C. Olive et al. 2010. Multiphoton microscopy of engineered dermal substitutes: Assessment of 3-D collagen matrix remodeling induced by fibroblast contraction. *Journal of Biomedical Optics* 15:056018.

Plotnikov, S. V., A. M. Kenny, S. J. Walsh et al. 2008. Measurement of muscle disease by quantitative second-harmonic generation imaging. *Journal of Biomedical Optics* 13:044018-1–044018-11.

Plotnikov, S. V., A. C. Millard, P. J. Campagnola, and W. A. Mohler 2006. Characterization of the myosin-based source for second-harmonic generation from muscle sarcomeres. *Biophysical Journal* 90:693–703.

Rabkin, E. and F. J. Schoen 2002. Cardiovascular tissue engineering. *Cardiovascular Pathology* 11:305–17.

Rice, W.L., S. Firdous, S. Gupta et al. 2008. Non-invasive characterization of structure and morphology of silk fibroin biomaterials using non-linear microscopy. *Biomaterials* 29:2015–24.

Rice, W. L., D. L. Kaplan, and I. Georgakoudi 2010. Two-photon microscopy for non-invasive, quantitative monitoring of stem cell differentiation. *Plos One* 5:e10075-1–e10075-13.

Ross, P., R. Mayer, and M. Benziman 1991. Cellulose biosynthesis and function in bacteria. *Microbiological Reviews* 55:35–58.

Schenke-Layland, K. 2008. Non-invasive multiphoton imaging of extracellular matrix structures. *Journal of Biophotonics* 1:451–62.

Schoen, F. J. 2004. Tissue engineering. In *Biomaterials Science*, eds. B. D. Ratner, A. S. Hoffman, F. J. Schoen and J. E. Lemons. San Diego: Elsevier Academic Press.

Sengupta, D. and S. C. Heilshorn 2010. Protein-engineered biomaterials: Highly tunable tissue engineering scaffolds. *Tissue Engineering Part B-Reviews* 16:285–93.

Shao, Z. F., J. Mou, D. M. Czajkowsky, J. Yang, and J. Y. Yuan 1996. Biological atomic force microscopy: What is achieved and what is needed. *Advances in Physics* 45:1–86.

Stoller, P., B. M. Kim, A. M. Rubenchik, K. M. Reiser, and L. B. Da Silva 2002. Polarization-dependent optical second-harmonic imaging of a rat-tail tendon. *Journal of Biomedical Optics* 7:205–14.

Svedberg, F., C. Brackmann, T. Hellerer, and A. Enejder 2010. Nonlinear microscopy with fiber laser continuum excitation. *Journal of Biomedical Optics* 15:026026-1–026026-4.

Tang, S., J. Liu, T. B. Krasieva, Z. P. Chen, and B. J. Tromberg 2009. Developing compact multiphoton systems using femtosecond fiber lasers. *Journal of Biomedical Optics* 14:030508-1–030508-3.

Werkmeister, E., N. de Isla, P. Netter, J. F. Stoltz, and D. Dumas 2010. Collagenous extracellular matrix of cartilage submitted to mechanical forces studied by second harmonic generation microscopy. *Photochemistry and Photobiology* 86:302–10.

Weska, R. F., G. M. Nogueira, W. C. Vieira, and M. M. Beppu 2009. Porous silk fibroin membrane as a potential scaffold for bone regeneration. *Key Engineering Material Volume 396–398 Bioceramics* 21:187–90.

Williams, R. M., W. R. Zipfel, and W. W. Webb 2005. Interpreting second-harmonic generation images of collagen I fibrils. *Biophysical Journal* 88:1377–86.

Wolf, K., S. Alexander, V. Schacht et al. 2009. Collagen-based cell migration models *in vitro* and *in vivo*. *Seminars in Cell & Developmental Biology* 20:931–41.

Yang, Y., F. Ding, H. Wu et al. 2007. Development and evaluation of silk fibroin-based nerve grafts used for peripheral nerve regeneration. *Biomaterials* 28:5526–35.

Yeh, A. T., B. S. Kao, W. G. Jung et al. 2004. Imaging wound healing using optical coherence tomography and multiphoton microscopy in an *in vitro* skin-equivalent tissue model. *Journal of Biomedical Optics* 9:248–53.

Yu, C. H., S. P. Tai, C. T. Kung et al. 2007. *In vivo* and *ex vivo* imaging of intra-tissue elastic fibers using third-harmonic-generation microscopy. *Optics Express* 15:11167–77.

Zaborowska, M., A. Bodin, H. Backdahl et al. 2010. Microporous bacterial cellulose as a potential scaffold for bone regeneration. *Acta Biomaterialia* 6:2540–47.

Zeugolis, D. I., S. T. Khew, E. S. Y. Yew et al. 2008. Electro-spinning of pure collagen nano-fibres—Just an expensive way to make gelatin? *Biomaterials* 29:2293–305.

Zhou, J. A., C. B. Cao, X. L. Ma, and J. Lin 2010. Electrospinning of silk fibroin and collagen for vascular tissue engineering. *International Journal of Biological Macromolecules* 47:514–19.

Zipfel, W. R., R. M. Williams, and W. W. Webb 2003. Nonlinear magic: Multiphoton microscopy in the biosciences. *Nature Biotechnology* 21:1369–77.

Index